Particle Technology Forum 2018

Core Programming Area at the 2018 AIChE Annual Meeting

Pittsburgh, Pennsylvania, USA
28 October - 2 November 2018

ISBN: 978-1-5108-7630-9

Printed from e-media with permission by:

Curran Associates, Inc.
57 Morehouse Lane
Red Hook, NY 12571

proceedings.com

Some format issues inherent in the e-media version may also appear in this print version.

Printed by Curran Associates, Inc. (2019)

AIChE
120 Wall Street, FL 23
New York, NY 10005-4020

Phone: (800) 242-4363
Fax: (203) 775-5177

www.aiche.org

Additional copies of this publication are available from:

Curran Associates, Inc.
57 Morehouse Lane
Red Hook, NY 12571 USA
Phone: 845-758-0400
Fax: 845-758-2633
Email: curran@proceedings.com
Web: www.proceedings.com

TABLE OF CONTENTS

Some abstracts in this book refer to images that were not made available by the publisher

535550 Development of Nanoparticle Alignment Regimes in Drying Cellulose Nanocrystal Droplet Suspensions for Additive Manufacturing

Abstract:

Nanoparticle suspensions have potential for a number of drop-on-demand applications, where they are deposited in droplet form and subsequently dried via evaporation of the suspending agent. The droplet drying process often results in deposition of ring patterns, typically referred to as the coffee ring effect. When drying picoliter-sized, aqueous, cellulose nanocrystal droplet suspensions, which are on the size scale of traditional drop-on-demand technologies, we observe alignment of these rod-like nanoparticles in multiple orientations. A radial alignment is formed near the center of the droplet, and tangential orientation is observed in the particles along the circumference of the droplet. Tuning particle orientation enables unique opportunities for control of nanocomposite structure required for next generation optical and electronic materials that can be generated using drop-on-demand or voxel based technologies. To successfully control the nanoparticle orientation during the drying process, a fundamental understanding of the alignment mechanism is necessary.

We have developed a 3-dimensional, continuum-based, finite element model implementing a deforming mesh to understand the impact of shear induced alignment driven by evaporative flow. Simulations of a single-phase, sessile water droplet suggest a relationship between velocity profile, shear stress, and particle orientation. We have explored the coupled effects of Marangoni flows, air convection, heat transport, and concentration dependent viscosity models on evaporation of single-phase picoliter-sized droplets leading to new insights into particle orientation in the evaporation of colloidal droplets that result from a complex interplay of initial concentration, substrate surface functionality, and particle-particle interactions. Velocity and shear stress profiles indicate the development of radial flows enacting shear stresses on the fluid driving the alignment of rod-like nanoparticles in the flow direction which varies with location in the confined flow field.

522259 Characterization of Poly(ether imide) Towards the Development of a Fused Filament Fabrication (FFF) Process Model

Abstract:

Fused filament fabrication (FFF) is the most common method of AM that relies on thermoplastic extrusion. As the AM field is transitioning from prototype visualization to functional prototypes and end product development, one of the greatest hurdles is the inherent z-axis (interlayer) weakness of the processed parts resulting from weak adhesion between deposited layers, producing anisotropic mechanical properties. This poor adhesion stems from insufficient polymer inter-diffusion between the layers, referred to as â€œself-healing.â€

To better understand the evolution of part strength during the FFF process, we are developing a model to describe the self-healing phenomenon that couples the complex thermal and geometric profiles with diffusion at the layer interface. This model is strongly dependent on the transient behavior of the polymer governed by the processing and environmental conditions present during the printing process. Additionally, we are examining the effect of adding a continuous composite fiber with anisotropic thermal conductivity. Our preliminary work has developed a realistic three-dimensional model of a printed part which will accurately represent the interface between individual layers. Previous models have utilized idealistic models based on circles and ovals of equal size. Our results prove the individual layers are flattened ellipsoids that vary in diameter depending on their height from the heated bed. We have also analyzed the complete re-entanglement time of the neat polymer to quantify the extent of self-healing, combined with characterizing the thermal conductivity of the composite filaments. The corresponding mechanical strength is driven by interlayer adhesion and characterized as a function of processing conditions. Our preliminary rheological findings of a poly(ether imide) amorphous thermoplastic indicate that times for complete re-entanglement range from 47 seconds to more than 22 minutes for isothermal temperatures of 375 °C and 345 °C, respectively. Since the cooling process is non-isothermal, and the layers remain above their glass transition temperatures for less than 3 seconds, this would limit the amount of chain re-entanglement at the layer interface and corresponding bond strength between layers of a printed part, and suggests that a modified printing process is required to create stronger parts with this material. The inclusion of the composite fiber does increase the time at which the temperature of the polymer is elevated and should help increase the amount of self-healing that occurs during the printing process.

537447 Additive Manufacturing of Core-Shell Microparticles Containing Thermosetting Resins

Abstract:

Cold spray deposition is a common additive manufacturing method for fabricating metal coatings that involves the spraying metal particles at high-speeds. However, this scalable, additive manufacturing process has not yet generated solid polymer coatings. Our study enables the cold spraying of solid polymer materials at room temperature. Here, we developed novel, core-shell thermoplastic microparticles containing a high concentration of thermosetting resins. Suspension polymerization was employed to produce the microparticles due to its many advantages, namely low conversion cost, mature scale-up processing, and a versatile tunability in terms of particle size and morphology. The thermal properties, size distribution, morphology, and composition of the particles were explored and fine-tuned in an effort to optimize the deposition of the particles at room temperature using cold spray technology. The outcome of this study is a novel form of sprayable thermosetting powders that deposit to form a coating with a high performance, easy operation processing, significant environmental benefits, and a commercial feasibility.

532737 Supersonic-Impaction Printing of Flame-Made Doped-Perovskite Nanoparticles

Abstract:

Perovskite nanoparticles find a range of applications in emerging electronics such as transparent conductors, solar cells, light-emitting diodes and sensors such as gas sensing. Such electroactive oxides typically have melting temperatures over 1000°C. Subsequently, their deposition, calcination and sintering also require extremely high temperatures. As a result, substrate compatibility with high temperature treatments is needed to facilitate usability and implementation of such materials. This is therefore a bottleneck for their application on substrates that are susceptible to high temperatures and oxidizing environments such as plastics or metals. More recently, the advent of additive manufacturing has shed new light on the deposition and growth of oxide materials. Among them, perovskite materials synthesized via sol-gel methods followed by inkjet and aerosol printing has shown promise for being able to be deposited on low temperature substrates. In this presentation, we will be showcasing a new concept and design for a deposition chamber intended for the supersonic impaction of aerosolized nanomaterials. Further, we will also discuss the implementation of a flame spray reactor as the source of synthesizing pure Barium Stannate ($BaSnO_3$) nanoparticles that are impacted using supersonic deposition to create electrically conducting perovskite thin films.

The principle of supersonic deposition is primarily based on acceleration of an aerosol-containing gas stream through a converging-diverging (de Laval) nozzle. This de Laval nozzle creates a high pressure drop across itself when downstream is pumped to moderate vacuum and upstream is maintained at a higher pressure. The pressure gradient across the nozzle accelerates the flowing aerosol to supersonic speeds and subsequently impact them onto the substrate. The kinetic energy of the particles is converted into heat upon impaction and thus the process can create a homogeneous thin film of sintered materials. Through computational fluid dynamics simulations, we found a number of parameters including the effect of gas velocities, nanoparticle density, size, trajectories and other physical properties to affect the efficiency of impaction. Unlike other precursors for competing additive manufacturing techniques such as inkjet, aerosol and electro-hydrodynamic jet printing, supersonic impaction does not require the use of any surfactant stabilized nanoparticle inks.

For this reason, flame spray pyrolysis was chosen as a scalable route for the synthesis of pure aerosolized ceramic nanoparticles without the use of any surfactants. More recently, flames are also being implemented for the synthesis of complex oxides such as tertiaries and perovskites. We found that the critical parameters that affect the flame temperature are the precursor and dispersion oxygen flow rates and enthalpy of the precursor material. A higher feed rate and precursors with high enthalpies ensure higher temperatures of the flame. Doped $BaSnO_3$ is an electroactive perovskite material and has a large indirect intrinsic band-gap of 3.2 eV. We will discuss our developments pertaining to the incorporation of electrical properties in the thin films of $BaSnO_3$ *via* vacuum annealing to induce oxygen vacancies and incorporation of external dopants that can substitute the Ba^{2+} site (n-type with La) or Sn^{4+} sites (p-type with K, Cs).

541761 Simulating Powder Handling Processes in Additive Manufacturing Using the Discrete Element Method

Abstract:

Additive Manufacturing is a rapidly advancing manufacturing area that enables users to produce complex parts in a short amount of time. One common approach to AM in industry involves the deposition and selective sintering/melting of successive layers of metallic powders. Through repeated build-up of powder layers, complex metal parts can be produced.

There are a wide range of metal powders available for AM production and each has its own characteristics. During printing a powder must be stored, transported, and processed within the printing equipment to produce the final part. Each of these stages need to be performed efficiently by the printer to ensure an optimum printing process.

The specific composition and behavioral traits of the powder used in AM is highly influential on printer set-up and resulting part manufacturing quality. Introducing a new powder into an AM process means that the set-up of the printing device needs to be configured to handle this new material effectively. Because of the variability of powder properties, flowability and susceptibility to environmental conditions, it is not straight-forward to make decisions as to how to reach the optimal configuration. At present, a common method to adjust printer set-up to account for the influence of a specific powder requires trial and error testing, or a reliance on know-how from previous work with the same printer and powder combinations. This takes time and extends the overall part production cycle.

Discrete Element Modelling (DEM) is a simulation method for recreating bulk and granular material behaviors in a virtual environment. It is a method used in other powder-focused industries, such as pharmaceutical, to provide engineering insight into powder handling and processing operations. By using DEM simulation, engineers can test design and process concepts and extract meaningful information from virtual versions of their material and equipment to help in the decision-making process.

The same opportunities exist in AM, and such a simulation method offers engineers an alternative to physical trial and error testing when trying to understand how key stages in part production will perform. In this presentation we will discuss the application of DEM, using EDEM software, to the additive manufacturing print process. Examples will be given to demonstrate how DEM can be used to help engineers understand the performance of powder delivery systems, as well as gaining insight into metrics such as powder porosity variation during the powder-layering process.

533694 3D Printed Complex Dosage Forms Embedded with Engineered BCS Class II Drug Particles

Abstract:

Recent work suggests complex structured dosage forms can be manufactured where drug particles can be integrated via 3D printing technology. However, incorporating engineered drug particles into 3D printed structures so that various intermediate and final product properties can be enhanced, for example, good mechanical properties and drug content uniformity of feed materials, has yet to be explored. It is hypothesized that engineered particles enable uniform distribution of drug into molten polymeric matrix via twin screw extruder (TSE). This distinct feature of engineered particles leads to obtaining enhanced dissolution profile from 3D structured dosage forms. The printable filaments containing varying drug loading (DL) %, and different extent of particle engineering as well as filament their mechanical properties were considered in this investigation. Griseofulvin (GF) and Hydroxypropyl cellulose (HPC) SL grade was used as the model BCS Class II drug and the polymeric matrix, respectively. The feed materials having four different DL% (5, 10, 15, and 25 wt.% GF) incorporating engineered and non-engineered particles were extruded by HME and 3D structured tablets from the feed materials were successfully created via FDM 3D printer. No drug degradation was observed in TGA results for any cases. All the drug particles kept their crystalline structure from HME extrusion (140 °C) as well as FDM printing (170 °C). The feed materials containing engineered particles exhibited higher tensile strength and better modulus of elasticity, regardless of the drug loading. Moreover, 3D printed tablets containing engineered particles showed enhanced drug release for all DL%. Overall, the 3D printed tablets embedded with engineered drug particles resulted in tablets without any degradation/recrystallization issues and demonstrated enhanced dissolution profiles for four different dosage amounts.

529744 3D Printing for Rapid Prototyping of Innovative Process Equipment for Pharmaceutical Crystallization

Abstract:

The development of pharmaceuticals involves various stages with preclinical and clinical trials. A short time-to-market is essential to maximize the patent protection and to avoid any loss in potential revenues.[1] Furthermore, most pharmaceuticals in development do not reach commercial manufacturing. However, for those pharmaceuticals that do reach market, efficient and novel processing equipment can provide a competitive edge. Hence, rapid prototyping of processing equipment during development is important to minimize time-to-market, while maximizing early learning on processing. 3D printing is an attractive technology to fabricate customized and novel processing equipment almost instantaneously. However, despite this potential, the use of 3D printing during early pharmaceutical process development is still in its infancy.

3D printing has found applications for the fabrication of micro-reactors and to some extent for mesoscale reactors as well.[2] In contrast to reactors, the use of 3D printing to fabricate innovative designs of separators has received limited attention. For example, for crystallization, which is a key unit operation in many pharmaceutical processes, no studies on the use of 3D printing for rapid prototyping have been reported to the best of our knowledge. Pharmaceutical crystallization is complicated by the need to optimize intrinsic crystal quality attributes such as size distribution, which affects the performance of downstream processes and possibly the efficacy of the final product. Optimization of crystallization requires control over various mixing processes. 3D printing can be a flexible and rapid method to fabricate innovative structures of crystallizers for optimized mixing at all relevant scales during pharmaceutical development.

The objective of this work is to develop and characterize 3D printed prototypes of innovative crystallization equipment with different structures for mixing. Two cases will be investigated. A common feature of both cases is that the mixing mechanism does not require moving parts, which makes fabrication with 3D printing attractive.

The first case involves a 3D-printed airlift crystallizer for protein crystallization. Protein crystallization in an airlift crystallizer has not been reported yet, but is attractive as protein crystals are generally more fragile in comparison to crystals of small-molecule active ingredients.[3] Therefore, mild and homogeneous mixing conditions are required, which can be provided by airlift columns due to the low maximum shear forces.[4] An airlift crystallizer was designed and fabricated with a desktop stereolithography (SLA) printer. The hydrodynamics of the column have been characterized for lysozyme as model protein. A sufficiently high upward velocity in the column demonstrates practical feasibility for protein crystallization in an airlift crystallizer. Next, the crystallization performance of the airlift crystallizer is characterized in this work and compared to a conventional stirred tank crystallizer.

The second case involves antisolvent and reactive crystallization in continuous flow. Rapid mixing is achieved by using static mixers. The effectiveness of kenics-type of static mixers has previously been demonstrated for anti-solvent crystallization of ketoconazole [5]. In this work, Y-mixers and komax-type mixers have been designed and fabricated using a desktop stereolithography (SLA) 3D printer. Preliminary studies on reactive crystallization of barium sulfate have been conducted, which demonstrate practical feasible and reduced agglomeration when using the komax mixer in comparison to when using a Y-mixer. The stability of many stereolithography resins used for 3D-printed structures is good for aqueous-based system. However, pharmaceutical crystallization often involves organic compounds. Therefore, this case particularly focuses on the use of various 3D printing techniques to address stability issues for organic solvents. In the long term, this research aims to develop a platform around 3D printing, which allows users to select, design and fabricate innovative crystallization equipment rapidly to achieve shorter time-to-market and intensified pharmaceutical crystallization.

Acknowledgement: The work described in this abstract was supported by a grant from the Research Grants Council of the Hong Kong Special Administrative Region, China (Project No. 16242916).

References

[1] G. Reklaitis, J. Khinast, and F. J. Muzzio, "Pharmaceutical engineering science—New approaches to pharmaceutical development and manufacturing," Chem. Eng. Sci., vol. 65, no. 21, pp. iv–vii, Nov. 2010.

[2] A. K. Au, W. Huynh, L. F. Horowitz, and A. Folch, "3D-Printed Microfluidics," Angew. Chemie - Int. Ed., vol. 55, no. 12, pp. 3862–3881, 2016.

[3] S. Tait, E. T. White, and J. D. Litster, "Mechanical Characterization of Protein Crystals", Part. Part. Syst. Charact., vol. 25, no. 3, pp. 266-276, Sep 2008.

[4] A. Soare, R. Lakerveld, J. Van Royen, G. Zocchi, A. I. Stankiewicz, and H. J. M. Kramer, "Minimization of attrition and breakage in an airlift crystallizer," Ind. Eng. Chem. Res., vol. 51, no. 33, pp. 10895–10909, Aug 2012.

[5] A. J. Alvarez and A. S. Myerson, "Continuous plug flow crystallization of pharmaceutical compounds," Cryst. Growth Des., vol. 10, no. 5, pp. 2219–2228, May 2010.

529577 Direct Write of UV Curable Polymer Bonded Magnets

Abstract:

Today, almost all commercial magnet producers are using the conventional sintering-based procedure to manufacture magnets. Polymer-bonded hard magnets have recently gained popularity as a novel magnet manufacturing technique due to its design freedom regarding shapes and magnetizing structures. Comparing to the conventional sintering process, the maximum energy product $(BH)_{max}$ of the polymer-bonded magnets is lower. However, the fabrication of polymer-bonded magnets does not require high temperature treatment and thus can be more efficient. Furthermore, no current mass production technologies can produce magnets with complex geometries.

This work utilizes extrusion-based direct write (DW), combined with UV curing, to fabricate polymer-bonded magnets at room temperature. DW is an additive manufacturing technique enabling the deposition of electronic components and functional or structural patterns using different types of materials, directly without utilizing masks or subsequent etching processes. Compared to the molding process, DW provides intricate details and smooth surface finish. In this work, Nd-Fe-B powder are directly mixed with photo-curable binders to form a printable DW ink. Experiments were carried out to understand the relationship between magnetic powder solid loading, processing, and mechanical strength. Magnetic performance of printed parts was compared to other polymer-bonded and sintered magnets. The study opens a new pathway for preparing polymer-bonded magnets at room temperature for various magnetic applications.

523345 Before You Click "Print": Regulatory Considerations for 3D Printed Oral Drug Products

Abstract:

Additive manufacturing/3D printing of oral drug products and medical devices has become more attractive to shift the manufacturing strategies to support individualized medicines of the future. The US Food and Drug Administration (FDA) is charged with protecting and promoting the public health by ensuring these products are safe and effective. To that end, FDA's recent approval of a 3D-printed drug product in August 2015 (SPRITAM®) introduced a new chapter in pharmaceutical manufacturing for solid oral dosage form. Although the first approved 3D-printed drug product is a solid oral immediate release product, the majority of 3D printing research for oral delivery has been focused on controlled release, targeting and precise delivery for extremely low dose drugs. Various principles of 3D printing of oral drug products could support production of complex and personalized medicines on demand. However, standardization of some aspects of the 3D printed dosage form and associated manufacturing process may be required to achieve this objective. In this presentation, technical and regulatory considerations for development and control of 3D printed oral products will be discussed for various 3D printing technologies.

DISCLAIMER

This presentation reflects the views of the presenter and should not be construed to represent FDA's views or policies.

531710 Two-Color Photo-Inhibited Systems for Rapid Additive Manufacturing

Abstract:

Additive manufacturing (AM), also known as 3D-printing, is exciting disruptive technology that may upend contemporary manufacturing practices. Production in AM generally proceeds by layer-by-layer addition of materialâ €"typically thermoplastic or photopolymerizable resinâ€"to gradually build up the desired geometry, allowing for the production of highly individualized parts with near limitless complexity. The achievable print speed is still one of the major limitations of the current technologies with layer-by-layer devices only able to achieve vertical print speeds of several millimeters per hour. Recently, a layer-less (or continuous) method for resin based AM, termed continuous liquid interface production, was developed. The method relies on a thin diffusion-controlled oxygen-inhibited dead zone to eliminate adhesion to the projection window during printing, allowing for an order of magnitude increase in the achievable printing speeds when compared with traditional SLA devices.

Photo-inhibitor molecules that inhibit polymerization in the presence of light may be an attractive alternative inhibition mechanism to oxygen-inhibited systems. The advantages of this type of system are that diffusion of oxygen no longer controls the thickness of the dead zone, and the limit of this zone is theoretically much larger. Larger dead zones allow faster polymer reflow under the printing part and, therefore, faster build speeds. This talk will present a novel method using a two-color irradiation scheme to spatially control polymerization to generate dead zones for rapid AM and enable additional functionality over conventional continuous additive manufacturing methods.

531942 Glass-Forming Polymer Networks for Shape-Memory Contact Printing

Abstract:

Current methods for microscale pattern transfer fail to combine high feature resolution, ability to pattern large areas, high throughput and cost effectiveness. Over the past years, several researchers have demonstrated how films of shape-memory polymers can exhibit topographical memory, leading to switchable adhesion or optical properties. We are designing and developing new stimuli-responsive materials to ultimately enable high fidelity pattern transfer by thermal modulation of the elastomeric contact printing process.

Different glass forming polymers that can withstand high elongation and store large amounts of elastic strain energy in compression are under investigation. Two systems will be described: (i) small molecule thiol-ene networks that are softened by addition of dithiol reagents, and (ii) glass forming t-butyl acrylates that are lightly crosslinked. Their shape-memory behavior in tension and in compression will be reported. Furthermore, a custom-built materials testing / contact printing apparatus will be used to set compressive strain between two soft surfaces while measuring stress and contact area. Contact mechanics experiments are modeled and interpreted using finite element analysis to offer deeper insight to the nature of contact and detachment.

531279 High Resolution Nanoparticle Sizing with Maximum a Posteriori Nanoparticle Tracking Analysis (MANTA)

Abstract:

Unbiased estimation of nanoparticle size distributions remains a challenge affecting numerous fields including nanotechnology and biopharmaceuticals. Current methods for particle sizing, such as dynamic light scattering, analytical ultracentrifugation, and field-flow fractionation, can suffer from a combination of intensity biases, difficult sample preparation, insufficient sampling, and ill-posed data analysis. The last issue, ill-posedness of the governing equations for these particle sizing techniques, guarantees that noise in the data will make reproducible measurements impossible without aggressive regularization. We have developed a new Bayesian method, Maximum A posteriori Nanoparticle Tracking Analysis (MANTA), for estimating the size distributions of nanoparticle samples from high-throughput single particle tracking experiments. In this method, unbiased statistical models are derived for two observable quantities in a typical nanoparticle trajectory (mean square displacement and the trajectory length), finite elements are used to discretize the particle size distribution, the weights of the finite elements are found via MAP estimation, and robust cross validation is applied to mildly regularize solutions to the MAP problem. MANTA infers nanoparticle size distributions with unprecedented resolution. The precision of the method is assessed by performing extensive Brownian dynamics simulations with synthetic particle size distributions that are known exactly. In experimental testing with gold nanoparticles, MANTA is capable of reliably resolving disparate populations of particles with relative variation in size as small as 30%.

537670 Core-Shell Graphene/Silicon Nanoparticles for Use As Lithium-Ion Battery Anodes

Abstract:

Although lithium-ion batteries have the highest specific energy capacity of any rechargeable battery chemistry in wide-spread use today the cycling stability and specific energy capacity of these batteries are still problematic for all-electric vehicles and a host of other high energy applications. Increasing the capacity of these batteries is crucial to keep pace with the demands of these new applications.

To decrease the weight of the anode, researchers have turned to materials with higher specific capacity than graphite, the most commonly used anode material in commercially available batteries. Silicon, with the highest known lithiation capacity of any material, has about 10 times the theoretical capacity of graphite. It has been noted that increasing the capacity of the anode up to about 1000 mAh/g greatly increases the total capacity of a lithium-ion battery cell.[2]

There are two well-known issues with utilizing silicon in anodes, both related to the fact that silicon undergoes a high expansion ratio upon lithiation (during the cycling of a battery). As the battery is charged (lithiated) expansion occurs in the anode, and as the battery is discharge (delithiated) contraction occurs in the anode. The volumetric expansion is only 10% percent for graphite, but is approximately 300% for silicon. The high volumetric expansion and contraction in the silicon can induce mechanical stress and leads to reformation and breakage of the solid electrolyte interface (SEI) during cycling producing loss of capacity and stability. The SEI layer is a polymeric boundary which naturally forms from the electrolyte at the surface of the anode and is known to further protect the electrolyte from further reactions at the surface of the anode. However, if the expansion ratio of the anode materials is sufficiently high, it appears that the SEI layer is broken and reformed to some extent during each cycle, thus leading to loss of efficiency and cycling stability.

Silicon has previously been used by a number of researchers, however simple admixture tend to have poorer cycling stability. Other more complicated techniques use hazardous materials or are difficult to scale up. Our approach consists of encapsulating silicon nanoparticles inside hollow graphene oxide membrane particles (HGOM) and thus, it avoids the uneven distribution of the silicon particles, where the HGOM particles act as nanoparticle carriers. When silicon nanoparticles are encapsulated inside these hollow sacks, they do not occupy the entire volume leaving some "void space" necessary to accommodate the physical change in volume of the silicon nanoparticles during cycling of the batteries. Such particles are often called "yolk-shell" particles to distinguish these from simpler core-shell particles in which there is no void space. In our method, the solid electrolyte Interface (SEI) forms around the graphene shells rather than around the silicon particles, which has a much smaller surface area providing larger stability and enhancement of capacity.

We have recently published a method to obtain these hollow graphene oxide membrane particles with controlled size and thus, controlled void space. Here, we propose a new facile method, for engineering the required void space and controlling the formation of the SEI around HGOM particles, which are assembled into anodes for Lithium ion batteries.

We employ a different technique based upon the newly discovered method of encapsulating silicon nanoparticles within graphene oxide shells, through the process of emulsification-diffusion, followed by sublimation of naphthalene and assembly into anodes. To our knowledge this method has never been accomplished before. This method has the potential to be scalable, while being able to control the void space within the core-shell structure. Although a sacrificial scaffold is used to create the void space, there is a great potential to recycle this material since no reactions occur during sublimation of the sacrificial scaffold.

Thermal reduction (pyrolysis) of the silicon/HGOM core shell particles yields highly conductive composites. Since the silicon nanoparticles have very high specific capacity, the resulting composite also has high specific capacity, thus increasing the specific capacity of lithium-ion batteries which employ these composites. We obtained an electrode with high cycling stability and a storage capacity greater than 1800 mA h/g after 200 cycles and greater than 1200 mA h g-1 after 450 cycles. The coulombic efficiency of the anode rapidly rose as cycling continued, eventually approaching 99.9%. The electrode materials were characterized via laser diffraction, zeta potential, EIS, cyclic voltammetry, FE-SEM, TEM, Raman Spectroscopy, BET, and, XRD.

Multiparameter Paramagnetic Particle Characterization by Dark-Field Imaging

Abhinav Sannidhi[1], Paul W. Todd[2], and Thomas R. Hanley[1]
[1]Department of Chemical Engineering, Auburn University, AL 36849
[2]Techshot, Inc., 7200 Highway 150, Greenville, IN 47124

Abstract

Magnetic particles widely varying in composition, size, and magnetization are currently in use in ferrofluids, biochemical separations, cell purification, cell labeling, cell physiology, endocytosis, endosome research, drug targeting, analytical biochemistry, environmental analysis, magnetic resonance imaging, hyperthermia, in vivo diagnostics and others. Most producers, whether research or commercial, are most interested in particle size uniformity and colloidal stability, but magnetization is never specified and is typically mentioned only in terms of percent iron oxide. Bulk measurements of magnetic susceptibility, using a squid magnetometer, vibrating sample magnetometer, Guoy's scale, Faraday's scale, inductive method or MR method provide an average value at best and do not account for the distributed nature of this variable or for particles with zero or very low susceptibility. By particle tracking velocimetry in dark field illumination, we have been able to measure multiple characteristics of several thousand individual particles per sample. The Hyperflux™ velocimeter is utilized to provide the quantitative video analysis of cells and particles using high definition camera/microscope and capture the images of the particle trajectories in an isodynamic field, the image analysis software converts the image data to the parameters of interest based on velocimetry calculations. By measuring calibrated magnetophoretic mobility and calibrated diameter of every particle we report magnetic susceptibility on a particle-by-particle basis and provide susceptibility distributions for numerous commercial products. In conclusion, the simultaneous determination of distributions of diameter, magnetophoretic mobility, and susceptibility using the commercial velocimeter provide good scope to know the detailed analysis of the sample on a particle-by-particle basis in different fields of science.

Keywords: paramagnetic particles, magnetophoretic mobility, characterization and quantification, dark field imaging, susceptibility distribution

Introduction

Magnetic particles have been extensively used in research and production since they are used in bioengineering and biomedical applications. The particle size and magnetic susceptibility play a prominent role in describing the quality of the sample. Different methods and devices have been recognized previously to characterize the magnetic particles such as particle size determination by Dynamic Light Scattering (DLS), observation of the particle microstructure by Scanning Electron Microscopy (SEM), Transmission Electron Microscopy (TEM), Atomic Force Microscopy (AFM), Magnetic force microscopy, determination of the elemental composition and crystalline phase information by X-Ray diffraction (XRD) and energy dispersive X-Ray spectrometer (EDX) respectively, magnetic susceptibility measurement by Superconducting quantum interference device (SQUID) [1], [2], Vibrating sample magnetometer(VSM) [3], Faraday microbalance [4], Guoy's scale [5], torque meter [6], The Bartington Magnetic Susceptibility System sensors [7], magnetic magnetoresistance sensors [8], and modified magnetization curves [9]. All these methods estimate either bulk average particle size or bulk average magnetic susceptibility. The characterization of magnetic susceptibility distribution on a particle by particle basis can be achieved by simultaneous assessment of both the particle size and the magnetophoretic mobility of the magnetic particles in the magnetic field [10]–[12]. Magnetophoretic mobility, which defines the key property of the magnetic particles, depends on the particle size, magnetic susceptibility, and viscosity of the medium. With the help of magnetophoretic mobility data, the magnetic susceptibility distribution can be easily extracted based on its relationship with the magnetophoretic mobility.

Hyperflux™ velocimeter is an instrument that has been used to measure the particle count, magnetophoretic mobility and particle size distribution on a particle by particle basis of the magnetic particles and magnetically labeled cells. This instrument provides twenty-three different characteristics of the sample along with particle size and magnetophoretic mobility [13]. It is also used in quantitative cell analysis as a surrogate to Fluorescence Activated Cell Sorters (FACS) [14] and in quantifying the nanoparticle endocytosis [15]. Individual particles' magnetophoretic mobility and diameter are determined using dark field image analysis within a few minutes by image analysis software. The optimum instrument settings which include threshold (intensity) settings, size gating for the particle size range to obtain accurate results have been found in the past research. The size calibration of the instrument is very crucial in determining the exact particle diameter of the sample as the dark field imaging technique over-estimates the particle size based on image pixel size [13].

In this present research, the extension of the size calibration capability of the instrument has been focused as there is a need to find the relationship between instrument reported diameter and actual particle diameter. The magnetic susceptibility distributions and particle size vs susceptibility plots of different magnetic beads were determined.

Theory

Magnetophoretic mobility is defined as the ratio of the terminal velocity of the particle, v, to the magnetic pressure, $\nabla\left(\frac{B_o^2}{2\mu_o}\right)$ (gradient of the magnetic field also denoted as S_m),

$$\mu_m = \frac{v}{\nabla\left(\frac{B_o^2}{2\mu_o}\right)} \qquad (1)$$

Units of magnetophoretic mobility are $\frac{m^3}{T\,A\,s}$ (cubic meter per tesla-ampere-second)

In the isodynamic field, S_m is constant. Then the value and direction of $\boldsymbol{v}$ are fixed. In other words, the same type of particles will move in a nearly uniform motion.

Magnetophoretic mobility derived from the force balance on a spherical particle in a viscous medium could be expressed as

$$\mu_m = \frac{2R^2\Delta\chi}{9\eta} \qquad (2)$$

Where R is the radius of the spherical magnetic particle, $\Delta\chi$ is the difference in magnetic susceptibility of particle and fluid, η is the viscosity of the fluid medium.

Materials and Methods

Calibration beads

Twenty-three calibration beads which include magnetic and non-magnetic beads with different vendor diameters were obtained and studied. Sixteen different magnetic beads (P1 – P16) having the size range from 0.68 μm to 5.01 μm, seven different non-magnetic beads (P17-P23) ranging from 0.43 μm to 9.61 μm were used, as shown in Table1.

Hyperflux Velocimeter

The magnetic particle velocimeter, obtained from IKOTECH, LLC, consists mainly of three components- a channel cell with magnetic assembly, video microscopy system, and image processing software. Its hardware includes a channel cell, a magnetic assembly which provides isodynamic magnetic field, a vacuum pump for sample changing and movement, a high sensitivity and high-resolution Grasshopper 2.0 MP CCD camera combined with 2X to 8X Edmund optics telecentric lens to capture dark field images of the magnetic particles, and a Dell computer with Linux Kubuntu operating system installed with software capable of capturing and analyzing the images of particle trajectories. The image data is analyzed by the software which generates 23 different characteristics of magnetic particles based on the intensity threshold and size limits.

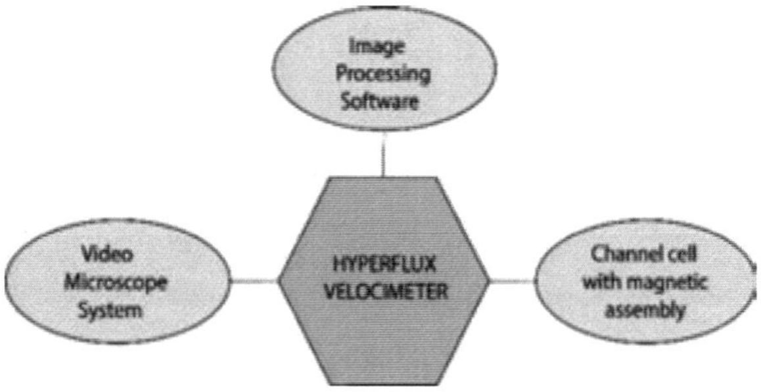

Figure 1- Components of Hyperflux velocimeter

Figure 2- Front view of the instrument

Results and discussion

Two forms of machine-based data were used to obtain diameter data from the images:

Cytotest Diameter Measurements**:** After running the sample beads of known vendor diameter through the channel cell, the image acquisition tab was used to collect images of the particles and then by default after successful completion of 10 two-second video sets at a rate of 30 frames per second, the project file was opened using Cytotest software which gives the magnetophoretic mobility distribution and the average diameter of the sample based on the collected dark field image data.

Peak Bin Diameter Calculation**:** The Magex software generates a "tracks.csv" spreadsheet which includes the twenty-three different characteristics of the particles tracked by the image analysis on a particle-by-particle basis. The 'tdiameter' data column, which gives the particle diameter analyzed from dark field images, was used to generate the diameter histograms shown in Figure 3 by varying the values of the bins based on their size (usually the diameter values estimated from the darkfield microscopy images are higher than the original diameter). The diameter histograms were generated using spreadsheet software with the average bin values of 0.8 μm to about 20 μm. Figure 3 depicts the diameter histograms for different beads. Identification of appropriate peaks in the noisy data is heuristic. All the peaks at bin value 2.2 or below were considered as debris, but the 0.7-micron particles were very close to this range. Then the next tall bin (I) above this was identified, then another tall bin some 1.5 X to 2.2 X times the previously selected bin value (I) was sought and considered to represent doublets, if such a pair exists, then the lower (on the x-axis) peak is declared as the peak bin diameter value for that particle type; otherwise the bin (I) should be considered as the desired peak diameter. This is easier for the larger (>2.5 μm) particles.

The two values of obtained mean diameters for all particles studied are given in Table 1.

Table 1- Different beads (magnetic and non-magnetic) used to extend the calibration capability of the instrument providing the vendor diameter, Cytotest diameter at threshold 35 and the peak bin diameter which is calculated from diameter histograms generated from the track's spreadsheet.

Particle Name	Vendor Diameter (micrometers)	CytoTest reported Diameter at threshold 35 (micrometers)	Peak Bin Diameter (micrometers)
MAGNETIC BEADS			
P1	0.68	4.48	3.2
P2	0.7	4.16	3
P3	0.87	5.73	3.7
P4	0.96	5.07	3.3
P5	1	4.48	3.5
P6	1	3.64	3.7
P7	1.1	5.68	3.4
P8	2.39	9.72	8.6
P9	2.6	10.63	8.5
P10	2.7	8.11	9.4
P11	2.7	9.03	7.4
P12	3	10.16	8
P13	3.12	7.1	11.7
P14	3.13	9.74	14.8
P15	3.27	10.92	8.9
P16	5.01	4.8	17.4
NON-MAGNETIC BEADS			
P17	0.43	3.78	3.6
P18	2.01	9.31	6.8
P19	2.9	11.4	8.3
P20	4.8	15.62	13.1
P21	4.993	15.27	14
P22	6.63	7.66	18.6
P23	9.61	7.47	20.3

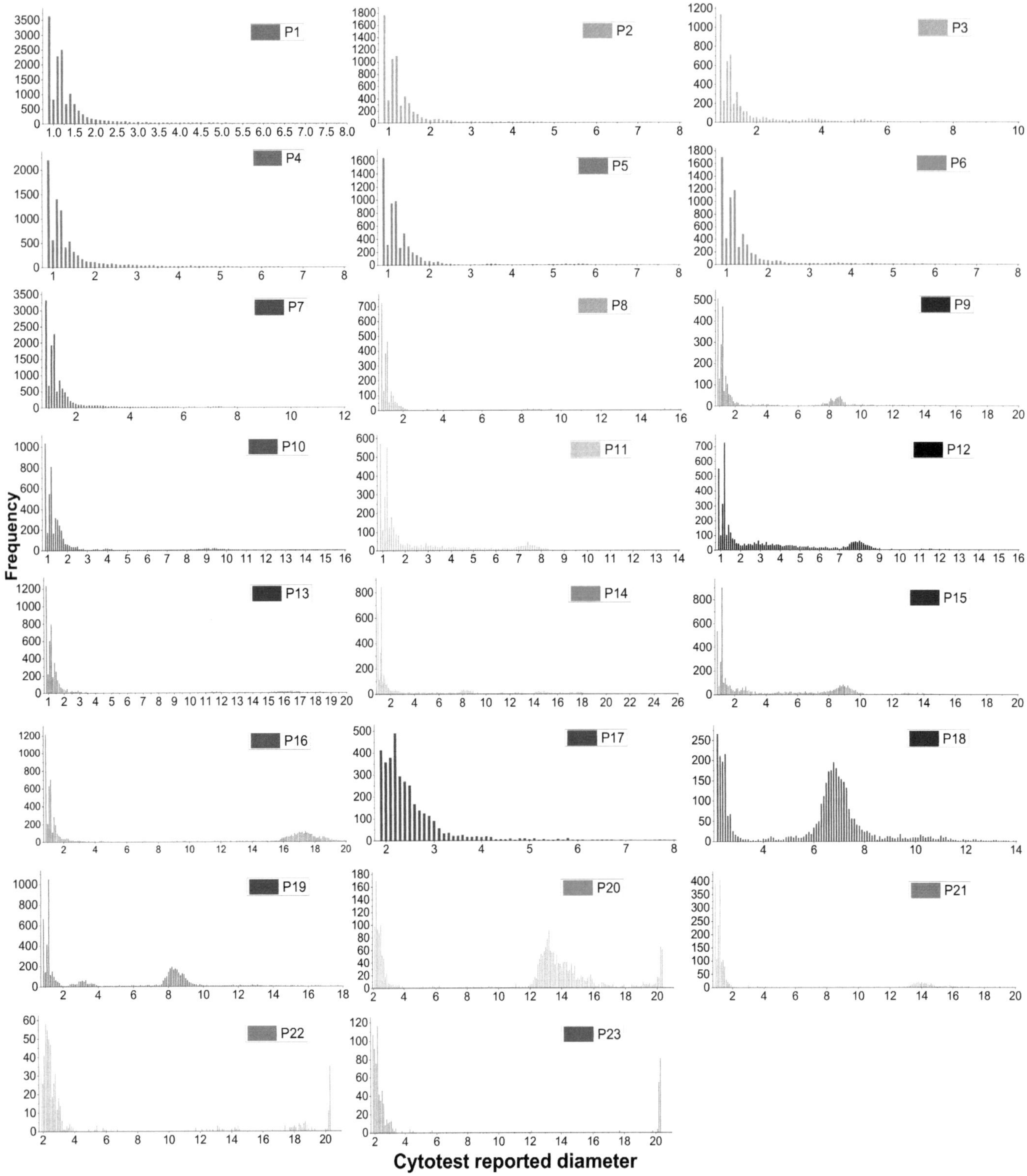

Figure 3- Histograms of diameters generated from "Track.csv" spreadsheets for 23 different beads, which were used to estimate the peak bin diameter

Calibration curves

Cytotest diameter vs Vendor diameter

The plots in Figure 4(A) show the relationship between the instrument automatically reported diameter ("Cytotest diameter") and the vendor diameter of the beads. This automatic calculation includes some debris when present. As Figure 3 shows there was a high debris count at or below 2.2 bin units for higher vendor diameters beads due to microbial contamination which was confirmed upon observing the tiny dots beside beads using Hemacytometer under an upright bright field microscope. So, the graph is considered inaccurate at high vendor diameter. The debris will "drag down" the average diameters calculated by the Cytotest program, and this can be seen in Table 1. Thus, by eliminating the three highest values of vendor diameter i.e., P16, P22, P23 a linear curve and a non-linear (second-degree polynomial) curve to the magnetic particles alone, a linear curve to the non-magnetic particles only, a linear curve to all the remaining points were generated (shown in Figure 4(A)). The curve equations, R-squared values, and Anova test p-values are shown in Table 2.

Cytotest diameter vs peak bin diameter:

The graph in Figure 4(B) depicts the relationship between the instrument reported diameter (Cytotest diameter) and the manually generated peak bin diameter. The points corresponding to the samples with high debris counts: P10, P13, P14, P16, P22, and P23 were eliminated. These tend to be the larger beads, whose diameters would be maximally "dragged down" by debris diameters. A linear and a quadratic function were applied to this dataset (shown in Figure 4 (B)) along with statistical analysis (shown in Table 2).

Peak bin diameter vs vendor diameter:

The relationship between the manually generated peak bin diameter and the vendor diameter is shown in Figure 4 (C). This is expected to be a better representation since the velocimeter data do not include debris as in the case of Cytotest diameter. A linear and a quadratic function was applied to this dataset (shown in Figure 4 (C)) along with statistical analysis (shown in Table 2). The linear coefficient of the calibration curve was positive as anticipated. Also, a linear fit with zero intercept was applied to the data.

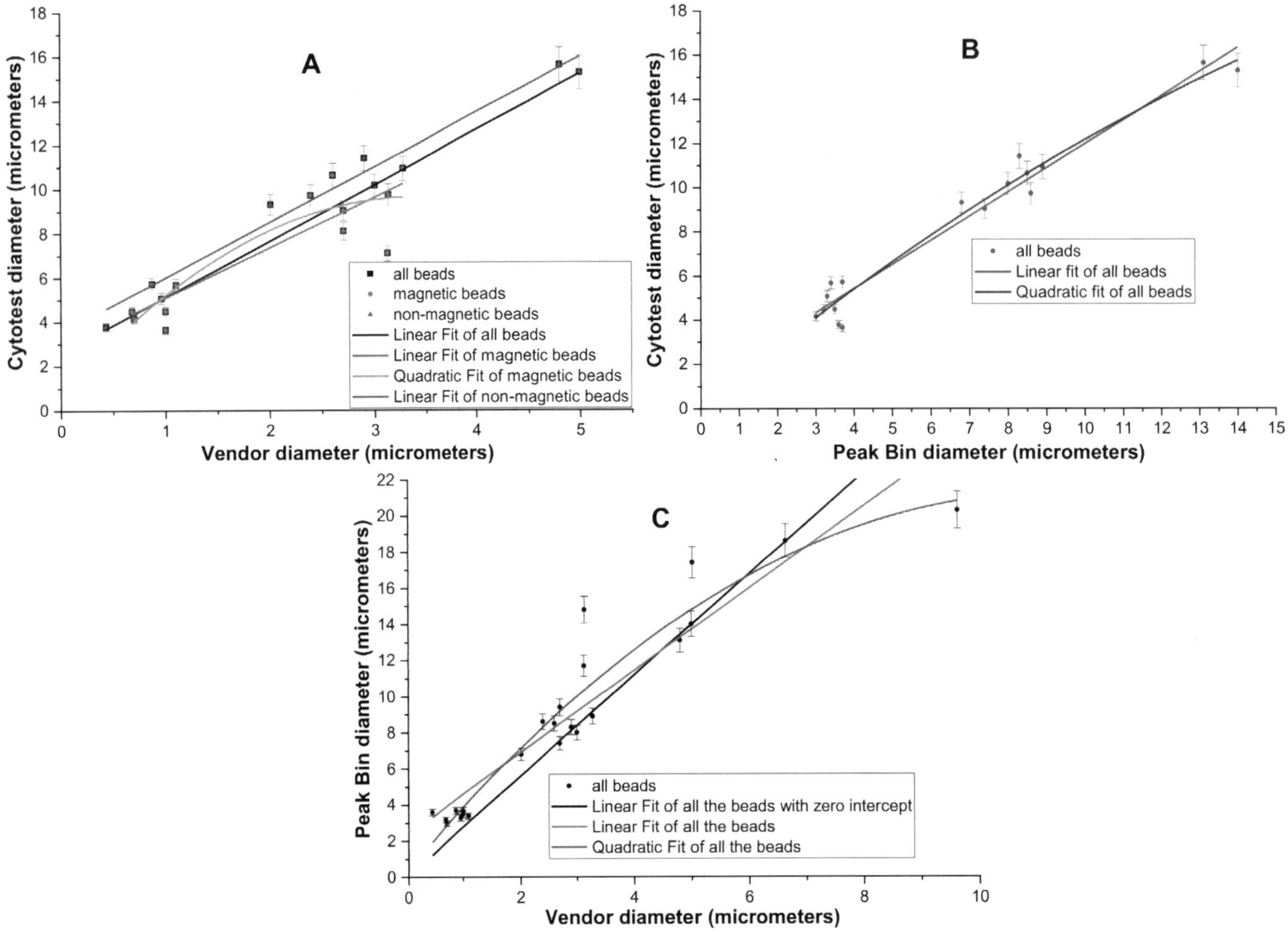

Figure 4- A: Relationship between Cytotest Diameter and Vendor Diameter a) for all the beads (linear) b) Magnetic Beads (Linear) c) Magnetic Beads (Polynomial) d) Non-Magnetic Beads (Linear). **B**: Relationship between Cytotest Diameter and Peak Bin Diameter for all the beads (linear fit, and quadratic curve fitting). **C:** Relationship between Peak Bin Diameter and Vendor Diameter for all the beads (linear fit, linear fit with zero intercept and quadratic curve fitting)

Table 2- Linear and Non-Linear Calibration curve equations along with R-squared values for the three calibration graph types applied to all beads and applied to magnetic and non-magnetic beads separately. From Anova test, at 95% confidence level (α=0.05), the slopes of all linear curves are significantly different from zero and non-linear curve equations are significant (p-value < α)

Type of Calibration Graph	Type of Beads	Curve Equation	R^2	Anova Test p-value
Cytotest Diameter vs Vendor Diameter	All beads (Linear)	$y = 2.5321x + 2.5863$	0.8926	3.71E-10
	Magnetic Beads (Linear)	$y = 2.2704x + 2.8207$	0.8072	5.37E-06
	Magnetic Beads (Poly)	$y = -0.7728x^2 + 5.2527x + 0.7268$	0.8257	2.80E-05
	Non-Magnetic Beads (Linear)	$y = 2.4997x + 3.5104$	0.9768	0.00151
Cytotest Diameter vs Peak Bin Diameter	All beads (Linear)	$y = 1.0908x + 1.0589$	0.9609	5.64E-12
	All beads (Poly)	$y = -0.0234x^2 + 1.4556x - 0.0546$	0.9661	5.11E-11
Peak Bin Diameter vs Vendor Diameter	All beads (Linear)	$y = 2.2764x + 2.3418$	0.8787	4.37E-11
	All beads (Linear with zero intercept)	$y = 2.7987x$	0.806	4.21E-06
	All beads (Poly)	$y = -0.1652x^2 + 3.7157x + 0.3486$	0.9228	7.54E-12

Magnetic susceptibility distribution

The apparent magnetic susceptibility distributions of the magnetic particles as shown in Figures 5 and 6 are calculated based on the relationship between susceptibility and magnetophoretic mobility (Equation 2) using the actual diameter calculated from the linear calibrated curve equation of peak bin diameter vs vendor diameter (Table 2), and assuming the viscosity of the medium (distilled water) to be 0.001 Pa.s (Pascal-second). Figure 5 shows the actual 2-parameter diameter vs susceptibility distribution, where the horizontal dashed line represents the vendor diameter. The graph confirms the presence of higher diameter particles with low susceptibility. As we move from P1 to P16 the formation of two populations can be observed with higher diameter value. Most of the debris and non-magnetic particles show negative or zero susceptibility, which can be eliminated by the logarithmic scale of the graph. There may be various causes of low susceptibility relative to magnetite, such as the presence of mixed iron oxides (FeO and γ-Fe_2O_3, having lower susceptibilities), the beads could be comprised of as low as 5% iron oxides, and the value of induced field, B_o, could be larger than the saturation magnetization of the beads. Figure 6 shows the single parameter apparent volume magnetic susceptibility distribution where most of the beads lie in the range of 10^{-4} to 10^{-3}.

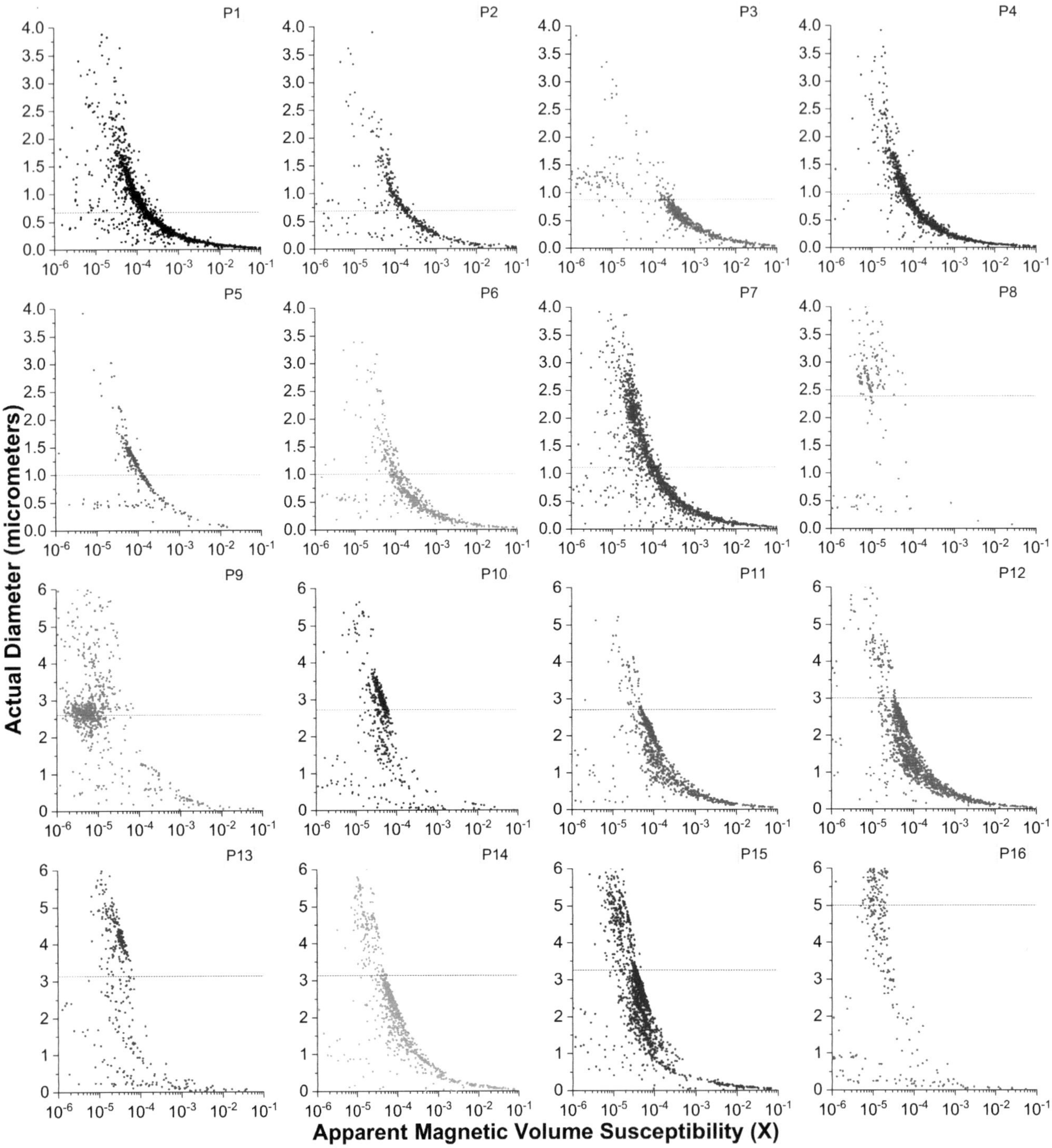

Figure 5- Two parameter Diameter vs Susceptibility distribution for 16 different magnetic particles, where horizontal dashed line represents vendor diameter

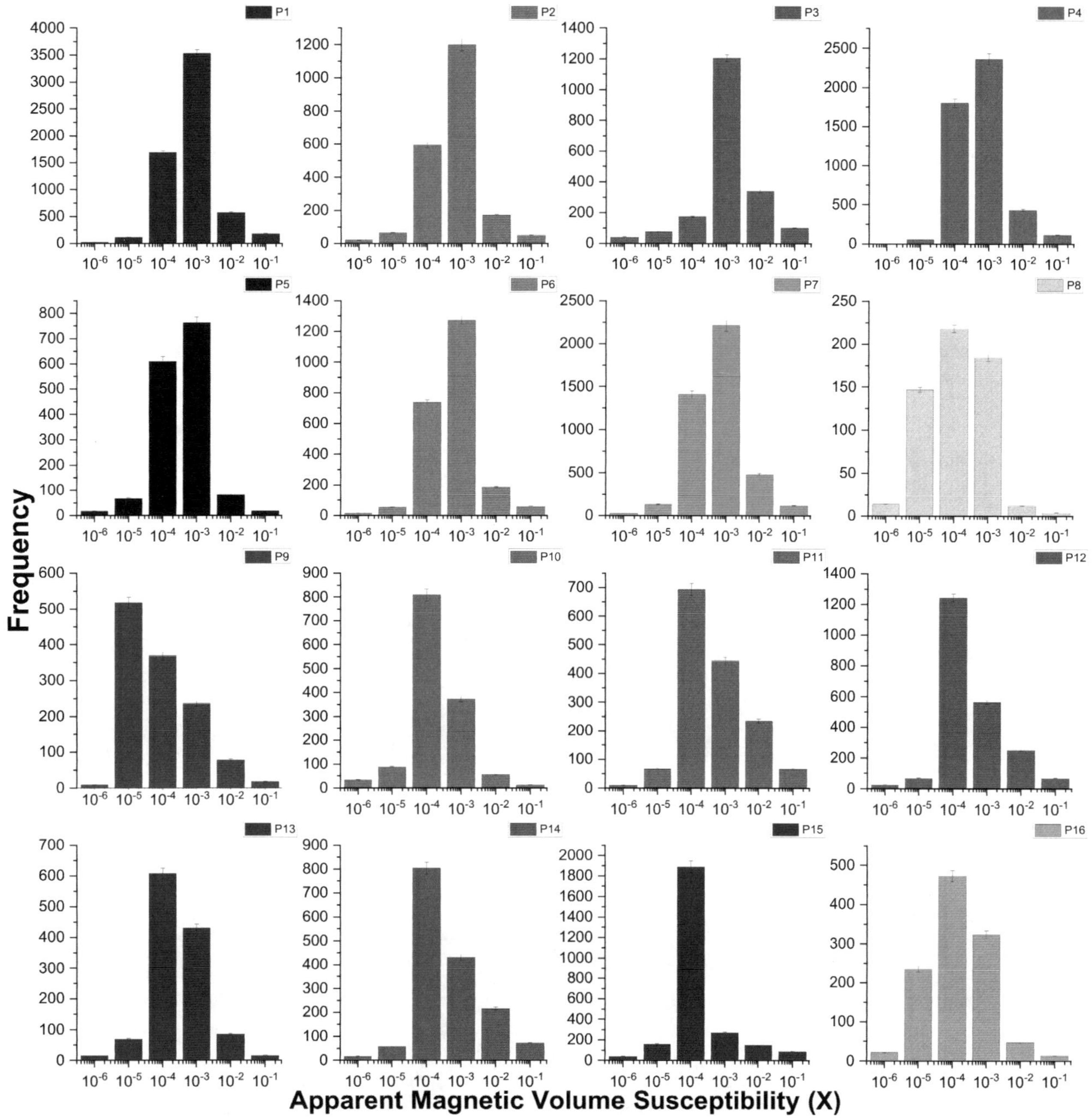

Figure 6- Single parameter susceptibility histograms of the 16 different magnetic particles

Conclusions

The size calibration capability of the Hyperflux velocimeter instrument has been extended by estimating the relationship between the vendor diameter, instrument reported diameter and peak bin diameter, using a wide range of differently sized magnetic and non-magnetic beads available commercially. The linear and non-linear calibration curves of particle size provide a means of determining each actual particle size. This study also benefits the researchers and vendors to find the exact size distribution of the sample without considering typical microbial contamination of magnetic beads, thus quantifying the quality of the particles. The simultaneous determination of actual particle size, magnetic susceptibility distribution along with particle size vs susceptibility plots on a particle by particle basis provides valuable information for product quality analysis and control.

References

[1] D. Hurt, S. Li, and A. Amann, "Versatile SQUID susceptometer with multiple measurement modes," *IEEE Trans. Magn.*, vol. 49, no. 7, pp. 3541–3544, 2013.

[2] C. Granata and A. Vettoliere, "Nano superconducting quantum interference device: A powerful tool for nanoscale investigations," *Phys. Rep.*, vol. 614, pp. 1–69, 2016.

[3] S.-H. Liao, H.-S. Huang, and Y.-K. Su, "Scaling characteristics of the magnetization increments of functionalized nanoparticles determined using a vibrating sample magnetometer for liquid magnetic immunoassays," *IEEE Sens. J.*, 2018.

[4] A. Laachir, V. Perrichon, A. Badri, J. Lamotte, E. Catherine, J. C. Lavalley, J. El Fallah, L. Hilaire, F. Le Normand, E. Quéméré, and others, "Reduction of CeO 2 by hydrogen. Magnetic susceptibility and Fourier-transform infrared, ultraviolet and X-ray photoelectron spectroscopy measurements," *J. Chem. Soc. Faraday Trans.*, vol. 87, no. 10, pp. 1601–1609, 1991.

[5] P. Marcon and K. Ostanina, "Overview of Methods for Magnetic Susceptibility Measurement," *PIERS Proc.*, no. 5, pp. 420–424, 2012.

[6] R. Kligfield, W. H. Owens, and W. Lowrie, "Magnetic susceptibility anisotropy, strain, and progressive deformation in Permian sediments from the Maritime Alps (France)," *Earth Planet. Sci. Lett.*, vol. 55, no. 1, pp. 181–189, 1981.

[7] J. A. Dearing, "Environmental Magnetic Susceptibility Using the Bartington MS2 System," *Interpret. A J. Bible Theol.*, p. 52, 1999.

[8] C. A. E. Little, J. Pellegrino, and S. E. Russek, "Microfluidic platform for magnetic nanoparticle trapping and detection," *IEEE Trans. Magn.*, vol. 49, no. 7, pp. 3402–3405, 2013.

[9] W. Liu, M. Zhou, and L. Kong, "Estimation of the size distribution of magnetic nanoparticles using modified magnetization curves," *Meas. Sci. Technol.*, vol. 20, no. 12, 2009.

[10] J. J. Chalmers, Y. Zhao, M. Nakamura, K. Melnik, L. Lasky, L. Moore, and M. Zborowski, "Instrument to determine the magnetophoretic mobility of labeled, biological cells and paramagnetic particles," *J. Magn. Magn. Mater.*, vol. 194, no. 1, pp. 231–241, 1999.

[11] M. Zborowski, J. Chalmers, and L. R. Moore, "Method for determining particle characteristics." Google Patents, 1999.

[12] M. Zborowski and J. J. Chalmers, *Magnetic cell separation*, vol. 32. Elsevier, 2011.

[13] C. Zhou, E. D. Boland, P. W. Todd, and T. R. Hanley, "Magnetic particle characterization—magnetophoretic mobility and particle size," *Cytom. Part A*, vol. 89, no. 6, pp. 585–593, 2016.

[14] C. Zhou, Z. Qian, Y. S. Choi, A. E. David, P. Todd, and T. R. Hanley, "Application of magnetic carriers to two examples of quantitative cell analysis," *J. Magn. Magn. Mater.*, vol. 427, no. November 2016, pp. 25–28, 2017.

[15] C. Zhou, Y. S. Choi, A. E. David, P. W. Todd, and T. R. Hanley, "Nanomaterial endocytosis: Estimation of particles per cell by magnetic measurement," *IEEE Magn. Lett.*, vol. 9, pp. 7–11, 2018.

516934 Numerical Study of the Evolution of Particle Size and Morphology in an Industrial Titanium Dioxide Reactor

Abstract:

Titanium dioxide is a ubiquitous industrial product with applications ranging from pigments to photocatalysts. Particle morphology, which plays a critical role in controlling optical properties, is determined by the interplay between complex particle processes such as inception, aggregation and sintering. There is therefore both scientific and commercial interest in understanding particle morphology. A stochastic population balance method is used to study the non-isothermal synthesis of particulate titanium dioxide via the chloride process under industrial conditions, with a multivariate particle model used to resolve fine-grained particle structure and inform surface related processes such as sintering and heterogenous growth. A network of ideal reactors is used as a modular approximation to the reactor dosing, reaction and cooling zones in which concentration and temperature gradients exist. Varying the network topology allows for simultaneous spatial and temporal resolution of particle aggregate morphology, including size distributions and connectivity of primary particles in the aggregates. Our model allows us to connect operating conditions directly to detailed morphology, for example illustrating how the aggregate collision diameters and the number of constituent primary particles increase as feed oxygen content increases. The industrial synthesis is performed under elevated temperatures and pressures, and the precursor concentration and particle number density are significantly higher than those occurring in laboratory studies. We discuss computational challenges encountered when "scaling" population balance modelling techniques to industrially relevant conditions and accounting for complex reactor geometries while incorporating multi-dimensional particle structure in the model. Adaptations to weighted particle methods are developed to address numerical issues encountered due to the high process rates, temperature exotherm and initial transience.

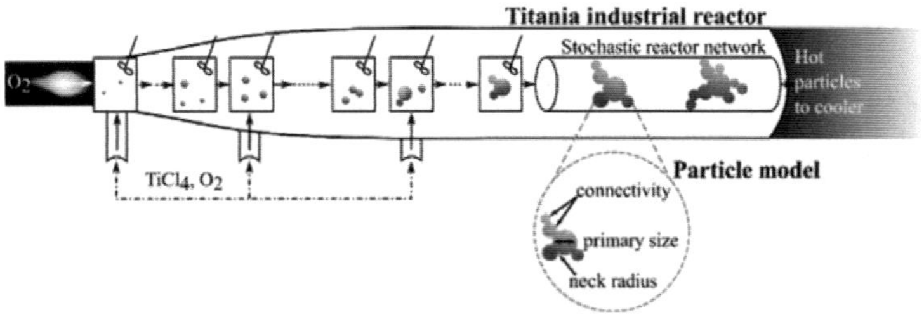

Figure 1. Industrial reactor modelled by a network of ideal reactors capable of evolving individual population balances and particle morphology distributions in each reactor

533067 Multiscale Modelling and Simulation of Particle Formation through Mono-Disperse Droplet Spray Drying

Abstract:

Multiscale Modelling and Simulation of

Particle Formation Through Mono-disperse Droplet Spray Drying

Xiao, J, Wu, WD and Chen, XD

*Presenting authorÕs e-mail: jie.xiao@suda.edu.cn

Suzhou Key laboratory of Green Chemical Engineering, School of Chemical and Environmental Engineering, College of Chemistry, Chemical Engineering and Materials Science, Soochow University, Suzhou, China, 215123

Abstract

A mono-disperse droplet spray dryer (MDSD) can offer uniform-sized droplets with well-controlled trajectory and drying history. It has demonstrated great potential in obtaining in-depth understanding of structured particle formation. Moreover, it has unique advantages in producing on-aim tailored particle products. Our group has been developing experimental and computational platforms for this promising technology.[1-4] In this talk, we will report our recent achievements on the development of a multiscale simulation platform for MDSD aiming at virtual manufacturing. Three new major contributions respectively at different scales will be presented.

At the dryer scale (macro), a comprehensive CFD modelling approach has been developed to characterize the multi-stage drying process. A level-set based free surface tracking model was coupled with a structural mechanics model of a piezo-ceramic nozzle, which allowed us to simulate droplet formation through atomization. A unique discrete phase model (DPM) was developed to track evaporating droplets flying through distinctively different sized chambers, i.e., the small dispersion chamber followed by the big drying chamber. For the first time, droplets in MDSD can be tracked from generation to the end of drying. At the single droplet scale (meso), a liquid-gas two phase flow model coupled with the moving mesh ALE method was developed to simulate droplet drying. The ability of quantifying the local evaporation rate and characterizing crust formation enabled us to track the shape evolution of a shrinking droplet. At the dropletsÕ surface layer scale (micro), a coarse-grained lattice Monte Carlo (LMC) approach was developed to simulate evaporation induced self-assembly of solute molecules. Evolution of the nanoscale surface morphology and chemistry under different drying conditions could then be systematically investigated. Furthermore, how to couple models at different scales aiming at multiscale structure based particle quality control will be discussed.

References:

1. **Xiao J***, Li Y, George OA, Li ZH, Yang SJ, Woo MW, Wu WD, Chen XD*. Numerical investigation of droplet pre-dispersion in a monodisperse droplet spray dryer. *Particuology* 2018; 38: 44-60.
2. George OA, **Xiao J***, Rodrigo CS, Mercade-Prieto R, Sempere J, Chen XD*. Detailed numerical analysis of evaporation of a micrometer water droplet suspended on a glass filament. *Chem. Eng. Sci.* 2017; 165:33-47.
3. George OA, Chen XD*, **Xiao J***, Woo M, Che L. An effective rate approach to modelling single-stage spray drying. *AIChE J.* 2015; 61(12): 4140-4151.
4. **Xiao J***, Chen XD*. Multiscale modeling for surface composition of spray-dried two-component powders. *AIChE J.* 2014; 60: 2416-2427.

537273 Atomic Structure and Stress Release Mechanism of Core-Shell Au-Pd Nanocubes

Abstract:

The internal structure of core-shell nanoparticles has remained difficult to investigate by imaging and diffraction techniques. We report large scale molecular dynamics (MD) simulations with reliable interatomic potentials to analyze the structure of an Au-Pd core-shell nanocube of 73 nm size in all-atom resolution. The cube was prepared by random atom deposition of Pd atoms onto an Au core, similar as in solution synthesis, which allows the shell to grow epitaxially onto the core. The final equilibrium structure was studied to analyze the internal structure, lattice strain, and stress release mechanism. The stress caused by the 4.8% lattice mismatch between Au and Pd results in stacking faults at the corners of the particle and long-range arcs along the faces and edges of the particle. The results of the simulation reveal three-dimensional strain on the nanoparticles in atomic detail that has not been previously analyzed and helps explain unique properties of core-shell nanoparticles. The currently feasible resolution by coherent diffractive imaging is 6.1 nm, which has been applied to this nanoparticle and is consistent with the data presented.

527342 Development and Application of a Computer Simulation Package for Wet Bead Milling of Nanoscale Pharmaceutical Particles Using Population Balance and Fundamental Principles

Abstract:

Nanometer scale particles have found increased use in pharmaceutical formulations and wet bead milling is one way to make these particles. The purpose of this work is to develop a computer modeling package that can simulate the entire wet bead milling process of making a nanometer scale pharmaceutical formulation. Fundamental equations or principles describing relationships of various milling components were obtained through either external publications or in-house studies. Raw material properties, equipment specifications, process parameters, and final product quality parameters (mainly particle size) were linked together by the computer modeling package using fundamental equations and principles. Compound and equipment specific coefficients in the fundamental equations were obtained through either readily available existing data or computer simulation. The computer modeling package can be used to optimize milling process, determine scale-up parameters, and answer submission related questions by simulating the effects of material, process, and equipment parameter changes on final product quality.

The computer modeling package consisted of a set of modular programs each of which performs a specific task. For example, one module was used to calculate particle size distribution change using population balance equation, while another module was used to link to equipment details including mill chamber volume and geometry, bead size, etc.. The package was designed in a way that more modules can be added to expand its capability without significant change on existing modules. Existing data were needed for determining equation coefficients. However, only readily available data were needed, that is, no measurement needed specifically for the purpose of developing the model. The coefficients for the equations in the model need to be determined before using the model for predictions. Based on the availability of the data, the coefficients were determined in two ways. Some data, such as temperature, volume of the chamber, initial concentrations, were used directly as equation coefficients. The rest of the coefficients were determined using simulations while using other indirect measurements as constraints.

The modeling package was successfully used to predict some particle size distribution changes of design of experiment batches under extrapolation conditions, help explain particle size distribution shapes and milling mechanisms, and perform scale-up related simulations.

537594 Quantitative Study of Conduction and Convection Heat Transfer Mechanisms in a Rotary Drum

Abstract:

Granular materials undergo process steps that include transportation, drying, heating and chemical or physical conversion. In several instances, these materials need to be heated or cooled during processing, and rotary drums (kilns) are the most commonly used process equipment for this purpose. It is not always possible to conduct experiments using different materials and consider all process conditions to study the thermal mechanisms due to the inherent complexity of these processes. Therefore, computational modeling and simulations have been extensively used to understand the heat transfer mechanisms in rotary drums. This work focuses on understanding the heat transfer mechanisms in the granular bed inside a rotating drum, using both simulations and experiments.

Experiments are performed using 4 mm diameter silica particles to investigate the granular flow and heat transfer mechanisms inside a 3-inch radius and 3-inch long stainless-steel rotary drum. The drum rides on two titanium wheels, precluding the direct contact of the drum wall with the rollers. One side of the drum is closed with a transparent quartz window, which can handle high temperatures. Another side of the drum is closed with a sapphire window, compatible with an IR camera, specifically chosen to give a high transmittance to the infrared light. The setup is designed to handle up to 1000^0 C, making it possible to study all modes of heat transfer via conduction, convection, and radiation. The wall of the drum is maintained at a desired high temperature using heat guns placed around the drum. A drum tumbler is used to a rotate the drum, and the temperature evolution in the particle bed is recorded using Omega thermocouples and a mid-wave infrared (MWIR) camera. The conduction and convection modes of heat transfer are quantified under varying operation conditions to establish a keen understanding of the heat transport. For this, both experiments and CFD-DEM (using MFIX-DEM, an open-source multi-solver suite) simulation techniques are used to analyze the thermal behavior.

543227 Development of a Continuum Model for Predicting Particle Segregation in Industrial Particulate Flows

Abstract:

As particulate solids flow they can segregate due to differences in particle size, density or other particle properties. Particle segregation has wide implication in various industrial processes such as hopper filling and discharge, mixing in rotating tumblers and chute transfer. The degree of particle segregation depends on flow conditions and differences between particles. We have recently developed a continuum model validated by experiments to quantitatively predict segregation. The model is based on an advection-diffusion equation for individual particle species with a semi-empirical percolation-induced segregation term. This continuum model incorporates three competing physical mechanisms for segregation â€" advection, percolation, and diffusion. The segregation term that we developed isolates the flow condition effect from the particle property effect on segregation, so that this segregation model can be applied directly to different flow geometries. We have applied this continuum model to several flow geometries including heap flows during hopper filling, hopper discharge, chute flows, and tumbler flows. We also extended this model to more practical conditions by considering particle polydispersity and three-dimensional flows. We continue to explore how the continuum model can be adapted to more complicated flow geometries and particle systems pertinent to industrial particle handling and processing systems.

526005 Effect of Cohesion on Gas Residence Time Distribution in Fluidized Beds

Abstract:

Effect of Cohesion on Gas Residence Time Distribution in Fluidized Beds J. Kolehmainen[1], A. Ozel[2], Y. Jiang[1], S. Sundaresan[1]

[1]*Department of Chemical and Biological Engineering, Princeton University, Princeton, New Jersey 08540, USA*

[2]*School of Engineering and Physical Sciences, Heriot-Watt University, Edinburgh EH14 4AS, UK*

Gas residence time distribution (RTD) is an important consideration in industrial fluidized bed reactors, owing to its influence on conversion and product selectivity [1]. Inhomogeneous flow behavior, which is a concern in fluidized beds, affects the RTD appreciably. Although inhomogeneities arise in fluidization of both cohesive and non-cohesive particles, the present study focuses on the effect of inter-particle cohesion on gas RTD in a model system.

In industrial applications such as fluid cokers [2,3], cohesion would come about through liquid bridge formed between particles [4]. Dissolution of small gaseous molecules in this liquid would affect their RTDs. In the present study, we have employed, for the sake of simplicity of numerical computations, inter-particle cohesion through van der Waals force as a simpler proxy for the more complex liquid bridge force. The strength of cohesion is characterized by the Bond number (Bo) representing the ratio of maximum cohesive force between two particles and the weight of a particle. In this study, we performed CFD-DEM simulations of a small fluidized bed. Changing Bo changes the flow behavior. In addition to the flow, we tracked the evolution of the concentration of an absorbing tracer. The rate of mass transfer between the particles and the gas is quantified through Gunnâ€™s correlation [5]. The partitioning of the tracer between the gas and the solid is quantified through a linear isotherm. By studying outlet concentration following a step change in feed concentration of the tracer, we determined the gas phase RTD. Through simulation campaign, we have stablished the effect of Bo and equilibrium constant on the RTD.

The mean residence time is essentially independent of the cohesion, but increases linearly with the equilibrium constant, which is consistent with one-dimensional models. For weakly absorbing particles, cohesion has only a mild effect on the higher moments of the residence time distributions. However, for moderately absorbing particles there is an appreciable increase in the higher moments when cohesion is increased. This increase in the higher moment results from a longer tail in the RTD, which is attributed to clustering that hinders the mass transfer between the particles and the gas.

[1] Gao, Yijie, Fernando J. Muzzio, and Marianthi G. Ierapetritou. "A review of the Residence Time Distribution (RTD) applications in solid unit operations." *Powder technology* 228 (2012): 416-423.

[2] Cui, H. P., et al. "Gas and solids mixing in a dynamically scaled fluid coker stripper." *Chemical engineering science* 61.2 (2006): 388-396.

[3] Song, Xuqi, et al. "Gas mixing in the reactor section of fluid cokers." *Industrial & engineering chemistry research* 44.16 (2005): 6067-6074.

[4] Boyce, C. M., et al. "Analysis of the Effect of Small Amounts of Liquid on Gasâ€Solid Fluidization using CFDâ€DEM Simulations." *AIChE Journal* (2017).

[5] Gunn, D. J. "Transfer of heat or mass to particles in fixed and fluidised beds." *International Journal of Heat and Mass Transfer* 21.4 (1978): 467-476.

536539 Drag Models and Their Validation: Unanswered Questions from the Past and Targeting Future Validation

Abstract:

Predicting accurate fluidized particle behavior with computational fluid dynamics (CFD) is potentially beneficial to industry via improving operation scaling up speed and reliability, fluidized bed operational efficiency, troubleshooting existing operations as well as exploring new equipment design, etc. Modeling the drag force, a major component of the fluid-particle momentum exchange, is essential to establishing a relevant toolbox for predicting fluidized particle behavior. Currently, a gamut of drag models exist, ranging from models that are based on the assumption of homogenous distributions of particles with each fluid grid to models that utilize sub-grid formulations to account for solids volume fractions gradients (i.e., heterogeneous particle distributions). A critical, comprehensive review of these models is presented with a focus on the parameter space relevant to development and validation. Key findings with respect to the limited range of experiments used for validating models, and the mismatch in drag model validation across different fluidization regimes are reported. Furthermore, the potential influence of neglecting, or incorrectly modeling, physical interactions between particles (e.g. cohesion, friction) are explored along with the effect of erroneously attributing all inaccuracies of the simulation to the drag model. Gaps associated with the experimental data available for model calibration and validation are discussed.

536328 Assessment of Mesoscale Solid Stress in Coarse Grid TFM Simulation of Geldart a Particles in All Fluidization Regimes

Abstract:

This study focused on assessing the effect of mesoscale solid stress in the coarse grid TFM simulation of gas-solid fluidized beds of Geldart Group A particles over a broad range of flow regimes, including bubbling, turbulent, fast, and pneumatic transport fluidization regimes. Particularly, the impact of mesoscale solid pressure, mesoscale solid viscosity, and mesoscale solid stress anisotropy were investigated by comparing six different coarse grid TFM settings. Compared with the available experimental data, it is found that both the kinetic theory based TFM with only drag correction and the filtered TFM can predict the flow behavior in all fluidization regimes. Mesoscale solid pressure and viscosity have the opposite impact on flow hydrodynamics; they compete and offset each other, which confirms the assumption employed in many previous studies that the mesoscale solid stress could be neglected in coarse grid TFM simulation.

530471 Structured Bubbling Fluidized Beds: Nucleation and Self-Arrangement Under Pulsation

Abstract:

Fluidized beds offer good mixing and transport rates in a wide range of gas-solid operations. The structure of the solids determines the transport resistances at a microscale, but the overall performance is often driven by the macroscopic flow patterns. The balance between the energy input due to drag and its dissipation in inelastic collisions originates very complex flows. Homogenous expansion, bubbling or channeling appear from the interplay of particle-particle interactions and the action of drag and gravity. The analysis of the momentum equations reveals the appearance of hydrodynamic instabilities [1], but not how they degenerate into chaotic hydrodynamics for which a numerical framework must be used. Kinetic models have had an undeniable success in describing the behavior of large scale granular systems, but they still rely on a simplified description of particle interactions and struggle to deal with systems dominated by dense flows [2].

While fluidized beds provide good mixing, their hydrodynamics are difficult to predictable and hard to scale. Alternatives reactors, such as fixed beds are easier to scale, however they suffer from heterogeneity, cannot deal with highly exothermic reactions and are difficult to control. In recent years, more complex designs have tried to incorporate new degrees of freedom to combine both attributes: good mixing with predictability and control [3,4]. Among them, the use of an oscillating gas is a promising route. Pulsation has been used for decades to improve the fluidization of cohesive powders. This work discusses how it can also be used to form a reproducible flow and control the dynamics of bubbling beds. An oscillatory flow introduces a temporal dynamic in the drag that, when properly balanced, synchronizes the nucleation of a bubble with the formation of intermittent granular structures that are repeated in every pulsation, hereby forming a recursive flow structure. Its reproducibility has been analysed with a pattern intensity index Λ that quantifies the level of order introduced in the bubble dynamic. Comparison across operating conditions, particle size and various bed geometries allows us to identify two operation regimes: unstructured and structured bubble flow. In the earlier, the oscillation does not suppress the hydrodynamic instability and the disposition of the bubbles remains largely chaotic and comparable to that of a constant flow. In the latter, the bubbles become perfectly ordered in a triangular tessellation; thus, they exhibit a narrower size and separation distribution, which is directly correlated to the frequency f and amplitude A of the oscillation used, and hereby may be controlled.

In this contribution, experimental data are compared with numerical studies under different flows, inter-particle friction factors, m, and heights. Comparison of the nucleation process predicted by continuous and discrete formulations showcases the limitations of the current closures of kinetic models (KTGF/TFM). In contrast, Eulerian-Lagrangian frameworks (CFD-DEM) explicitly solve the local granular rheology and reproduce the experimental bubble pattern [5], but also raise interesting questions in relation to the prediction of the energy dissipation. The energy dissipated through friction appears to be key in the formation of clustered structures between the gas bubbles and in their wake. The flow around these regions generate compressive stresses that maintain particles together and lock in frictional contact. These regions act as "pivots" that restrict the mobility of the solids across the wakes and prevent any long-range solid circulation. As the flow becomes compartmentalized in small regions between neighboring wakes, the nucleation in a subsequent pulse becomes a tighter function of the position of the existing bubbles. A continuous numerical framework based on the Kinetic Theory of Granular Flow KTFG is unable to track this phenomenon, because the closures for the frictional stress cannot deal with such rapid changes in rheology, due to clustering, and they overpredict the solids mobility. In contrast, a CFD-DEM frame is able to track correctly the nucleation mechanism. Some interesting discrepancies however remain between CFD-DEM and the experimental bubble nucleation in the history of the bubble shape and its rising velocity that stress the importance of other numerical factors in the stabilization of a pattern in DEM, such as the calibration of mechanical properties or the effect of polydispersity.

<u>References</u>

[1] Batchelor G.K. Secondary instability of a gas-fluidized bed, J. Fluid Mech. 251 (1993), 359-311.

[2] Chialvo S., Sun J., Sundaresan S. Bridging the rheology of granular flows in three regimes Physical Review E 85, 021305 (2012) 1-8.

[3] Coppens M.O., Ommen J.R.van, Structuring chaotic fluidized beds, Chem. Eng. J. 96 (2003) 117-124.

[4] Niyogi K., Torregrosa M.M., Pantzal N., Heynderickx G.J., Marin G.B. (2017). Experimentally validated numerical study of gas-solid vortex unit hydrodynamics. Powder Technology. 305, 794-808.

[5] Wu K, De Martin L., Coppens M-O. (2017). Pattern formation in pulsed gas-solid fluidized beds – The role of granular solid mechanics. Chemical Engineering Journal, 329, 4-14.

532295 Cluster-Induced Deagglomeration in Unbounded Fluidization of Cohesive Particles

Abstract:

The dilute gas-solid flows of cohesive particles in unbounded fluidization are studied via CFD-DEM simulations, focusing on two categories of heterogeneities in particle concentration: particle clusters of hydrodynamic origin (instabilities) and agglomerates of cohesive origin. The former refers to the regions of higher solid concentrations than surroundings, while the latter is defined as particles held together by cohesive forces. Both clustering and agglomeration have significant impacts on reaction rates, momentum, heat and mass transfer in multiphase flow applications. By tracking enduring contacts in simulations, we isolate agglomerates from the overall heterogeneities (resulting from both clusters and agglomerates) in the system, which is quantified by local particle number density fluctuations. In simulations with increasing domain volume, the degree of clustering increases (as is the norm for non-cohesive systems) while agglomeration decreases, contrary to the behavior in granular (no fluid) systems, where clustering enhances agglomeration due to the reduced relative velocities of particles in clusters. Based on statistical and theoretical analyses of particle velocities, higher levels of clustering in larger systems are shown to result in increased relative velocities between the gas and solid phases. This increased relative velocity between the phases serves as an added source of granular temperature (collision velocity between particles), thereby enhancing deagglomeration. We coin this newly-identified mechanism as â€œcluster-induced deagglomerationâ€ and demonstrate its robustness for systems with increasing cohesion. Specifically, as larger cohesion leads to higher energy dissipation during particle collisions that enhances both clustering and agglomeration, the cluster-induced deagglomeration is able to explain the surprising saturation of agglomeration level with increasing cohesion, as opposed to the monotonic behavior in granular systems.

533362 Discrete Particle Model for Non-Spherical Large Objects in Dense Gas-Solid Flows

Abstract:

In a number of applications of fluidized bed such as combustors, gasifiers and incinerators, large solid objects are included in the beds in addition to small medium particles for fluidization. The large solid objects are usually far from spherical and the prediction of non-spherical object motions in the beds is important for robust and efficient operations and bed designs. Depending on its orientation, large objects with various shapes experience different fluid and contact forces respectively from surrounding gas and particles and its motions become quite complex.

In the present study, to enable reliable numerical prediction of large non-spherical solid objects in dense gas-solid flows such as fluidized beds, Super-ellipsoid (SE) model that is a model to represent non-sphericity of particles in DEM framework was combined with Fictitious particle (FP) model that is a model to represent large solids in DEM-CFD framework. After a number of fundamental studies to check its accuracy and validity, the proposed model was compared against experiments in that positions, orientations and dynamic forces working on single non-spherical objects in a three-dimensional fluidized bed were measured non-invasively using a magnetometer and an acceleration sensor installed inside of the objects. Both the calculations and experiments showed characteristic orientation changes of the objects during fluidization and the efficacy of the proposed model was well demonstrated.

537017 Influence of Model Parameters on Runtime and Accuracy of CFD-DEM Simulations of a Prismatic Spouted Bed

Abstract:

Influence of Model Parameters on CFD-DEM Simulation of a Prismatic Spouted Bed

Thomas Eppinger, Leonard Becker, Felix Klippel, Oleh Baran, Ravindra Aglave

Spouted beds are very effective gas–solid or liquid–solid reactors or contactors and are widely used for drying, granulation, coating or chemical reactions in the pharmaceutical, agricultural, food or chemical industry. In the prismatic, slot-rectangular configuration they show some advantages compared to conventional axisymmetric configurations [1].

In this contribution we are investigating these kind of reactors/contactors by means of a coupled CFD–DEM approach with a focus on runtime acceleration techniques since runtime is a major issue for applying CFD-DEM simulations on industrial scale. In a first step the numerical results are compared against experimental results [1] in terms of pressure drop, pressure fluctuation frequencies and solid particle distribution. Then the effect of geometric simplifications is discussed and finally we show how different level of coarse graining affects the time-averaged and instantaneous results as well as the total runtime.

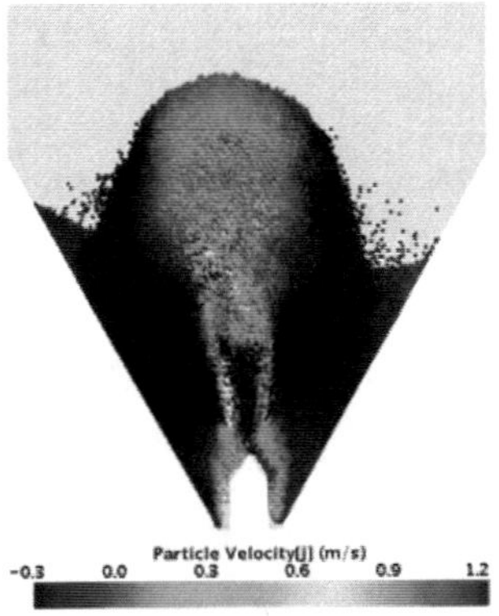

Figure 1: Particle velocity in a dense spouting bed

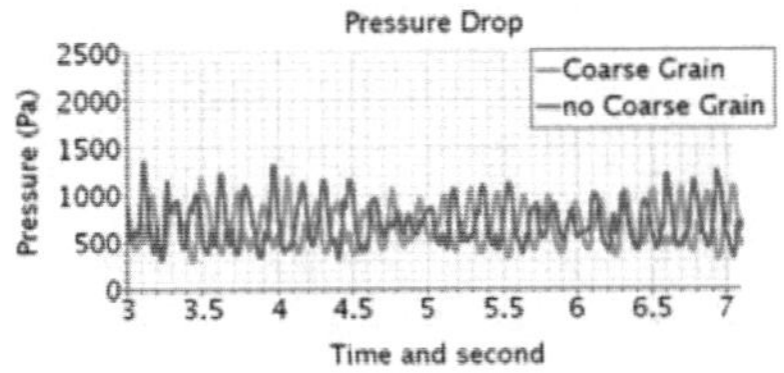

Figure 2: Pressure drop fluctuations for a simulation with and without coarse graining

[1] Salikov V. et al., Characterization and CFD-DEM modelling of a prismatic spouted bed (2015), Pow.Tech (270), 622-636

537144 Flow Behavior of Particulate Pine Forest Residues and Corn Stover: A Comparison of Experiments and Simulations

Abstract:

Feeding and handling problems have been recognized as a major contributor to low production of renewable cellulosic ethanol in the US, which achieved only 7% of nameplate capacity in 2016. In this work we compare the flow behavior of pine forest residues and corn stover at various particle sizes and moisture contents in instrumented pilot-scale hopper tests and instrumented shear testers to continuum finite element method (FEM) and discrete element method (DEM) models. The shear testers include a Schulze ring shear tester, a custom large Peschl-style tester, and a modified Johanson direct shear tester. In addition to measurements of normal force and torque, the shear testers are also instrumented with arrays of pressure transducers. Modeling approaches include a Drucker-Prager model, a micropolar hypoelasticity model, and DEM simulations employing non-spherical, deformable particles. Results from physical experiments indicate that continuum elastoplastic models, such as Drucker-Prager models, do not account for the full complexity of biomass flow, and that more advanced rheological models are better able to predict nuances of the flow behavior, especially the compressive stress field.

522827 Flow Rate Interference Effects in a Silo with Two Openings

Abstract:

Here we study the mass flow-rate of grains from a flat-bottomed rectangular silo with two symmetrically-spaced openings, each of width W. In recent DEM and experimental studies [1,2] it was found that as the spacing between these two openings was increased the flow rate decreased monotonically from a maximum (when the spacing was zero, i.e. a single large opening of size 2W) to a minimum when the openings are far apart. However, in an analogous system, the evacuation of pedestrians from a room with two exits, as the separation between exits doors was increased the flow rate dropped, reached a minimum value, and then rose again to an equilibrium rate.
Using a load-cell measuring device we have experimentally measured the mass flow rate of grains from our silo, modifying the spacing between the openings. Our results were performed with mustard seeds (d = 2.15mm), amaranthus (d = 1.08mm), and screened sand (d = 0.46 mm). We found that we also recovered non-monotonic behaviour. Initially, with a single opening of size 2W the flow rate was at a maximum. Similarly to the pedestrian case [3], as the spacing, L, between the openings was increased, the flow rate rapidly dropped, reached a minimum, then rose to a rate close to twice that of a single opening of size W. This is the first experimental evidence of such behaviour in a granular silo system.

Our experiments were then simulated using DEM (MFix). For values of low inter-granular friction monotonic behaviour was observed (as in [1,2]), but for larger values, the flow-rate behaved as in our physical experiment and the pedestrian dynamics simulation [3]. These observations illustrate the importance of granular friction in the drainage dynamics of these systems.

[1] Zhang, X., Zhang, S., Yang, G., Lin, P., Tian, Y., Wan, J. F., & Yang, L. (2016). Investigation of flow rate in a quasi-2D hopper with two symmetric outlets. Physics Letters A, 380(13), 1301-1305.

[2] Xu, C., Wang, F. L., Wang, L. P., Qi, X. S., Shi, Q. F., Li, L. S., & Zheng, N. (2018). Inter-orifice distance dependence of flow rate in a quasi-two-dimensional hopper with dual outlets. Powder Technology.

[3] Sticco, I. M., Frank, G. A., Cerrotta, S., & Dorso, C. O. (2017). Room evacuation through two contiguous exits. Physica A: Statistical Mechanics and its Applications, 474, 172-185.

534954 Dissolution of Polymer Particulate Systems: Population Ensemble Modeling

Abstract:

The present study addresses the dissolution of cellulosic biomass particles in conditions emulating large-scale processing where the particles exhibit a distribution of size and degree of crystallinity. To this end, we have developed a model in which the behavior of a population of particles is obtained from an ensemble of individual polymer particle dissolution models. The dissolution of individual semicrystalline polymer particles is based on the relevant transport phenomena and kinetics and reveals decrystallization and disentanglement as two important and potentially rate-determinant steps in the process. [Ghasemi, M.; Singapati, A. Y.; Tsianou, M.; Alexandridis, P., Dissolution of semicrystalline polymer fibers: numerical modeling and parametric analysis. AIChE Journal 2017, 63 (4), 1368-1383. DOI: 10.1002/aic.15615] The average value or the number distribution of any intra-particle property captured by the individual particle model can be determined by simulation of a sufficient number of individual particles such that ensemble averages are independent of the particle numbers and the computed particle distributions are acceptably smooth. Using this population ensemble model, various particle size distributions and crystallinity distributions are analyzed for different dissolution parameters. The findings from this study would be useful for the rational design and optimization of biomass pretreatment processes to reduce the particle size and degree of crystallinity. [Ghasemi, M.; Tsianou, M.; Alexandridis, P., Assessment of solvents for cellulose dissolution. Bioresource Technol. 2017, 228, 330-338. DOI: 10.1016/j.biortech.2016.12.049] [Ghasemi, M.; Alexandridis, P.; Tsianou, M., Dissolution of cellulosic fibers: effect of crystallinity and fiber diameter. Biomacromolecules 2018, 19 (2), 640-651. DOI: 10.1021/acs.biomac.7b01745]

527833 Benchmarking a Novel 0-D Model Against Data from Two-Fluid Model Simulations of a Wet Fluidized Bed

Abstract:

Processes relying on wet fluidized beds (FBs) are widely used in various industries such as the petrochemical and the food & pharmaceuticals sector, for several purposes: i) to improve the particle flowability, taste, and appearance; ii) to modify the particle density, or the particle size distribution; and iii) to protect the particles against humidity, light, oxygen etc. Consequently, it is required to consider wetting, drying, particle shaping and size enlargement effects, as well as the homogenization of the product. Fluidized beds allow integrating all of the above purposes into a single process step, mainly due to the high rate of heat and mass transfer in these devices.[1]

Various phenomena occur in wet fluidized beds that need special consideration: i) deposition of droplets on the particle surface; ii) evaporation of liquid from the particle surface; iii) evaporation of free-flowing droplets; and iv) particle agglomeration (not considered in this study). Since contributions of these exchange rates are difficult to be experimentally investigated, the numerical study of such system can be beneficial.

In our present study, a Two-Fluid Model (TFM)-based code (MFiX[2]) was extended to consider gas-particle-droplet interactions in a small-scale wet fluidized bed. Since the computational cost is a big challenge in simulation of industrial beds on a system level, a novel 0-D model was developed. The accuracy of this simplified model was examined by quantifying the uniformity of the particlesâ€™ LoD (Loss on Drying) distribution in the bed as simulated by the TFM. The results of TFM simulation showed that evaporation and deposition phenomena are mainly limited to the spray zone, while drying happens in the dense bed. This leads to an almost uniform distribution of LoD in the bed in case no droplets are washed out of the FB with the fluidization air.

To quantify the degree of uniformity of LoD, the standard deviation of LoD was computed in the bed based on TFM data. It was demonstrated that this standard deviation is 2 orders of magnitude smaller than the total change of LoD during the flow time. As a consequence, the bed can be assumed to well-mixed as long as no droplet loss happens. Several sets of simulations were performed considering various spray rates and initial bed temperatures using both the TFM and the 0-D model. The macroscopic bed characteristics from TFM simulation were imported to the 0-D model simulations. It was proven that the key process parameters (i.e., LoD, gas humidity, and gas temperature) could be well predicted by the 0-D model. However, in case droplet loss occurs as a consequence of a too small droplet injection velocity, the 0-D model fails to predict the outcome of our TFM simulations. Specifically, the largest deviation between TFM and 0-D model prediction was observed for bed temperature and particle LoD. Also, deviations between the model predictions for very shallow beds were unveiled by our simulations. This highlights the important effect of the solid flow pattern on the mixing rate in the bed, and hence the ability to assume a well-mixed particle phase in FBs in the context of system level models.

[1]Mörl L, Heinrich S, Peglow M. Fluidized bed spray granulation. *Handbook of powder technology*. 2007;11:21-188.

[2]Syamlal, Madhava, William Rogers, and Thomas J. OBrien. *MFIX documentation theory guide*. No. DOE/METC--94/1004. USDOE Morgantown Energy Technology Center, WV (United States), 1993

537410 Scale up Studies of Dry Catalyst Impregnation for Improved Content Uniformity Using Simulations and Experiments

Abstract:

Dry catalyst impregnation of active metals onto a porous catalyst support is an important step in the preparation of heterogeneous catalyst. In a typical dry impregnation process, metal solutions are sprayed over a particulate bed in a mixing vessel until the pore volume is reached. The inter-particle variability of the impregnated liquids inside the particles and metal content may significantly affect the activity and selectivity of the resulting catalyst. Current scale-up practices lead to poor fluid distribution and inhomogeneity in metal content. The aim of this work is to understand the dynamic behavior of the particles under the spray nozzle, which is essential for desired content uniformity, and to develop a scale-up model for the dry impregnation process.

Four dimensionless numbers were considered to characterize the scale up of the system. All these quantities were derived taking into account that the system should have geometric and dynamic similarity. They were kept constant for different scales.

$\Pi_1=L/D$, $\Pi_2=A_{wetted}/L^2$, $\Pi_3=Fr=\Omega^2D/2g$, and $\Pi_4=C_Q=Q/(\Omega\times D^3)$, where L is the length of the cylinder, D is the diameter of the cylinder, A_{wetted} is the wetted area of the cylinder, Fr is the Froude Number, C_Q is the Flow Rate Number, Ω is the angular velocity (or rotational speed), and Q is the flow rate.

For the scale up study of our system, water impregnation was performed in 2 different cylindrical sizes: 1) small cylinder (D_1=19.3cm and L_1=29cm) and 2) large cylinder (D_2=40cm and L_2=60cm). Both systems have two spray nozzles located close to each other, generating a continuous wetted area at the center of the vessel. The rotational speed and spray rate were determined by keeping Fr and C_Q numbers constant for the 2 scales. The water content of the particles was compared for different times and locations. Relative standard derivation (RSD) is calculated from the axial water content, and is plotted as a function of the number of revolutions. It is shown that the two RSD curves collapse into one between the 2 different scales, indicating the dimensionless numbers are good scale-up parameters.

In addition, Discrete Element Method (DEM) simulations coupled with a novel algorithm allowing the transfer of metal solution to and between particles were used in combination with geometrically equivalent experiments to model dry catalyst impregnation. DEM simulations were performed for a cylindrical slice in three different sizes, varying from D=10cm, 20cm, and 30cm. Note that the slices are used in order to save computational time. Simulation results showed similar performance in all different size of the vessel and confirmed that the dimensionless numbers can be used to scale-up the impregnation process. We also performed water impregnation simulations in a cylindrical vessel with full length (L=29cm) and diameter (D=19.3cm), keeping a "one to one" size correspondence with experiments (i.e. we kept the exact geometrical dimensions and nozzle configuration as compared to the ones used in experiments) with the goal of validating the parameters chosen in the DEM simulations. These simulations are computationally very expensive due to the size of the vessel. Also notice that these simulations have axial dispersion as opposed to the cylindrical slice shapes. In addition, we explored the value of optimal Froude number. Rotational speeds were varied from 6rpm, 9rpm, 12rpm, and 15rpm. An optimal Fr value was found when the RSD is minimal, which gives the best content uniformity.

537953 Dense Packing of Cell Monolayers: Jamming of Deformable Polygons

Abstract:

Collective motion in dense packings of cells occurs in wound healing, embryonic development, and cancerous tumor growth. Most current computational models of dense cell packings either treat the system as collections of spherical particles or assume that the system is confluent, with no extracellular space. We have developed a new model for dense cell packings in two and three spatial dimensions, where the cells are modeled as deformable particles that have a preferred area and perimeter in 2D and a preferred area and volume in 3D. We measure the packing fraction, Ï†$_j$ at jamming onset as a function of the asphericity, Î±, which is the ratio of the perimeter square to the area of the particle i. We find that the jammed packing fraction increases monotonically with Î± and that the system becomes confluent with Ï†$_j$ = 1 for Î± > 1.16. Using surface Voronoi analysis, we show that this value for Î± corresponds to the case when the cells completely fill their Voronoi-tesselated regions. We also demonstrate that the free area per cell obeys a k-gamma distribution, which has been found for jammed packings of non-deformable particles. Finally, we will describe results from our model concerning the mobility of deformable particles subjected to applied forces, as well as diffusion of deformable particles subjected to active forces to discuss the effect of geometry in active jamming of deformable particles.

525805 Blend Uniformity Prediction Based on Discrete Element Method

Abstract:

Prediction of Blend Uniformity by utilizing Discrete Element Method simulation

Shuichi Tanabe[1, 2, 3], Srikanth R. Gopireddy[1], Shuichi Ando[2], Hidemi Minami[2], Nora A. Urbanetz[1], Regina Scherlieb[3]

1 Pharmaceutical Development, Daiichi Sankyo Europe GmbH, Pfaffenhofen 85276, Germany

2 Formulation Technology Research Laboratories, Daiichi Sankyo Co., Ltd., Hiratsuka 2540014, Japan

3 Department of Pharmaceutics and Biopharmaceutics, Kiel University, Grasweg 9a, 24118 Kiel, Germany

This work presents the prediction of the probability density distribution of blend uniformity of a binary granular blend using Discrete Element Method (DEM). Sample Blend Uniformity (BU), i.e., %SD of the active ingredient concentration in samples taken from various locations of bulk blend, was evaluated to assure the quality of finished drug product such as content uniformity. There exists three key factors in BU analysis (BUA) for assuring homogeneity of the bulk blend, and these include (1) particle size of components in the blend, (2) the sampling regimen that defines sampling locations and the number of samples taken from each sampling location, and (3) the acceptance criteria. The acceptable ranges of the particle size of the components at a given sampling regimen and acceptance criteria can be confirmed based on experiments. However, as resources are limited it is practically difficult to conduct comprehensive experiments to identify the physical propertiesf allowable ranges in commercial scale, which makes the gedge of failureh setting difficult and conservative in manufacture. With the increasing computational capabilities over the last years, computationally intensive in-silico experiments using DEM are becoming an important tool to understand production processes such as blending. The advantage of DEM compared to other simulation techniques is that it is able to capture the trajectory of each and individual particle in the system through Newton's equation of motion, by calculating the new positions and velocities of the particles based on the forces acting on it at a defined time step. The forces acting on each particle which may be due to particle-particle or particle-wall contacts or the non-contact forces such as gravity, cohesion, etc. enter as source terms in the equation of motion (Zhu et al., 2007). Herein this study aims at predicting the sample BU at a given combination of key factors based on DEM simulation, and thus selecting the appropriate sampling regimen.

The blending process of a binary granular blend (active : placebo = 14 : 86 w/w%) using a 50 L bin blender is considered. After filling of more than 50% of total placebo granules the active granules are filled into the blender. Following to the active granule filling the remaining portion of the placebo granules are filled, result in a sandwich like filling. Blender rotation speed is 6 rpm, the Froude number is 0.02. Both granules are mono-sized (180 micrometre in diameter) having a spherical shape and no cohesiveness. In order to keep the simulation time reasonably low, upscaling of particle size and downscaling the blender geometry were applied. This decreases the computational time as less particles are involved. At a given blending time, the blend is divided into *M* equivalent mass spaces as sampling locations. Each of these *M* sampling locations is sub-divided into *N* equivalent mass spaces. The mass present in one of those *M* x *N* mass spaces is assumed to be a sample. Sample BU is calculated using only those samples taken according to a sampling regimen, i.e., 1 sample each from 10 ($M = 10$) locations (10x1). The 10x1 sampling regimen at $N = 8$ will have 10^8 ways of sampling resulting in 10^8 sample BU values giving the probability density distribution of the sample BU. In total, 14 DEM simulations were performed having different particle size expansion level (F_{PS} = 3, 5, 7), blender geometry reduction level (F_G = 0.05, 0.1, 0.2), and number of samples in a sampling location (*N = 4 and 8*) of a same blend. The equation below has been derived that allows the prediction of the sample BU of an actual blend in a given blender based on the DEM simulation which was performed with enlarged particles and reduced blender geometry.

$$BU_{S1}/BU_{S2} = (PN_{S2}/PN_{S1})^{0.5}$$

Here PN_{S1} and PN_{S2} are the number of particles in a sample of the DEM condition 1 and 2, respectively. Root mean square error normalized by the mean (nRMSE) was calculated based on the equation by considering the sample BU at (F_{PS}, F_G) = (5, 0.1), $N = 8$, 25% fill level as a reference.

Qualitative and quantitative prediction of sample BU probability density distribution in a binary blend was successfully demonstrated, which have not been addressed previously. The sample BU was decreased gradually until 60 to 80 sec, and reached a plateau state after 80 to 100 sec (8 to 10 rotations) of blending. The powder blending dominated by diffusion was also observed visually in the snapshots of the time-series DEM simulation. The blending pattern shown at 100 s (10 rotations of blending), shows that blending is complete, and any further blending may not be necessary. At the plateau state the probability density distribution of the sample BU was normally distributed, which is a reasonable approximation of a binomial distribution for a complete random mixture of a binary blend. The mean sample BU became small as the number of particles in a sample increase as reported in the previous study (Muzzio et al., 2002). The sample BU probability density distribution with 10x1, $N = 4$ having ca. 3200 particles in a sample was smaller than those with 1x10, $N = 8$ having ca. 1600 particles in a sample, while they have overlapping portion. The relative standard deviation (RSD) of the sample BU was constant regardless of the mean sample BU in a given blend. The sample BU of the blends was successfully predicted where the nRMSE was 0.2, suggesting a feasible prediction accuracy. The number of particles in a sample was considered to be a critical parameter in predicting the mean sample BU based on DEM results. Therefore, quantitative sample BU probability density distribution prediction of a given blend having different physical properties of granules should be possible at a similar prediction accuracy considering the principle of DEM. Further verification studies will be performed for blends having a particle size distribution and cohesiveness.

With the proposed DEM simulation a sophisticated in-silico design of experiments for setting proven acceptable ranges can be performed, which has been considered practically difficult to perform in experiment due to limited resources. The DEM simulation will provide not only the mean sample BU that will be obtained in experiments but also the probability density distribution, which will provide a forecast of the possibility to pass/fail the acceptance criteria for sample BU. A detailed visualization enabled by simulation also reveals intriguing blending development, and helps e.g. to identify the critical regions of blending thereby optimizing the sampling locations. These in-silico experiments would provide comprehensive understanding of the allowable particle size range of components at a given sampling regimen and acceptance criteria, which helps in setting a reliable gedge of failureh specification.

References

Muzzio, F.J., Sudah, O.S. Coffin-Beach, D, 2002. Effects of blender rotational speed and discharge on the homogeneity of cohesive and free-flowing mixtures. Int. J. Phram. 247, 57-68

Zhu, H.P., Zhou, Z.Y., Yang, R.Y., Yu, A.B., 2007. Discrete particle simulation of particulate systems: theoretical developments. Chem. Eng. Sci. 62, 3378–3396.

544942 One-Way Coupled CFD-DEM Analysis of Particulate Flows in a Monodose Dry Powder Inhaler

Abstract:

In this study, we perform a one-way coupled computational fluid dynamics (CFD) and discrete element method (DEM) simulation of a spinning, filled capsule in a monodose dry powder inhaler (DPI). Commercial software packages ANSYS Fluent and Rocky DEM are used for performing, respectively, the CFD and DEM computations. The coupled multiphysics model is used to primarily study the mechanisms by which powder is discharged from the capsule as well as factors that lead to in-capsule and in-device powder retention. Model results suggest that powder discharge from the capsule is influenced by two opposing forces: those that result from the centrifugal acceleration that drives the powder to the capsule ends, and those that result from the high-velocity air at the capsule ends that impedes powder from discharging from the capsule. In addition, air flow into the capsule, the amount of which increases as the capsule becomes more empty, combined with the dome shape of the capsule ends, causes a small amount of powder to be inevitably retained within the capsule. In addition, we demonstrate that the model may be used to perform parametric studies to determine solid formulation properties critical to reducing in-capsule powder retention, as well as device design modifications that may minimize in-device powder retention. The limitations of the one-way coupled approach, determined through comparing quantities of interests with those predicted from a two-way coupled model, will also be discussed.

540924 Nano- and Microfabricated Hydrogels for Regenerative Engineering

Abstract:

Engineered materials that integrate advances in polymer chemistry, nanotechnology, and biological sciences have the potential to create powerful medical therapies. Our group aims to engineer tissue regenerative therapies using water-containing polymer networks, called hydrogels, that can regulate cell behavior. Specifically, we have developed photocrosslinkable hybrid hydrogels that combine natural biomolecules with nanoparticles to regulate the chemical, biological, mechanical and electrical properties of gels. These functional scaffolds induce the differentiation of stem cells to desired cell types and direct the formation of vascularized heart or bone tissues. Since tissue function is highly dependent on architecture, we have also used microfabrication methods, such as microfluidics, photolithography, bioprinting, and molding, to regulate the architecture of these materials. We have employed these strategies to generate miniaturized tissues. To create tissue complexity, we have also developed directed assembly techniques to compile small tissue modules into larger constructs. It is anticipated that such approaches will lead to the development of next-generation regenerative therapeutics and biomedical devices.

Commercial Scale Manufacturing of a Pharmaceutical Product Using Powder-Liquid 3D Printing Technology

Timothy S. Tracy, PhD, Aprecia Pharmaceuticals, Blue Ash, OH

Abstract

Three dimensional printing (3DP) opens the possibility to manufacture pharmaceuticals differently from today and provides the opportunity to create novel dosage forms for patients. Despite significant advances in the field, an obstacle to commercialization of 3DP pharmaceuticals has been the difficulty in being able to manufacture at commercial scale. Recently, utilizing powder-liquid 3DP technology a pharmaceutical (levetiracetam, Spritam™) has been developed through the 505(b)(2) regulatory pathway at commercial quantities and has received FDA approval for manufacture and sale. In addition, the product is a novel dosage form that utilizes fast-melt technology (ZipDose™) that provides for rapid dispersion in the mouth (< 10 seconds) with only a sip of liquid, while containing doses of levetiracetam up to 1000mg. Products that rapidly disperse in the mouth are particularly useful for patients with dysphagia or difficulty swallowing, particularly of large tablets. This rapid dispersion at high doses was possible because of the selective porosity and hardness of the dosage form created by the 3DP layering process. To produce this unique dosage form, processes were developed to assure consistent and precise jetting of print fluid and careful control of powder deposition across up to 30 layers at commercial production speed to produce tablets that were highly accurate and reproducible with respect to dose of active drug. The regulatory approval and commercialization of this product demonstrated that unique pharmaceutical dosage forms could be created through 3DP technology that helped address unmet needs.

540925 Closing the Circle on Design, Hardware, and Materials for Additive Manufacturing

Abstract:

3D printing is a subset of additive manufacturing (AM) and represents an evolving set of tools and capabilities to prototype, and in some cases produce, new products in much faster iteration cycles than traditional subtractive processes. For 3D printing to reach its full potential as a viable large-scale manufacturing option, new advancements are still required across the three major segments of the industry â€“ design, hardware, and materials â€“ and to accelerate these advancements, highly collaborative efforts between the segments will provide 3D printing users with the quickest and best solutions for their production needs. The presentation will highlight the current state-of-the-art in materials for 3D printing, known existing property gaps, Dowâ€™s vision for the future development and performance of 3D printable materials, and how the broader industry can collaborate to shorten the path to wider scale adoption of these tools in 21st century manufacturing.

541541 Additive Manufacturing and Architected Materials

Abstract:

Material properties are governed by the chemical composition and spatial arrangement of constituent elements at multiple length-scales. This fundamentally limits material properties with respect to each other creating trade-offs when selecting materials for specific applications. For example, strength and density are inherently linked so that, in general, the more dense the material, the stronger it is in bulk form. We are combining advanced microstructural design, using flexure and screw theory as well as topology optimization, with advanced additive micro and nanomanufacturing techniques to create new material systems with previously unachievable property combinations â€" mechanical metamaterials. The performance of these materials is fundamentally controlled by geometry at multiple length-scales rather than chemical composition. We have demonstrated designer properties of these mechanical metamaterials in polymers, metals, ceramics and combinations thereof. Properties include ultra-stiff lightweight materials, negative stiffness, and negative thermal expansion. Our additive micro-manufacturing techniques include Projection Microstereolithography (PmSL), Direct Ink Writing (DIW), and Electrophoretic Deposition (EPD). These tools are capable of generating the designed structures which are highly three-dimensional micro- and nano-scale architectures with multiple constituent materials in the same structure.

527980 A DEM Study of Shear Flows of a Binary Mixture of Non-Spherical Particles

Abstract:

Discrete element method (DEM) simulations of shear flows of a binary mixture of large irregular glued-sphere particles and small spheres are conducted in order to investigate the effect of particle shape and solid fraction ratio on the behavior of binary granular flows. The normal and shear stresses first decrease and then increase with total solid fraction. The normal and shear stresses increase with the increasing solid fraction ratio of large particles. However, for dense regions, the stresses for bidisperse systems are lower than those for monodisperse systems, which is due to the lubrication effect of small spheres in the bidisperse system. The orientation distributions of large particles also show that, in dense regions, the bidisperse system with a higher solid fraction ratio of small spheres exhibits a more uniform alignment for large particles. By incorporating the effective particle projective area in the plane perpendicular to the flow plane, the conventional kinetic theory of binary spherical systems is found capable of predicting the stresses for binary systems of non-spherical particles. Stresses for binary systems with various solid fraction ratios, in the dilute and moderately dense regions, collapse into a single line by further incorporating the root-mean-cubed diameter, but this scaling is not adequate for dense flows.

533945 Comparisons of Continuum FEM Models of Hopper Flow of Particulate Materials to Experimental Measurements

Abstract:

Finite element analysis with elasto-plastic models such as the Mohr-Coulomb and Drucker-Prager models have been shown to agree well with steady-state continuum theories for symmetric hopper geometries. For isotropic cohesion-less particulate materials with constant bulk properties, experimental observations of hopper velocity and stress profiles match closely with continuum theories and the mass flow rate agrees with empirical correlations. In this work, finite element modeling is applied to cohesive and variable property particulate materials flowing through symmetric and asymmetric hoppers and quantitative comparisons are made to experimental results. Based on these comparisons, limitations of FEM modeling with elasto-plastic models are reported.

533387 Development of a Coupled CFD – DEM Simulation Method for a Tablet Coating Process

Abstract:

Tablets are the most common form of drug administration and they are often coated to mask the API's taste, to add protection functionality, to provide a modified release of an active pharmaceutical ingredient (API), to add a second API (active coating) or to provide differentiation through coloring. Especially, with respect to active coating, ensuring content uniformity - and thus coating uniformity - is a challenging task. Consequently, intra-tablet and inter-tablet coating variability are critical quality attributes.

Simulation of the tablet coating process can help to understand the underlying principles of the tablet coating process more deeply. This includes the tablet movement inside the tablet coater including mixing inside the tablet drum, the tablet surface velocity as well as the residence time inside the spray zone (or other zones inside the tablet coater) and the cycle time between two distinct spray zone visits and can calculate coating variability from this data.

Mechanistic models are developed to help find the optimal process conditions in regards to the heat and mass transfer as well as the drying mechanics during the tablet coating process [1]. These models can help to optimize an existing coating process. But to test new processes in silico more sophisticated simulation tools are necessary. Over the last couple of years CFD-DEM simulation has become more complex [2]. DEM nowadays allows the realistic simulation of industrial scale applications [3] as well as incorporating realistic particles shapes. Tablet coating is an attractive application of this modelling approach due to the reasonable number of tablets and the necessity to incorporate both fluid and tablet motion to adequate capture process performance.

In this work a model to realistically simulate biconvex tablets will be coupled to a CFD simulation. The momentum exchange will be done via a drag model that includes the non sphericity of the tablets as well as corrections for the changing void fractions surrounding the tablets. We will also show a method to include the heat and mass transfer in the CFD-DEM simulations. The proposed models will be verified and validated against existing theoretical models as well experimental results.

References:

[1] D. Niblett, S. Porter, G. Reynolds, T. Morgan, J. Greenamoyer, R. Hach, S. Sido, K. Karan, and I. Gabbott, "Development and evaluation of a dimensionless mechanistic pan coating model for the prediction of coated tablet appearance," *Int. J. Pharm.*, vol. 528, no. 1–2, pp. 180–201, 2017.

[2] D. Jajcevic, E. Siegmann, C. Radeke, and J. G. Khinast, "Large-scale CFD–DEM simulations of fluidized granular systems," *Chem. Eng. Sci.*, vol. 98, pp. 298–310, Jul. 2013.

[3] P. Boehling, G. Toschkoff, K. Knop, P. Kleinebudde, S. Just, A. Funke, H. Rehbaum, and J. G. Khinast, "Analysis of large-scale tablet coating : Modeling , simulation and experiments," *Eur. J. Pharm. Sci.*, vol. 90, pp. 14–24, 2016.

534611 Implementation of a Non-Local Granular Fluidity Model in Openfoam for Simulation in Arbitrary 3D Geometries

Abstract:

Unlike simple liquids and gases, the bulk flow and transport of granular materials remain poorly understood by physicists and pose many problems for engineers. Discrete element methods (DEM) are currently considered the state-of-the-art for simulating the flow of granular materials, but DEM is limited in system size to about a few million particles due to high computational costs, even when run on current high-performance computing architectures. This limitation prevents the simulation of flows in industrial-scale vessels and equipment, e.g., grain silos and coal-ash disposal piles, where particle counts can easily approach 100 million to a billion. There are several notable and ongoing efforts to develop continuum constitutive models to describe the emergent dynamic behavior of deforming granular materials, including the so-called inertial rheology model [*Nature*, 441(7094), 727â€“730]. Many of these models are unable to reproduce nonlocal behavior that arise when granular phenomena occur on length scales that are near the scale of the system geometry, e.g., jamming of hopper outlets. More recently, an extension to the inertial rheology model has been proposed, called the nonlocal granular fluidity (NLGF) model, that is able to quantitatively reproduce flow phenomena that depend on system length scales [*PNAS*, 110(17), 6730â€“6735]. While inertial-rheology models have been implemented and demonstrated in volume-of-fluid (VOF) CFD solvers, according to our knowledge the NLGF model has been so far only solved analytically for simple cases and implemented for defined strains in FEM software. In this work, we implement the NLGF model in a VOF solver developed on the OpenFOAM CFD framework. An evolution equation for a scalar *fluidity* variable is solved along with the Navier-Stokes equations for each volume element and at each time increment, where the viscosity is a function of the fluidity and pressure. A regularization of the viscosity is required due to yield-stress behavior that would otherwise result in infinite viscosity values. Our new NLGF-VOF solver is demonstrated for a few simple flow geometries, including flow down an incline and out of a silo, and the simulation results are compared to those found in the literature.

530158 Multi-Scale Modelling of Biomass Gasification: The Effects of Intraparticle Transfer on Syngas and Biochar Production

Abstract:

Biomass gasification is an efficient and environmentally friendly technology to realize energy conversion from biomass. In this work, a coupled representative particle and fuel-bed model was developed based on a multi-scale approach. In this model, the reactor was discretized in the flow direction of inlet gasifying agent and a fuel-bed model was constructed to characterize gas-solid field in the gasifier reactor. In each cell, a representative particle model was developed to predict the effects of intraparticle transfer on the flow field and biomass conversion rate. In addition, a three-region approach (i.e. natural, mixed and forced convection region) was applied in this model in terms of different fluid velocity profiles inside the gasifier reactor. The molecular, particle and reactor levels were all considered in the multi-scale modelling approach. The effects of particle shape and size, intraparticle transfer, different gasification agents and air inlet locations on the gasification performance were investigated in this work. Experimental results were used to validate the model and the deviation between the model prediction and experimental data was found to be less than 15%.

529693 Coarse-Grained Discrete Element Model for Powder Shear Flow

Abstract:

A discrete element method (DEM) is widely used for numerical simulations of various powder handling processes. However, the DEM has a big issue of high calculation cost. Therefore, in the DEM simulation, the number of particles is limited. In order to overcome this issue, a coarse grained (CG) model has been proposed. In the CG model, a group of original particles is replaced to a single CG particle which is larger than the original particle. Thereby, a large-scale DEM simulation can be performed using much smaller number of particles than a number of original particles. However, the CG model for a powder shear flow is not proposed yet. In this study, a new CG model for a powder shear flow was proposed, which is maned as CGSF: coarse-grained method for granular shear flow.In the CGSF, the translational velocity distribution and frictional force generatd in a powder shear flow were considered in the coarse graining. The CGSF was applied in a rotating drum mixer. As a result, the CGSF was able to simulate powder flow of the original particles accurately.

530025 Numerical and Experimental Studies of Granular Materials in the Quasi-Static Regime

Abstract:

In the chemical and process industries, a large percentage of granular materials are handled and produced on a daily basis. Indeed, such materials are widely used in powder-based unit operations such as granulation and drying; fluid catalytic cracking processes; coal gasification [1]. Those materials exhibit a broad range of intricate behaviours and might vary in size and shape; in addition, moisture can be present which might lead to particles agglomeration. In our work, we propose a numerical simulation at the particulate scale together with an experimental study that will allow us to understand the effects of those parameters on the flow behaviour of granular materials. This study will ideally help industrial practitioners handle and produce particulates in an efficient and less costly way.

Among all the methodologies available for the measurement of granular materials' flow properties; the shear test is a preferred method due to the relative simplicity of its procedure [2]. In this test, the granular material is subjected to a given applied normal pressure (normal stress *s*) and the stress needed to generate shear (shear stress *t*) leading to either compaction or dilation of the granular bed is evaluated. In this study, the shear cell apparatus of an automated powder rheometer (FT4 powder rheometer), developed by Freeman Technology Ltd. (Castlemorton Common, Worcestershire, UK), is used in order to generate experimental shear test data (*s* versus *t*) obtained with dry monodisperse spherical glass beads (0.5mm diameter). We also perform the discrete element method (DEM) simulations of simple shear flows of glass beads in a periodic domain by applying Leeds-Edwards boundary conditions [3,4]. In these simulations, the particle-particle friction coefficient is a lump parameter, which is calibrated using the shear test data. This will serve as a base case for the modelling of the shear response of cohesive particles.

Once the DEM simulation is calibrated and validated, constitutive models for particles stresses will be proposed in order to simulate granular flows in industrial-scale applications. These proposed models will be revisited by undertaking further experimental studies of granular materials with variable morphological properties (size and shape) and moisture content.

[1] F. J. Muzzio, A. Alexander, C. Goodridge, E. Shen, and T. Shinbrot, "Solids mixing part A: fundamentals of solids mixing," *Handb. Ind. Mix. Sci. Pract.*, pp. 887–985, 2004.

[2] Y. Wang, S. Koynov, B. J. Glasser, and F. J. Muzzio, "A method to analyze shear cell data of powders measured under different initial consolidation stresses," *Powder Technol.*, vol. 294, no. Supplement C, pp. 105–112, Jun. 2016.

[3] S. Khamseh, J.-N. Roux, and F. Chevoir, "Flow of wet granular materials: A numerical study," *Phys. Rev. E Stat. Nonlin. Soft Matter Phys.*, vol. 92, no. 2, p. 022201, Aug. 2015.

[4] M. Badetti, A. Fall, F. Chevoir, and J.-N. Roux, "Shear strength of wet granular materials: macroscopic cohesion and effective stress," *SciRate*, Feb. 2018.

533870 Experimental Observation of the Initial Stages of Localized Fluidization

Abstract:

Localized fluidization is the process where fluidization is restricted to a section of the bed while the surrounding areas remain static. Localized fluidization is of importance for the operation of conical and spouted beds, as well as the understanding of soil erosion and ground water movement. Although fluidization has been extensively studied in the literature, there are fewer studies dealing with localized fluidization. In particular, there is very little experimental information on the initial stages of the formation of localized fluidization. Therefore, the objective of this paper was to conduct an experimental investigation of the early stages of localized fluidization.

The experiments were carried out in a 30 cm x 20 cm x 10 cm transparent rectangular box where different particle diameters, injection ports and bed heights were tested. To observe the formation of the fluidization, a combination of Planar Laser Induced Fluorescence (PLIF) and Refractive Index Matching (RIM) was used. RIM renders the medium transparent and PLIF allowed for the observation of a particular plane inside the bed away from wall interactions. This setup permitted the observation of the growth of the fluidized zone from the initial localized particle movement within a cavity to the fully fluidized state.

The results include the identification of two very distinct regimes for the expansion of the fluidized cavity: a regular regime that has been previously observed, and a newly found ultra-slow regime that requires much longer time to achieve full fluidization. It was also found that the duration of the expansion from cavity to fluidized chimney depends strongly on the injection port size in the ultra-slow regime, while it only depends weakly in the regular regime. In addition, the diameter of the injection port has a significant influence on the evolution of the fluidization in the area close to the injection, while having almost no effect in the region closer to the top of the bed. Furthermore, a parametric study of the particle sizes, diameter of the injection port, and bed height was performed for the regular regime. An empirical expression for this regime was developed that allows the collapse of the expansion time for all the different variables studied.

537225 Bubbles Distribution in a 0.6-m-Diameter Disk and Donut Fluidized Bed Stripper

Abstract:

The disk and donut stripper has been widely used in the fluid catalytic cracking (FCC) process to strip out product hydrocarbon vapors entrained by the catalyst before regeneration. The stripper consists of a number trays of central conical hats and inclined annular rings. Steam introduced at the bottom of the stripper displaces hydrocarbon vapors from the downwardly flowing emulsion phase into the rising steam bubbles or voids. Disk and donut strippers haveexcellent stripping efficiencies, but they are very prone to flooding. The objective of this study was to determine the cause of flooding through the measurement of stripper density and local bubble void fraction. Tests were conducted in a cold-flow 0.6-m-diameter, 7.6 m tallunit that had seven trays of disk and donut spaced 0.46 m apart. Bubble void fraction profiles were measured with optical fiber probes at five axial locations for stripper solids fluxes of zero to about 70 kg/s.m^2. Bubble probes were traversed in the 133 mm space between the apex of the cone and the bottom of the donut above it, and in the 31.8 mm space between the top of the donut and bottom of the cone. The stripper bed density was found to decrease linearly with increasing solids flux until a solids flux was reached at which it dropped precipitously. The solids flux at the sudden density drop was taken as the stripper flooding point. The flooding solids flux was determined for fluidizing gas velocities of 0.15, 0.3 and 0.46 m/s. The geometry of the disk and donut trays had a significant impact on the radial distribution of bubbles in the stripper. The radial bubble void fraction profiles had a characteristic, M-shape, profile. The bubble void fraction increased outward from a low value at the center reaching a peak at the radius corresponding to the edge of the cone and then decreased to low values close to the wall. The void fraction at the wall increased more with increasing superficial gas velocity than that in the core causing the M-shape profile to nearly become an inverted-V shape at the highest gas velocity. Similar profiles were observed at no solids outflow (Gs = 0). The bubble void fraction both in the core region and near the wall for solids outflow conditions at any given superficial gas velocity was, however, higher than at zero solids flux. It appears that, large pockets of gas accumulated momentarily underneath the conical hats and donuts before escaping, while the donuts â€˜funneledâ€ solids through the core of the stripper causing the core to have fewer bubbles. The only escape route of the rising gas bubbles was through the annular gap between the cone edge and the wall. As the solids flux increased it became increasingly more difficult for the gas voids to rise against the downflowing solids, which eventually led to flooding. It is likely that the width of the annular gap and, therefore, the column diameter play a role in the flooding of gas fluidized bed strippers.

530332 Magnetic Resonance Imaging of Injected Bubble and Jet Dynamics in Fluidized Beds

Abstract:

The behavior of bubbles and jets in fluidized beds is key to gas-solid contact and solids mixing in industrial systems. The behavior or these voids has fascinated scientists and engineers for decades because of their similarities to voids seen in gas-liquid systems (*1–3*). These similarities exist despite a lack of surface tension in gas-solid systems and the ability for gas to flow through the "interface" between a bubble and the surrounding particulate phase.

Due to scientific and industrial interest, a variety of theoretical, experimental and computational studies have been conducted to better understand bubbles and jetting behavior in fluidized beds. A frequent difficulty in understanding bubble behavior is the difficulty in visualizing dynamics inside 3D opaque systems coupled with the fact that bubbles behave differently is pseudo-2D fluidized beds than 3D beds. For this reason, tomographic imaging has been used to visualize bubble dynamics using techniques such as electrical capacitance tomography (*4*), X-ray (*5–7*) and magnetic resonance imaging (MRI) (*8–11*). An advantage of MRI over other tomographic imaging techniques is that it can image the flow field of particles in addition to the local concentration of particles; however, MRI has traditionally been limited by low temporal resolution and small fluidized bed sizes as compared to other tomographic imaging techniques. Recently, multichannel MRI measurements conducted in medical MRI scanners have been introduced (*12*), enabling imaging of particle concentration and flow field in a fluidized bed 190 mm in diameter with a temporal resolution of 18 ms.

In this study, we have utilized this rapid, large-scale MRI capability to image the dynamics of bubbles and jets injected into an incipiently fluidized bed. The imaging capabilities have enabled new insights into the evolution of bubbles, the interaction of multiple bubbles and how the flow of particles is affected by these interactions. This data has been used to quantify the leakage of air from injected bubbles into interstitial flow as well as the volume of particles carried up in the wake of bubbles under different particle and injection conditions. Additionally, anomalous flow phenomena have been uncovered, such as the "disappearance" of bubbles with a neighboring bubble taking up the air from a disintegrated bubble as well as a "zipper-like" instability forming between two interacting jets. The flow phenomena and quantitative data obtained serve as important challenge problems for computational models.

References:

1. R. D. Toomey, H. F. Johnstone, Gaseous fluidization of solid particles. *Chem. Eng. Prog.* **48**, 220–226 (1952).
2. J. F. Davidson, D. Harrison, *Fluidised Particles* (Cambridge University Press, Cambridge, 1963).
3. J. R. Grace, The viscosity of fluidized beds. *Can. J. Chem. Eng.* **48**, 30–33 (1970).
4. W. Warsito, L.-S. Fan, Measurement of real-time flow structures in gas–liquid and gas–liquid–solid flow systems using electrical capacitance tomography (ECT). *Chem. Eng. Sci.* **56**, 6455–6462 (2001).
5. P. N. Rowe, R. Matsuno, Single bubbles injected into a gas fluidised bed and observed by X-rays. *Chem. Eng. Sci.* **26**, 923–935 (1971).
6. F. Fischer *et al.*, An ultra fast electron beam x-ray tomography scanner. *Meas. Sci. Technol.* **19**, 094002 (2008).
7. R. F. Mudde, Time-resolved X-ray tomography of a fluidized bed. *Powder Technol.* **199**, 55–59 (2010).
8. T. Pavlin *et al.*, Noninvasive Measurements of Gas Exchange in a Three-Dimensional Fluidized Bed by Hyperpolarized 129Xe NMR. *Appl. Magn. Reson.* **32**, 93–112 (2007).
9. D. J. Holland, C. R. Müller, J. S. Dennis, L. F. Gladden, A. J. Sederman, Spatially resolved measurement of anisotropic granular temperature in gas-fluidized beds. *Powder Technol.* **182**, 171–181 (2008).
10. C. M. Boyce *et al.*, Magnetic resonance characterization of coupled gas and particle dynamics in a bubbling fluidized bed. *Phys. Rev. Fluids.* **1**, 074201 (2016).
11. H. T. Fabich, A. J. Sederman, D. J. Holland, Study of bubble dynamics in gas-solid fluidized beds using ultrashort echo time (UTE) magnetic resonance imaging (MRI). *Chem. Eng. Sci.* **172**, 476–486 (2017).
12. A. Penn *et al.*, Real-time probing of granular dynamics with magnetic resonance. *Sci. Adv.* **3**, e1701879 (2017).

536726 Experimental Validation of Indirect Conduction Model and Biot Number Analysis for Wall-to-Particle Heat Transfer

Abstract:

Heat transfer between particles and a hot surface is the governing step in many unit operations, such as rotary dryers, heat exchangers, packed beds, etc. In such dense-phase processes, conductive heat transfer from the hot surface (wall) to particle is often the dominating mechanism. Particulate conduction can further be divided into two categories: (i) conduction at the direct contact between the particles and wall (direct conduction), and (ii) conduction through the interstitial fluid surrounding the particles (indirect conduction). The relative importance of the two mechanisms is gauged by the ratio $k_g R_p / k_{eff} R_c$, where k_g is the thermal conductivity of the interstitial fluid, R_p is the radius of the particle, k_{eff} is the effective thermal conductivity for the particle and wall, and R_c is the radius of the contact area between the particle and wall; indirect conduction dominates if the ratio is greater than unity. Hence, for particles having exceedingly small contact area and relatively low thermal conductivity â€" sand, glass, steel, etc. â€" indirect conduction is important. The indirect conduction theory developed by Rong and Horio [1] is being widely used for MFIX-DEM simulations, though experimental validation for the above model is lacking in the literature. The present work is focused on experiments to assess the validity of the indirect conduction model for wall-to-particle heat transfer. To achieve this goal, experiments were carried out using glass and steel particles, where indirect conduction is expected to dominate. These experimental results were compared to DEM simulations performed with the aforementioned indirect conduction model, with good agreement for steel but not for glass. The mismatch for glass particles was ultimately traced to a non-uniform temperature gradient within the particle due to indirect conduction. Accordingly, a modified Biot number which includes the effect of indirect conduction is introduced.

Reference

[1] D. Rong, M. Horio, DEM simulation of char combustion in a fluidized bed, Second International Conference on CFD in the â€¦. (1999).

535775 Heat Transfer Measurement and Modelling in a Fluidized Bed with Pulsed Gas Flow

Abstract:

To gain a better understanding on heat transfer properties of light-weight, cohesive biomass particles during fluidization, bed-to-surface heat transfer coefficients of biomass were measured in a fluidized bed with pulsed gas flow. Effects of flow rate, pulsation frequency, particle size distribution, fines and vibration on heat transfer were investigated. Higher gas flow rates generally yielded higher heat transfer coefficients. Natural frequency was found to be optimum as it offered ample bed movement and internal solid circulation without too much inactivity between pulsation cycles. Heat transfer was also heavily influenced by the interaction between gas convective and particle convective heat transfer, which was verified by the proposed heat transfer model. Two mechanisms, one of which treated the flow-on and flow-off period within a pulsation cycle individually, the other utilized the actual bubble rise velocity obtained via velocimetry were identified to account for the significantly different flow behavior below and above natural frequency. Good agreement was observed between experimental data and modeled results.

Behavior of Nanosized TiO_2 Catalyst in a Microjet and Vibration Assisted Fluidized Bed

Keju An, Jean M. Andino, School for the Engineering of Matter, Transport & Energy

Arizona State University, Tempe, AZ, United States

Introduction

Gas–solid fluidization is beneficial for enhancing the availability of material surface area and minimizing the formation of large agglomerates. The minimization of agglomeration is particularly helpful when considering nanosized particles and the large interparticle forces (e.g. electrostatic and Van der Waals forces) that support agglomeration.[1] However, the fluidization of nanosized particles is challenging and may result in exceptionally restricted bed expansion and large bubble formation when the minimum fluidization velocity is attained.[2]

Numerous techniques have been employed to fluidize nanoparticles. Microjet assisted fluidization of agglomerate particulate fluidization (APF)- and agglomerate bubbling fluidization (ABF)-type nanosized powders has previously been reported as a particularly effective method[3,4]. Results indicated that the quality of gas–solid fluidization of nanoparticles was greatly enhanced by adding a high-velocity jet produced by a micronozzle pointing vertically downward. Bed expansions for an APF-type nanopowder were up to 50 times the original bed height, and difficult-to-fluidize ABF- type nanopowders exhibited behavior similar to APF type, although with lower bed height expansion. Although previously published work was successful in fluidizing nanoparticles, past studies using ABF-type nanoparticles such as titanium dioxide utilized chemicals (e.g. alcohol solutions) in order to minimize electrostatic forces.[5-7] The addition of an alcohol to the gas flow to significantly reduces the particle-to-particle and particle-to-reactor surface electrostatic charges, thereby minimizing agglomeration and improving fluidization quality.[8] Although the use of chemicals improves the fluidization quality, these chemicals limit the opportunities to use a microjet assisted fluidization system for environmental remediation or chemical reaction engineering work.

In this study, nanosized titanium dioxide (P25 TiO_2), a photocatalyst of ABF type, was selected as the test particle, due to its usefulness in air pollution control applications. For the first time, we have studied the dual effect of vibration and a downward microjet on the fluidization of nanosized P25 TiO_2. The fluidization quality, as measured by the parameters of nondimensional bed height, nondimensional pressure drop, and minimum fluidization velocity were determined. This work shows the usefulness of the new MVA method over conventional fluidization, thereby enabling new applications.

Experimental Methods and Materials

The experiments were performed in a column reactor with an inner diameter of 76 mm and a height of 800mm made of clear cast acrylic. Nanosized powders were fluidized with high purity nitrogen at a superficial velocity range from 0.005 to 0.035 m/s, supplied to the fluidized bed through a gas distributor at the bottom consisting of a sintered quartz plate 2mm thick having a pore size of 20 microns. The nitrogen leaving the compressed gas cylinder passed through a water bubbler which was connected to the bottom of the column. The exhaust gas was filtered through a HEPA filter to collect any elutriated nanoparticles. The pressure drop was measured using a differential pressure manometer between two taps, one at the top of the column and the other located 2 cm above the gas distributor.

The column was attached to a wooden plate with four springs to isolate the vibration from the ground. The electromagnetic vibrator was mounted below the wooden table that supported the transparent acrylic reactor. Electromagnetic vibration was controlled by a signal generator that provided vertical sinusoidal motions with controlled amplitude and frequency, and the frequency range used was from 40 to 70 Hz. The amplitude was monitored using a vibration meter attached to the table. From previous work using a system of similar dimensions, placing the downwards pointing nozzle at a height of ~10 cm above from the distributor showed enhanced fluidization performance.[9-14] Thus, this distance was used. The nozzle diameter used in this work was 500 microns, which is identical to the nozzle size used by Quevedo at el.[3] The overall schematic of the experimental setup is shown in Figure 1.

Dry nitrogen was supplied from a compressed gas cylinder, and a mass flow controller (MFC) with a range from 0 to 15 L/min was used to control the gas flow rate at the bottom distributor, as shown in Figure 1. The mass flow controller enabled facile adjustment of the flowrate and hence the superficial velocity through the reactor. Two pressure regulators were used to produce flow through the nozzle, delivered at 10 psi (68.9 kPa) gage, as well as gas flow through the mass flow controller. The vibrator was turned on before the microjet was started in order to avoid large bubble formation and channeling. The dynamic behavior of the nanoparticles was characterized by the nondimensional bed height, nondimensional pressure drop, and minimum fluidization velocity. In this study, the characteristics of fluidization were investigated by increasing the superficial gas velocity in small steps (0.1 cm/s), starting from 0.005 m/s (by changing the flowrate). After each velocity increase, the bed pressure drop and height expansion ratio were recorded after 1 to 2 min to allow the fluidized bed behavior as well as the pressure drop value to stabilize. The minimum fluidization velocity (U_{mf}) was

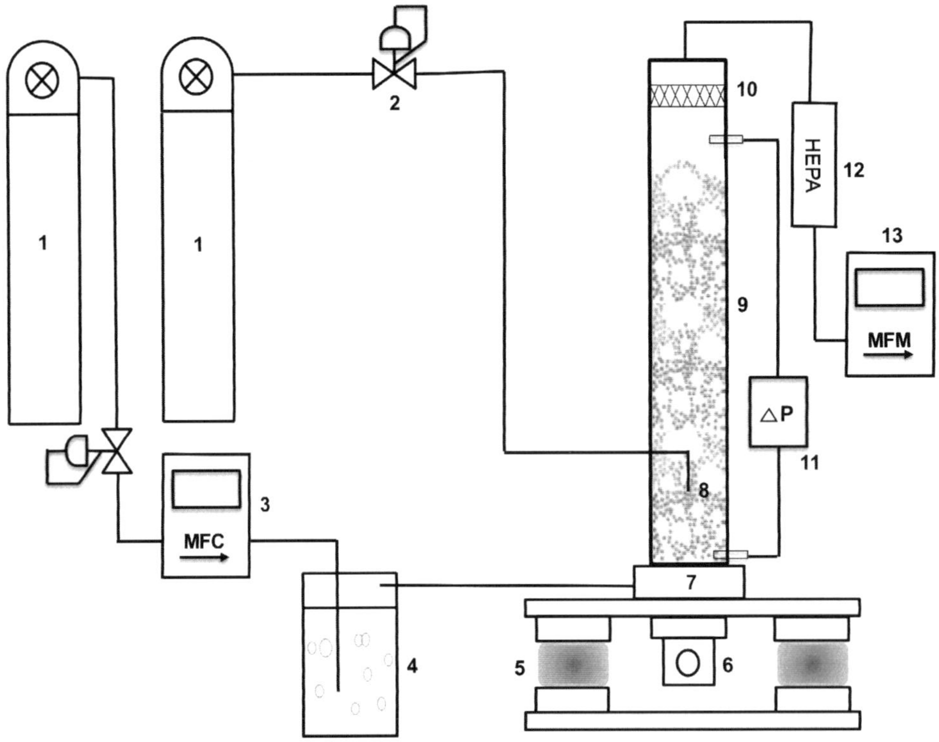

Figure 1. Schematic diagram of the experimental setup
1: Compressed N_2 cylinder, 2: Pressure regulator, 3: Mass flow controller, 4: Water bubbler, 5: Vibration isolators, 6: Magnetic vibrator, 7: Gas distributor 8: Microjet, 9: Fluidization reactor,10: Pre-filter, 11: Pressure manometer, 12: HEPA filter, 13: Mass flow meter

determined using a plot of the nondimensional pressure drop against superficial gas velocity. The experiments were performed under two different conditions. Firstly, we varied the superficial gas velocity while the vibration frequency was held constant. Secondly, we performed the experiments by varying the vibration intensity while a constant superficial gas velocity was maintained. All experiments were repeated multiple times to ensure the reproducibility of the results.

The commercial ABF type nanopowder, TiO_2 P25, manufactured by Evonik-Degussa was used for this study. Before all experiments, the nanopowder was sieved using a 500 μm sieve placed on a vibration shaker, in order to remove large agglomerates that may have been gradually enlarged during packing, storage, and transportation. The properties of the TiO_2 P25 that were used are shown in Table 1. In order to compare our results to those that were obtained previously, we conducted background experiments using the ABF-type TiO_2 nanopowder. We employed simple microjet assisted (MA) fluidization as well as vibro-fluidization (VF) in our experiments using microjet and vibration assistance without alcohol solution. The MA, VF, and combined microjet and vibration assisted (MVA) fluidization methods were compared with the results of previous fluidization studies[3] that employed a microjet as a secondary flow with a dilute alcohol solution. When using TiO_2 nanopowder, Quevedo et al.[3] observed a bed expansion of up to 6 times the original bed height.

Table 1. TiO_2 P25 Nanopowder properties

Type	**Fluidization Behavior**	**Material**	**Primary Particle Size, [nm]**	**Surface Area, [m²/g]**	**Tap density, [kg/m³]**
TiO_2 P25	ABF	TiO_2	21	50	130

Results and Discussion

As seen in Figure 2, using only microjet assistance to fluidize the particles in the absence of an alcohol solution, severe channeling and powder fluctuation problems occurred. This result supported the need for additional fluidization assistance methods(e.g. vibration) for the nanopowder that was used in this work. Applying only vibration assistance resulted in smooth fluidization, but relatively s mall bed expansion (as is shown subsequently). Combined MVA fluidization resulted in smooth fl uidization and significant bed expansion.

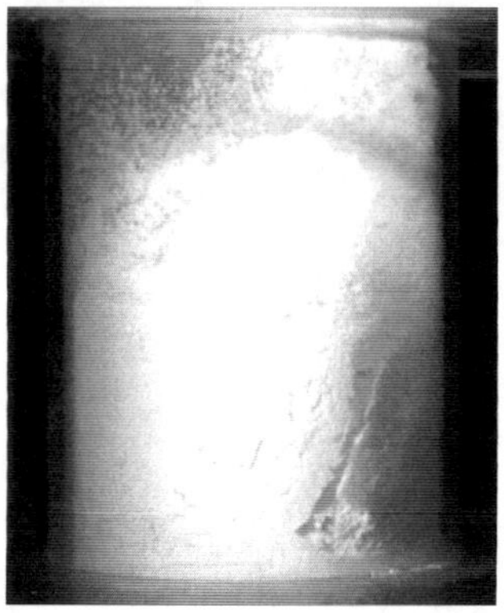

Figure 2. Microjet assisted fluidized bed(MA) channeling

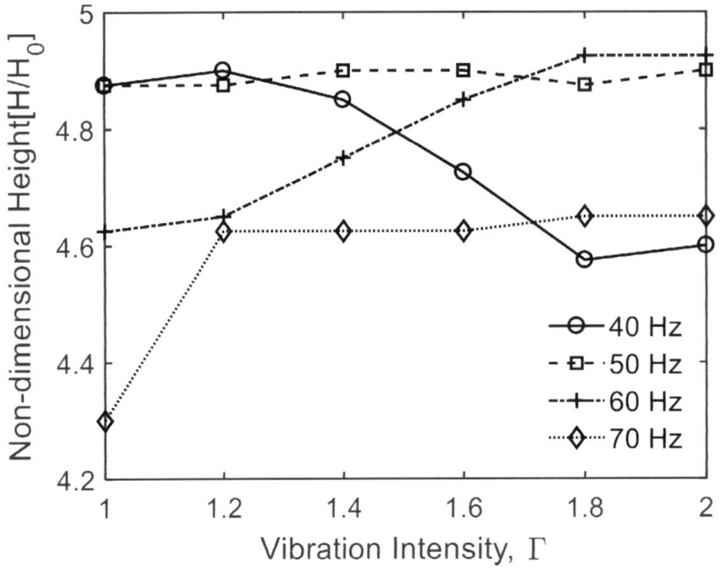

Figure 3. Nondimensional height versus vibration intensity in a fluidized bed employing microjet and vibration assistance.

The experimental results of the MVA fluidized bed that appear in Figure 3 show the evolution of the nondimensional height of the fluidized bed against vibration intensity with fixed frequencies of 40 to 70 Hz. The vibration intensity is defined as the ratio of vibrational acceleration to gravitational acceleration, $\Gamma=(Aw^2)/g$, where A is the amplitude of vibration, $w=2\pi f$, and f is the vibrational frequency.[15] Nondimensional bed height is obtained by dividing the stable fluidized bed height (H) by the initial packed bed height of the bed (H_0) when the sieved TiO_2 powder has been just loaded. From the experimental results, nondimensional height showed independent behavior from the vibration intensity when the frequency was 50Hz. When the frequency was 70 Hz, the nondimensional height shows the lowest values in the vibrational intensity range that was tested. These preliminary results suggest that there is an optimal operational frequency of 50 Hz within the custom designed MVA fluidization reactor system utilized in the current work.

As indicated previously, MA fluidization alone did not result in significant bed expansion, whereas the vibration assisted and MVA systems resulted in smooth fluidization and bed height expansion. As seen in Figure 4, as the gas velocity was increased with a fixed frequency of 50 Hz and a vibration intensity of 2, the nondimensional height increased and then reached a plateau for both the vibro-fluidized bed (VFB) and MVA fluidized bed. The VFB system only showed a bed expansion of 3 times the initial bed height, and fluctuation was observed at the top of the bed when gas velocity was more than 0.025 m/s. The maximum bed expansion of the MVA fluidized P25 TiO_2 was approximately 5 times the initial bed height. The MVA fluidized bed height did not expand further and the bed height remained constant when the superficial gas velocity reached 0.02 m/s. It is important to note that the MVA fluidized bed's nondimensional height was similar to the previously published microjet assisted fluidized bed that employed an alcohol solution.[3] The current result is significant since it allows the MVA system to be employed in scenarios where nanosized particle fluidization would be desirable, but where added chemical contamination would be detrimental to the process.

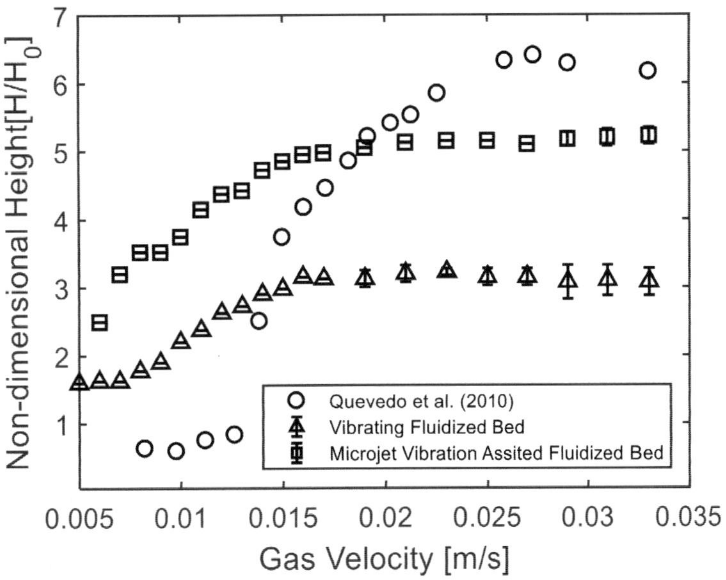

Figure 4. Nondimensional height versus superficial gas velocity for P25 TiO_2 in a vibrofluidized bed (VFB) and a microjet and vibration assisted (MVA) fluidized bed observed in the current work in the absence of an alcohol solution as compared to previously published data source for a microjet assisted (MA) fluidized bed in the absence of an alcohol solution that was used to minimize electrostatic effects.

Nondimensional pressure drops were obtained by dividing the actual measured pressure drop by the apparent weight of the bed(weight of powder divided by the cross-sectional area of the column). The nondimensional pressure drop parameter is useful in indicating whether the gas flow suspends a portion of powder's mass. A nondimensional pressure drop close to unity suggests that the flow suspends most of the entire mass of powder. The nondimensional pressure drop plotted against gas velocity is shown in Figure 5 for the VFB and MVA fluidization systems. The figure shows that when microjet assistance is applied along with vibrational assistance, the TiO_2 nanopowder is suspended in the gas phase more than when the VFB system is employed. Moreover, the minimum fluidization velocity (defined as the velocity where the pressure drop becomes constant) of 0.009 m/s for MVA fluidization is smaller than the 0.013 m/s for the VFB system. Therefore, MVA fluidization allows for more of the particles to become suspended at a lower superficial gas velocity than with the VFB system.

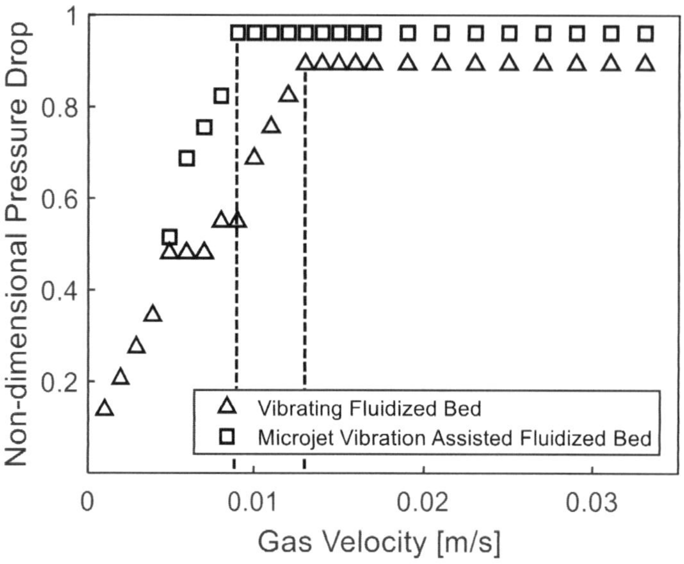

Figure 5. Nondimensional Pressure Drop vs Gas Velocity

Conclusions

The fluid dynamic characteristics and fluidization quality of the MVA system at vibration intensities of 1 to 2 and frequencies of 40 to 70 Hz were investigated. We found that the optimum performance was obtained at a frequency of 50 Hz with the MVA system over a wide range of vibrational intensities. As compared to other fluidization methods, MVA fluidization requires a relatively low minimum fluidization gas velocity in order to achieve a relatively high nondimensional height. The TiO_2 nanopowder bed height in the MVA system expanded several times more than in the VFB and microjet only assisted bed systems, and showed smooth fluidization without an alcohol solution, even at higher superficial gas velocities. The results that were obtained in this work are similar to previously published work for TiO_2 nanosized powders fluidized in a microjet system with the aid of an alcohol solution that was used to minimize electrostatic interactions. The MVA approach is, however, more useful when considering environmental remediation or reaction engineering work where the use of additional chemicals may be undesirable. Additional studies on a wider range of nanoparticle materials and conditions is required in order to fully characterize fluidization behavior in the MVA system.

References

1. Seville, J. P. K.; Willett, C. D.; Knight, P. C., Interparticle forces in fluidisation: a review. *Powder Technology* 2000, 113, (3), 261-268.
2. Quevedo, J. A.; Pfeffer, R., In Situ Measurements of Gas Fluidized Nanoagglomerates. *Industrial & Engineering Chemistry Research* 2010, 49, (11), 5263-5269.
3. Quevedo, J. A.; Omosebi, A.; Pfeffer, R., Fluidization Enhancement of Agglomerates of Metal Oxide Nanopowders by Microjets. *Aiche Journal* 2010, 56, (6), 1456-1468.
4. Pfeffer, R.; Quevedo Jose, A.; Flesch, J., Fluidized bed systems and methods including micro-jet flow. United States Patent No. US 8118243 B2, 2012.
5. Tahmasebpoor, M.; Abadi, R.; Noupoor, Y.; Badamchizadeh, P., Model Based on Electrostatic

Repulsion and Hydrogen Bond Forces To Estimate the Size of Nanoparticle Agglomerates in Fluidization. *Ind. Eng. Chem. Res.* 2016, 55, (50), 12939-12948.

6. Amjadi, O.; Tahmasebpoor, M., Improving fluidization behavior of cohesive Ca(OH)2 adsorbent using hydrophilic silica nanoparticles: Parametric investigation. *Particuology* 2018, 40, 52-61.
7. Fotovat, F.; Bi, X. T.; Grace, J. R., Electrostatics in gas-solid fluidized beds: A review. *Chemical Engineering Science* 2017, 173, 303-334.
8. Pfeffer, R.; Quevedo Jose, A., Systems And Methods For Reducing Electrostatic Charge In A Fluidized Bed. United States Patent No. US 7905443 B2, 2011.
9. Clift, R.; Filla, M.; Massimilla, L., Gas and particle motion in jets in fluidized beds. *International Journal of Multiphase Flow* 1976, 2, (5), 549-561.
10. Hong, R.; Li, H.; Li, H.; Wang, Y., Studies on the inclined jet penetration length in a gas—solid fluidized bed. *Powder Technology* 1997, 92, (3), 205-212.
11. Hong, R. Y.; Guo, Q. J.; Luo, G. H.; Zhang, J. Y.; Ding, J., On the jet penetration height in fluidized beds with two vertical jets. *Powder Technology* 2003, 133, (1), 216-227.
12. Chyang, C. S.; Chang, C. H.; Chang, J. H., Gas discharge modes at a single horizontal nozzle in a two-dimensional fluidized bed. *Powder Technology* 1997, 90, (1), 71-77.
13. Vaccaro, S., Analysis of the variables controlling gas jet expansion angles in fluidized beds. *Powder Technology* 1997, 92, (3), 213-222.
14. Hong, R.; Ding, J.; Li, H., Fluidization of fine powders in fluidized beds with an upward or a downward air jet. *China Particuology* 2005, 3, (3), 181-186.
15. Nam, C.; Pfeffer, R.; Dave, R.; Sundaresan, S., Aerated vibrofluidization of silica nanoparticles. *American Institute of Chemical Engineers. AIChE Journal* 2004, 50, (8), 1776-1785.

Identification and characterization of meso-scale flow structure in the dense gas-solid flow in a fluidized bed

Li Niu, Mengxi Liu*, Zhimin Chu

Stake Key Lab. of Heavy Oil, China University of Petroleum, 102249, Beijing, China

*Email:*mengxiliu@sina.com*

Abstract–Experiments were carried out using fluidized catalytic cracking (FCC) catalyst and silica power mixed particles under different conditions. Transient signals were registered in a fluidized bed utilizing optical probe. A three-phase structural model consisting of a bubble phase, an emulsion phase and particle agglomerates was developed and the criterion for defining bubbles and particle agglomerates was proposed. The volume fractions of the three flow structures vary with operation pattern and spatial positions. Hydrodynamics characteristics of bubbles and particle agglomerates were obtained.

Introduction

Gas-solid fluidized bed reactors are of great importance in the chemical and petrochemical processes, facilitated by the advantages of simple structure, wide range of fluidized particle size distribution and high mass and heat transfer efficiency. In many cases, the gas-solid fluidized bed were not well designed, leading to inefficient interphase contact and limited mass and hear transfer. One of the important reasons is the lack of scientific and quantitative description of the meso-scale flow structure.

The description of dense gas-solid flows in fluidized beds tends to follow the classical two-phase model. It assumes that both meso-scale structures of bubble and emulsion phase are uniform, which is not always consistent with experimental results. Liu proposed that the meso-scale structure of dense gas-solid flow can be described by a three flow structure model, involving bubbles with particles inside, emulsion phase and particle agglomerates, where agglomerate refers to the particle groups having a solid holdup greater than the solid holdup at initial fluidization. Particles inside the agglomerates are not fluidized, but as a group, they are fluidized with emulsion phase.

Experimental setup

In the present work, transient bed density and velocity signals were measured by a specially designed optical probe. The signals were decoupled by statistical analysis methods to identify the meso-scale flow structures. The experimental apparatus is shown schematically in Figure 1. The column was 286 mm in inner diameter and 5460 mm in height. Air was humidified to avoid electrification in the bed.

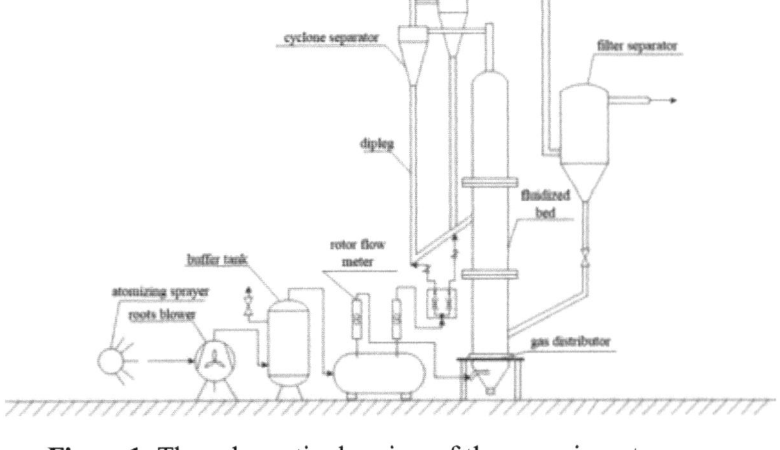

Figure1. The schematic drawing of the experiment process

The bed material was a kind of mixed particle of FCC catalyst (95 wt. %) and silica powder (5 wt. %, $\bar{d}_p$ =10μm) with mean particle sizes of 74μm. The particle size distribution is presented in Figure 2. The solid holdup at the minimum fluidization was 0.51, determined by the plot of pressure drop versus air flux. The superficial gas velocity varied from 0.1 m/s to 0.5 m /s, among which the superficial gas velocity from bubbling fluidization to turbulent fluidization is 0.4 m /s. The static bed height was 0.8 m for all experiments. Measurements were conducted at various radial positions on the cross section at heights of 112 mm, 312 mm, 512 mm and 712 mm, respectively.

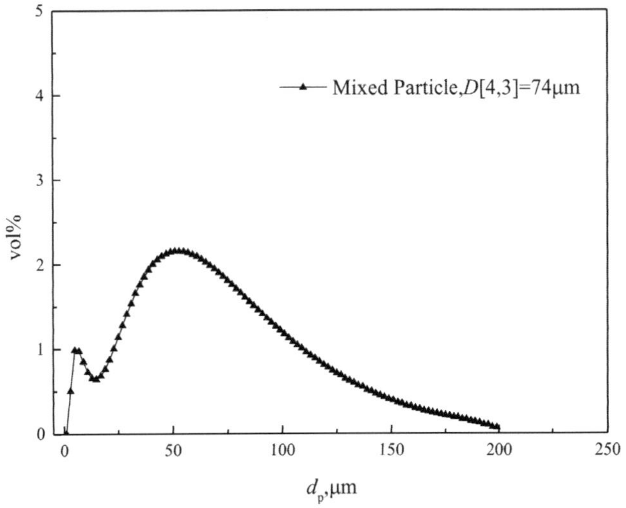

Figure2. Particle size distributions for mixed powder (95 wt. %FCC and 5 wt. %silica powder)

A PV-6D Particle Velocity Analyzer (Institute of Process Engineering, Chinese Academy of Sciences) connecting with a small probe was used to determine the local solid holdup and particle velocity simultaneously. The probe consists of two optical fiber tips with a vertical distance of 3.88 mm and a diameter of 0.8 mm for each tip. One fiber emits light and the other measures the intensity of the light reflected from the particles. The receiving optical fibers transmit visible light reflected by particles to a photomultiplier, and the signals are then logged into a computer via an A/D converter. The sampling rate is 20 kHz with a total sampling time of 196.68 s. Due to the relationship between the output voltage and bed density is nonlinear, an experiment was conducted to obtain the relationship. Then the calibration curve was fitted as

$$\varepsilon_s=0.040e^{0.64V} \text{ (95wt.\%FCC+5wt.\% silica powder)} \quad (1)$$

Analysis method

The whole flow structures can be characterized by signal moment estimations, such as mean solids holdup ε_s, standard deviation σ, skewness S, kurtosis K. For analysis, the four moments was directly calculated from the experimental series. Then, as shown in Eq.s (2) to (5), the four moment values was used to back calculate the three key parameters, namely the solid holdup of the dense phase ε_{sd}, the solid holdup of the bubble phase ε_{sd} and the fraction of the dense phase f_d.

$$\bar{\varepsilon}_s = f_d\varepsilon_{sd}+(1-f_d)\varepsilon_{sb} \quad (2)$$

$$\sigma=\sqrt{(\varepsilon_{sd}-\bar{\varepsilon}_s)^2 f_d+(\varepsilon_{sb}-\bar{\varepsilon}_s)^2(1-f_d)} \quad (3)$$

$$S=\frac{1}{\sigma^3}\left[(\varepsilon_{sd}-\bar{\varepsilon}_s)^3 f_d+(\varepsilon_{sb}-\bar{\varepsilon}_s)^3(1-f_d)\right] \quad (4)$$

$$K=\frac{1}{\sigma^4}\left[(\varepsilon_{sd}-\bar{\varepsilon}_s)^4 f_d+(\varepsilon_{sb}-\bar{\varepsilon}_s)^4(1-f_d)\right] \quad (5)$$

Meanwhile, by choosing a threshold of the bubble phase, the three key parameters (ε_{sd}', ε_{sb}', f_d')

can be statistically obtained from the probability density distribution (PDD) curves of the solid holdup signals. Based on traversing method, the precise threshold can be determined when the following condition was satisfied.

$$\begin{cases} \left| \varepsilon_{sb} - \varepsilon_{sb}' \right| \to 0 \\ \left| \varepsilon_{sd} - \varepsilon_{sd}' \right| \to 0 \\ \left| f_d - f_d' \right| \to 0 \end{cases} \tag{6}$$

Figure 3 illustrates a typical PDD curve which was characterized by a typical bimodal distribution and divided by a threshold. The peak of low solids holdup represents the bubble phase, which can be fitted by a log-normal distribution. (Eq.7) The other peak of high solids holdup corresponds to the dense phase which follows a Gaussian distribution. (Eq.8)

$$f(\varepsilon_{sb}) = \frac{1}{\sigma_{sb}\varepsilon_{sb}\sqrt{2\pi}} \exp\left[-\frac{1}{2\sigma_{sb}^2} (\ln\varepsilon_{sb} - \mu_{sb})^2 \right] \tag{7}$$

$$f(\varepsilon_{sd}) = \frac{1}{\sigma_{sd}\sqrt{2\pi}} \exp\left[-\frac{1}{2} \left(\frac{\varepsilon_{sd} - \mu_{sd}}{\sigma_{sd}} \right)^2 \right] \tag{8}$$

The PDD curve then was divided into three parts by the threshold and the initial solids holdup $\varepsilon_{s,mf}$. The three parts represent three meso-scale flow structures as bubble phase, emulsion phase and particle agglomerates, respectively.

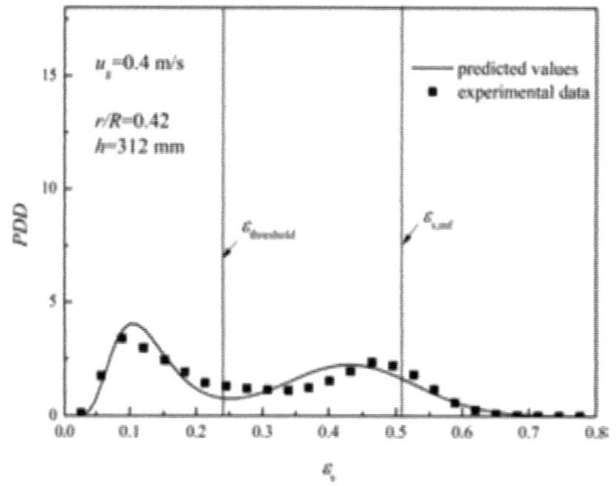

Fig.3 Probability density distribution of solids holdup

Figure 4 illustrates the radial distribution of the volume fraction of the bubble phase, emulsion phase and particle agglomerates at a bubbling bed (0.2 m/s) and a turbulent bed (0.5 m/s) respectively. The sum volume fraction of the three parts is 1. The volume fraction of bubble phase keeps stable at the center of the fluidized bed and decreases at the wall. As the operating gas velocity increases, the volume fraction of bubble phase increases while the particle agglomerates fraction decreases.

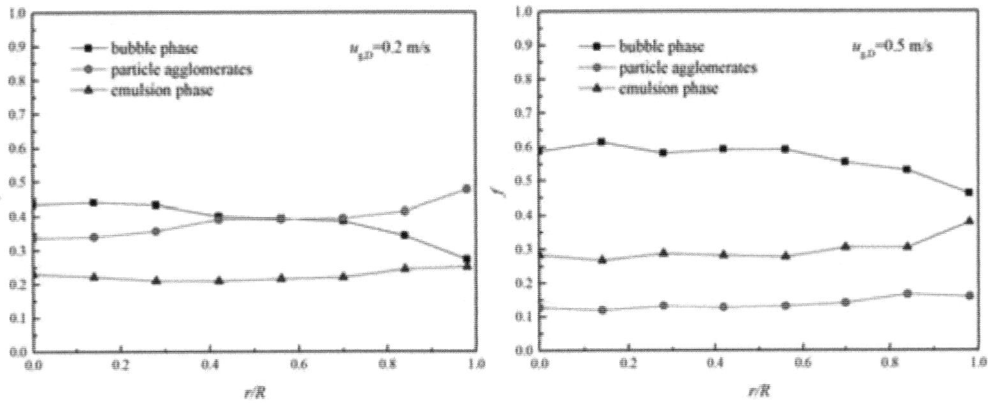

Fig.4 Radial distribution of the volume fraction of the bubble phase, emulaison phase and particle agglomerates at different flow pattern

After determining the threshold of bubbles and particle agglomerates, we need to decouple the corresponding signals from the original optical signals. The method is as follows. (Taking the identification of bubbles as an example)

(1) Find the bubble peak along the time axis of the signals. When the value of the solids holdup signal is less than the threshold value, consider this point A to present the start of the bubble x, and then search backward from this point to find the first point B reach the threshold, if the following 20 consecutive points are all greater than the threshold, point B is considered to be the end point of the bubble, and the bubble peak number x, start and end times T_{x1} and T_{x2} are recorded.

(2) Find the minimum position of the bubble x in channel 1 and channel 2. n_{x1} is the number of data points between the lowest point of the bubble x and A in the statistical channel 1, and n_{x2} is the number of data points between the lowest point of the bubble x and B. Take n_{x1} and n_{x2} respectively on both sides of the lowest point of the bubble peak in channel 2.

(3) When bubbles move in the direction of the two fibers, two reflected signals with similar shapes and a certain delay in time will be generated. For the same bubble, the generated fluctuating signals should be similar. So we compared the bubble peaks obtained from the two channels. It was found that for a correlation coefficient greater than 0.7, the pair of bubbles from the two channels correspond to the same bubble.

Results and discussion

Figure 5 shows the flow structure of the bubble phase. The fraction of the bubble phase increases with the increase of operating gas velocity. At the side wall, as the number of particle agglomerates increases, it decreases due to the wall effect. In the center of the fluidized bed, the distribution of solids holdup seems flat, while it shows an increasing trend at the side wall. The average solids holdup decreases as the operating gas velocity increases in a bubbling bed. When the operating velocity continues to increase to more than 0.3 m/s, the average solids holdup of changes little with the increase of operating velocity. The standard deviation of solid holdup of the bubble phase shows an increasing trend in radial direction at each gas velocity. With the increase of operating gas velocity, it decreases at each radial position. As the operating gas velocity increases, a large amount of gas becomes bubbles and the bubble frequency increases. When the gas velocity is greater than 0.4 m/s, that is, the turbulent bed, the frequency of bubbles appears to increase significantly. At the side wall, the frequency of bubbles is significantly reduced due to the side wall effect. As the operating gas velocity increases, the bubble velocity also tends to increase and sharply decreases at the side wall. The bubble chord varies from 10-43 mm and is less affected by changes in operating gas velocity.

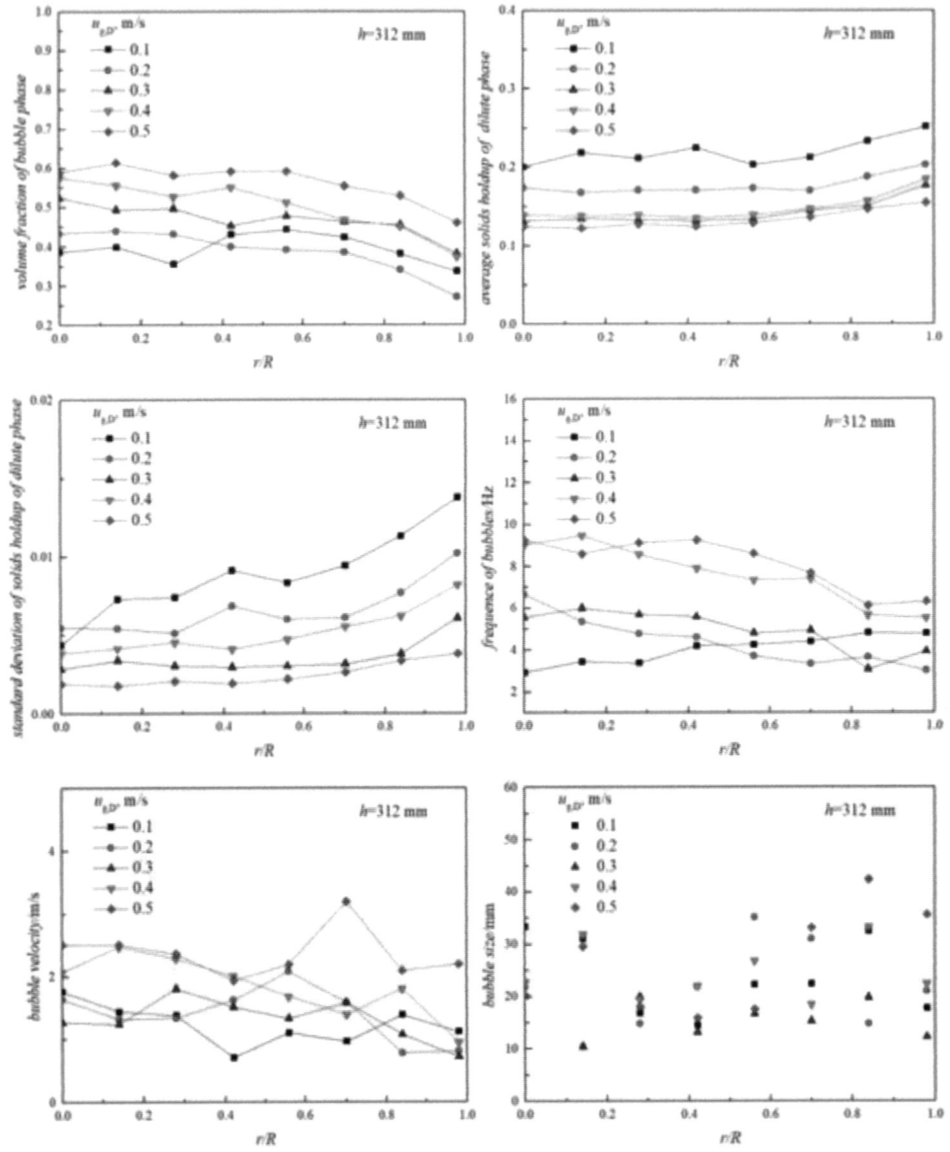

Fig.5 Flow structure of dilute phase

Figure 6 shows flow structure of the dense phase. In a bubbling bed, the average solids holdup is evenly distributed along the radial direction. As the operating gas velocity increases, more gas enters the dense phase and the average solids holdup of the dense phase decreases. After entering the turbulent bed, the solids holdup does not change much with the gas velocity. The standard deviation of the solids holdup shows a decreasing trend from the center of the bed to the side wall. This is due to the fact that there are more bubbles in the center than in the side walls, gas exchange is more frequent, and fluctuations are greater. The fluctuation of the average solids holdup of the particle agglomerates is very small, basically between 0.54 and 0.59. With the increase of operating gas velocity, the frequency of agglomerates tends to increase. In the center of the fluidized bed, it is relatively stable, but increases at the side wall. The speed of particle agglomerates increases with increasing operating gas velocity while it gradually decreases at the side wall. The agglomerates size increases significantly for the gas velocity greater than 0.3 m/s. The agglomerates chord length is 0.5-1.5 mm in a bubbling bed and is 1.5-2.9 mm in a turbulent bed.

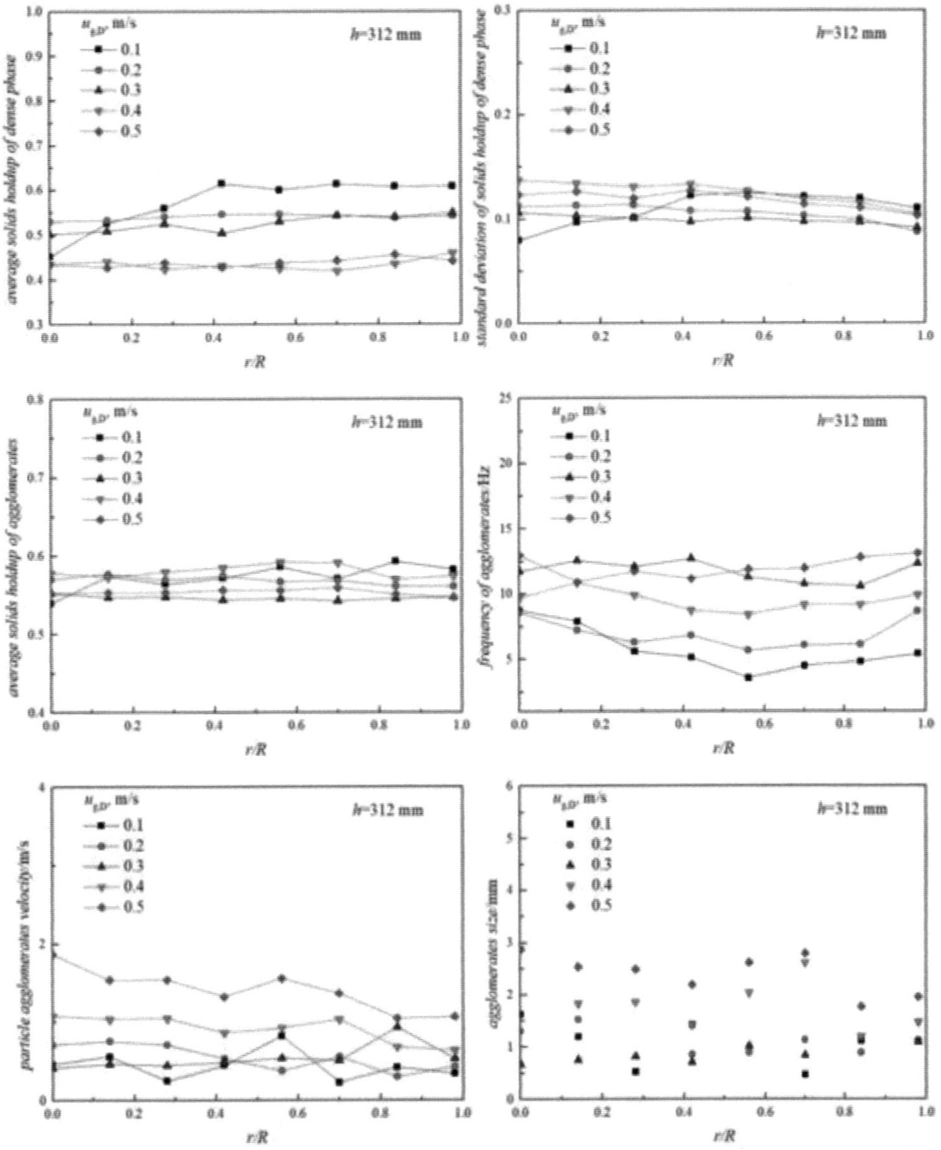

Fig.6 Flow structure of dense phase

Conclusion

A new method of decoupling optical fiber fluctuation signals is presented in this work, suggesting an accurate approach to identify the bubble and the dense phases from original signals, and proposing the definition and the distinguish basis of the agglomerates. By using the method, meso-scale flow structures such as solids holdup, velocity, size and frequency are quantitatively characterize.

Reference

[1] Liu M.X, Lu C.X. Micro-scale Two-Phase Flow Structure in a Modified Gas−Solid Fluidized Bed. Ind. Eng. Chem. Res. 2014, 53, 13475−13487.

[2] Bi H T, Su P C. Local phase holdups in gas-solids fluidization and transport. AIChE Journal, 2001, 47(9): 2025-2031.

[3] Jesse Zhu, Maozhan Qi, Shahzad Barghi. Identification of the flow structures and regime transition in gas-solid fluidized beds through moment analysis. AIChE Journal, 2013, 59(5):

1479-1490.
[4] Lin Q, Wei F, Jin Y. Transient density signal analysis and two-phase micro-structure flow in gas-solid fluidization. Chem. Eng. Sci., 2001, 56, 2179-2189.
[5] Cocco R., Shaffer F., Hays R., Reddy K.S.B, Knowlton T. Particle clusters in and above fluidized bed. Powder Technol., 2010, 203, 3-11.

Kutta-Joukowski force: the Radial Distribution of Particles Concentration in a Riser

Yiping Fan, Chen Li, Mengxi Liu, Chunxi Lu

Stake Key Lab. of Heavy Oil, China University of Petroleum, 102249, Beijing, China

*Email: *fanyipin2002@yahoo.com*

Abstract–The annular-core structure of radial particle concentration distribution in riser was widely reported. In these years, the occurrence of the core-annular was usually merely contributed to the velocity gradient in the boundary layer or the drag force. In this paper, the aerodynamic theory is "transplanted" into the two-phase flow analysis. A transverse force, the Kutta-Joukowski force, is used to explain and quantify the formation of the core-annular in a riser. The radial profile of the Kutta-Joukowski force is proposed by large cold-model riser experiment. Furthermore, for engineering design/operating purpose, the empirical formula for predicting the Kutta-Joulowski force and the coefficient of the concentration difference force K are given, respectively.

INTRODUCTION

CFB riser is widely used in many processes such as FCC unit and CFB boiler. The most desirable gas-solid distribution in a riser is uniform in both axial and radial directions. However, the particle concentration presents quite dense in the wall area while dilute comparatively in the riser center. In other words, it assumes a typical annular-core structure. The annular-core is obviously disadvantageous to the contacting of gas-solids two phases in the riser. It is necessary to explore the occurrence mechanism of the annular-core in order to further improve the gas-solid contacting condition.

Reports covering the occurrence mechanism of the annular-core are quite limited. Except for some qualitative description, the occurrence of the core-annular structure was usually merely contributed to the velocity gradient in the boundary layer or the drag force. However, these researches do NOT analyze the force(s) exerted on the particle(s) in the riser.

In the annular-core structure, a great number of particles concentrate in the riser wall region while they do NOT disperse/diffuse towards the central area. It must closely relate to a transverse force which is towards the riser wall. Magnus force and Saffman force are two typical transverse forces.

The expression of Magnus force is $F_M = \rho_g v\Gamma$. Its magnitude is related to the particle rotation. Some researchers have reported that the average instantaneous angular velocity of the 400~500 μ m glass beads even reached 200~300 r/s by observation with high speed camera. In the past researches, the Magnus force caused by individual particle rotation with 200~300 r/s is probably underestimated, which was also supported by some CFD simulations. The computed results reported that the radial particle distribution would be more non-uniform if introducing the rotations of particles; in other words, the annular-core is more evident when considering the rotation of particles. The Magnus force exerted on individual particle is approximately $3\times10\text{-}10^{-1}\times10^{-9}$ N. However, the Magnus force CANNOT be the main cause of annular-core because it is absolutely impossible that all the particles in a riser rotate in one direction. On the other hand, Saffman force is caused by the velocity gradient and its direction is towards the riser center, thus it does not contribute to the annular-core structure.

Therefore, it is necessary to introducing a more general transverse force to quantitatively interpret the annular-core.

Fan et al (2010) have "transplanted" the aerodynamic theory into the two-phase flow analysis and revealed a force named Kutta-Joukowski force acted on the particles in the feedstock zone of FCC riser. The Kutta-Joukowski force has a similar mechanism to the Magnus force. However, it has two differences comparing with the Magnus force. First, it is the curl that used in the Kutta-Joukowski force rather than the actual rotation speed. Second, the Kutta-Joukowski force act on a cluster of particles (or particles group) but not individual particle.

THEORETICAL ANALYSIS

In a riser, the Kutta-Joukowski force exerts on a cluster of particles (or some particles) per unit length is expressed by Eq.(1)

$$F_{K-J} = \rho_g (v_{g,z} - v_{p,z})\Gamma \tag{1}$$

Here $v_{g,z} - v_{p,z}$ means the local velocity vector difference between the gas and the particles in any enclosed area (corresponding to the slip velocity in a riser). Γ is the velocity circulation around the area; its value equal to the integral around the closed curve. It is difficult to directly compute the velocity circulation, Γ.

In classical hydrodynamics, the Stokes Theorem has been used for any area A that is enclosed by a closed curve. For a fluid with a constant density, the velocity circulation around the curve, Γ, is equal to the vortex strength through the area A.

$$\Gamma = I \tag{2}$$

Consider the relation of the vortex strength I to the angular velocity Ω:

$$I = 2\Omega A = 2A \times [\frac{1}{2}(\frac{\partial v_{p,r}}{\partial z} - \frac{\partial v_{p,z}}{\partial r})] \tag{3}$$

Eq.(3) establishes the relation of transverse force (Kutta-Joukowski force) to the velocity gradient in ANY closed area A in a multiphase flow.

In hydrodynamics, the clockwise velocity circulation (or angular velocity) is negative, while the anticlockwise is positive, vice versa.

The Kutta-Joukowski force can be used to explain the generation of the annular-core flow in a riser. As shown in Fig.1, consider the direction of the slip velocity, the relative velocity between the particles and the gas is illustrated by the arrow CD. In the fully developed zone of a riser, the gradients of particle velocity conform to $\frac{\partial v_{p,r}}{\partial z}$=0 and $\frac{\partial v_{p,z}}{\partial r}$<0, then $\Omega = \frac{1}{2}(\frac{\partial v_{p,r}}{\partial Z}) - \frac{\partial v_{p,z}}{\partial r}$>0. The vortex strength Ω is positive, corresponding to an anti-clockwise velocity circulation as illustrated by the circle 1 in Figure 1. Therefore, the direction of the Kutta-Joukowski force on particles is towards the riser wall. The annular-core structure is then formed by this force.

As shown in Fig.2, consider a differential element which encloses a cluster of particles (or some particles), the Kutta-Joukowski force act on the middle section, ABCD, is,

$$\mathrm{d}F_{K-J} = \rho_g (v_{g,z} - v_{p,z})(-\frac{\partial v_{p,z}}{\partial r})(r + \frac{\mathrm{d}r}{z})\ \mathrm{d}r\mathrm{d}z\mathrm{d}\theta \tag{4}$$

Thus the Kutta-Joukowski force on per unit area is,

$$F_{K-J} = \rho_g (v_{g,z} - v_{p,z})(-\frac{\partial v_{p,z}}{\partial r})r \tag{5}$$

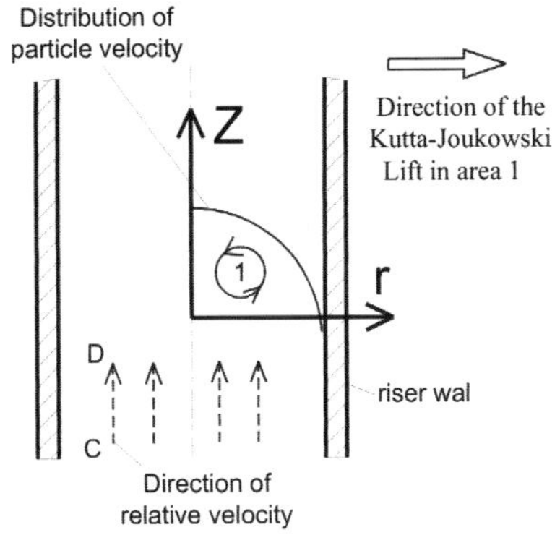

Fig.1 Kutta-Joukowski force on particles in a riser

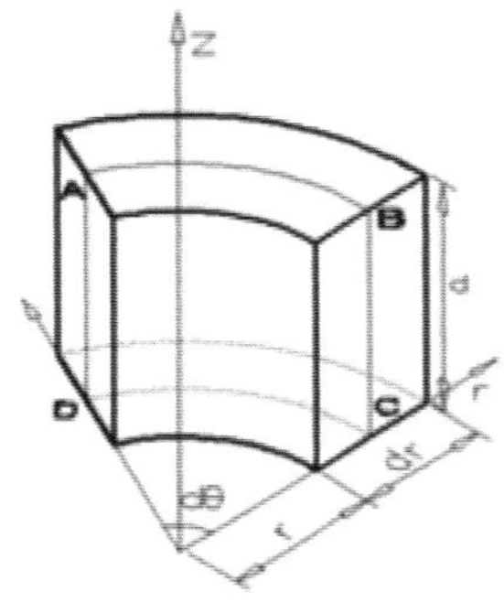

Fig.2 a differential element in polar coordinates

On the other hand, in an annular-core structure, the solids concentration near the riser wall is much higher than that in the center area. This concentration difference is likely to cause a tendency that the particles diffuse from the high concentration area to the low area.

Suppose the force caused by the concentration difference is,

$$F_\rho = K\rho A \tag{6}$$

Here the dimension of coefficient K is $[\frac{\mathrm{L}^2}{\mathrm{T}^2}]$. It is obviously a function of both diffusion coefficient and velocity gradient. ρ denotes the local density whose value can be obtained via $\rho = \rho_\mathrm{p}\varepsilon_\mathrm{s}$.

Thus the concentration difference force exerting on the differential element is

$$F_\rho = K(\rho + \frac{\partial \rho}{\partial r} r)\mathrm{d}r\mathrm{d}z\mathrm{d}\theta \tag{7}$$

This force is towards the riser center.

If the differential element shown in Fig.2 is in the fully development zone of a riser, the particle average velocity keeps almost unchanged. The Kutta-Joukowski force F_{K-J} balances the concentration difference force F_ρ.

$$F_\rho = F_{K-J} \tag{8}$$

It is,

$$K(\rho + \frac{\partial \rho}{\partial r} r) = \rho_g (v_{g,z} - v_{p,z})(-\frac{\partial v_{p,z}}{\partial \mathrm{r}})r \tag{9}$$

The left side of Eq.(9) represents the force caused by the local density gradient while the right side is the Kutta-Joukowski force which is brought about by the particle velocity gradient $\frac{\partial V_{p,z}}{\partial \mathrm{r}}$ and the slip velocity $v_{g,z} - v_{p,z}$.

In this paper, the particle velocity gradient was obtained with fiber-optical probe. As to the slip velocity $v_{g,z} - v_{p,z}$, the formula proposed by Yang (1993) was used,

$$v_{slip} = 10.8(1-\varepsilon)^{0.292} v_t \tag{10}$$

EXPERIMENTAL SETUP AND MEASURING METHOD

The large-scale cold-model riser is shown in Fig.3. The riser is a plexiglass pipe with an ID of 100 mm and its height is 11m. The lift gas and the fluidizing gas are atmospheric air and are all supplied by the Roots blowers. Their flowrates are controlled by a series of rotameters. The superficial velocity of the riser ranges from 6.02~12.03m/s; while the solid flux in the riser is controlled in a range from 0~155kg/m^{2}s.

The particles are typical FCC catalysts. The particle density ρ_p is 1500kg/m^3, while the mean diameter d_p is 60μm.

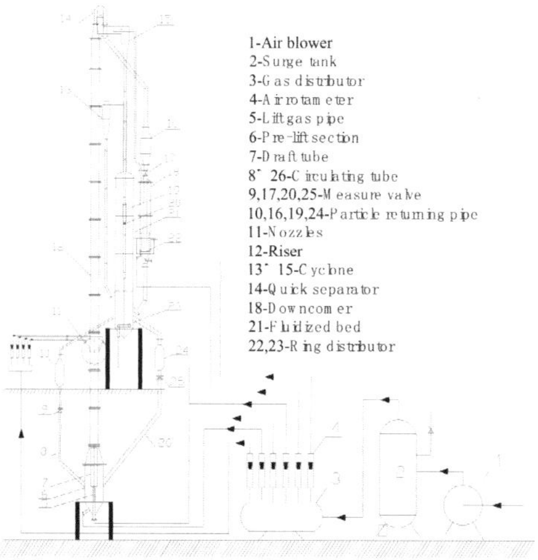

Fig.3 Schematic diagram of experimental apparatus

A PV-6 Particle Velocity/Concentration Analyzer connecting with a small probe was used to determine the local solids volume fraction and particle velocity simultaneously. The probe consists of two optical fiber sensors with a vertical distance of 2 mm. The delay time (τ) of particles passing through the tip of these two sensors was measured via the sampled signal. Then the local velocity of particles moving in front of the probe tip was obtained by dividing the distance (l=2 mm) between the two sensors by the delay time. As to the local solids volume fraction measuring, when a beam of light irradiates a cluster of particles, part of the light will be reflected while the other part will be absorbed. The intensity of reflected light depends on the concentration of the particle cluster irradiated. The local solids volume fraction in the riser was obtained by analyzing the reflected light signal.

EXPERIMENTAL RESULTS AND DISCUSSION

Based on the measured distributions of the local particle velocity and the concentration, the Kutta-Joukowski force on the particles per unit area is then computed. The distribution of the Kutta-Joukowski force Vs. that of the local density (solids hold-up) is given in Fig.4.

It is clear seen that the local density has close relationship to the Kutta-Joukowski force. ε_s assumes a evident monotonic functional relation to F_{K-J}. Kutta-Joukowski force increases with r/R and reaches a maximum near the riser wall. In the area r/R=0~0.5, the values of both F_{K-J} and ε_s are relatively low while they increase sharply in r/R=0.5~1 with a similar gradient.

It is noted that the expression of Kutta-Joukowski force by Eq.(5) is the function of the slip velocity and the particle velocity gradient. Fig.4 demonstrated that the Kutta-Joukowski force directly contributed to the annular-core structure.

Based on the experimental results, the empirical correlations for Kutta-Joukowski force and the coefficient in Eq.(6) are proposed as below.

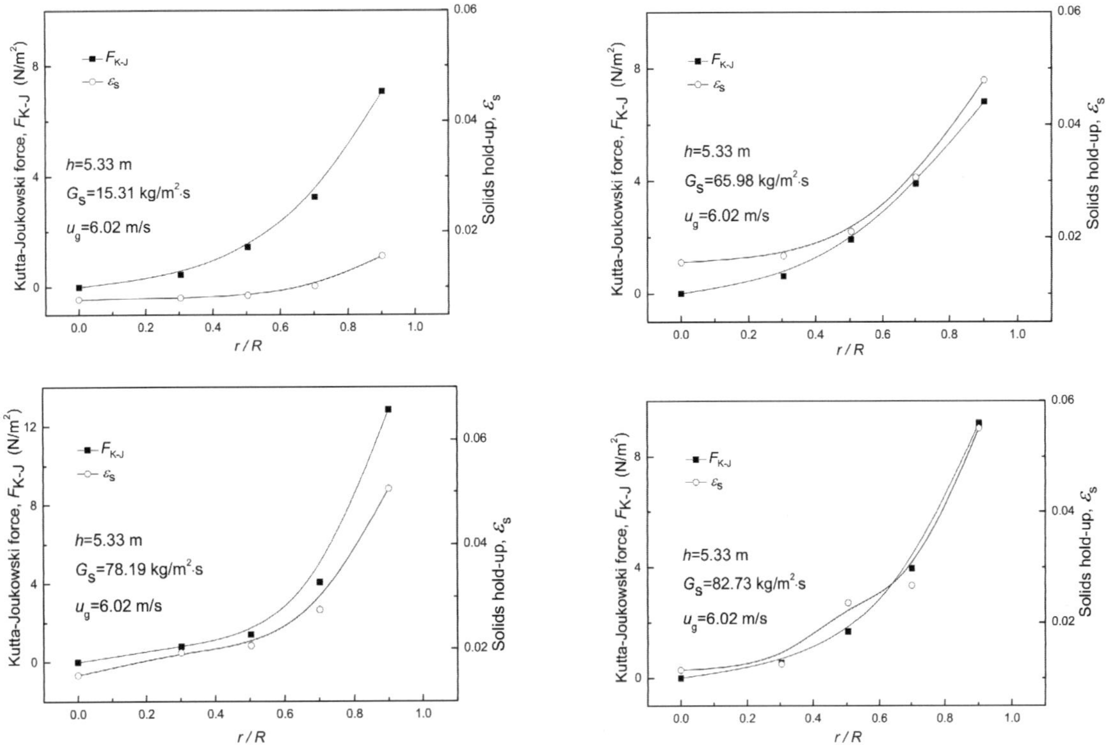

Fig.4 Radial profiles of Kutta-Joukowski force Vs. local density (solids hold-up)

$$F_{K-J} = \rho_g U_g^{\ 2} \varepsilon_s^{0.5558} (\frac{v_p}{U_g})^{0.1493} (\frac{r}{R})^{0.8964} (\frac{h}{H})^{0.361} \tag{11}$$

For $r/R \le 0.5$, $K = 1.4679\times10^{-11} u_t^2 (\mathrm{Re}_{ap})^{2.0111} (\mathrm{Re}_{ag})^{1.1876} (1-r/R)^{-1.16392} (h/H)^{0.1807}$ (12)

For $r/R > 0.5$, $K = 3.0046\times10^{-11} u_t^2 (\mathrm{Re}_{ap})^{-0.1729} (\mathrm{Re}_{ag})^{2.1946} (1-r/R)^{-0.0606} (h/H)^{0.7436}$ (13)

Fig.5 compares the Kutta-Joukowski force & coefficient K between the calculated and theoretical data. The error can be acceptable for engineering design/operating purpose.

CONCLUSION

In this paper, the aerodynamic theory is "transplanted" into the two-phase flow. A transverse force, namely the Kutta-Joukowski force, is used to explain and quantify the formation of the core-annular structure in a riser.

The PV-6 optical fiber probe was applied to investigate the local particle concentration and the local particle velocity distribution in a large-scale cold-model riser. Furthermore, the method to calculate local Kutta-Joukowski force is proposed through theoretical analysis. This force is the function of the particle velocity curl, the slip velocity and the gas density.

Based on the experimental data, the distribution of Kutta-Joukowski force in the riser is given. On the other hand, the particle concentration difference force is introduced and quantified. This force balances the Kutta-Joukowski transverse force in the riser. The empirical formula for predicting the Kutta-Joulowski force and the coefficient K are given, respectively.

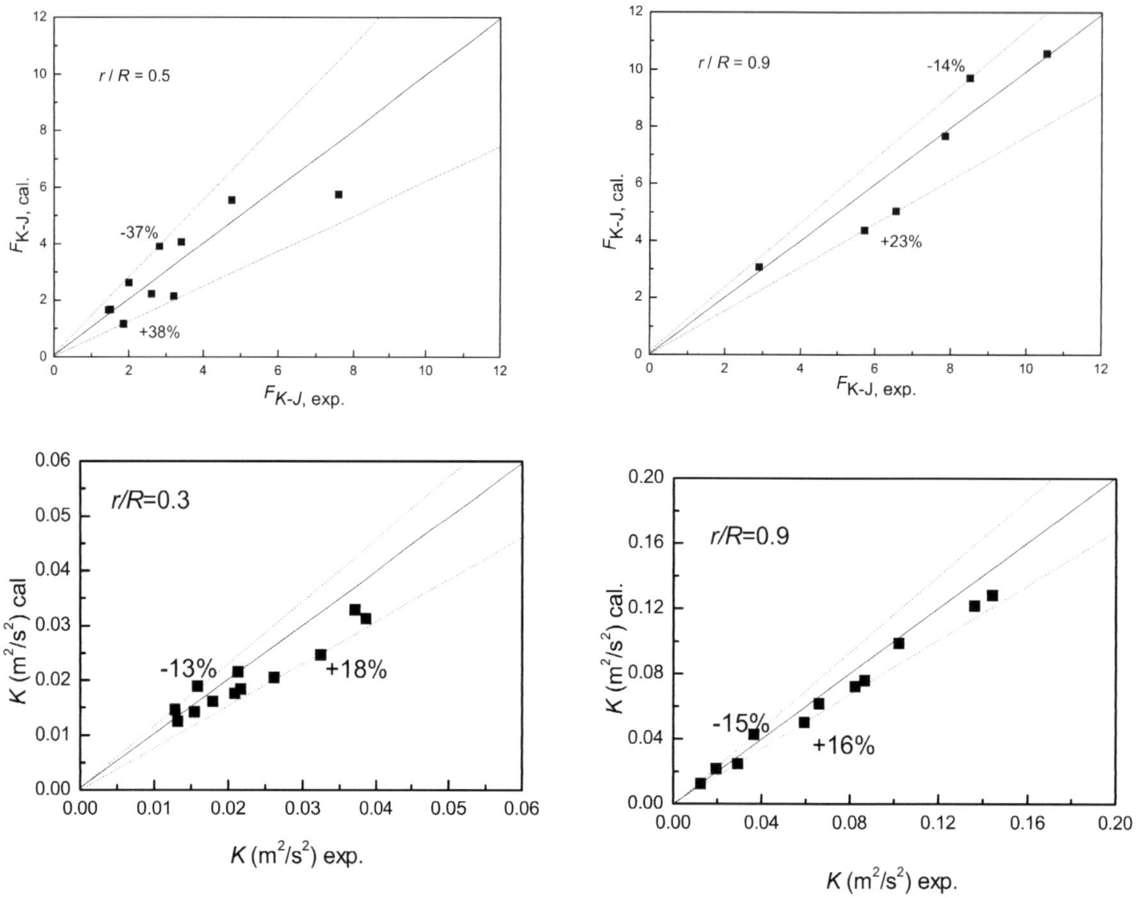

Fig.5 Comparison of the Kutta-Joukowski force & coefficient K between calculated and theoretical data

NOTATION

A	area， m^2
d_p	Particle diameter， m
$F_{K\text{-}J}$	Kutta-Joukowski force， N/m^2
F_ρ	Concentration difference force， N/m^2
G_s	Solids flux， $kg/(m^2 \cdot s)$
h	Axial position， m
H	Height of riser， m
I	vortex strength， m^2/s
K	Coefficient of Concentration difference force， m^2/s

r	Radial position， m
R	Radius of riser， m
u_t	Particle terminal velocity， m/s
U_g	Superficial gas velocity， m/s
$v_{p,r}$	Radial component of local particle velocity，
$v_{p,z}$	Axial component of local particle velocity， m/s
v_{slip}	Slip velocity， m/s
$v_{g,z}$	Radial component of local gas velocity， m/s
ε	Voidage
ε_s	Local solids volume fraction
ρ_g	Gas density， kg/m^3
ρ	Local density， kg/m^3
ρ_p	Particle density， kg/m^3
Ω	Curl， s^{-1}

REFERENCES

1. Bi, H. (2002). Some issues on core-annulus and cluster models of circulating fluidized bed. Canadian Journal of Chemical Engineering 80, pp. 809-817.
2. Clift, R., Grace, J., Weber, M. (1978). Bublbles, Drops, and Particles, Academic Press, San Diego Douglas, J., Gasiorek, J., Swaffield, J. 1992. Hydrodynamics, China Higher Education Press, Beijing.
3. Fan, Y., E, C., Shi, M. (2010). Diffusion of feed spray in fluid catalytic cracker riser. AIChE Journal 56, pp. 858-868. Gidaspow, D. 1994. Multiphase flow and fluidization continue and kinetic theory descriptions, Academic Press, San Diego.
4. Jin, S., Raneine, B. (2006). Hydrodynamic modeling of particle rotation for segregation in bubbling gas-fluidized beds. Chemical Engineering Science 61, pp. 1470-1479.
5. Liu J. (2001). Particle and gas dynamics of high density circulating fluidized beds, Doctor thesis, University of British Columbia, Vancouver.
6. Martin, M., Turlier, P., Bernard, J. (1992). Gas and solid behavior in cracking circulating fluidized beds. Powder Technology 92, pp. 249-258.
7. Poon, E., Ooi, A., Giacobello, M. (2013). Hydrodynamic forces on a rotating sphere. International Journal of Heat and Fluid Flow 42, pp. 278-288.
8. Wang C. (2013). High-density gas-solid circulating fluidized bed riser and downer reactor, Doctor thesis, University of Western Ontario, London.
9. Yang Y., Jin Y., Yu Z. (1992). Investigation on slip velocity distribution in the riser of dilute circulating fluidized bed. Powder Technology 73, pp. 67-73.
10. Yang Y., Jin Y., Yu Z. (1993). Local slip behavior in the circulating fluidized bed. AIChE Journal 89, pp. 81-90.

526540 Insights from Microhydrodynamic Modeling of Nanomilling in a Wet Stirred Media Mill

Abstract:

Owing to their large surface area and unique physico-chemical and mechanical properties, materials with nanoscale features have found use in a multitude of products. Wet stirred media milling (WSMM) has been used for the production of nanoparticles of various active agents such as poorly water-soluble drugs, agrochemicals, nutraceuticals, etc. in suspension form as it is a scalable, solvent-free, environmentally benign process. However, the process is energy intensive, costly, time-consuming, and can cause contamination of the product due to bead wear if the process is not developed properly. The development of a cost-effective, robust wet media milling process that also avoids significant product contamination entails a good understanding of the underlying particle change mechanisms such as breakage and aggregation, and their relation to process and formulation parameters. Unfortunately, a great majority of the previous investigations especially in various high-value product industries, where WSMM has been used to manufacture active nanoparticles, the development approaches are usually very empirical, which is devoid of any fundamental understanding. In fact, majority of the effort has been directed toward formulation development to ensure that the nanosuspensions are physically stable over time. It is no surprise that in many product-value industries such as pharmaceuticals, WSMM has not become a platform technology due to the aforementioned engineering challenges.

In this presentation, insights from recent microhydrodynamic modeling studies [1–3] will be presented to explain the impact of various process parameters such as rotor speed, bead loading/size, and drug loading on the breakage kinetics in a unified framework. We used griseofulvin (GF), a poorly water-soluble hydrophobic drug, as a model active agent and prepared its aqueous nanosuspension with hydroxypropyl cellulose (HPC) and sodium dodecyl sulfate (SDS). The combined HPC–SDS provided excellent electrosteric stabilization and prevented particle aggregation [4], thus allowing us to solely discern the breakage kinetics and the impact of process parameters on the kinetics. Milling experiments were performed using a Netzsch Microcer wet stirred media mill. For various processing conditions, time-wise evolution of GF particle size was measured during milling using laser diffraction offline, and the process time constant for each set of processing conditions was calculated via an exponential decay model. Additionally, average power consumption as well as density–apparent shear viscosity of the milled suspensions were measured. Using a modified version of the microhydrodynamic model developed earlier by Eskin et al. [5], we calculated various microhydrodynamic parameters such as the granular temperature, the average bead oscillation velocity, the frequency of single-bead oscillations, the maximum contact pressure at the center of the contact circle of the colliding beads, the average frequency of drug particle compressions, and a milling intensity factor. The experimental data and calculated characteristic process time constants were analyzed in view of the calculated microhydrodynamic parameters. Most significant findings can be summarized as follows: upon an increase in rotor speed, more mechanical energy was imparted and all microhydrodynamic parameters increased monotonically. Higher rotor speed led to more frequent and energetic/forceful bead–bead collisions and more frequent drug particle compressions, which explains the experimentally observed faster breakage. An increase in volumetric bead concentration led to counteracting effects; higher bead loading led to more bead–bead collisions and drug particle compressions, but less energetic/forceful collisions/compressions. Overall, the experimentally observed positive impact of the bead loading, i.e., faster breakage of the drug particles, was explained by an increase in the milling intensity factor. An increase in drug loading led to a slight, almost linear decrease in all microhydrodynamic parameters, except the milling intensity factor, which exhibited a sharper decrease, thus explaining the reduced breakage rate at higher drug loading in the experiments. The process time constants from all different sets of processing conditions were uniquely predicted by a single exponential decay-type correlation of a single microhydrodynamic parameter, i.e., the milling intensity factor. Finally, in the experiments, there was an optimal bead size that achieved the fastest breakage at each rotor speed and that it shifted to a smaller size at higher speeds. Calculated microhydrodynamic parameters reveal two counteracting effects of bead size: more bead–bead collisions with less energy/force upon a decrease in bead size. The optimal bead size exhibited a negative power-law correlation with the specific energy consumption and each microhydrodynamic parameter. Hence, the microhydrodynamic model rationalizes the use of smaller beads for more energetic wet media milling. Overall, this presentation reveals significant insights from recent microhydrodynamic studies in a unified framework, which will guide streamlined development of more robust WSMM processes.

References:

1. Li, P. Alvarez, E. Bilgili, A microhydrodynamic rationale for selection of bead size in preparation of drug nanosuspensions via wet stirred media milling, International Journal of Pharmaceutics, 524 (2017) 178–192.
2. Li, N. Yaragudi, A. Afolabi, R. Dave, E. Bilgili, Sub-100 nm drug particle suspensions prepared via wet milling with low bead contamination through novel process intensification, Chemical Engineering Sciences, 130 (2015) 207–220.
3. Afolabi, O. Akinlabi, E. Bilgili, Impact of process parameters on the breakage kinetics of poorly water-soluble drugs during wet media milling: A microhydrodynamic view, European Journal of Pharmaceutical Sciences, 51 (2014) 75–86.
4. Bilgili, A. Afolabi, A combined microhydrodynamics–polymer adsorption analysis for elucidation of the roles of stabilizers in wet stirred media milling, International Journal of Pharmaceutics, 439 (2012) 193–206.
5. Eskin, O. Zhupanska, R. Hamey, B. Moudgil, B. Scarlett, 2005b. Microhydrodynamic analysis of nanogrinding in stirred media mills. AIChE Journal. 51 (2005) 1346–1358.

546543 Recent Advances in the Modeling of Non-Linear Particle Breakage

Abstract:

Milling is an important unit operation in the production of raw materials, specialty chemicals, and value-added products. Population balance models (PBMs) have been used as a quantitative tool to model the evolution of particle size distribution (PSD) during milling processes. Most PBMs used in the last 70 years assumed the linearity of the breakage rate, i.e., first-order breakage kinetics. In this presentation, I will present some fundamental experimental studies that provide strong evidence for the emergence of non-linear breakage: deviations from first-order breakage kinetics, significant impact of initial PSD on the breakage kinetics, anomalies in binary breakage tests, etc. The non-linear breakage appears to be prevalent in dense milling systems such as in dry ball mills where enduring mechanical multi-particle interactions occur among particles. Then, I will give highlights from recent theoretical developments that address this complex phenomenon of non-linear particle breakage. The traditional linear PBM, the time-variant PBM, and the non-linear functional model [1] have been critically analyzed in view of experimental data. Extensive numerical simulations of dry ball milling systems [2,3] and particle bed breakage experiments [4] suggest that the non-linear functional model can serve as a unified framework by which non-linear particle breakage in a multitude of comminution systems can be systematically studied and quantified at the process length scale. Finally, the presentation will highlight recent efforts in exploring ***the origin of non-linear breakage*** [5] via multi-scale DEM–PBM models and how to best use such models judiciously to predict the evolution of PSD in non-linear milling processes [6].

References:

[1] E. Bilgili, J. Yepes, B. Scarlett, "Formulation of A Non-Linear Framework for Population Balance Modeling of Batch Grinding: Beyond First-Order Kinetics," Chemical Engineering Science, Vol. 61, 2006, pp. 33–44.

[2] E. Bilgili, B. Scarlett, "Population Balance Modeling of Non-Linear Effects in Milling Processes," Powder Technology, Vol. 153, 2005, pp. 59–71.

[3] M. Capece, E. Bilgili, R. Dave, "Identification of the Breakage Rate and Distribution Parameters in a Non-Linear Population Balance Model for Batch Milling," Powder Technology, Vol. 208, 2011, pp. 195–204.

[4] E. Bilgili, M. Capece, "A Rigorous Breakage Matrix Methodology for Characterization of Multi-Particle Interactions in Dense-Phase Particle Breakage," Chemical Engineering Research and Design, Vol. 90, 2012, pp. 1177–1188.

[5] M. Capece, R. N. Dave, E. Bilgili, "On the Origin of Non-linear Breakage Kinetics in Dry Milling," Powder Technology, Vol. 272, 2015, pp. 189–203.

[6] M. Capece, R. N. Dave, E. Bilgili, "A Pseudo-coupled DEM–Non-linear PBM Approach for Simulating the Evolution of Particle Size during Dry Milling," Powder Technology, Vol. 323, 2018, pp. 374–384.

524051 Surface Dynamics of Micronized Active Particles

Abstract:

Surface Dynamics of Micronized Active Particles

Vibha Puri[1], Jagdeep Shur[2], Robert Price[2] & Ajit Narang[1]

[1]Small Molecule Pharmaceutical Sciences, Genentech, Inc., 1 DNA Way, South San Francisco, CA 94080, USA

[2]Nanopharm Ltd., Cavendish House, Newport NP10 8FY, UK

ABSTRACT

Purpose

The performance of pharmaceutical powders in drug products, such as dry powder inhalations (DPIs), is predominantly governed by the interfacial interactions of the particles in multi-component powder blends. The particle size targeted for lung delivery of drugs is in the range of 1 to 5 micron. Air jet micronization is commonly used for particle size reduction of synthetic, crystalline APIs. However, this high energy process can induce structural and surface disorder in the micronized API. Furthermore, this mechanically activated API on storage can undergo relaxation (surface stabilization) that can alter the material's surface properties. Often a *conditioning or resting* step is introduced to allow the energized, micronized API to stabilize. Hence, there is a need for better understanding of the surface of micronized API and development of a conceptual framework for selection of conditioning environment for the micronized API.

The objective of the study was to investigate the molecular and particle level changes occurring during storage of a model micronized API (compound A), and their impact on interparticle interactions. A multitude of techniques were used to investigate surface structural disorder, surface energetics, and interparticle interactions.

Materials and methods

Micronized API (Compound A) produced by air jet milling was stored under different temperature and relative humidity conditions of 20°C/30%RH, 60°C/30% RH, 20°C/80%RH, and 60°C/80% RH, and characterized at 0, 3, 6, and 9 week time period. The structural disorder in the micronized API was measured by RH-perfusion microcalorimetry (using a thermal activity monior (TAM), TA Instruments, USA). Surface free energy (SFE) was measured by inverse gas chromatography (using surface energy analyser, SMS Instruments, USA). The interparticle forces between drug-drug (cohesive) and drug-lactose (Respitose SV003, DFE Pharma, USA) (adhesive) particles were measured by atomic force microscopy (AFM) (Nanoscope IIIa, CA, USA). The slope of plots of force of cohesion versus adhesion was denoted as the cohesive-adhesive balance (CAB). Changes in the specific surface area, particle size, and crystallinity were also monitored.

Results and discussion

Compound A is non-hygroscopic with <0.3% water uptake at 25°C/90% RH. Complementary use of melting peak characterization (by differential scanning calorimeter) and thermodynamics of sorption-desorption isotherm (by TAM) showed that the structural disorder due to micronization was predominantly at the surface. The TAM profiles indicated that all stored samples underwent structural relaxation over time (Figure 1). The effect of high temperature (60°C/30% RH) and high relative humidity (20°C/80%RH) in lowering the structural disorder was comparable. Maximum relaxation was achieved under combined conditions of high temperature and relative humidity (60°C/80%RH).

The freshly micronized API showed total, dispersive, and specific (acid-base) surface energies of 64.0±1.2, 58.9±0.9, and 5.1±0.5 mJ/m^2, respectively, at the surface coverage of 0.01 n/n_m. The dispersive component of the SFE was the major contributor to the total SFE, and showed quantitative lowering in the stored samples (Figure 2). At the 9-week time point, all the samples showed lowering of SFE to similar extent. However, the kinetics was significantly fast at high temperature and high relative humidity condition (60°C/80%RH), and the samples showed greater heterogeneity.

The SFE profiles determined for API and lactose monohydrate were collectively used to compute work of cohesion (drug-drug) and adhesion (drug-lactose). The freshly micronized API had IGC-CAB of 1.16, suggesting the material was cohesive. The SFE change was most rapid at extreme stress condition of 60°C/80% RH and the CAB value lowered to 1.04, both, for the 3- and 9-week samples. However, the remaining three storage conditions showed a staggered fall over the 3- and 9-week periods, with CAB value lowering to about 1.11, followed by reduction to 1.04.

The AFM studies provided a direct measurement of interparticle forces (Figure 3). The freshly micronized API sample displayed a CAB value of 1.35 suggesting that the drug-drug cohesion was higher than the drug-lactose adhesion. On storage, the CAB values showed a trend to reduce and fall below 1. At the stress condition of high temperature-high humidity (60°C/80%RH), the CAB value reached 0.65 at 9-week time point, which indicated that the drug-lactose adhesion forces were ~1.5 fold higher than the drug-drug cohesion forces. Comparing the percent change in CAB across the four storage conditions, the low temperature-low humidity (20°C/30%RH) environment showed least change (CAB of 1.18).

The powder samples showed lowering of the specific surface area over the period of storage, with maximum reduction seen for high temperature-high relative humidity condition (60°C/80%RH).

Conclusion

The study highlights the use of surface specific techniques to quantitatively measure changes in the micronized API. Micronization of pharmaceutical powders can induce structural disorder in the constituent particles. These materials continue to undergo surface re-construction on storage, which can then impact their physicochemical properties.

Compound A changed in molecular and particle level properties upon storage under different temperature and humidity conditions. Surface structural disorder, surface energy, and specific surface area lowered on storage. The CAB of micronized API changed from being cohesive (CAB>1) to adhesive (CAB<1) on exposure to stress conditions over time.

The study shows that the extent of change in structural disorder, surface energy, particle features (such as specific surface area), and the overall interparticle forces were at different scales for the different environmental conditions. The higher stress conditions showed faster kinetics of change in material properties along with the greater heterogeneity. Hence, the selection of conditioning environment for the micronized API can impact the overall drug product performance.

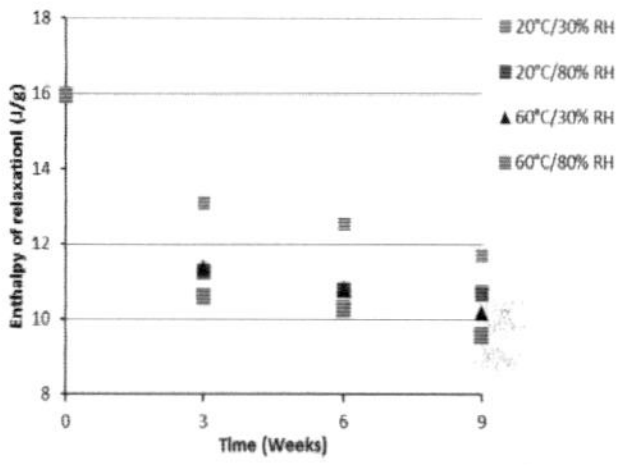

Figure 1. Plot of enthalpy relaxation of the micronized API powder samples as a function of storage time.

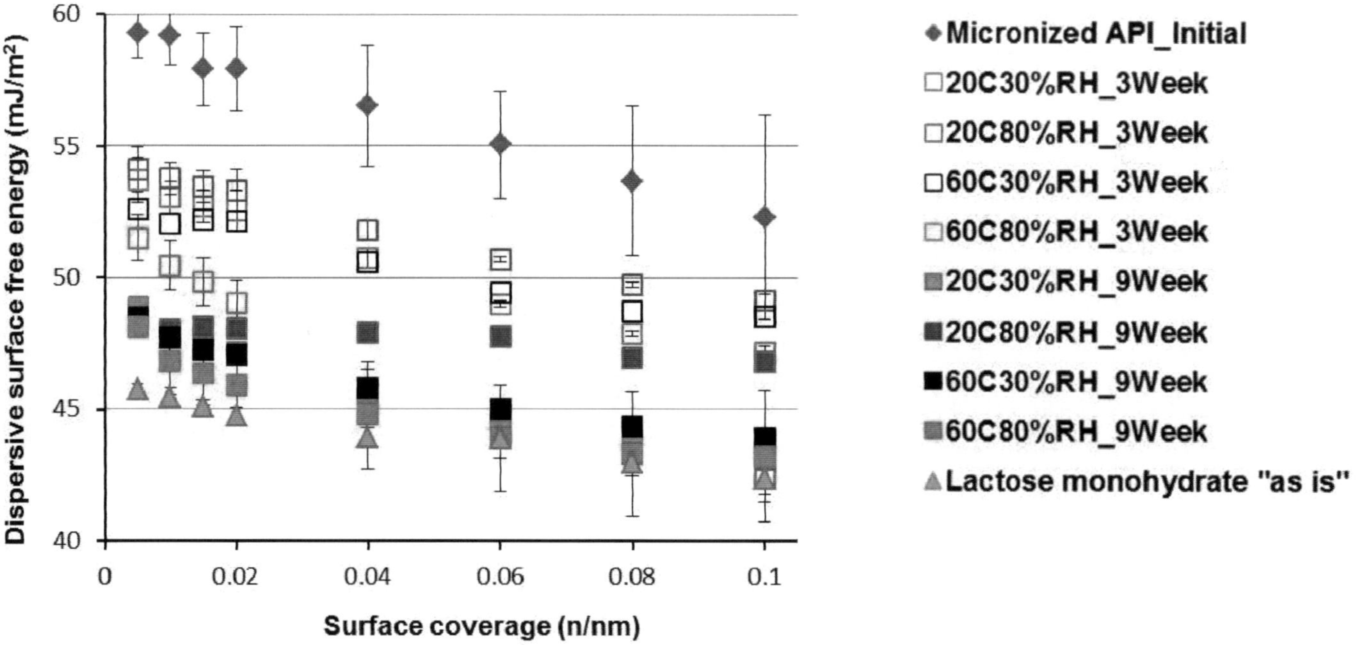

Figure 2. Plot of dispersive surface energy of micronized API and lactose samples.

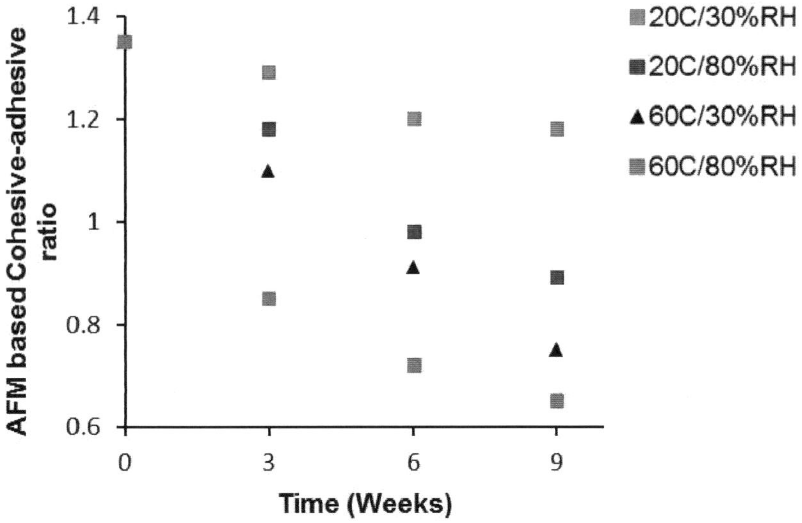

Figure 3. Plot of AFM based CAB ratios of the micronized API powder samples as a function of storage time.

537911 An Investigation into the Performance of an Industrial-Scale Roll Mill

Abstract:

Prediction of mill performance is a difficult task. In order to predict the performance of a roll mill, it is useful to determine the selection and breakage functions of a mill for a particular bulk solid as a function of mill parameters such as roll gap, roll differential speed, and mill feed rate. This enables modeling of the grinding circuit under changing mill and feed conditions and estimation of the product particle size. In this investigation, a production-sized roll mill was fed tightly sized fractions of particles which yielded the selection function and daughter particle size distribution (PSD) directly. In addition, the mill was also fed a polydisperse PSD in order to investigate how well the exit PSD was modelled using the data from the tightly sized fractions. Lastly, the experimental breakage matrix was compared with the predictions of several breakage matrix models.

548145 Are Low Molecular Weight Polymers Really Effective for Stabilization of Wet-Milled Drug Suspensions?

Abstract:

Owing to their large surface area and supersaturation capability, drug nanoparticles prepared via wet stirred media milling have been widely used to address inadequate bioavailability of poorly water-soluble drugs [1,2]. Unfortunately, stabilization of drug nanoparticles in aqueous medium poses challenges as nanoparticles tend to aggregate, leading to loss of surface area available for dissolution. Thus, drug nanosuspensions necessitate the use of various dispersants/stabilizers. According to some published work [3,4] and prevailing notion among pharmaceutical formulators, low MW polymers are usually more effective than and preferred over high MW polymers in the production and stabilization of drug nanosuspensions. This study subjects this commonly-held notion to serious scrutiny. In this study, we explore the impact of various polymers and their molecular weight (MW) on the aggregation state and rheology of wet-milled suspensions of itraconazole (ITZ), a poorly water-soluble drug. ITZ suspensions with SSL, SL, and L grades of hydroxypropyl cellulose (HPC) having MW of 40, 100, and 140 kg/mol, respectively, hydroxypropyl methyl cellulose (HPMC E3 with 10 kg/mol), polyvinylpyrrolidone (PVP K30 with 50 kg/mol), sodium dodecyl sulfate (SDS, surfactant), and HPC SL–SDS were wet-milled in a stirred media mill. Laser diffraction, viscometry, scanning electron microscopy, and electrophoretic light scattering (ELS) were used to characterize the milled suspensions.

Laser diffraction results show that 2.5% HPC SL–0.2% SDS led to the finest ITZ nanosuspension, whereas without SDS, only 4.5% HPC with SL/L grades ensured minimal aggregation. Rheological characterization reveals that aggregated suspensions exhibited pronounced pseudoplasticity, whereas stable suspensions exhibited near Newtonian behavior. Wet-milled stable 10% ITZ nanosuspensions exhibiting near-Newtonian flow behavior were prepared with 4.5% HPC SL/L (100 and 140 kg/mol MW, respectively). At this concentration, HPC SSL (40 kg/mol), HPMC E3 (10 kg/mol), and PVP K30 (50 kg/mol) could not suppress ITZ nanoparticle aggregation, leading to significant pseudoplastic behavior. Contrary to previous studies that highlight the favorability of low MW polymers [3,4] (specifically <50 kg/mol for ITZ–HPC [3]), our study demonstrated that higher MW HPC (100 and 140 kg/mol) is more favorable for ITZ nanosuspension stabilization, which in line with the thermodynamic considerations of polymer adsorption. Unfortunately, studies like [3,4] that used low-energy mills do not truly reflect the impact of different polymers or MW on steric stabilization and reduction of drug aggregate formation. The use of higher MW polymer and/or higher polymer concentration causes more pronounced viscous dampening [5], which slows down the breakage in low-energy mills more profoundly than in high-energy mills like wet stirred media mills used in this study. Hence, such studies drew somewhat confounded conclusions about the impact of polymer MW due to pronounced impact of viscous dampening. Not only did high-energy, wet stirred media milling enabled faster production of ITZ nanoparticles than low energy mills used in ref. [3] for ITZ–HPC, but also it allowed us to elucidate the role of polymer MW in stabilization of drug nanoparticles, which in line with the thermodynamic consideration of polymer adsorption. Although higher MW HPC led to better stabilization of the ITZ nanosuspensions here, the use of HPC with MW >140 kg/mol could cause more pronounced viscous dampening and ensuing slower breakage. Hence, overall our results point out the criticality of choosing optimal polymer MW in the stabilization of wet-milled drug nanoparticles and invalidity of the commonly-held notion that low-MW polymers are more effective than high-MW polymers for the stabilization of drug nanosuspensions.

References:

1. Li, M. Azad, R. Davé, E. Bilgili, Nanomilling of drugs for bioavailability enhancement: a holistic formulation–process perspective, Pharmaceutics, 8 (2016) 27.
2. Bhakay, M. Rahman, R. N. Dave, E. Bilgili, Bioavailability enhancement of poorly water-soluble drugs via nanocomposites: Formulation processing aspects and challenges, Pharmaceutics, 10 (2018) 86.
3. -Y. Choi, C. H. Park, J. Lee, Effect of polymer molecular weight on nanocomminution of poorly soluble drug, Drug Delivery, 15 (2008) 347–353.
4. Sepassi, D. Goodwin, A. Drake, S. Holland, G. Leonard, L. Martini, M. Lawrence, Effect of polymer molecular weight on the production of drug nanoparticles, J. Pharm. Sci., 96 (2007) 2655–2666.
5. E. Bilgili, A. Afolabi, A combined microhydrodynamics–polymer adsorption analysis for elucidation of the roles of stabilizers in wet stirred media milling, Int. J. Pharm., 439 (2012) 193–206.

536943 Attrition in Submerged Jetting Region in Fluidized Bed

Abstract:

One of major sources of attrition in fluidized bed and circulating fluidized bed is the jets from the gas distributor. Particles are generally picked up by the jet at the gas distributor. They accelerate along with the jet penetration length, and slam into the relatively slow moving particles. This is how attrition occurs at the gas distributor.

Particle attrition in the submerged jetting region depends strongly on orifice velocity and orifice size. Over the years, the attrition works in open literature were mostly concerned with higher gas jet velocities and smaller orifice size than are commonly found in industrial scale equipment. Furthermore, attempts on attrition model development were mainly based on empirical fitting to the experimental data with a limited number of test materials and narrow range of operating conditions. Attrition rates are typically expressed with a power-law dependence for gas jet velocity and hole diameter. A wide range of power indices are reported in open literature; for example, dependence of jet velocity on attrition rate has been reported to vary from 1 to 5.8.

In this work, PSRI conducted a thorough and systematic study of submerged jets in order to assess attrition in the jetting region of a large scale fluidized bed; a commercially relevant grid nozzle design and experimental conditions were used. Grid attrition rate was found to be proportional to the square of shroud velocity.

One of the major new findings was to have a better understanding on the gas jet orientation on attrition. It was generally known that that downwardly directed jets always produced more fines than the upwardly pointed jets. However, this was found to be not always the case. This result will change our knowledge on the effect of gas jet orientation on attrition.

539672 Dust Dispersion Particle Breakage: Classification Based on Brittleness Index

Abstract:

Dust explosions are a frequent problem in the process industries causing loss of life and property. Standards such as ASTM, ISO etc. provide guidelines for determining the explosion parameters P_{max} and K_{St}, (the maximum pressure and rate of pressure rise, respectively) in a dust explosion. These parameters are found in standard 20-L or 1-m^3 devices and are used to perform explosion risk assessments and mitigation. In these devices, it is assumed the particle size distribution of the dispersed dust remains constant throughout the testing procedure. Recent studies show a significant particle size reduction can result from dispersion in the standard 20-L apparatus, which can affect the explosion parameters. The reduction in particle size due to the dispersion process depends on the dust and the explosion apparatus. Therefore, it is necessary to quantify the particle size reduction of different dusts due to dispersion in various explosion apparatuses (20-L, 36-L and 1-m^3) and correlate it to the dust properties that influence particle breakage.

This research studied particle size reduction for a range of materials in the 20-L, 36-L and 1-m^3 dust explosion apparatus. The dispersion-induced particle breakage was correlated to the properties of the dusts, which were measured using Nanoindentation. Based on the particle breakage, dusts were categorized into three categories, with a brittleness index range associated with each category. In addition, the effect of dust concentration on particle breakage and behavior of nanomaterial was examined.

Dispersion results show the 20-L and 36-L dust explosion apparatus cause significantly more particle breakage than the 1-m^3 apparatus. It also shows that not all materials undergo particle breakage. In addition, the 1-m^3 apparatus caused particle breakage only in material with high brittleness index. The results show that the brittleness index correlates directly to the particle breakage trend. Also, results indicate particle breakage is inversely related to dust concentration. Nanomaterial dispersion result shows particle breakage followed by subsequent re-agglomeration post-dispersion.

This research will help process industries identify dusts that are susceptible to breakage based on their brittleness index and prone to give misleading explosion results during standard dust explosion testing, permitting a proper dust explosion risk assessment.

535757 Compartmental Modeling and Simulation Study of Wet Twin Screw Granulator

Abstract:

A twin-screw wet granulator (TSWG) is an emerging technology for continuous wet granulation that achieves desired level of mixing through a combination of appropriate screw configuration and a range of process settings (e.g. feed rate, screw speed, etc.). The wet granulation process in TSWG can be divided into several stages. A number of different mechanisms, including nucleation, growth, aggregation, and breakage, which ultimately determine the characteristics of the produced granules, typically drive the dynamics of wet granulation. In this study, a laboratory scale TSWG is used to granulate Î±-Lactose monohydrate as a function of the screw speed, powder throughput and liquid to solid (L/S) ratio. The experimental results are used to construct and validate a Population Balance Model (PBM) of the granulator using the concept of compartmental modeling. The system is divided into three compartments based on the dominating process occurring in the respective compartments, i.e., first, second and third compartment are dedicated to aggregation, aggregation with breakage and breakage respectively. A recently developed discretization ([1] and [2]) is adapted and implemented to solve this model whose idea is based on the conserving the important properties of the system. The proposed model is analyzed by varying the liquid to solid (L/S) ratio as well as the rate of mass exchange between the compartments to track the particle size distribution. We find that the proposed mathematical model is capable of tracking the particle size distribution accurately for various values of L/S ratio and discuss the effect of these parameters on the performance of the granulator.

Keywords: Population Balance Equation, Twin Screw Wet Granulator, Finite volume Scheme.

References:

[1] Kumar, J., Kaur, G. and Tsotsas, E. (2016) An accurate and efficient discrete formulation of aggregation population balance equation. *Kinetic and Related Models* 9(2), 373-391.

[2] Saha, J., Kumar, J., Buck, A. and Tsotsas, E. (2016) Finite volume approximations of breakage population balance equation. *Chemical Engineering Research and Design* 110, 114-122.

535224 Developing a Formulation Dependent Mechanistic Kernel to Predict the Granule Size Distribution in a Two Component High Shear Wet Granulation Process

Abstract:

Wet granulation process is often a complex multi-component unit operation in pharmaceutical industry. The complexity of this process lies in understanding the effects of process design, operating conditions and material properties on the granule product quality. A holistic understanding of the granulation rate processes will greatly help in predicting such granule product attributes. Although an actual pharmaceutical granulation process involves more than one solid component (Active Pharmaceutical Ingredient (API), binder, surfactant, disintegrant etc.) there were few models reported to evaluate the functional dependency of the constituent materials. Most of the population balance modeling studies were focused on the representation of granulation process with single solid component and granule product attributes were predicted using the effect of individual volumes of solid, liquid and gas [1,2,6,7].

The purpose of this study was to develop a formulation dependent mechanistic kernel and to provide an experimental basis for incorporating constituent material property such as contact angle. The preferential wetting of constituent materials with liquid binder was expected to be the governing factor for the dynamics of granulation process and hence, two materials (Ibuprofen and Micro crystalline cellulose (MCC-101)) with wide difference in contact angle were chosen. Other factors such as particle segregation during dry mixing [5] and drug migration during drying of granules [3] were avoided as Ibuprofen and MCC-101 were of similar particle size and both were insoluble.

This study was performed in a high shear wet granulator with percentage of formulation varying from 40% to 60% for a range of operating conditions (liquid to solid ratio and wet massing time), by face-centered central composite design. The analysis of variance study indicated a significant effect of percent API on the granule size distribution (GSD). Though, it was expected to have a narrow GSD with increase in hydrophobic material [4], it was observed that with the increase in hydrophobic material there was an increase in granule d50 as well as a wider GSD. Also a change from induction growth to steady growth was observed with increase in percentage composition of Ibuprofen. This provided an additional insight that apart from the availability of binder liquid around the granules other material properties like material yield strength and maximum pore saturation, that affect the granule deformation and uptake of powder into the liquid droplet, also play an important role in granulation rate processes. Finally, a composition dependent mechanistic kernel was proposed considering these findings and this was validated against the experimental results obtained.

References

1. Barrasso, D., Hagrasy, A. E., Litster, J. D., Ramachandran, R., 2015. Multi-dimensional population balance model development and validation for a twin screw granulation process. Powder Technology 270, Part B, 612 – 621.
2. Chaudhury, A., Kapadia, A., Prakash, A. V., Barrasso, D., Ramachandran, R., 2013. An extended cell-average technique for a multi-dimensional population balance of granulation describing aggregation and breakage. Advanced Powder Technology 24 (6), 962 – 971.
3. Katari, A., 2018. An analysis of drug migration during drying of granules as an underlying cause of content non-uniformity in wet granulation. Ph.D. thesis, Rutgers, the State University of New Jersey.
4. Nguyen, T. H., Shen, W., Hapgood, K., 2010. Effect of formulation hydrophobicity on drug distribution in wet granulation. Chemical Engineering Journal 164 (2), 330 – 339, pharmaceutical Granulation and Processing.
5. Oka, S., Smrka, D., Kataria, A., Emady, H., Muzzio, F., tpnek, F., Ramachandran, R., 2017. Analysis of the origins of content non-uniformity in high-shear wet granulation. International Journal of Pharmaceutics 528 (1), 578 – 585.
6. Poon, J. M.-H., Immanuel, C. D., Francis J. Doyle, I., Litster, J. D., 2008. A three-dimensional population balance model of granulation with a mechanistic representation of the nucleation and aggregation phenomena. Chemical Engineering Science 63 (5), 1315 – 1329
7. Ramachandran, R., Barton, P. I., 2010. Effective parameter estimation within a multidimensional population balance model framework. Chemical Engineering Science 65 (16), 4884 – 4893.

524280 A Model-Based Design of Experiment (MB-DOE) Approach Towards Scale-up of High Shear Wet Granulation Operation

Abstract:

Several dimensionless numbers and their correlations have been proposed in literature (e.g. Power number correlation [1], Regime Map & Dimensionless Spray Flux [2]) which can aid in high shear wet granulation (HSWG) scale-up. Each of these dimensionless quantities and their correlations capture certain process/granule attributes and granulation rate process (i.e., liquid addition & nucleation, granule growth & consolidation, layering and breakage). If the granulation process is operated in the right growth and nucleation regime, the final granules will have properties as desired. Since HSWG is a combination of several rate processes happening at the same time, it is necessary to have a robust scale-up strategy. It is common to observe product quality failure at the commercial scale if the scale-up strategy doesnâ€™t account for the effect of all the critical process parameters (CPPs) on these rate processes.

In this work, a predictive data driven hybrid process model of the HSWG operation has been developed by combining all the available correlations among dimensionless numbers and relationships between other relevant geometric, quality and operational parameters, such that the effect of all the CPPs on the granule attributes can be quantified.

This process model has been further implemented within an optimization framework for designing experiments such that the space covered by the output variables (i.e., granule quality attributes) is maximized, while obtaining an orthogonal bracketing study in the process input parameters, within the desired design space. The resulting problem is a multi-objective optimization that seeks to fulfill the lower level problem, which solves the hybrid process model, while maximizing the optimality criterion written in the upper problem [3].

This work will illustrate a model-based DOE approach where appropriate experiments have been designed for the initial ranging study at the pilot scale granulator based on the observations made at the small-scale granulator. This also aided in identifying the range of operating parameters at the pilot scale such that the granulation has properties in the desired space.

In addition, the hybrid process model, implemented within the optimizer has been further used to check that the proposed operating conditions for the very first commercial scale study will yield granulation with property in the desired space, prior to the actual experimental run.

References

[1] Faure et al., 1999. European Journal of Pharmaceutical Sciences, 8, 85-93

[2] J.D. Litster, 2003, Powder Technology, 130, 35-40

[3] Chen et al., 2015, AIChE Annual Conference, November 9, 2015, Salt Lake City, UT

527781 A Population Balance Based Rheological Model for Fresh Cement Paste

Abstract:

A Population Balance based Rheological Model for Fresh Cement Paste

Juan Pablo Gallo-Molina[1,2], Ingmar Nopens[1] and Karel Lesage[2]

[1]BIOMATH, Department of Mathematical Modelling, Statistics and Bio-informatics, Ghent University, Ghent, Belgium.

[2]Magnel Laboratory for Concrete Research, Department of Structural Engineering, Ghent University, Ghent, Belgium.

Due to the fact that it is the largest material volume used in the world, its critical importance for infrastructure development and the environmental issues associated with the large masses involved, concrete has been attracting increasing research attention since the last decades [1]. For hardened concrete, civil engineers already possess a series of engineering tools that allow them to satisfactorily predict properties of interest (e.g. deformation, stress, etc.) as a function of material characteristics. However, when this material is encountered in a fluid state, the picture is completely different: only some empirical tests are capable of relating some material variables (e.g. concrete composition) with relevant properties (i.e. flow speed, flow spread, etc.). Although this practice is longstanding, the increasing complexity of concrete mixtures - introduced by the growing use of mineral additions, chemical admixtures and non-traditional cement chemistries - has invoked the need for developing mechanistic models with predictive capabilities.

The size range of the particles found in a typical concrete can encompass several orders of magnitude, which implies that numerous kinds of interactions and phenomena are involved. For this reason, a bottom-up modelling approach is often aimed for in the modelling of fresh concrete [2]. Consequently, adequately modelling pure cement pastes is deemed as a first step towards the mentioned goal. These pastes are, in turn, particulate systems consisting of a mixture of mineral particles suspended in an aqueous medium. As the particle size distribution and its evolution in time due to aggregation significantly impact rheological properties, Population Balance Models (PBM) are a valuable tool in the developing of models capable of predicting the rheology of fresh cement paste.

As an initial approach, in this work we present a population balance equation based rheological model [3]. This model is implemented for cement pastes under simple torsional flow and it is applied to a parallel plate rheometer geometry. In order to consider the effect of vertical position on shear rate, we introduce one spatial dimension in the model. The objective of this modelling effort is to better understand both rheological and particle size distribution measurements of these systems and to further develop the understanding of complex rheological behavior resulting from aggregate interactions. This work seeks to contribute in the developing of a suitable model capable of correlating cement properties and the rheological behavior of its paste.

References

[1] R. J. Flatt, N. Roussel, and C. R. Cheeseman, "Concrete: An eco material that needs to be improved," *J. Eur. Ceram. Soc.*, vol. 32, no. 11, pp. 2787–2798, 2012.

[2] N. Roussel, A. Lemaître, R. J. Flatt, and P. Coussot, "Steady state flow of cement suspensions: A micromechanical state of the art," *Cem. Concr. Res.*, vol. 40, no. 1, pp. 77–84, 2010.

[3] M. Mohtaschemi, A. Puisto, X. Illa, and M. J. Alava, "Rheology dynamics of aggregating colloidal suspensions," *Soft Matter*, vol. 10, no. 17, p. 2971, 2014.

527765 Agglomeration-Driven Product Selection in a Continuously Operated Fluidized-Bed Crystallizer

Abstract:

A fluidized-bed crystallizer is continuously operated with a specified crystal-size distribution. Under selected experimental conditions of temperature-dependent supersaturation and varied flow conditions the crystal size distribution changes dominantly by agglomeration. A flow-through microscope QICPIC is used to measure and quantify the changes in shape, size and agglomeration over time.

Corresponding model calculations for the fluid flow as well as for the crystal size and shape changes are used to deepen our understanding of the important process phenomena and the influencing process parameters.

The application of the fluidized-bed crystallizer in bigger process scheme of a continuous production of specified crystals will be the prospect of the presented work.

534985 Roller Compaction Modelling: Pharmaceutical Application

Abstract:

Roller Compaction Modelling: Pharmaceutical Application

R. Sousa[1], L. Bardet[2], S. Doktorovova[1], V. Sainz[1], P.C. Valente[1], J.R. Authelin[2]

[1] R&D Drug Product Development, Hovione Farmacência S.A, Estrada do Paço do Lumiar, 1649-038 Lisboa, Portugal;

[2]Sanofi-aventis, Research and Development, Physical Quality, 94403 Vitry-sur-Seine, France

We present the comparison of different roller compaction models that predict the ribbon density from the process parameters and material constitutive laws benchmarked against experimental data. Two models from the literature Johanson's Model (Johanson, 1965) and Thin Layer Model (Peter, Lammens, & Steffens, 2010; Reimer & Kleinebudde, 2018) were compared with a new model, which consists in an analytical solution to predict the pressure at gap. The three models consider two regions between the rolls, a slip region where the roll is faster than the powder and a nip region where the material is compressed. Johanson's Model finds the nip angle that separates the two regions by intersecting the pressure gradient curves in both regions, considering the log density as a linear function of the log stress in the nip region (Johanson, 1965). The Thin Layer Model was modified to solve the nip angle numerically to match the specific roll force and relates the stress as a function of the natural strain (Yu, 2012). Regarding the new model, the nip angle is calculated as an analytical function of the densification factor. Then, the nip angle expression is used in the force integral to derive the maximum pressure, which can be used to estimate the density at gap. All the calculations explained in the new model result in a single expression that relates the at gap density with the material constitutive laws (a constant and the compressibility factor), the specific roll force, gap width and roll dimensions.

Ribbon relative density has been shown to be a key predictor of granule properties post milling. Densification should be as high as necessary to achieve the desired granule flowability, but as low as possible to avoid the loss in compactibility (Kleinebudde, 2004). Therefore, in order to save time and material during formulation development, it is advantageous to have a predictive model based on easily accessible parameters and possible to calibrate with few experiments.

We present different methodologies to calibrate these models based on roller compaction experiments or die uniaxial simulations. The three models were calibrated for three different formulations with one third of the samples used for external validation. The new model was as reliable as the Johanson or Thin Layer models, with a prediction error of approximately ±0.1 g/mL. In conclusion, these models are important tools to obtain the desired granule properties with minimum loss in development time and material. For instance, a roller compaction experiment requires a minimum of 100 g of material, while a die compaction simulation would only need a few milligrams to estimate the material properties. Furthermore, they facilitate the scale up and should be easily transferred between different roller compactors.

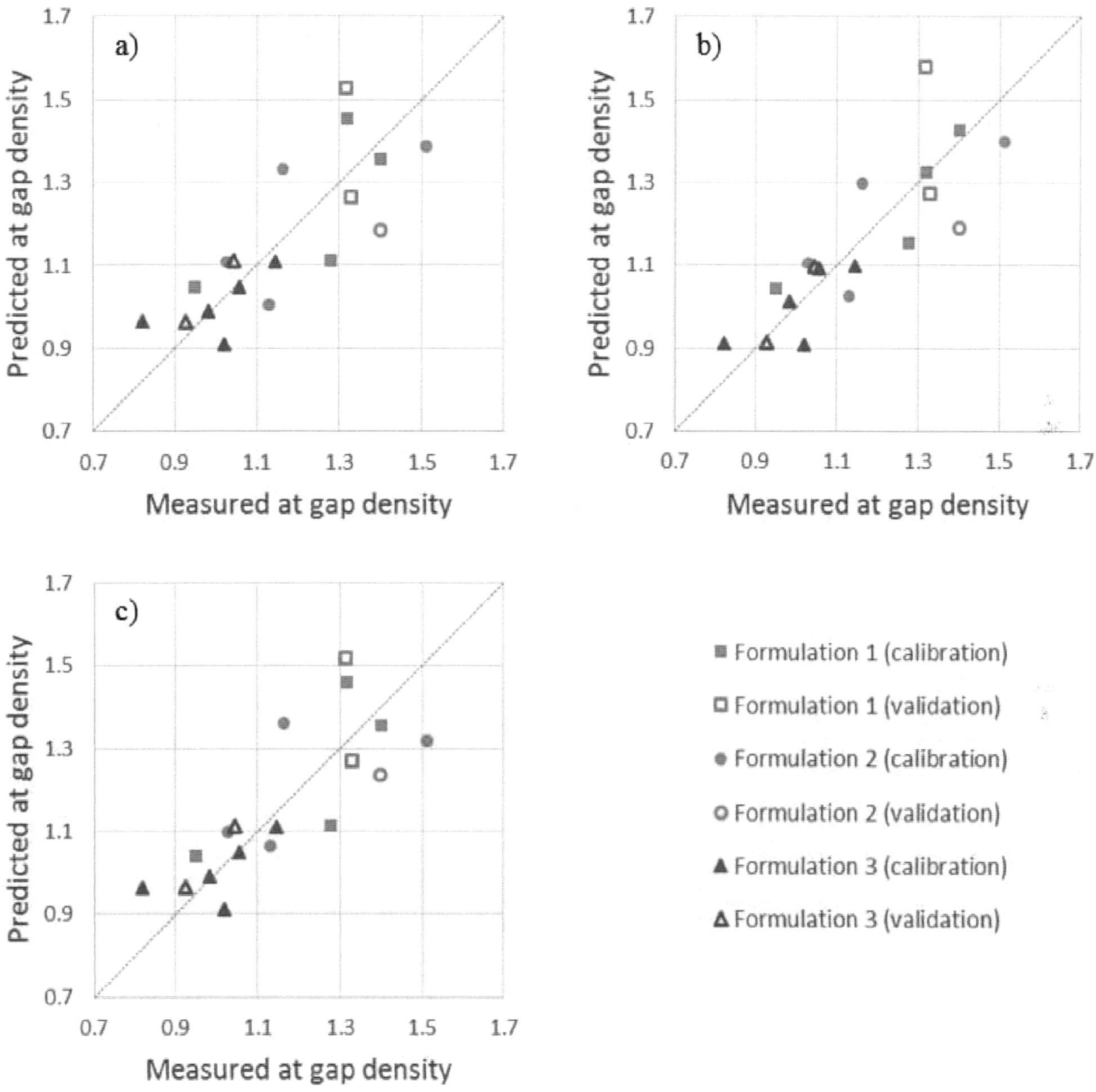

Figure 1 – Comparison of the measured and predicted at gap densities for three different formulations. a) Thin Layer Model; b) Johanson's Model; c) New model.

We dedicate this work to prof. Noel Midoux who wrote all equations for the new model.

References

Johanson, J. R. (1965). A Rolling Theory for Granular Solids. *Journal of Applied Mechanics*, *32*(4), 842–848. https://doi.org/10.1115/1.3627325

Kleinebudde, P. (2004). Roll compaction/dry granulation: Pharmaceutical applications. *European Journal of Pharmaceutics and Biopharmaceutics*, *58*(2), 317–326. https://doi.org/10.1016/j.ejpb.2004.04.014

Peter, S., Lammens, R. F., & Steffens, K. J. (2010). Roller compaction/Dry granulation: Use of the thin layer model for predicting densities and forces during roller compaction. *Powder Technology*, *199*(2), 165–175. https://doi.org/10.1016/j.powtec.2010.01.002

Reimer, H. L., & Kleinebudde, P. (2018). Hybrid modeling of roll compaction processes with the Styl'One Evolution. *Powder Technology*, #pagerange#. https://doi.org/10.1016/j.powtec.2018.02.052

Yu, S. (2012). ROLL COMPACTION OF PHARMACEUTICAL EXCIPIENTS by SHEN YU A thesis submitted to The University of Birmingham for the degree of School of Chemical Engineering The University of Birmingham October 2012, (October).

530202 Anisotropic Mechanical Properties of Compacted Powders with Cohesive Contacts

Abstract:

In this work the multi-particle finite element method (MPFEM) is used to simulate the mechanical properties of compacted powders with cohesive contacts. The mechanical properties under consideration are the elastic properties (Youngâ€™s modulus, shear modulus and Poison ratio) and the yield properties (uniaxial yield strength, yield surface). The MPFEM model consists of a representative volume element (RVE) of monodisperse, spherical, deformable particles which are compacted during simulation. A simple contact model is used to describe the interaction of particles. The contact model includes repulsive forces, friction forces and cohesion forces and is therefore suitable to describe tensile strength of powders in addition to the compression strength.

The whole MPFEM simulation is divided in the compaction step and the testing step. In the compaction step, a RVE of randomly closed-packed spheres is compacted to a defined packing density in the range of 0.65 to 0.95. Different compaction modes are used to evaluate the influence of the strain path during compaction. In addition to the two most common compaction modes isostatic and closed die, five extra modes with different strain paths are considered. The strain path during compaction is described by means of the ratios of the three main principal strains and the average strain (, ,). The strength of the contact cohesion is varied to investigate the whole range from non-cohesive powder to completely bonded powders. After compaction the powder is unloaded to obtain the reference state for the testing step.

During the testing step the RVE of compacted powder is reloaded to determine its mechanical properties. First of all, the RVE is loaded in particular normal and tangential directions while the deformation is recorded to determine the elastic properties. Then the uniaxial yield strength for compression and tension is determined in x-, y- and z-direction. Finally, the RVE is stressed in many different arbitrary directions to yield, in order to generate the whole yield surface of the compacted powder.

The results of the simulations show the mechanical properties of compacted powder as function of the packing density after compaction, the cohesion strength of the contacts and the strain path. Algebraic equations are derived to condense the simulation results of the parameter study. A continuous mathematical description of the mechanical properties is given as function of the relative density, the contact cohesion strength and the three strain ratios during compaction. The equations can be used to calculate the mechanical properties of compacted powder for arbitrary packing densities, contact cohesion and strain paths.

The MPFEM model in this work mechanistically describes the anisotropic properties of compacted powder. The derived equations for the mechanical properties can be used in a future work as input data in a conventional FEM model of powder compaction. In this way the spatial distribution of the powder properties inside a whole part of compacted powder can be investigated. This way, the stability of parts can be investigated for different powder properties and different compaction conditions.

532439 Development of CFD-DEM Coupling Model for Particles-Liquid-Gas Flow

Abstract:

Particles-liquid-gas flow can be encountered in many environmental phenomena such as debris flow and riverbed erosion as well as many industrial applications including powder dispersion in a stirred tank and liquid imbibition in powders during a wet granulation process. The flow can be extremely complex due to the interactions between the phases, e.g. surface tension / capillary forces, wetting of particles surface and momentum exchanges. In the present work, a novel meso-scopic coupling model of CFD and DEM is developed which takes into account these interactions. The Diffuse Interface (DI) method is used to capture liquid and gas phases in CFD whilst DEM is used to track the movement of particles. Smooth mapping of particle-scale quantities to continuous fields, e.g. particle solid fraction, is achieved by using the diffusion-based coarse graining [1,2] so that the numerical cell size can be chosen independently of the particle size, which can be critical to treat the interfacial phenomena properly. Several simulation case studies are performed to show the validity of the model developed.

[1] R. Sun, H. Xiao, Diffusion-based coarse graining in hybrid continuum-discrete solvers: Theoretical formulation and a priori tests, Int. J. Multiph. Flow. 77 (2015) 142–157. doi:10.1016/j.ijmultiphaseflow.2015.08.014.

[2] R. Sun, H. Xiao, Diffusion-based coarse graining in hybrid continuum – discrete solvers : Applications in CFD – DEM, Int. J. Multiph. Flow. 72 (2015) 233–247.

514842 Meso-Scale Nonequilibrium Characteristics in a Bubbling Fluidized Bed

Abstract:

To understand the meso-scale, far-from-equilibrium behavior in fluidization [1], we investigate, both numerically and experimentally, the nonequilibrium features in a pseudo 2D bubbling fluidized bed. In experiment, velocities of individual particles are measured by using a particle tracking velocimetry (PTV) method, and void fractions are obtained with Voronoi tessellation. A bimodal shape of probability density function (PDF) for particle vertical velocity is found in not only time-averaged but also time-varying statistics, which is caused by the transition between the dense and dilute phases and breaks the local-equilibrium assumption in continuum modeling of fluidized beds [1]. The results of time-varying radial distribution function and voidage distribution also confirm this finding. Moreover, analysis of voidage, velocity of particle, granular temperature and turbulent kinetic energy of particles shows that there is no scale-independent plateau over the interface, and it seems hard to find a scale-independent plateau to separate the micro- and meso-scales of fluidized beds, which require sub-grid meso-scale modeling for continuum or coarse-graining methods of gas-fluidized systems [2]. In numerical simulation, dense discrete particle method is used with the energy minimization multi-scale (EMMS) drag. Simulation results generally agree with the experiment.

References

1. W. Wang, Y. Chen, Mesoscale modeling: Beyond local equilibrium assumption for multiphase flow, Adv. Chem. Eng. 47(2015) 193â€"277.
2. Y. Tian, J. Geng, W. Wang, Structure-dependent analysis of energy dissipation in gas-solid flows: Beyond nonequilibrium thermodynamics, Chem. Eng. Sci. 171 (2017) 271â€"281.

530886 Development of Drift Velocity Transport Equation for Filtered Drag Force Model

Abstract:

Division: Particle Technology/ Fluidization and Fluid-Particle Systems

Topic: Computation Modeling and Validation for Fluidization Processes

Title: Development of Drift Velocity Transport Equation for Filtered Drag Force Model

Authors: Yundi Jiang[1], Ali Ozel[2], Jari Kolehmainen[1], Yannis Kevrekidis[1,3], Sankaran Sundaresan[1]

[1]Department of Chemical and Biological Engineering, Princeton University, Princeton, New Jersey 08540, USA

[2]School of Engineering and Physical Sciences, Heriot-Watt University, Edinburgh EH14 4AS, UK

[3]Department of Chemical and Biomolecular Engineering, Johns Hopkins University, Baltimore, MD, 21218, Maryland, USA

Introduction: Gas-particles flows in industrial vessels such as circulating fluidized beds are inherently unstable and exhibit structures of various temporal and spatial scales. Filtered Euler-Euler models are introduced to simulate such flows so that we do not have to resolve the fine-scale flow structures, which are impractical to resolve due to computing limitations [1]. It solves filtered transport equations for both solid and gas phases and the fine-scale flow structures that are unresolved are taken into account through sub-grid modeling [2,3,4]. It was found that, among all the terms in the filtered transport equations, the sub-grid contribution of the drag term is the most significant [3,4]. Previous research has established that sub-grid drift velocity, defined as the difference between the filtered gas velocity and the filtered gas velocity seen by the particles, is an essential marker to model the sub-grid contribution of the drag term [3,4], and similarly in Reynolds-averaged kinetic theory models [5]. Although drift velocity can capture the sub-grid contribution of the filtered drag term, the accessibility of this sub-grid variable is lacking in coarse simulation and the estimation of it remains a challenge. In the present study, we seek to develop a transport equation to calculate drift velocity.

Methods and Results: A transport equation for drift velocity is derived from filtered and unfiltered solid continuity and gas momentum equations, yielding thirteen unclosed sub-grid scale terms. We perform budget analysis by taking the domain and time average of each term and four out of the thirteen terms are significant. These four terms are further combined and reduced to two: a sub-grid gas pressure gradient and a sub-grid void fraction scaled drag term. Constitutive models are required to close these two terms. Finding an explicit model with the guidance of physical intuition has been challenging, so we use data mining techniques to close the equation. Neural network models are built on the filtered variables computed from fine-grid Euler-Euler simulation results. Correlations coefficients of 0.8 and 0.91 are obtained for the two terms. After closing the transport equation, we assess its predictability by performing a coarse-grid Euler-Euler simulation of a 3-D dense fluidized bed. A comparison of the axial direction average solid volume fraction profile of fine- and coarse-grid simulation indicates that the sub-grid contribution of the drag term can be well captured by evolving the transport equation.

Conclusion: Sub-grid modeling of the drag term requires an accurate estimation of drift velocity, which needs a transport equation due to the significance of temporal effects. A transport equation is derived theoretically, simplified with budget analysis, closed with neural network models, and assessed with a coarse-grid simulation. The results from the coarse simulation indicate that by including a drift velocity evolution, we can effectively correct the drag term.

References:

[1] Agrawal, K., Loezos, P. N., Syamlal, M., & Sundaresan, S. (2001). The role of meso-scale structures in rapid gas–solid flows. *Journal of Fluid Mechanics, 445*, 151-185.

[2] Igci, Y., Andrews, A. T., Sundaresan, S., Pannala, S., & O'Brien, T. (2008). Filtered two fluid models for fluidized gas particle suspensions. *AIChE Journal*, *54*(6), 1431-1448.

[3] Ozel, A., Fede, P., & Simonin, O. (2013). Development of filtered Euler–Euler two-phase model for circulating fluidised bed: high resolution simulation, formulation and a priori analyses. *International Journal of Multiphase Flow, 55*, 43-63.

[4] Parmentier, J., Simonin O., & Delsart,O. (2012) A functional subgrid drift velocity model for filtered drag prediction in dense fluidized bed." *AIChE Journal* 58.4: 1084-1098.

[5] Fox, R. O. (2014) On multiphase turbulence models for collisional fluid–particle flows. *Journal of Fluid Mechanics* ,742, 368-424.

525850 Numerical Study on a Gas-Solid Flow in an Arbitrary Shape Boundary Including Thin Plates

Abstract:

Gas-solid flow is widely encountered in the chemical engineering such as a fluidized bed. Coupled Discrete Element Method (DEM) - Computational Fluid Dynamics (CFD) method[1] has been often employed in various solid-fluid flow systems in industries so far and its adequacy was proven in the previous studies[2,3]. In DEM-CFD method, the size of CFD grid should be sufficiently larger than the size of DEM particle diameter according to the local volume average technique[4]. Although this technique provides a great advantage on the representation of the macroscopic behavior of a gas-solid flow, it is of great difficulty to calculate the solid-fluid multiphase flow systems including thin plates. This is because the thin wall cannot be represented by the existing DEM-CFD method due to the resolution.

In order to solve this problem, we developed a Refined Grid Model in DEM-CFD method, where the CFD grid size could be given independently of the particle diameter. This approach is indeed useful in many systems in industries. In this study, the Refined Grid Model was applied to the fluidized bed with a narrow gap channel to show the adequacy of the model. Fluidization conditions in a narrow channel were successfully performed by using this model, though the existing method could not simulate this system. Consequently, it is shown that this model makes it possible to perform a gas-solid flow in a powder system including thin plates.

Acknowledgement

This study was financially supported by JSPS KAKENHI (17J02825).

References

[1] Y. Tsuji, T. Kawaguchi, and T. Tanaka, "Discrete particle simulation of two-dimensional fluidized bed," *Powder Technol.*, vol. 77, no. 77, pp. 79–87, 1993.

[2] M. Sakai, H. Takahashi, C. C. Pain, J. P. Latham, and J. Xiang, "Study on a large-scale discrete element model for fine particles in a fluidized bed," *Adv. Powder Technol.*, vol. 23, no. 5, pp. 673–681, 2012.

[3] M. Sakai, M. Abe, Y. Shigeto, S. Mizutani, H. Takahashi, A. Viré, J. R. Percival, J. Xiang, and C. C. Pain, "Verification and validation of a coarse grain model of the DEM in a bubbling fluidized bed," *Chem. Eng. J.*, vol. 244, pp. 33–43, 2014.

[4] T. B. Anderson and R. O. Y. Jackson, "A Fluid Mechanical Description of Fluidized Beds," *Ind. Eng. Chem. Fundam.*, vol. 6, no. 4, pp. 527–539, 1967.

536673 Catalytic Propane Oxidative De-Hydrogenation with High Propylene Selectivity in a Downer Fluidized Bed Reactor: Kinetics and CPFD Simulation

Abstract:

Propylene is one of the major building block in the chemical industries. The oxidative dehydrogenation (ODH) process is a low-temperature process offering potentially high olefin yields and extended catalyst life given the low coke deposition. But ODH is still hindered by low alkane conversion and poor olefin selectivity. This prevents the ODH process from being applied in large scale processes. ODH kinetics involve a complex reaction network. In the presence of gas phase oxygen, olefin combustion and correspondingly lower olefin selectivity are favored.

To demonstrate the catalytic propane oxidative dehydrogenation (PODH) with high propylene selectivity, runs were performed in a mini-fluidizable chemical reactor engineering center (CREC) Riser Simulator [1] using a novel catalyst 7.5 wt. % vanadium supported on a ZrO_2-γAl_2O_3 (1:1 wt. %) [2]. Reactions were carried out at 500-550°C, 10-20 seconds, atmospheric pressure under oxygen-free gas phase conditions. Given that, the ODH circulating fluidized bed system requires a limited degree of catalyst re-oxidation, the ODH catalyst was evaluated using consecutive propane injections (e.g. 10 cycles). There was no catalyst regeneration in between the 10 cycles. Propylene selectivities up to 94% with a 25% propane conversion were obtained.

These data were employed for establishing a kinetic model. This kinetic was based on a Langmuir-Hinshelwood rate equation and a parallel-series reaction network [3,4]. The 6-independent intrinsic kinetic parameters were calculated via numerical regression. The high propylene selectivity led to a much larger 2.82×10^{-5} mol.gcat^{-1}s^{-1} frequency factor for propylene formation versus the 1.65×10^{-6} mol.gcat^{-1}s^{-1} frequency factor for propane combustion. Calculated energies of activation (55.7 kJ/mole for propylene formation and 33.3 kJ/mole for propane combustion) appear however, to moderate this effect, with frequency factors influence prevailing. Furthermore, propylene conversion into carbon oxides (CO_x) oxidation appears as a non-favoured reaction step, given the 98.5 kJ/mole activation energy and 4.80×10^{-6} mol.gcat^{-1}s^{-1} frequency factor.

To address the issue of ODH process development, our research group at the CREC-University of Western Ontario, Canada has led the implementation of a PODH circulating process involving a downer unit and a dense phase fluidizable regenerator. This reactor systems overcome the issues of fixed-bed reactors, namely non-isothermal conditions and mass transfer limitations. So, propylene selectivity is highly influenced by a suitable catalyst, reactor configuration, operation parameters and proper kinetics.

As far as we are aware of, this is the first contribution where a large-scale simulation of PODH process in a circulating fluidized bed reactor is reported. Regarding the selected downer, it has 10 cm diameter and 10 m length, with two inlets in the top cyclone, one is for the partially reduced catalyst particles and a second one is for the oxidized catalysts (10/1 ratio). After leaving the cyclone, propane is fed into annular feeding section and the reaction starts in earnest with all chemical changes accounted via the developed kinetic model. Catalyst particles with a 87.13 µm mean diameter, a 3357 kg/m^3 particle density and 20 seconds total reaction time are considered.

Computational particle-fluid dynamics (CPFD) is employed to demonstrate the main features of process, based on an energy minimum multi-scale (EMMS) drag model coupled with CPFD. This Hybrid Barracuda CPFD model uses the Eulerian-Lagrangian approach called multi-phase particle-in-cell (MP-PIC). The software's numerical methodology considers a direct element method wherein solids are modeled as discrete particles with both size and density distributions, and the fluid is modeled as a continuum. Barracuda CPFD results are reported including gas and particles in the downer feeding section, particle-fluid flow in the stabilized region and gas compositions at different axial locations. Results are useful to establish downer performance, including propane ODH conversions in the 20% range with 92-94% propylene selectivity.

It is anticipated that thus research allows: (a) to gain insights on PODH performance (propane conversion and propylene selectivity) at various catalyst particle loadings; (b) to show the anticipated favourable flow dynamics in downer units; (c) to evaluate the mixing of partially reduced and oxidized particles streams at the top cyclone-feeding section.

References:

1. de Lasa, H. Riser Simulator. **1992**.
2. Rostom, S.; de Lasa, H. I. Propane Oxidative Dehydrogenation Using Consecutive Feed Injections and Fluidizable $VO_x/\gamma Al_2O_3$ and VO_x/ZrO_2–γAl_2O_3 Catalysts. *Ind. Eng. Chem. Res.* **2017**.
3. Al-Ghamdi, S.; Moreira, J.; de Lasa, H. Kinetic Modeling of Propane Oxidative Dehydrogenation over VO_x/γ-Al_2O_3 Catalysts in the Chemical Reactor Engineering Center Riser Reactor Simulator. *Ind. Eng. Chem. Res.* **2014**, *53*, 15317–15332.
4. Hossain, M. M. Kinetics of Oxidative Dehydrogenation of Propane to Propylene Using Lattice Oxygen of $VO_x/CaO/\gamma Al_2O_3$Catalysts. *Ind. Eng. Chem. Res.* **2017**, *56*, 4309–4318.

534003 An Orthogonal Recursive Bisection (ORB) Based Time Advancement Algorithm for CFD-DEM Solvers

Abstract:

The time integration of the granular phase in coupled computational fluid dynamics (CFD) â€“ discrete element method (DEM) simulations presents a unique computational challenge brought about by the large variations in particle collisional time scales. Particles in the dilute regions of the computational domain can be advanced with large time steps while dense regions require much smaller time increments. However, the time step size in most solvers is globally set as the limit for accuracy and stability imposed by the collisions and is typically orders of magnitude less than that required away from collisions. This work addresses this precise issue and provides a strategy to avoid the use of a global conservative small time step size for the entire set of particles.

A novel time stepping algorithm for CFD-DEM solvers using a partitioning approach using orthogonal recursive bisection (ORB) that allows for variable time steps among particles is described and its computational performance is compared against baseline explicit methods, typically used in several CFD-DEM solvers. ORB has advantages of being relatively quick and easy to update incrementally and has the required heuristic behavior (i.e., it will split the region in half with a cluster on each side) when groups of particles are well separated (clustered). The algorithm presented in this work uses a local time stepping approach to resolve collisional time scales for subsets of particles that are present at the leaves of the ORB, thereby resulting in substantial reduction of computational cost. The parallel implementation of this method where a ``knapsackâ€ algorithm is used in tandem with ORB for effective load-balancing is also presented, where a best possible partitioning is obtained based on number of particles and local time-stepping costs. The algorithm is tested against benchmark problems with varying particle distributions that include fluidized bed and riser flow scenarios. Preliminary results indicate that the approach is 2-3X faster than traditional explicit methods for problems that involve both dense and dilute regions, while maintaining the same level of accuracy.

529646 Effect of Collision Angle on Particle-Particle Adhesion of Colliding Particles through Liquid Droplet

Abstract:

In powder handling, a particle adhesion mediated by a liquid bridge is one of the quite important phenomena. In an actual process, the liquid bridge shows dynamic motion due to continuous motion of the particles. Therefore, understanding of the particle adhesion phenomenon by a dynamic liquid bridge is essential to adequately and precisely control powder handling processes. This study presents a direct numerical simulation of the particleâ€“particle adhesion by a dynamic liquid bridge. Collision of a dry particle and a wet particle was simulated at various collision angles. In particular, translational and rotational motions of the particle at different collision angles were discussed through comparison with a conventional static liquid bridge force model. As a result, it was found that both translational and rotational motions were largely different between simulation results of the direct numerical simulation and static liquid bridge force model, especially at the tangential collision. To understand these results, we focused on the rotational behavior of the particle and deformation of the liquid bridge. It was concluded that the non-slip behavior of the liquid bridge on the particle surface is a key phenomenon for the particle-particle adhesion by the dynamic liquid bridge at the tangential collision.

Reference: H. Kan, H. Nakamura, S. Watano. Adv. Powder Technol., in press.

533140 Spatially-Averaged Models for Heat Transfer in Gas-Solid Flows

Abstract:

Gas-solid flows, like fluidised beds or risers, are used in industry for a wide range of applications, from food drying and polymer production to the direct reduction of iron ore. While for small scale fluidised beds, simulation is quite feasible, the occurring physical phenomena cannot be fully numerically analysed in large scale reactors yet, due to a large range of involved scales. In fluidised bed reactors of several meters, the smallest stable structures are usually only several particle diameters wide [1]. Thus, the microscopic behaviour of individual particles, i.e. gas-solid drag, intraphase and interphase heat transfer between the particles and the gas as well as collisions cannot be fully resolved due to computational limitations. However, the unresolved heterogeneous structures have a significant influence on the flow properties. The velocity fluctuations around particle clusters give rise to turbulence, i.e. cluster induced turbulence [6, 7]. An overestimation of the gas-solid drag force by not accounting for this turbulence leads to an overestimation of the bed expansion [5].

A model accounting for the unresolved terms can be derived by spatially averaging the kinetic theory based two-fluid model equations [4]. The filtered gas-solid drag can be, for example, approximated by the filtered drag coefficient times the filtered slip velocity corrected by a drift velocity [2, 3]. The drift velocity can be seen as the gas-phase velocity fluctuations seen by the particles. Additionally, it is also shown in [3] that the filtered drag force depends on the turbulent kinetic energies of both phases, as well as strongly on bulk density fluctuations. Closures for the turbulent kinetic energies and the bulk density fluctuations are derived [4, 8].

The spatially averaged two-fluid model will be extended to include a model for heat transfer and, therefore, filtered energy equations are derived. The effective heat transfer coefficient divided by the solid volume fraction is approximated by its zeroth order Taylor series expansion about the filtered variables. This gives rise to a similar construct as the drift velocity. That is that the temperature difference between both phases is corrected by a 'drift temperature' stemming from phase averaging. Closure models for this 'drift temperature' as well as for the other unresolved terms in the filtered equations for the thermal energy balance are derived following the concepts outlined in [4]. Finally, a comparison shows fairly good agreement of the presented closure models with the fine grid data.

References

[1] Agrawal K, Loezos PN, Syamlal M, Sundaresan S. The role of meso-scale structures in rapid gas-solid flows. *J. Fluid Mech.* 2001;445:151–185.

[2] Ozel A, Gu Y, Milioli CC, Kolehmainen J, Sundaresan, S. Towards filtered drag force model for non-cohesive and cohesive particle-gas flows. *Phys. Fluids* 2017;29:103308.

[3] Schneiderbauer S, Saeedipour M. Approximate deconvolution model for the simulation of turbulent gas-solid flows: An a priori analysis. *Phys. Fluids* 2018;30:023301.

[4] Schneiderbauer S. A spatially-averaged two-fluid model for dense large-scale gas-solid flows. *AIChE J.* 2017;63(8):3544-3562.

[5] Schneiderbauer S, Puttinger S, Pirker S. Comparative analysis of subgrid drag modifications for dense gas-particle flows in bubbling fluidized beds. *AIChE J.* 2013;59(11):4077-4099.

[6] Fox RO. On multiphase turbulence models for collisional fluid-particle flows. J. *Fluid Mech.* 2014;742:368-424.

[7] Capecelatro J, Desjardins O, Fox RO. Strongly coupled fluid-particle flows in vertical channels. I. Reynolds-averaged two-phase turbulence statistics. *Phys. Fluids* 2016;28(3):033306.

[8] Capecelatro J, Desjardins O, Fox RO. Numerical study of collisional particle dynamics in cluster-induced turbulence. *J. Fluid Mech.* 2014;747:R2.

537995 Effect of Baffles on the Rate of Heat Transfer in Rotating Drums

Abstract:

Rotating drums are one of the most common devices used for industrial processes that involve thermal treatment of granular materials, such as drying and calcination. Adding baffles/lifters in the rotating drums is one of the approaches used to enhance the time required to heat the particles and improve the uniformity of the final product. The effect of the baffles on the heat transfer depends on the shape, the size, and number of baffles. We use discrete element method (DEM) simulations, in order to investigate the effect of baffles on heat transfer of granular materials in rotating drums. The simulations and analysis are limited to conductive heat transfer mechanism. We investigated the effect of baffles by changing the operating conditions, such as the fill level, speed of rotation, and baffle size and thermal properties of the particles and the drum. We found that the increase in the rate of heat transfer is directly related to the size and number of baffles. Particularly, we investigated the effect of baffles on the scale up of heat transfer based on the recent model developed by Yohannes et al (2016)* for granular flows in rotating drum with no baffles. We found that the scale up equation can be modified to account for the presence of baffles. In this talk, we will present results from our simulations and comparison of the results to prior models and experimental results. *B. Yohannes, H. Emady, K. Anderson, I. Paredes, M. Javed, W. Borghard, F. J. Muzzio, B. J. Glasser, and A. M. Cuitiño (2016), Physical Review E 94(4) 042902.

529883 A Novel Approach for Radiative Thermal Exchange in Coupled Particle Simulations

Abstract:

Radiative thermal heat exchange in particulate systems is of major importance in a variety of industrial applications, ranging from chemical reactors, thermal insulation, powder metallurgy, solar-thermal receivers, laser sintering, combustion processes and many other domains. Various experimental studies (see e.g. Chen et al. [1], Goshayeshi et al. [2]) show, that thermal transport through radiation becomes the leading heat transfer mechanism at high temperatures, i.e., above 750°C. In the contrary, numerical studies on thermal transport in dense gas-particle suspension flows are often limited to conductive and convective thermal transport (Forgber and Radl [3]).

The goal of our present work is to introduce a novel approach to predict radiative energy exchange rates in Discrete Element Method (DEM) simulations. Previously, computationally expensive approaches, e.g. Monte Carlo or inline ray-tracing (Dayal [4]), were used to study this mode of energy exchange. Unfortunately, such approaches limit the application to small particle arrangements. We introduce a novel, efficient method for thermal exchange through radiation and extend the capabilities of the open-source particle simulator LIGGGHTS (Kloss et al. [5]). The method is based on a per-particle view factor evaluation which utilizes standard discrete element method routines (see Forgber and Radl [6]). Subsequently, the energy transport in dense sheared particle beds is investigated numerically, and the relative importance of radiation compared to other heat transfer mechanisms is quantified. Most important, our present contribution further explores the influence of boundaries. Therefore, we (i) present a simplified setup which includes heated walls, (ii) detail on the numerical model, and (iii) discuss the formulation of the boundary conditions that needs to be used. Our results are helpful to decide whether radiative wall-particle heat transfer must be considered in a numerical model or not.

References

[1] J. C. Chen, J. R. Grace, and M. R. Golriz. Heat transfer in fluidized beds: design methods. Powder Technology, 150(2):123–132, 2005.

[2] A. Goshayeshi, J. R. Welty, R. L. Adams, and N. Alavizadeh. Local Heat Transfer Coefficients for Horizontal Tube Arrays in High-Temperature Large-Particle Fluidized Beds: An Experimental Study. Journal of Heat Transfer, 108(4):907–912, 1986.

[3] T. Forgber and S. Radl. Heat transfer rates in wall bounded shear flows near the jamming point accompanied by fluid-particle heat exchange. Powder Technology, 315:82-193, 2017.

[4] R. Dayal. Numerical Modelling of Processes Governing Selective Laser Sintering. PhD thesis, Technische Universität Darmstadt, 2014.

[5] C. Kloss, C. Goniva, A. Hager, S. Amberger, and S. Pirker. Models , algorithms and validation for opensource DEM and CFD-DEM. Progress in Computational Fluid Dynamics, 12:140–152, 2012.

[6] T. Forgber and S. Radl. A novel approach to calculate radiative thermal exchange in coupled particle simulations. Powder Technology, 323:24-44, 2018.

538333 Experimental Studies of Thermal Properties of Packed Powder Beds

Abstract:

Knowing the thermal properties of a given powder/material is important for many industrial applications such as drying and calcination. One such property is the bulk thermal diffusivity, which controls the rate of heat transfer in the powder. Even though the thermal diffusivity is a crucial factor to determine the required time to heat up the powder bed for industrial processes that involve thermal treatment, there is no reliable data of the thermal diffusivity of commonly used powders. We experimentally measured the temperatures inside a packed bed of powders in order to 1) determine the thermal diffusivity of the powders 2) to study the factors, such as density and moisture content, that affect the thermal diffusivity and 3) to relate the powder bed thermal diffusivity (macro scale) to the particle thermal diffusivity (micro scale). Our experiments were performed using a powder filled metal pipe where the temperature of the powder bed could be monitored as it was heated. We collected temperature data using wireless thermocouple transmitters and a receiver; the temperature of the powder bed was recorded from each thermocouple every second with an accuracy of 1°C. We calculated the thermal diffusivity of the powders based on the measured temperature and Heisler charts. For all the materials we tested, the thermal diffusivity varied from 0.5E-07 to 1.3E-07 (m^2/s). In this talk, we will present details of the experimentally measured temperature, factors affecting thermal diffusivity, and comparisons with particle scale heat transfer simulations.

530494 Particle Dynamic Simulations of Heat Transfer in a Bladed Mixer: Effect of Material and Process Parameters

Abstract:

Drying of active pharmaceutical ingredients (APIs) is an important step in the production of pharmaceuticals that controls the moisture content of the drug crystals. The drying step is carried out after crystallization and filtration of the API. Insufficient drying can lead to microbial growth and polymorph transformations. Often the API is heat sensitive so the temperature during drying must be carefully controlled. While a variety of dryers are used in the pharmaceutical industry, one of the most popular types is an agitated filter dryer. Such dryers are fairly simple in form in that they can be approximated as a cylinder with a rotating impeller (bladed mixer) with heated walls. During the drying step, the API particles undergo simultaneous transfer of mass and heat such that moisture content is reduced to a desired level. In this work, we focus on understanding heat transfer in a bladed mixer. Due to the complex flow dynamics associated with particles, heat transfer through a particle bed in agitated systems remains difficult to understand and optimize. As a result, the process can prove to be challenging to adapt from the laboratory scale to the pilot plant or even manufacturing scale. In this work, we make use of the discrete element method (DEM) to examine both heat transfer and flow of granular material in a bladed mixer geometry. We capture two important modes of heat transfer: conduction and dry granular convection. The first mode involves heat conduction through physical contacts between particles, while the other one describes heat transfer due to movement of particles. We observe that these modes of heat transfer can be expressed *via* two relevant time scales. The first time scale is a conductive time scale that describes the time needed for a single particle in contact with a heated surface to attain the temperature of said surface. The second time scale is a mixing time scale and defines the rate at which thermal energy is distributed throughout the bulk material. By looking at the relative significance of these competing time scales, we witness and characterize different regimes for heat transfer. Furthermore, we evaluate the influence of material properties such as conductivity and operating conditions such as agitation rate on heating performance. Developing such fundamental understanding of heat transfer in a bladed mixer provides an insight into how the performance of agitated filter dryers and their scale-up can be optimized.

536541 Synthetic and Compositional Control of Multicompontent Copolymers to Promote Drug Solubility and Bioavailability

Abstract:

Synthetic polymers have enabled amorphous solid dispersions to emerge as a strategy for overcoming poor solubility and bioavailability of intractable drugs via oral administration. In such formulations, polymeric carriers generate and maintain supersaturation of drugs via noncovalent interactions, which serve to increase the apparent drug solubility upon dissolution and enhance gastrointestinal absorption and oral bioavailability. In this work, we use controlled polymerization to create a family of multicomponent copolymers containing numerous functional groups that promote hydrogen bonding and hydrophobic interactions with several drug structures such as phenytoin and nilutamide. Through systematic characterization and screening we show that macromolecular design parameters with comomoners significantly improve efficacy by using: i) one monomer to serve as a precipitation inhibitor that discourages drug recrystallization and ii) another comonomer to provide hydrophilicity and solubility of the hierarchical system. Systems that maintained drug supersaturation in amorphous solid dispersions were identified with molecular-level understanding of noncovalent interactions using NOESY and DOSY NMR spectroscopy. Through this approach, effective polymers have been discovered that promote high drug solubilization and in vivo efficacy such as a Poly(N-isopropylacrylamide-co-N,N-dimethylacrylamide) (poly(NIPAm-co-DMA)). In vitro dissolution tests and in vivo pharmacokinetics of select spray-dried dispersions created via this approach have shown polymer chemical composition and solubilization performance relationships, which will be discussed in detail.

524288 Nanocellulose Gels As a Flexible, High Surface Area Material for Crystallizing Pharmaceuticals

Abstract:

Cellulose is a highly abundant natural polymer that, being a plant-based material, exhibits renewability and biodegradability. It is frequently used in the form of microcrystalline cellulose as a filler and binder in pharmaceutical tablets. A nanoscale form of cellulose can be prepared by removing much of the amorphous cellulose material and leaving small crystalline particles with a large concentration of hydrogen bonding groups on the surface. Due to its strong hydrogen bonding tendency and the presence of charges on the nanocellulose surface, it has the potential to bind with drug molecules and act as a release inhibitor, a stabilizer or a site for heterogeneous crystallization of the drug itself. Beyond hydrogen bonding and charge, however, the nanocellulose surface can be readily altered to display a variety of functional groups, leading to flexibility for rational design of nanocellulose-drug composites. We use nanocellulose functionalization, such as carboxylation and sulfonation, in combination with surfactants such as octadecylamine to prepare gels of nanocellulose. Addition of pharmaceutical compounds to the gels results in changes in drug form, such as polymorph transitions or amorphous form formation due to molecular interactions between the drugs and the nanocellulose gels. The high surface area of the nanocellulose provides a high concentration of interaction sites and the small size of the nanocellulose excipient particles has interesting performance in promoting assembly and packing of the composite in drug products such as compressed tablets and dense free standing films.

528646 Controlling the Particle Morphology of Spray Dried Poly(methacrylic acid-co-methyl methacrylate) (Eudragit L100) Polymer

Abstract:

Title: Controlling the particle morphology of spray dried poly(methacrylic acid-co-methyl methacrylate) (Eudragit L100) polymer

Authors: Kimberly B. Shepard and Michael M. Morgen

Purpose: Poly(methacrylic acid-co-methyl methacrylate) (PMMAMA, or trade name Eudragit L100) is an enteric random co-polymer of methacrylic acid and methyl methacrylate with a high glass transition temperature (~190°C). Spray dried amorphous dispersions (SDDs, left figure) which use PMMAMA as a matrix polymer leverage these properties and often have excellent physical stability, limited gastric release, and fast dissolution rate at intestinal pH. In previous work, we have demonstrated that SDDs of PMMAMA, HPMC and a rapidly crystallizing drug can be formulated to achieve rapid dissolution and sustained supersaturation during *in vitro* dissolution testing. However, during scale-up of PMMAMA SDD manufacture, poor powder properties lead to very low yields, which complicate formulation and process development. In particular, clusters of string-like structures were found throughout the SDDs (see center Figure). In this work, we investigate a process and formulation space in which string formation can be minimized or eliminated.

Methods: PMMAMA SDDs were prepared on small scale (1-100g) or lab scale (100g-10kg) custom spray drying equipment from either 97/3 acetone/water or pure methanol solutions, with solids loading varying from 3-12%wt. Process parameters, such as inlet and outlet temperature (80-140°C, and 35-65°C), liquid atomization pressure (100-600psi), atomization gas pressure (5-30psi), nozzle type (e.g. 2-fluid or pressure swirl), solution temperature, were varied. SDDs were characterized via scanning electron microscopy (SEM).

Results: During a typical pressure-swirl atomization process, droplets are generated when the liquid sheet exiting the nozzle first breaks up to form long one-dimensional "filaments," which in turn de-stabilize and break up to form droplets. During spray drying, liquid droplets dry into particles via a skinning process, i.e. solvent evaporates rapidly from the surface of the droplet, leaving behind a solid-like polymer skin on the surface of the droplet that occurs at a critical concentration. We hypothesize that PMMAMA string structures form when polymer skinning occurs before the filaments fully atomize into droplets. SEM images show examples of "bead on a string" structures, where skinning occurred in the midst of droplet breakup (see right Figure). In these experiments, string formation was reduced by either increasing the time for droplets to skin (e.g. decreasing drying kinetics or decreasing solution concentration) or decreasing the time to droplet formation (e.g. increasing atomization pressure).

We have identified a robust processing space on both small-scale and lab-scale spray dryers where string formation is avoided. Decreasing the outlet temperature from 65°C to 35°C results in adequate drying while eliminating strings. By spray drying from a more dilute solution, the time to skinning can be significantly increased. A solids loading of 3% and 5% in 97/3 acetone/water shows negligible string formation. Increased atomization pressure reduces the time to droplet breakup and subsequent string formation for pressure-swirl and 2-fluid nozzles. Utilizing a slower-drying solvent, such as methanol, or reducing the temperature of the spray solution can also reduce string formation.

Conclusions: By controlling the spray solution solids loading, drying temperature, atomization conditions and other parameters, we identified a process space where PMMAMA SDDs can be manufactured without string formation and with acceptable particle properties (i.e. lack of strings). Stability and performance implications of this processing strategy are currently being evaluated. This process is applicable to other polymer systems in which string formation during spray drying is observed.

537276 Application of Fundamental Relationships and Models to Predict Spray-Dried Dispersion Particle Size

Abstract:

Purpose:

Atomization models can assist spray drying formulators in estimating ranges of droplet size that may be achieved for a given spray solution formulation and drying process. Droplet size can then be related to the final spray-dried dispersion (SDD) particle size by accounting for the fundamentals of droplet solidification and final particle density [1]. Specifically, for film-forming polymeric SDD systems, droplet drying kinetics and thermodynamics must also be considered due to their resulting particle morphological variations [2]. By combining these fundamental relationships with empirical correlations, a rational process development strategy can be employed early in the product lifecycle and give initial insight into future scale-up constraints or risks. Using basic solution properties as model inputs, multiple empirical approaches to predicting droplet size were evaluated. This study showed excellent correlation of predicted droplet size to the resulting final SDD particle size. In addition, use of a simplified porous particle approach that relies on the viscous gel point to predict particle size from droplet size was effectively demonstrated.

Methods:

Viscosity, surface tension and density were evaluated for a range of solids concentrations using two different dispersion polymers, HPMCAS-M (1-12 wt%) and PVP/VA (1-20 wt%) prepared in either acetone or methanol. Additionally, the viscous gel point (skinning concentration) was determined using a gravimetric method. Droplet size was measured sing a phase Doppler particle analyzer (PDPA) to study the effects of solids composition, nozzle geometry, and atomization parameters. Spray-drying experiments were then conducted at lab and feasibility scale to obtain final SDD particle size and evaluate potential impact of this approach on scale-up. SDDs were characterized for particle size distribution using light scattering methods, bulk/tap density, and morphology by scanning electron microscopy.

Results:

Droplet size predictions were conducted using an empirical model developed by Lefebvre:

These predictions correlated well with measured droplet size (PDPA) and measured SDD particle sizes (light scattering) for the formulations studied [3]. HPMCAS-M showed a steeper response of particle size to droplet size than PVP/VA possibly due to higher viscosities observed in solution at similar concentrations. Differences were observed between the relationship of predicted droplet size and SDD particle size based on the spray solvent used for both PVP/VA and HPMCAS-M polymers. A shrinking sphere model was initially applied to understand the relationship between droplet size and final particle size however, this approach was not found to demonstrate the desired predictability. Therefore, a porous particle approach was evaluated that accounts for the differences in skinning concentration of the polymers in various solvent systems. This was found to significantly improve the predictability of final SDD particle size.

Conclusions:

Application of simple models and fundamental relationships was shown to provide a high level of accuracy for predicting both droplet size and final SDD particle size. The utility of this approach can be further expanded to aid in scale-up predictions to rationally choose a droplet size to target a desired SDD particle size during initial process development and optimization on lab-scale spray dryers. Key differences between the dispersion polymer or solvent chosen were explained through either solution properties or relationship of polymer solidification.

[1] Singh, A. and Van den Mooter, G. Spray drying formulation of amorphous solid dispersions. *Adv. Drug Deliv. Rev.* **2015**.

[2] Vehring R. Pharmaceutical particle engineering via spray drying. Pharmaceut Res. 2008;25(5):999-1022.

[3] Lefevbre, Arthur. Atomization and Sprays. CRC Press, 1988.

538532 Improving Spray Drying Processing through Modeling and Characterization

Abstract:

Advances in engineering spray dried dispersion (SDD) for enhancing the oral bioavailability of drugs enable us to develop robust solid dispersion drug products. To this end, research in pharmaceutical spray drying has focused on coupling characterization of particles, solution and the spray with predictive modeling such as using computational fluid dynamics. Here, I will highlight two approaches we have used to gain fundamental understanding of spray drying process and the SDD particles.

To understand the microscopic characteristics of SDDs, we analysed the morphology and chemical distribution of the drug within SDD particles. Our work revealed the existence of local amorphous amorphous phase separation of drug from the polymer within particles. Using theoretical analysis, we elucidated the role of the complex drying kinetics and the thermodynamics of the process resulting in the observed SDD spatial and chemical microstructure. We also highlight the importance of probing the microscopic origins in such a system.

To understand the transport phenomena during spray drying, we turned to computational modelling of spray drying using solvents. We studied the variation in drying kinetics as a function of various processing conditions and the most influential processing parameters. We highlight the similarities and differences in drying when moving from a pilot to a commercial scale dryer. I will also briefly discuss the computational model of polymeric droplets to describe the specific drying trajectories, accounting for the formation of a polymeric skin on the droplet surface, the variation of the solvent diffusivity throughout drying.

533575 Mechanistic Approach to Predict Amorphous Solid Dispersion Thermal Degradation in Spray Drying Processes

Abstract:

Mechanistic approach to predict ASD thermal degradation in spray drying processes

G. Poeiras[1,3], T. Porfírio[1,2*], C. Sá Couto[1], J. Pereira[1], R. Silva[1], I. Duarte[1], M. D. Afonso[3], J. Vicente[1]

[1]Hovione Farmaciência SA, Sete Casas, 2674-506 Loures, Portugal; *tporfirio@hovione.com

[2]LAETA, IDMEC, Mechanical Engineering Department, Instituto Superior Técnico, Universidade de Lisboa, Av. Rovisco Pais 1, 1049-001 Lisbon, Portugal

[3]CeFEMA, Chemical Engineering Department, Instituto Superior Técnico, Universidade de Lisboa, Av. Rovisco Pais 1, 1049-001 Lisbon, Portugal

The number of drugs with solubility limitations is increasing, being the limited aqueous solubility a major challenge in the development of oral-dosage forms, as it may affect oral bioavailability. Several solubilization strategies have been developed, being the amorphous solid dispersions (ASD) one of the most common. Spray drying (SD) process is one of the most widely used technologies for ASD production allowing the control of the product attributes, viz. particle size, density and physical state. SD is a solvent based process that often includes a secondary drying (post-drying stage) to reduce residual solvent content to values below ICH guidelines (International Conference on Harmonization of technical requirements for registration of pharmaceuticals for human use). A key processing phenomena in both equipment is the particles deposition on the walls and the subsequent build-up that affects process operation, safety and may impact the product quality through the deposited particles degradation, Figure 1 [1]. For that reason, it is crucial to assess the product chemical and physical quality, which have to be guaranteed throughout the process.

The degradation is intrinsically dependent on the process conditions in which the product is exposed, such as temperature and saturation. Computational Fluid Dynamics (CFD) may be used to model the temperature profile throughout the chamber and provide an estimation of the product exposure during a manufacturing batch [2]. Moreover, CFD simulations forecast possible zones of wall accumulation. The output of such analysis supports the assessment of the accumulated product quality. Subsequently, the goal is to build a general model capable of predicting the ASD chemical degradation kinetics for any combination of temperature and relative saturation.

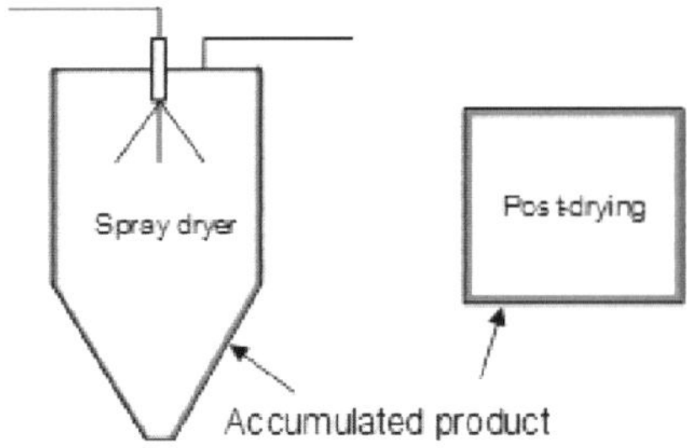

Figure 1 – Process exemplification including product accumulation.

As for the physical stability, the thermodynamic properties and molecular mobility inherent to the amorphous state are responsible for the formation of crystalline material over time, compromising the bioavailability and performance of the solid dosage form. Molecular mobility is more or less pronounced according to the process conditions that the product is subjected to, namely temperature, humidity and saturation. The experimental determination of these molecular motions is related to the study of structural relaxation processes, physical aging or annealing, which consist in the continuous evolution of the amorphous product towards a more stable equilibrium state at a given temperature. Structural relaxation can be predicted from the sample thermal history and its degree of non-equilibrium [3].

In sum, this work developed a mechanistic approach to predict ASD stability based on (Figure 2):

- CFD to model and predict temperature and saturation profiles inside the spray drying chamber and/or post drying equipment;
- Estimation of the degree of chemical degradation using a kinetic model;
- Physical stability using structural relaxation studies.

A case study comprising the spray drying and/or post-drying steps using a model drug to form an ASD will be presented following the proposed approach. In this case study, the operating conditions ranges, namely temperature and relative saturation, were defined based on the results provided by the CFD modelling and also aiming to stress the ASD chemically and physically.

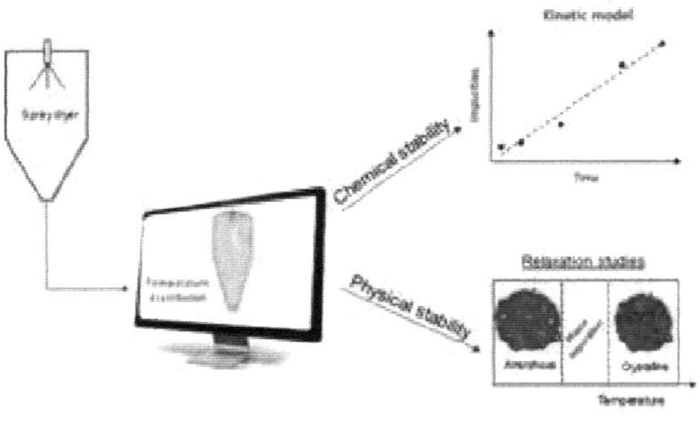

Figure 2 – Proposed approach for ASD thermal degradation study.

-

References

[1] S. Keshani, W. R. W. Daud, M. M. Nourouzi, F. Namvar, and M. Ghasemi, "Spray drying: An overview on wall

deposition, process and modeling", Journal of Food Engineering, pp. 152–162, 2015.

[2] L. Huang and A. S. Mujumdar, "Simulation of an Industrial Spray Dryer and Prediction of Off-Design Performance", Drying Technology, vol. 25, pp. 703–714, 2007.

[3] R. Hilfiker, Polymorphism in the Pharmaceutical Industry, pp. 260-281, 2006.

Downstream Processing of a Ternary Amorphous Solid Dispersion: the Impacts of Spray Drying and Hot Melt Extrusion on Powder Flow, Compression and Dissolution

Mark T. Davis[1*], Catherine B. Potter[1] & Gavin M. Walker[1]

1. The Synthesis and Solid State Pharmaceutical Centre (SSPC), Bernal Institute, The University of Limerick, Ireland
*email: mark.davis@ul.ie; phone: +353 61 213096

Abstract

Downstream processing aspects of a stable form of amorphous itraconazole exhibiting enhanced dissolution properties were studied. Preparation of this ternary amorphous solid dispersion by either spray drying or hot melt extrusion led to significantly different powder processing properties. Particle size and morphology was analysed using scanning electron microscopy. Flow, compression, blending and dissolution were studied using rheometry, compaction simulation and a dissolution kit. The spray dried material exhibited poorer flow and reduced sensitivity to aeration relative to the milled extrudate. Good agreement was observed between differing forms of flow measurement, such as Flow Function, Relative flow function, Flow rate index, Aeration rate, the Hausner ratio and the Carr index. The stability index indicated that both powders were stable with respect to agglomeration, de-agglomeration and attrition. Tabletability and compressibility studies showed that spray dried material could be compressed into stronger compacts than extruded material. Blending of the powders with low moisture, freely-flowing excipients was shown to influence both flow and compression. Porosity studies revealed that blending could influence the mechanism of densification in extrudate and blended extrudate formulations. Following blending, the powders were compressed into four 500 mg tablets, each containing a 100 mg dose of amorphous itraconazole. Dissolution studies revealed that the spray dried material released drug faster and more completely and that blending excipients could further influence the dissolution rate.

Keywords: Amorphous, itraconazole, flow, compression, tabletting, dissolution

1. *Introduction*

Amorphous solid dispersions (ASD) have received a surge in research interest over recent decades, due to the large and increasing number of poorly soluble drugs on today's market(Brouwers et al., 2009; Leuner and Dressman, 2000). Although several examples of commercial ASDs exist(Huang and Williams III, 2017), the technology has sometimes lost out to other bioavailability enhancement strategies, often due to problems with recrystallisation or perceived difficulties in downstream processing. Although publications addressing recrystallisation issues are manifold, far less research has been reported addressing the issues of flow and compression(Demuth et al., 2015).

Efficiency of flow and compression can affect processing times and tablet quality. Bridging, arching or rat-holing for instance, can lead to increased processing times, inaccurate die fill and unacceptable dose variation. Compression issues can cause sticking, picking or flaking of tablets and ultimately reduction in patient compliance(Jain, 1999; Leuenberger and Rohera, 1986; Patel et al., 2006). In an age of greater focus on quality by design (QbD), global manufacturing competition and 'right first time' demands, the need to resolve or minimise such issues at the research stage is stronger than ever.

The purpose of this study was to assess the impacts of spray drying (SD) and hot melt extrusion (HME) on the flow, compression and dissolution characteristics of an ASD containing itraconazole (ITZ). Comparisons between properties of amorphous dispersions prepared by SD and HME are popular,(Davis and Walker, 2018) although most of these do not specifically deal with downstream processing issues. We have recently reported the preparation of a novel ternary ASD using both SD(Davis et al., 2017) and HME(Albadarin et al., 2017). Ternary dispersions have become increasingly popular(Davis and Walker, 2018) in the struggle to meet the sometimes competing demands of ASDs: increased bioavailability, adequate stability, efficient processing, the use of existing plant equipment and bearable costs. The formulations, containing drug and excipient polymers Soluplus® and HPMCP, displayed excellent stability and enhanced dissolution in both studies(Albadarin et al., 2017; Davis et al., 2017). The following is an account of the impact of preparation on the downstream processing of one of those formulations.

2. *Materials and Methods*

2.1 Materials

ITZ (>99%) was purchased from Xi'an Lyphar Biotech Ltd. Accurate analytical standard of itraconazole (99.8%) for HPLC calibration was purchased from Sigma Aldrich. Polymers Soluplus® & HPMCP HP-55 (HPMCP) were donated by BASF and Shin Etsu respectively. Excipients Avicel 101 & Avicel PH 200 LM were donated by FMC and Mannogem granular was donated by SPI Pharma. Solvents and chemicals were purchased from Sigma-Aldrich and were of HPLC and reagent grade.

2.2 Spray drying

Spray drying was conducted using a Büchi B-290 mini spray dryer connected to a B-295 inert loop. A 2-fluid pneumatic nozzle (0.7 mm) was fitted and liquid feedstock was pumped at 4 mL/min (15%). Drying nitrogen was set at maximum by fixing the aspirator at 100% (35 m^3/h). A solution of ITZ-Soluplus®-HPMCP (30-40-30 w/w) in dichloromethane-methanol 7-10 (v/v) (1.0 L, 10.0 % w/v) was pumped through the nozzle at inlet temperature 100 °C, producing an outlet temperature of 65 ± 2 °C. High yields (88.6%) were achieved by using high spray gas rates (670 L/h) and attaching the high efficiency cyclone option. Powdered product was immediately transferred into a stainless steel pan and dried overnight in a vacuum oven at 40 °C and 10^{-2} mBar to ensure complete removal of residual levels of solvent. Powder was then stored in a sealed glass bottle in a desiccator over anhydrous molecular sieves.

2.3 Hot melt extrusion and milling

Itraconazole, Soluplus® and HPMCP 30:40:30 (w/w) (50g) were pre-mixed by hand in a polythene bag, having first removed larger particles of HPMCP, by passing through a 435 µm sieve. The pre-mixed powders were then manually fed into a Three-Tec twin-screw extruder (Three-Tec GmbH, Germany) with a barrel diameter of 12 mm and an L/D ratio of 40:1. The set points of the six heating zones, from feed to die, were 80, 110, 120, 140, 150 and 150 °C, respectively and screw speed was maintained at 15 rpm throughout. Following extrusion the material was milled for 1 min at 20 Hz using a 25 mL stainless steel jar attached to a Retsch Mixer Mill MM 400 ball mill (Retsch GmbH, Germany). Milled extrudate was sieved through 435 and 90 µm sieves and the 90 – 435 µm and < 90 µm fractions were stored separately in sealed glass bottles in a desiccator over anhydrous molecular sieves.

2.4 Scanning electron microscopy (SEM)

Drug polymer formulations were attached to double sided carbon tape and sputter coated with a thin layer of gold followed by imaging using a Joel CarryScope JCM-5700 scanning electron microscope. Micrographs were recorded at various magnifications, using a beam acceleration voltage of 5 kV, a spot size of 40 and a working distance of 12.

2.5 Powder Rheometry

Powder flow was analysed using an Freeman FT4 powder rheometer in a manner previously reported(Freeman, 2007). Stability and Variable flow rate was performed in the 25 mm vessel, Aeration in the 25 mm vessel and Shear in the 10 mL and 1 mL vessels.

2.6 Compaction simulation

Powders were tableted using a Gamlen R-series tablet press containing a 500 kg load cell and 6 mm punch and die. The material was subjected to uniaxial compression tests to form flat-face cylindrical tablets with a target weight of 90 mg. Punch load and displacement were measured during compression and the effect of compaction forces from 0.5 to 5 kN, equivalent to pressures of 17 to 173 MPa, was studied. Tablet hardness, defined as the diametral force required to break the compact, was investigated using a PharmaTest hardness tester (Pharma Test Apparatebau AG, Germany) immediately after tabletting.

2.7 True density

The true density of the extruded and spray dried products and powder blends was measured under ambient conditions using an helium pycnometer (AccuPyc II 1340, Micromeritics Inc., Norcross, GA). The volume of the pycnometer was verified with a standard before carrying out analysis. During sample analysis five purge cycles were recorded for samples and the reported result is an average of measurements made on three different days.

2.8 *Formulation and tabletting*

ASD products and excipients were accurately weighed as described in Table 4 and the individual components were vortex mixed for 2 min in 10 mL vessels. The resultant prototype tablet formulations were pressed according to scalable parameters determined through small scale compaction simulation performed on 90 mg material. Blends were weighed into a 0.5 inch (12.7mm) diameter die and then pressed at 1000 kg pressure on a manual hydraulic tablet press to form flat-face cylindrical tablets with a target weight of 500 mg.

2.9 *In-Vitro Dissolution Studies*

Drug release studies were carried out using non-sink conditions, in a USP type II apparatus, PharmaTest dissolution tester (Pharma Test Apparatebau AG, Germany), with 900 mL 0.1 *N* HCl buffer (pH 1.2) per well, a paddle speed of 100 rpm and a temperature of 37 ± 0.2 °C. Formulated tablets of mass 500 mg, equivalent to 100 mg of drug were added to the dissolution vessel (n = 3), with start times staggered by 2 min. 5 mL aliquots of media were withdrawn at predetermined intervals (0.25, 0.5, 0.75, 1, 2, 3, 4, and 6 h), filtered through a 0.45 µm polytetrafluoroethylene (PTFE) syringe filter and then the first 1 mL of filtrate was discarded. 500 µL of the filtrate was subsequently added to 500 µL of HPLC mobile phase in

2 mL HPLC vials. 5 mL of fresh medium was immediately replaced following each withdrawal.

2.10 High performance liquid chromatography (HPLC)

Dissolution samples were analysed using an Agilent (Agilent, Little Island, Cork) 1260 Infinity II high performance liquid chromatography system, comprising of quaternary pump G1311B and diode array detector G1315D set at wavelength 263 nm. The thermostated column compartment G1316A was set at 25.0 °C and equipped with a Kromasil 100 C18 5 µm 250 x 4.6 mm (Kromasil, Mainz, Germany) RP-HPLC column. The system was operated under isocratic flow at 1 mL/min using, a mobile phase of acetonitrile:water:diethanolamine (69.95/30/0.05, v/v/v). Samples were injected from autosampler G1329B in volumes of 20 µL and data was collected and analysed by Agilent OpenLAB CDS Chemstation software. Standards of ITZ (99.8% purity) were prepared from a stock solution of 1 mg/mL in methanol, to quantify the levels of drug in the dissolution media.

3. Results and Discussion

3.1 Preparation of Amorphous Solid Dispersions by Spray Drying and Hot Melt Extrusion

In our previous studies we discovered that it was possible to prepare novel ternary amorphous dispersions of BCS class II drug itraconazole by hot melt extrusion(Albadarin et al., 2017) and by spray drying(Davis et al., 2017). Following rational polymer selection(Davis et al., 2015), these blends proved to have excellent stability and enhanced dissolution profiles. As design of experiments (DoE) had previously been performed during formulation design, only one blend was selected for downstream processing. A ternary ratio of itraconazole-Soluplus®-HPMCP 30-40-30 (w/w) was selected to test a high drug load and simultaneously facilitate HME processing, by choosing a Soluplus® ratio that was slightly higher than that of HPMCP.

Particle size is a critical factor in powder flow. Often the particles from spray drying processes are very small (< 10 µm)(Vehring, 2008), which can be excellent for dissolution, but challenging for powder flow. It is possible to increase particle size by changing nozzle type to a pressure nozzle(Beyerinck et al., 2010), processing parameters, drier geometry or feed stock concentration. Of these options, the most directly available was an increase in feedstock concentration, although it is likely that a pressure nozzle would be more effective. More concentrated feed solutions produce more viscous droplets and hence larger particles(Vehring, 2008). For this reason the feed was prepared at 10% (w/v), following optimisation experiments on the dichloromethane-methanol solvent system(Davis et al., 2017), leading to an optimal ratio of 7-10 (v/v). In addition, the yield of the spray dried powder was optimised by tuning processing parameters. In particular, when the spray gas rate was increased from 25 mm (301 $L.h^{-1}$) to 55 mm (670 $L.h^{-1}$) on the flow meter, the yield increased from 51% to 89%.

With hot melt extrusion, it was less difficult to prepare the quantities of material required (at least 100 g), as the materials are not diluted in liquid before feeding to the equipment. However, it was found that sieving, pre-mixing and manually feeding the ternary materials was favourable, to promote adequate mixing and prevent system overload by excess torque. Early runs lead to the production of opaque amber material and excessive torque. The torque issue was ameliorated by modifying the ternary blend mixture from itraconazole-Soluplus®-

HPMCP 30-35-35 to 30-40-30 (w/w), as the HPMCP was difficult to extrude, but the Soluplus® processed very well. In addition, the adjustment of zone temperatures by increasing zones 4 and 5 from 130 °C and 140 °C to 140 °C and 150 °C respectively, resulted in production of glassy yellow material. After cooling, the glassy extrudate was milled, sieved and stored.

Cleaning of the extruder was also challenging, requiring addition of pure Soluplus® as a 'cleaning polymer', followed by hot removal of thermocouples, hot removal of barrels from screws, cooling, separation of barrel blocks using levers and finally immersion in warm 1% surfactant with sonication.

3.2 Particle size and morphology

The impact of particle size and morphology on powder flow is well known and can drastically effect the efficiency of pharmaceutical powder processes, such as die filling(Gallo et al., 2011; Hancock et al., 2002; Nokhodchi et al., 2007; Sinka et al., 2004; Sivasankaran et al., 2010; Wu and Cocks, 2004; Xie and Puri, 2006). Variation in powder composition or flow can lead to unacceptable deviations in tablet weight and content uniformity, possibly leading to the rejection of a batch in extreme circumstances. Poor flow can potentially also lead to inefficient or even dangerous processes on the plant, caused by bridging or caking of powder masses during storage in containers or silos and the potential for powder flooding. Because spray dried powder contains particles mostly < 10 μm and milled extrudate < 90 μm (or 90 – 435 μm, for the larger sieved extrudate fraction), the cohesion and flow of these powders will differ profoundly and require differing hopper angles, orifices or alternative storage design for safe and efficient handling.

Scanning electron microscopy (SEM) was performed on the materials to provide insight into particle size and morphology. Itraconazole was received as relatively flat, wedge-like crystals (Figure 1a) approximating 40 μm length by 10 μm width, with a mostly smooth surface that nevertheless displayed several brittle shards. Soluplus® was received as much larger, approximately spherical granules (Fig 1b) with a diameter approximating 250 μm. HPMCP (Fig 1c) was received as pod- or cocoon-like elongated particles approximating 350 μm by 70 μm, with a smooth but broken surface. The physical mix was a combination of the above, although the difference in size and shape between the small itraconazole particles and the large HPMCP particles was particularly apparent (Fig 1d). This shaped the decision to sieve the HPMCP before powder blending and extrusion.

The extruded particles were ball milled and sieved prior to SEM analysis. The particles appeared as irregularly shaped granules, comprising considerably more material per granule than the spray dried particles. There were many rough spots on the surface of each individual granule, representing possible areas for particle interlocking and cohesion. Particles were typically 50 μm x 55 μm in the < 90 μm sieved fraction (Fig 1e) and 240 μm x 350 μm in the 90 - 435 μm sieved fraction (Fig 1f). The spray dried sample appeared as a combination of some hollow spheres with diameters of up to 30 μm (Fig 1g), some burst hollow spheres (Fig 1h) of similar diameter, but mostly crumpled spheres of less than 10 μm, that had grouped into agglomerates of various shapes, with 40 - 60 μm in total diameter. The particles size of the spray dried material in this study represented an increase to that of our previous work, where a lower feed concentration of 0.5% (w/v) typically led to particle diameters of 1.5 - 2.4 μm, depending on formulation(Davis et al., 2017). Although increasing the feed concentration did increase the particle size, the spray dried particles were still very small.

While this is advantageous for dissolution, the powders are expected to be quite cohesive during flow analysis.

3.3 Powder Flow

3.3.1 Powder stability during flow

Because unknown powders may be prone to attrition, agglomeration or caking, it is important to begin by measuring whether the powder will change during the course of testing. To discover if the flow properties alter under the stresses imposed during flow,(Iveson et al., 2001; Salman et al., 2002) the 'stability program' runs the powder through a series of identical conditioning and test cycles: if the powder is stable then the results from each measurement are similar. This is observed as a straight line from points 1 through to 7 in Figure 2. The stability index, defined in equation 1, is the ratio of energy in test 1 to energy in test 7:

$$Stability\ index, SI = Energy\ Test\ 7/Energy\ Test\ 1 \qquad (1)$$

Generally, a stable powder will have an SI approximating 1, powders with SI > 1 are unstable, affected by de-aeration, agglomeration, segregation, moisture uptake or electrostatic charge and powders with SI < 1 are unstable, affected by attrition, de-agglomeration, over-blending or coating of the vessel and blade by an additive(Freeman, 2007). The data in Figures 2 and 3 demonstrated that milled extrudate (HME & BM), physical mixture, itraconazole and spray dried ASD were largely stable to flow rate changes, with SI values of 0.96 ± 0.06, 1.08 ± 0.11, 0.97 ± 0.18 and 1.24 ± 0.26, respectively. Only excipients Soluplus® (1.91 ± 0.35) and HPMCP (1.35 ± 0.17) were unstable with respect to flow stress, with the stability results for HPMCP particularly inconsistent (data not shown). The underlying cause for the instability may be connected to static charge accumulation in HPMCP during testing.

3.3.2 Powder variable flow rate during flow

The rate at which powders are handled and transferred varies and the sensitivity of a powder to this is an important factor. Often, cohesive powders are more sensitive to changes in flow rate than non-cohesive or granular materials. During the variable flow rate test (Figure 2), the flow rate is reduced from 100 mm/s to 10 mm/s over tests 8-11. The flow rate index (FRI) is given by equation 2:

$$Flow\ rate\ index\ (FRI) = Energy\ Test\ 11/Energy\ Test\ 8 \qquad (2)$$

Usually, a high flow rate sensitivity (usually for very cohesive powders) is depicted as FRI > 3.0; an average flow rate sensitivity (including most powders) lies between 3.0 > FRI > 1.5; flow rate insensitive materials (with large PSD or surface coating) have FRI ≈ 1.0 and pseudoplastic or Neutonian Flow rates (such as powders containing flow enhancers like magnesium stearate) are found with FRI < 1.0(Freeman, 2007).

As depicted in Figures 2 and 3 the experimental FRI values recorded were as follows: itraconazole 1.89 ± 0.03, Soluplus® 1.28 ± 0.06, HPMCP 1.63 ± 0.24, HME/BM 1.40 ± 0.02; physical mix 1.68 ± 0.11 & spray dried 3.73 ± 0.35. So, most of the materials can be classified as having average flow rate sensitivity, with the Soluplus® and milled extrudate edging towards being insensitive. The clear exception was the spray dried material, with a high FRI value of 3.73, as would be expected for a cohesive powder comprised of very small particles. The difference between the spray dried ASD and the milled extrudate ASD was 3.73 – 1.40 = 2.33 units. It would therefore be reasonable to expect a significantly different sensitivity of these two ASDs to flow rate during processing, despite their identical chemical compositions.

3.3.3 Powder Aeration Studies

The bulk properties of all powders are to some extent influenced by air because the inter-particulate spaces themselves contain air. The volume of air impacts interparticular interactions and flow. Some powders are readily aerated and require only modest quantities of air to transform into a fluidised bed. Others are cohesive and therefore not easily aerated, although the flow properties may still improve with increasing air entrainment. The aeration ratio (AR) is given by the equation:

$$Aeration\ Ratio, AR = Energy\ (Air\ velocity\ 0)/Energy\ (Air\ velocity\ 10) \qquad (3)$$

Powders with AR $\approx$ 1 are not sensitive to aeration and are generally very cohesive. An average sensitivity to aeration with 2 < AR < 20 is found for most powders. Very sensitive powders with AR >> 20 will most likely become fluidised easily. Referring to Figure 4, the three raw materials and the two ASD powders were measured on the aeration program and produced the following AR results: HPMCP = 1.38 – 1.68 (not sensitive, very cohesive); Soluplus® = 0.45 – 0.54 (not sensitive, special case); itraconazole = 1.62 – 3.51 (cohesive to average); milled extrudate = 45.9 – 58.2 (very sensitive, non-cohesive); spray dried = 1.73 – 2.02 (not sensitive, very cohesive)(Freeman, 2007).

These results can be rationalised by considering the particle size and morphology of the powders (Figure 1 & section 3.2). HPMCP particles are elongated along one axis and typically several hundred µm long, so a preference for horizontal stacking may add to cohesion. Soluplus is spherical and around 300 µm in diameter, the spherical shape is quite aerodynamic and allows a laminar flow around the spheres, leading to the very low AR value of 0.48. This may also explain the unexpected requirement for more energy as the blade flows through the sample at higher aeration velocities. Itraconazole comprises straight, flat crystalline wedges of approximately 10 µm x 40 µm, so it is unsurprising that this material is also cohesive and AR insensitive. The test on the milled extrudate was performed prior to sieving. The particles contain a significant fraction of larger particles (> 300 µm), with various shapes and sizes, so the material was non-cohesive and very sensitive to aeration. Finally, the spray dried material contains hollow, burst and crumpled spheres, with many particles < 10 µm. The material is very cohesive due to the tiny particle size and therefore AR insensitive

The average AR values for the milled extruded and spray dried samples were 50.23 ± 6.89 and 1.88 ± 0.15 respectively, a difference of 48.35 AR units, representing a significant difference in sensitivity towards aeration. The result has implications for understanding Würster coating and other processes that intentionally use air or for flow by gravity (Bruni et al., 2007; Freeman, 2007; Leturia et al., 2014). In particular, die filling during tabletting requires both that the powder fill the die by gravity and also that the air in the die exits easily through the particles as they ingress(Sinka et al., 2004; Wu and Cocks, 2004). As speed of tabletting increases this becomes more important.

3.3.4 Powder shear testing

Shear cell testing is perhaps the most studied and understood flow test performed using powder rheometry(Fitzpatrick et al., 2004; Jenike, 1964; Schulze et al., 2008; Schwedes, 2003; Walker, 1966). The procedures and testing sequences have been reported elsewhere(Freeman, 2007), but briefly consist of powder conditioning, consolidation, vessel splitting, re-consolidation and finally shear testing. The tests were performed in triplicate to increase accuracy. FT4 software enabled automatic construction of Mohr circles, providing valuable data such as cohesion (C), unconfined yield strength (UYS) , major principle stress (MPS), minor consolidation stress (MCS), angle of internal friction (AIF), Peschl's Relative Flow Function (Rel_P) (Peschl and Colijn, 1977) and Jenike's Flow function (FF of ffc)(Ennis, B J; Witt, W; Weinekoter, R; Sphar, D; Gommeran, E.; Snow, R H; Allen, T; Raymus, GJ; Litster, 2007; Freeman, 2007; Jenike, 1964). The equations for FF and Rel_P are given as follows:

$$Flow\ Function, FF = MPS\ /UYS \tag{4}$$

$$Rel_P = (MPS - MCS)/UYS \tag{5}$$

In Figure 5, the yield loci (plot of applied normal stress vs shear stress) of itraconazole, the physical mixture and the two ASD powders were plotted with the corresponding shear parameters reported in Table 1. The most frequently reported value derived from shear cell data to describe flowability is the Jenike Flow Function (FF of ffc)(Freeman, 2007; Jenike, 1964). The FF values for itraconazole (2.03 ± 0.03) and the spray dried ASD (3.45 ± 0.18) indicated that they were cohesive, the physical mixture was very cohesive (1.98 ± 0.25) and the milled extruded ASD was free-flowing (57.01 ± 53.62). The low standard deviation values indicated that repeatability was excellent for the cohesive powders but rather poor for the free-flowing extrudate, demonstrating that this test was more accurate for cohesive powders(Freeman, 2007). Similarly (but with subtle differences in cohesive ranking), the Relative Flowability (Rel_P)(Peschl and Colijn, 1977) values indicated that itraconazole and the physical mix were extremely cohesive, the spray dried ASD was very cohesive and the milled extrudate was cohesionless (Tables 1 & 3).

The highest values of Unconfined Yield Strength (UYS) were found in the physical mixture (11.83 kPa) and itraconazole (10.75 kPa), with much lower values for the spray dried (4.14 kPa) and milled extrudate (2.98 kPa) ASDs, respectively. This was expected, as more cohesive powders will generally have higher values for UYS(Freeman, 2007). In addition, a steeper yield locus often represents a powder that is more sensitive to increasing levels of consolidation. The highest AIF value was for the physical mixture (44.09 ± 8.75°), followed by itraconazole, roughly equal to the extruded/milled ASD and finally the lowest value was for the spray dried ASD (27.75 ± 0.61°). Therefore the pure drug was most sensitive to increasing consolidation and the spray dried ASD was least sensitive overall and less sensitive than the milled extrudate (by 8.64° on average). Consolidation results have implications for packaging, shipping and processing considerations.

3.3.5 A comparison of powder flow testing parameters

To better understand the accuracy and precision of the various parameters used to measure flow, simple mass/volume measurements were taken to calculate the Hausner ratios and Carr's indices for the powders. Although these parameters have little theoretical basis, they are easily obtained and have been used to describe powder flow for many decades, so their inclusion seems worthwhile. The FT4 rheometer conditions the powder prior to recording bulk density and this imparted a high degree of repeatability (accuracy) on the values obtained (Table 2), with standard deviations (n = 3) between 0.01 and 0.08 g/mL for the 5 powders included. Conversely, the 'measured' values were simply taken as a recording of pouring the powders from their containers in whatever state they happened to be resting in (raw materials were stored in 1 kg bags (HPMCP), 5 kg bottles (Soluplus) and 1 kg bottles (itraconazole) and ASDs (100 g) in sealed glass jars in a desiccator) into a pre-weighed graduated cylinder then re-weighing to get the density and this was performed once. Although the repeatability of bulk density values for Soluplus® and HPMCP from the FT4 and direct measurements was excellent (± 0.01 g/mL) and those of spray dried and milled extrudate were comparable (± 0.05 & 0.10 g/mL respectively), the value for itraconazole differed wildly (± 0.19 g/mL, or 37%) (Table 2). The net effect of these differences is demonstrated by Carr index values of 43.39 and 10.71 for itraconazole depending on the bulk density value taken - this clearly demonstrated the requirement for a more consistent method for measuring powder flow, particularly for powders like itraconazole that are both cohesive (C = 2.55 ± 0.17 kPa) and sensitive to consolidation (UYS = 10.75 ± 0.56).

To explore these points further, the various flow parameters obtained from the FT4 rheometer were tabulated with the Hausner ratio and Carr's index in Table 3. For itraconazole, interpretations of values for the Flow Function, Relative Flowability, Flow rate index, Aeration ratio, Hausner ratio and Carr's index were in good agreement with each other: cohesive, extremely cohesive, average flow rate sensitivity, average aeration sensitivity, extremely poor flow and extremely poor flow, respectively. Similar agreement was observed for Soluplus®, HPMCP, physical mixture, spray dried ASD and the milled extrudate. The high level of agreement between results suggests precision in the flow parameters from the instrument. In general, the accuracy of the measurements from the FT4 was excellent. However, while analysing the milled extrudate sample, although the FRI measurement was very accurate (SD ± 0.02), the results for Rel_P and FF were very inaccurate (± 40.07 and ± 53.62 respectively), reflecting the lack of sensitivity of shear testing towards larger particles.

3.3.6 *The effect of additional excipients on ASD powder flow*

To see if flow could be improved by blending with appropriate excipients, the spray dried and the milled extrudate ASDs were blended (separately) with Avicel PH 200 LM (Av200) and Mannogem granular (ManG) to give four formulations (Table 4). In order to enhance powder flow, these excipients had relatively larger particle sizes (on average 200 µm and 400 µm, respectively), although the difficulty in homogenously mixing powders of differing sizes is well documented. Furthermore, both excipients were specially treated commercial forms of MCC and mannitol respectively, both known for their excellent compression characteristics. Thirdly, due to the moisture-sensitive nature of amorphous dispersions, both excipients were selected for their low moisture ('LM') or non-hygroscopic (mannitol) properties. To achieve a 100 mg drug dose in the final tablet (the standard itraconazole dosage) required 333.3 mg (66.6% w/w) of the ASD, allowing only 150 mg (30% w/w) of excipient for the 500 mg total, with disintegrant and lubricant accounting for the remaining 3.4% (w/w).

In order to analyse the effect of these excipients on powder flow, further shear studies were performed using the 1 mL annular shear cell (due to lower powder quantities). The yield locus plots are depicted in Figure 7 with shear data in Table 5. Considering the spray dried ASD and formulations, addition of Av200 reduced cohesion by 0.46 kPa and increased FF by 1.01 units. Conversely, addition of ManG increased cohesion by 0.25 kPa and decreased FF by 0.25 units. Considering milled extrudate the picture was dissimilar: addition of Av200 reduced cohesion by 0.24 kPa and increased FF by 3.56 units, but addition of ManG *further reduced* cohesion by 0.46 kPa and further increased FF by 14.18 units. Clearly the enhancement or hindrance of flow by addition of further excipients is a combination of many complex factors, is specific to the precise set of interparticulate forces acting between the given excipient and dispersion and is therefore difficult to predict or simulate.

3.4 *Compression*

3.4.1 Tabletability and Compactability of raw materials and ASDs

In order to better understand the effects of ASD preparation on tabletability (compaction pressure vs tensile strength) and compactability (solid fraction vs tensile strength), the materials were examined using a Gamlen tablet press (GTP) and a PharmaTest hardness tester. The data was converted into plots according to equation 6:

$$\sigma = 2F / \pi DH \tag{6}$$

Where, σ is tensile strength, F is the load required to fracture the tablet, D is the tablet diameter and H the tablet height(Banker et al., 2002; Endicott et al., 1961; Fell and Newton, 1970).

The spray dried material produced a stronger compact than milled extrudate at similar compaction pressures (Figure 6a). This was attributed to the bonding area, i.e. the effective surface area taking part in physical interparticulate bonding, which is greater for smaller particles (Olsson and Nyström, 2001). Tablet tensile strength is related to both bonding area and bonding strength with the overall strength due to an interplay between the two (Osei-Yeboah et al., 2016; Sun, 2011). Assuming that a minimal amount of brittle fracture occurs in the powders during the initial stages of compression, the greater mechanical strength of spray dried compacts can be partly attributed to the smaller particle size produced by spray drying relative to milled extrudate.

The strongest compact was formed with pure HPMCP (slightly stronger than pure Avicel 101) and the weakest with Soluplus® (Figure 6a). It is also possible that differences in excipient hygroscopicity impacted compact strength, by introducing moisture mostly during ambient storage or during weighing and compression, causing more hygroscopic materials (such as HPMCP) to form stronger compacts(Malaj et al., 2010; Sun, 2011). Confirmation of this would require further testing.

The compactability (tensile strength vs solid fraction) profiles (Figure 6b) showed a range in tablet tensile stress with solid fraction. Soluplus® was very compacted at the tensile strengths studied, with values over 0.95. Milled extrudate fractions were comparable to values obtainable for a typical filler like Avicel 101 (MCC), but spray dried ASD had lower solid fraction ranges, up to 0.80. HPMCP had the lowest values, up to 0.68, showing the possibility for further compaction at higher pressures.

3.4.2 Tabletability and Compactability of Prototype Tablet Formulations

Formulation of spray dried powders with Av200 (Formulation 1) improved tabletability, however inclusion of ManG (Formulation 2) produced softer tablets (Figure 8a). Neither Av200 nor ManG improved the tabletability of extruded powders, with both Formulation 3 and Formulation 4 producing softer tablets than the milled extrudate alone. Addition of either excipient to the milled extrudate ASD produced a more linear relationship between compaction pressure and tensile strength above pressures of 115 MPa, and this was more marked with the ManG formulation.

Compactability values were higher for all extruded material formulations, although inclusion of both Av200 and ManG resulted in enhanced compactability values for all of the spray dried and extrudate formulations (Figure 8b).

3.4.3 Compressibility of Prototype Tablet Formulations

Tensile strength measurement is carried out on the tablet following compression and is an out-of-die method, however analysis of the powder in-die during compression provides further information, especially on elastic deformation. The non-linear behaviour observed between tensile strength and compression pressure of the extruded powder, Formulation-3 and Formulation-4 provided the opportunity to investigate whether a change in the mechanism of densification from plastic to elastic deformation had occurred. Figure 9 compares the behaviour of extruded powder and extrudate formulations 3 & 4 in-die, illustrating both the change in porosity with compression pressure (dotted line) and the resulting Heckel plot, based on the logarithmic change in porosity (full line). Porosity was calculated by determining the solid fraction using the density of a compact from its mass and volume then dividing by the true density (equations 7 and 8).

$$Solid\,fraction = \frac{Tablet\ Density}{True\ Density} = \frac{Mass/(\pi \cdot \left(\frac{Diameter}{2}\right)^2 \cdot Thickness)}{True\ Density} \quad (7)$$

$$Porosity = 1 - Solid\,fraction \quad (8)$$

For each of the three Heckel plots, an initial non-linear region was observed, typically attributed to particle rearrangement (Heckel, 1961; Shapiro, 1995) or particle fragmentation(Duberg and Nyström, 1982). This was followed by a linear region indicating plastic deformation(Heckel, 1961), then a further non-linear region signifying elastic deformation of the powder bed(Sun and Grant, 2001). While all three regions are present in each of the plots, the nature of the regions varied from powder to powder. The extruded material (Figure 9a) exhibited a shortened plastic deformation region with elastic deformation shown to be the mechanism of compression for pressures greater than 70 MPa. This is in agreement with the plateau in the tensile strength observed for these powders at higher compression pressures (Figure 8a). Similar behaviour is shown by Formulation-3 (Figure 9b), but with a lesser degree of elastic deformation. Formulation-4 (Figure 9c), which maintained a linear relationship between tablet tensile strength and compression pressure for the pressures studied (Figure 8a), behaved close to linearly, even at the highest pressures. Previous studies have shown that while it is difficult to completely separate the contributions of plastic and elastic deformation mechanisms to the strength of a compact following compression, strength increases are mainly due to the formation of physical bonds between particles due to plastic deformation(Hiestand, 1997). Much of the elastic deformation is temporary and disperses once the load is released(Picker, 2001) however it seems that in this case some tensile strength increase due to elastic deformation is permanent, indicated by the fact that the tensile strength of extruded powders in Figure 8 continues to increase at higher pressures and does not completely plateau.

3.5 Dissolution studies

Dissolution profiles of tablets 1, 2, 3 and 4 are plotted in Figure 10. As expected, the formulations containing spray dried dispersions performed better than those containing milled extrudate(Albadarin et al., 2017; Davis et al., 2017). Drug concentrations at 2 h approximated 50 $\mu g.mL^{-1}$ and 30 $\mu g.mL^{-1}$, respectively. After 6 h, average drug concentrations increased to

82.3 μg.mL^{-1} in tablet 1 (spray dried with Av200), 69.1 μg.mL^{-1} in tablet 2 (spray dried with ManG), 44.2 μg.mL^{-1} in tablet 3 (milled extrudate with Av200) and 44.5 μg.mL^{-1} in tablet 4 (milled extrudate with ManG). The highest drug release observed was from one of the spray dried and Av200 tablets, achieving a 91.0 μg.mL^{-1} concentration after 6 h, a drug release of 81.9%.

The most immediate factors facilitating the higher performance of the spray dried material were the smaller particle size, narrower particle dimensions and larger particle surface area, relative to milled extrudate (Figure 1). Only the smaller sieved fraction (< 90 μm) of milled extrudate was brought though for dissolution studies, as this was closer in particle size to the spray dried formulation and therefore more comparable. For an absolute comparison an identical particle size would be best, although this would require a very small sieve fraction from the milled extrudate (or use of a jet mill) and a further enhancement of spray dried particle size by a factor of 5 to 10. This could be achieved using alternative spray drying parameters, such as a pressure nozzle(Beyerinck et al., 2010).

Interestingly, the Av200 tablets performed better than the ManG tablets on average, however the results were not simplistic and at certain times the performance of ManG tablets exceeded Av200 tablets. For example, the Av200 formula out-performed the ManG formula during almost all of the milled extrudate tests, apart from the first half hour, where it was marginally higher and the final test point (6 h), where it was also marginally higher. Additionally, for the spray dried tests, the ManG formula out-performed the Av200 formula during the first 2 hours, but this changed between 2 h and 3 h and the Av200 formulation then released more drug through to 6 h (Figure 10). Depending on the type of formulation required (e.g. immediate or sustained release) either of these results could be seen as a favourable direction for development. It is worth noting that although the ASD and 'filler' are present in each tablet at 66.6% and 30% (w/w) respectively, the impact of 'filler' excipient on each formulation is significant. For example, the highest recorded drug release (at 6 h) for spray dried with avicel was 91.0 μg.mL^{-1} and the lowest for spray dried with mannogem was 67.6 μg.mL^{-1}, a difference of 23.5 μg.mL^{-1}.

Clearly, formulation of either the spray dried or the extruded ASD into an optimised tablet is a very complex task and difficult to predict. Firstly, further studies are recommended for formulation optimisation, such as examining ASD level, filler type, filler level, lubricant level and other factors impacting dissolution rate; including the potential for gelling polymer networks (GPN)(Demuth et al., 2015). Secondly, optimisation of tablet processing factors such as the effect of compression pressure on dissolution, the presence/absence of crystalline drug caused by compression, reduced particle surface area, disintegration time and wettability are also recommended. Finally, optimisation of processing parameters during ASD preparation, such as nozzle type, feed concentration, outlet temperature (for spray drying), or screw size, screw speed and sieve fraction (for extrusion) would further enhance control over critical parameters such as yield, processing time and particle size. Because an exhaustive trial of all possible permutations is impractical, a well-constructed design of experiment (DoE) approach appears the most straightforward and beneficial, given the potential for statistically relevant information based on fewer experimental trials. Optimisation of the formulation aspect of the ternary dispersions themselves using DoE has already been accomplished in this group and was recently reported in two separate publications (Albadarin et al., 2017; Davis et al., 2017).

4. *Conclusions*

The differences in downstream processing properties between amorphous solid dispersions prepared *via* hot melt extrusion or spray drying are significant. Preparation choice substantially impacted flow, compression and dissolution. The fundamental differences can be attributes to variances in particle size and morphology generated from the alternative preparation methods. Spray dried powders had smaller particles sizes, more complex morphology, were more cohesive and had poorer flow than milled extrudate. Good agreement was observed between various rheometric and density based parameters used to measure flow, such as Flow Function (FF), Relative flow function, Aeration ratio (AR), Flow rate index (FRI) and also the more rudimentary methods such as the Hausner ratio and Carr's index. Spray dried materials produced stronger compacts during tabletability studies and had lower solid fractions during compactability studies. Porosity studies revealed differences in the mechanism of densification in extrudate and blended extrudate formulations. It was possible to alter the flow and compression properties of the amorphous dispersions by blending with specific excipients. Dissolution studies on four prototype tablets containing 100 mg of amorphous itraconazole, revealed that drug release was faster and more complete from the spray dried samples and could be further influenced by selective excipient blending.

Acknowledgements

This work was funded under Science Foundation Ireland grant "Synthesis and Solid State Pharmaceutical Centre (SSPC)" - (12/RC/2275). Special thanks are due to our colleagues Dr Barbara Schaller, Dr Patrick Cronin and Dr Javed Iqbal for useful discussions on compression, flow and HPLC, respectively.

References

Albadarin, A.B., Potter, C.B., Davis, M.T., Iqbal, J., Korde, S., Pagire, S., Paradkar, A., Walker, G., 2017. Development of stability-enhanced ternary solid dispersions via combinations of HPMCP and Soluplus® processed by hot melt extrusion. Int. J. Pharm. 532, 603–611. doi:10.1016/j.ijpharm.2017.09.035

Banker, G.S., Siepmann, J., Rhodes, C., 2002. Modern Pharmaceutics, Fourth Edition, Drugs and the Pharmaceutical Sciences. CRC Press.

Beyerinck, R.A., Ray, R.J., Dobry, D.E., Settell, D.M., 2010. Method for making homogeneous spray-dried solid amorphous drug dispersions using pressure nozzles.

Bruni, G., Barletta, D., Poletto, M., Lettieri, P., 2007. A rheological model for the flowability of aerated fine powders. Chem. Eng. Sci. 62, 397–407. doi:10.1016/j.ces.2006.08.060

Davis, M., Walker, G., 2018. Recent strategies in spray drying for the enhanced bioavailability of poorly water-soluble drugs. J. Control. Release 269, 110–127. doi:https://doi.org/10.1016/j.jconrel.2017.11.005

Davis, M.T., Egan, D.P., Kuhs, M., Albadarin, A.B., Griffin, C.S., Collins, J.A., Walker, G.M., 2015. Amorphous solid dispersions of BCS class II drugs: A rational approach to solvent and polymer selection. Chem. Eng. Res. Des. doi:10.1016/j.cherd.2016.04.008

Davis, M.T., Potter, C.B., Mohammadpour, M., Albadarin, A.B., Walker, G.M., 2017. Design of spray dried ternary solid dispersions comprising itraconazole, soluplus and HPMCP: Effect of constituent compositions. Int. J. Pharm. 519, 365–372.

doi:http://dx.doi.org/10.1016/j.ijpharm.2017.01.043

Demuth, B., Nagy, Z.K., Balogh, A., Vigh, T., Marosi, G., Verreck, G., Van Assche, I., Brewster, M.E., 2015. Downstream processing of polymer-based amorphous solid dispersions to generate tablet formulations . Int. J. Pharm. . doi:10.1016/j.ijpharm.2015.03.053

Duberg, M., Nyström, C., 1982. Studies on direct compression of tablets. VI. Evaluation of methods for the estimation of particle fragmentation during compaction. Acta Pharm. Suec. 19, 421–436.

Endicott, C.J., Lowenthal, W., Gross, H.M., 1961. New instrument and method for evaluating tablet fracture resistance. J. Pharm. Sci. 50, 343–346. doi:10.1002/jps.2600500415

Ennis, B J; Witt, W; Weinekoter, R; Sphar, D; Gommeran, E.; Snow, R H; Allen, T; Raymus, GJ; Litster, J.D., 2007. Solid-Solid Operations and Processing, in: Perry, D.W.G.R.H. (Ed.), Perry's Chemical Engineers' Handbook. McGraw-Hill Professional.

Fell, J.T., Newton, J.M., 1970. Determination of tablet strength by the diametral-compression test. J. Pharm. Sci. 59, 688–691. doi:10.1002/jps.2600590523

Fitzpatrick, J.J., Barringer, S.A., Iqbal, T., 2004. Flow property measurement of food powders and sensitivity of Jenike's hopper design methodology to the measured values. J. Food Eng. 61, 399–405. doi:10.1016/S0260-8774(03)00147-X

Freeman, R., 2007. Measuring the flow properties of consolidated, conditioned and aerated powders — A comparative study using a powder rheometer and a rotational shear cell. Powder Technol. 174, 25–33. doi:http://dx.doi.org/10.1016/j.powtec.2006.10.016

Gallo, L., Llabot, J.M., Allemandi, D., Bucalá, V., Piña, J., 2011. Influence of spray-drying operating conditions on Rhamnus purshiana (Cáscara sagrada) extract powder physical properties. Powder Technol. 208, 205–214. doi:10.1016/j.powtec.2010.12.021

Hancock, B.C., Carlson, G.T., Ladipo, D.D., Langdon, B.A., Mullarney, M.P., 2002. Comparison of the mechanical properties of the crystalline and amorphous forms of a drug substance. Int. J. Pharm. 241, 73–85. doi:10.1016/S0378-5173(02)00133-3

Heckel, R., 1961. Density-pressure relationships in powder compaction. Trans Met. Soc AIME 221, 671–675.

Hiestand, E.N., 1997. Principles, tenets and notions of tablet bonding and measurements of strength. Eur. J. Pharm. Biopharm. 44, 229–242. doi:10.1016/S0939-6411(97)00127-6

Huang, S., Williams III, R.O., 2017. Effects of the Preparation Process on the Properties of Amorphous Solid Dispersions. AAPS PharmSciTech. doi:10.1208/s12249-017-0861-7

Iveson, S.M., Litster, J.D., Hapgood, K., Ennis, B.J., 2001. Nucleation, growth and breakage phenomena in agitated wet granulation processes: A review. Powder Technol. 117, 3–39. doi:10.1016/S0032-5910(01)00313-8

Jain, S., 1999. Mechanical properties of powders for compaction and tableting: An overview. Pharm. Sci. Technol. Today 2, 20–31. doi:10.1016/S1461-5347(98)00111-4

Jenike, A.W., 1964. Storage and Flow of Solids, Bulletin of the University of Utah. University of Utah.

Leturia, M., Benali, M., Lagarde, S., Ronga, I., Saleh, K., 2014. Characterization of flow properties of cohesive powders: A comparative study of traditional and new testing methods. Powder Technol. 253, 406–423. doi:10.1016/j.powtec.2013.11.045

Leuenberger, H., Rohera, B.D., 1986. Fundamentals of Powder Compression. I. The Compactibility and Compressibility of Pharmaceutical Powders. Pharm. Res. An Off. J. Am. Assoc. Pharm. Sci. 3, 12–22. doi:10.1023/A:1016364613722

Malaj, L., Censi, R., Gashi, Z., Di Martino, P., 2010. Compression behaviour of anhydrous and hydrate forms of sodium naproxen. Int. J. Pharm. 390, 142–149. doi:10.1016/j.ijpharm.2010.01.036

Nokhodchi, A., Maghsoodi, M., Hassan-Zadeh, D., Barzegar-Jalali, M., 2007. Preparation of agglomerated crystals for improving flowability and compactibility of poorly flowable and compactible drugs and excipients. Powder Technol. 175, 73–81. doi:10.1016/j.powtec.2007.01.030

Olsson, H., Nyström, C., 2001. Assessing tablet bond types from structural features that affect tablet tensile strength. Pharm. Res. 18, 203–210. doi:10.1023/A:1011036603006

Osei-Yeboah, F., Chang, S.-Y., Sun, C.C., 2016. A critical examination of the phenomenon of bonding area - Bonding strength interplay in powder tableting. Pharm. Res. 33, 1126–1132. doi:10.1007/s11095-016-1858-8

Patel, S., Kaushal, A.M., Bansal, A.K., 2006. Compression physics in the formulation development of tablets. Crit. Rev. Ther. Drug Carrier Syst. 23, 1–65.

Peschl, I.A.S.Z., Colijn, H., 1977. NEW ROTATIONAL SHEAR TESTING TECHNIQUE. J Powder Bulk Solids Technol 1, 55–60.

Picker, K.M., 2001. Time dependence of elastic recovery for characterization of tableting materials. Pharm. Dev. Technol. 6, 61–70. doi:10.1081/PDT-100000014

Salman, A.D., Biggs, C.A., Fu, J., Angyal, I., Szabó, M., Hounslow, M.J., 2002. An experimental investigation of particle fragmentation using single particle impact studies. Powder Technol. 128, 36–46. doi:10.1016/S0032-5910(02)00151-1

Schulze, D., Schwedes, J., Carson, J.W., 2008. Powders and bulk solids: Behavior, characterization, storage and flow, Powders and Bulk Solids: Behavior, Characterization, Storage and Flow. doi:10.1007/978-3-540-73768-1

Schwedes, J., 2003. Review on testers for measuring flow properties of bulk solids. Granul. Matter 5, 1–43. doi:10.1007/s10035-002-0124-4

Shapiro, I., 1995. Compaction of powders XII. Evaluation of published compaction equations for modeling purposes, in: Advances in Powder Metallurgy and Particulate Materials. p. 2/97-2/114.

Sinka, I.C., Schneider, L.C.R., Cocks, A.C.F., 2004. Measurement of the flow properties of powders with special reference to die fill. Int. J. Pharm. 280, 27–38. doi:10.1016/j.ijpharm.2004.04.021

Sivasankaran, S., Sivaprasad, K., Narayanasamy, R., Iyer, V.K., 2010. An investigation on flowability and compressibility of AA 6061100-x-x wt.% TiO2 micro and nanocomposite powder prepared by blending and mechanical alloying. Powder Technol. 201, 70–82. doi:10.1016/j.powtec.2010.03.013

Sun, C., Grant, D.J.W., 2001. Influence of crystal structure on the tableting properties of sulfamerazine polymorphs. Pharm. Res. 18, 274–280. doi:10.1023/A:1011038526805

Sun, C.C., 2011. Decoding powder tabletability: Roles of particle adhesion and plasticity. J. Adhes. Sci. Technol. 25, 483–499. doi:10.1163/016942410X525678

Vehring, R., 2008. Pharmaceutical particle engineering via spray drying. Pharm. Res. 25, 999–1022. doi:10.1007/s11095-007-9475-1

Walker, D.M., 1966. An approximate theory for pressures and arching in hoppers. Chem. Eng. Sci. 21, 975–997. doi:10.1016/0009-2509(66)85095-9

Wu, C.-Y., Cocks, A.C.F., 2004. Flow behaviour of powders during die filling. Powder Metall. 47, 127–136. doi:10.1179/003258904225015617

Xie, X., Puri, V.M., 2006. Uniformity of powder die filling using a feed shoe: A review. Part. Sci. Technol. 24, 411–426. doi:10.1080/02726350600934663

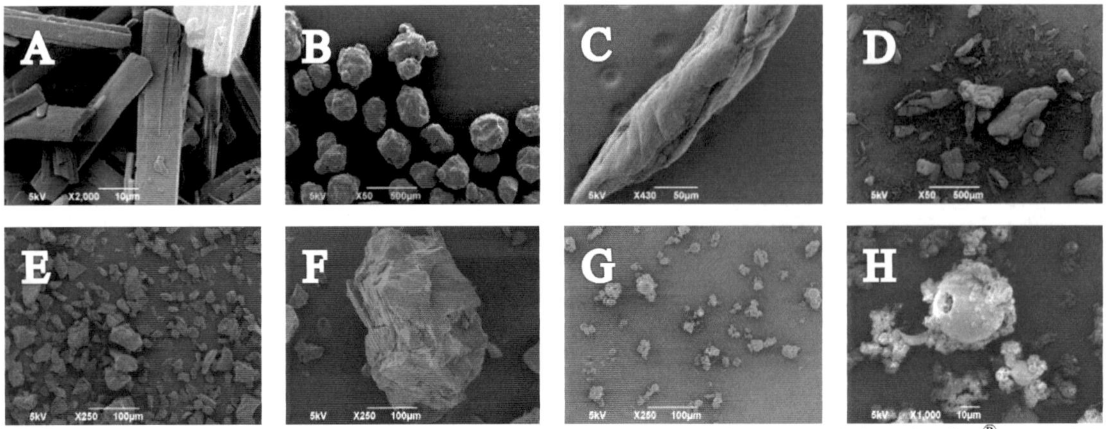

Figure 1: Scanning electron micrographs of (A) itraconazole x2000 (B) Soluplus® x50, (C) HPMCP x430, (D) physical mix x50, (E) milled extrudate < 90 μm sieve fraction x250, (F) milled extrudate 90 – 450 μm sieve fraction x250, (G) spray dried x250 & (H) spray dried x1000.

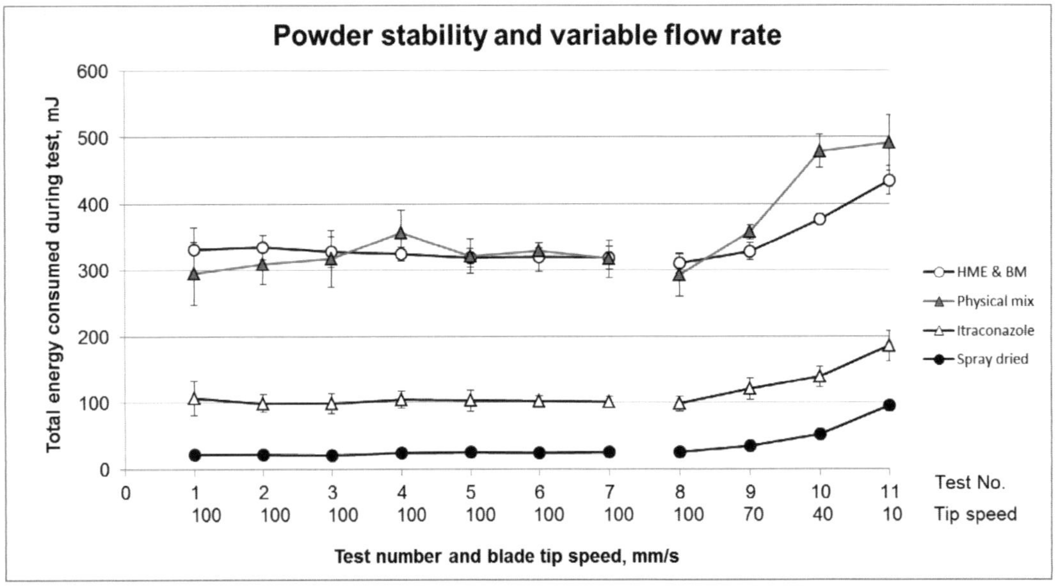

Figure 2: Stability and variable flow rate data recorded on an FT4 powder rheometer for ASD powders of itraconazole-Soluplus®-HPMCP 30-40-30 (w/w/w), physical mix and pure drug. Tests 1-7 are performed at constant blade tip speed and measure stability, a stable powder will show an approximately flat straight line. Tests 8-11 feature a reduction in blade tip speed and measure the sensitivity of powder to flow rate, typically more energy is consumed as the blade passes at a slower speed. Measurements were recorded in triplicate.

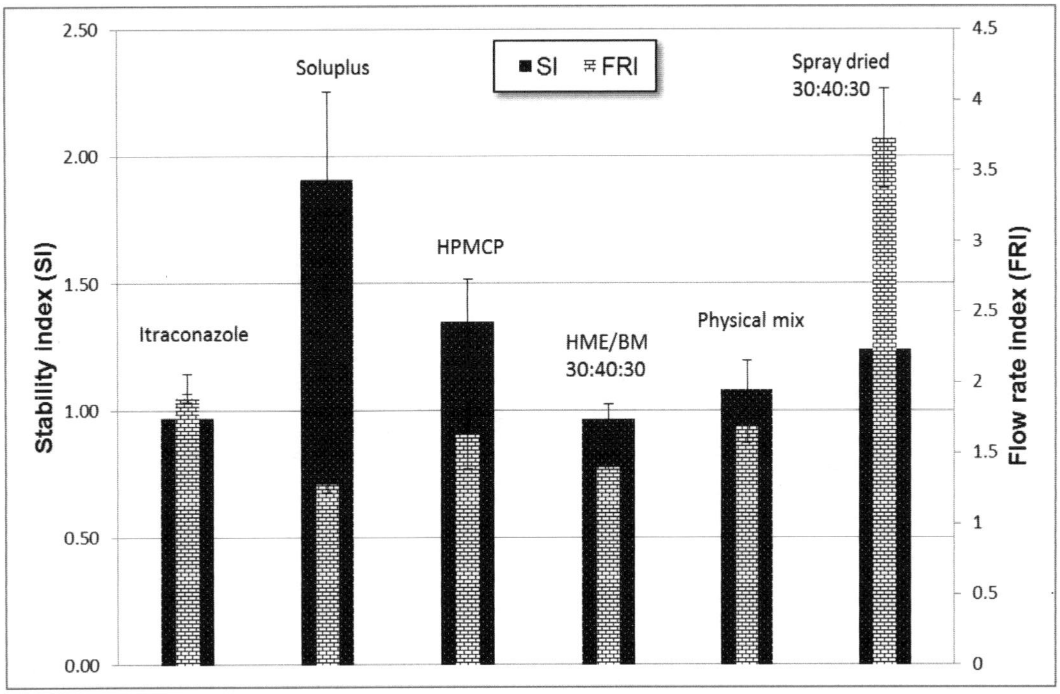

Figure 3: The stability index (SI) and flow rate index (FRI) values recorded in triplicate for itraconazole, Soluplus®, HPMCP, physical mix, HME/BM powder and spray dried powder by the FT4 powder rheometer. An SI value of 1.0 ± 0.1 indicates the powder is not sensitive to flow stress, for example the HME/BM result. An FRI > 3.0 indicates that the powder is highly sensitive to flow rate and most likely cohesive, such as the value of the spray dried sample.

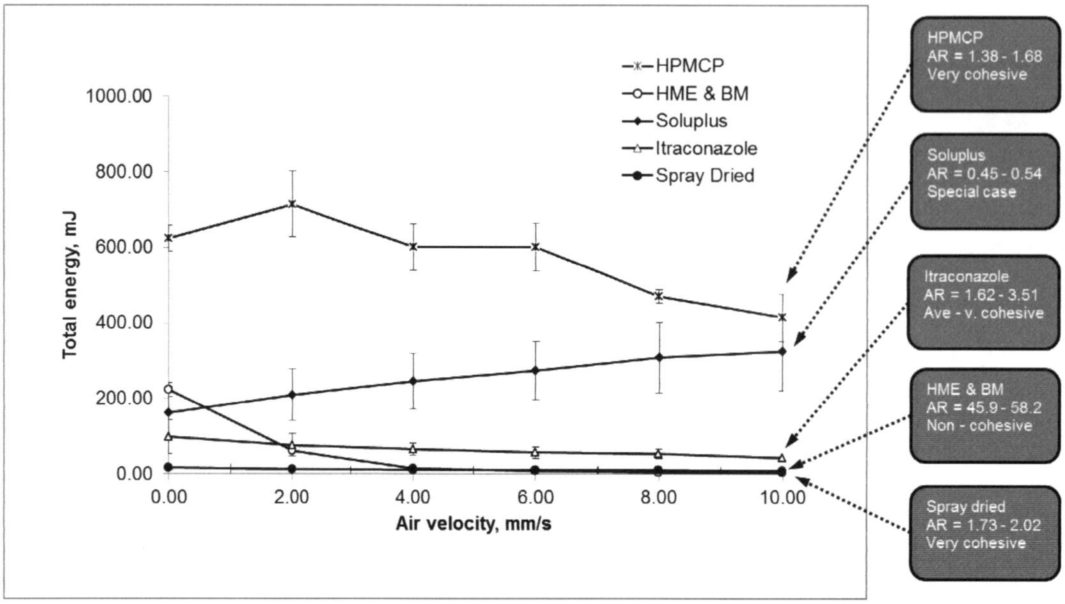

Figure 4: Aeration studies on raw materials itraconazole, Soluplus®, HPMCP and ASD powders prepared by hot melt extrusion then ball milling (HME & BM) or by spray drying. Data was recorded in triplicate on an FT4 powder rheometer. During the test, a blade is passed at 100 mm/s tip speed through a bed of powder while air is introduced at the base at progressive velocities. The extent by which the flow energy is reduced is a function of the physical properties of the powder (such as cohesion) and provides the aeration ratio (AR).

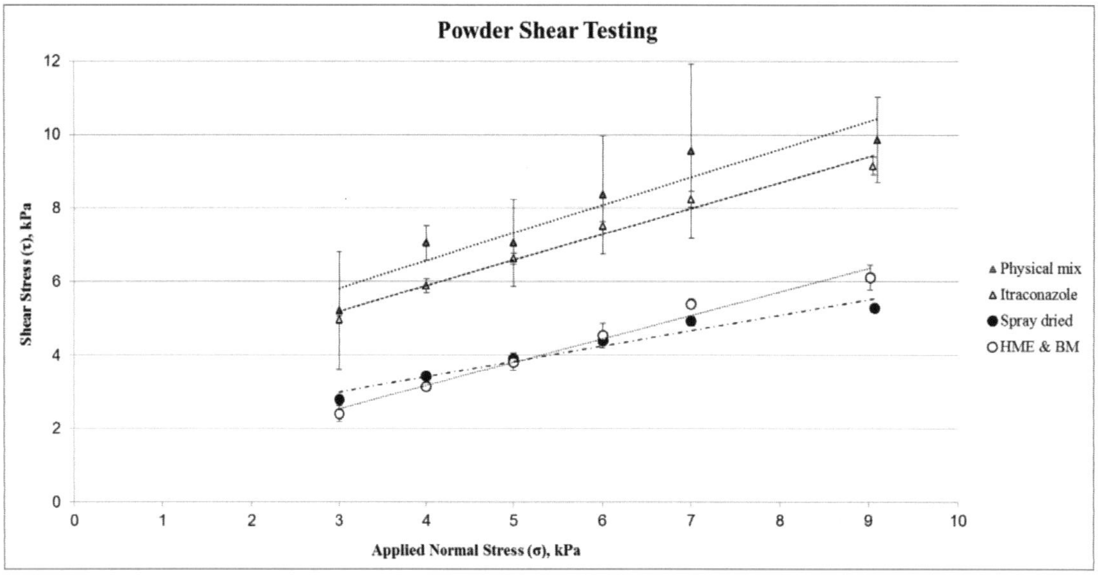

Figure 5: Yield locus plots for itraconazole, physical mixture and ASD powders prepared by spray drying and by hot melt extrusion (followed by ball milling), recorded in triplicate on an FT4 powder rheometer. Corresponding shear parameters are recorded in Table 1.

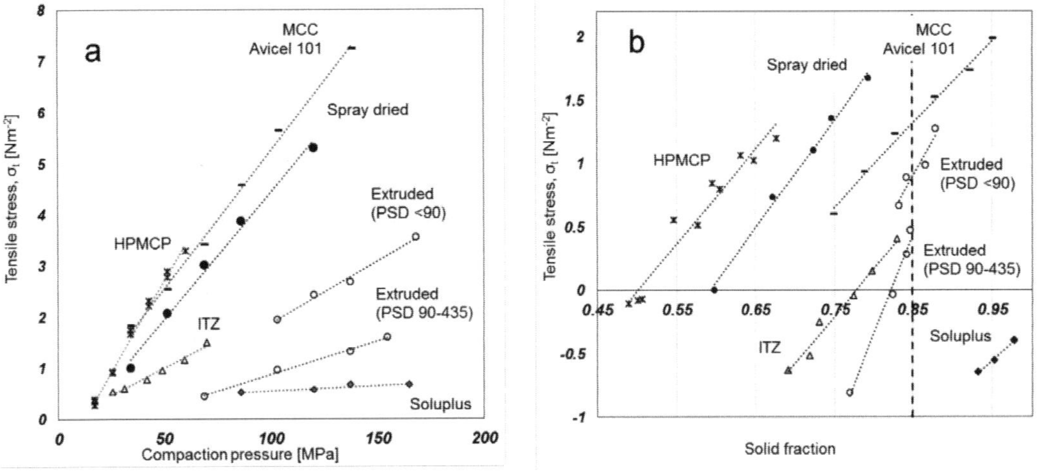

Figure 6: Tabletability (a) and Compactability (b) plots of raw materials, spray dried ASD and milled extrudate. ITZ = itraconazole, MCC = microcrystalline cellulose, PSD = particle size distribution. Two sieve fractions were measured for extrudate: < 90 µm & 90 – 435 µm.

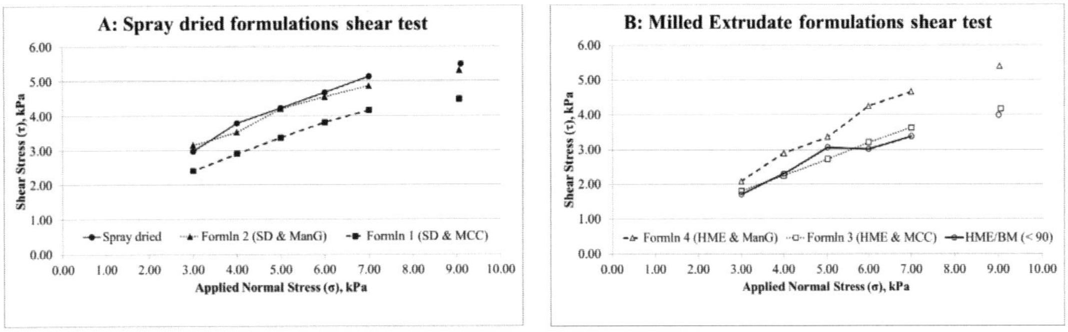

Figure 7: Yield locus plots for prototype formulations, comprising spray dried (SD) powder with Avicel PH 200 LM microcrystalline cellulose (MCC) (Formulation 1) and Mannogem granular mannitol (ManG) (Formulation 2) and milled extrudate (HME/BM <90) with MCC (Formulation 3) and ManG (Formulation 4). For complete formulations see Table 4, for shear data see Table 5 and for discussion see Section 3.3.6.

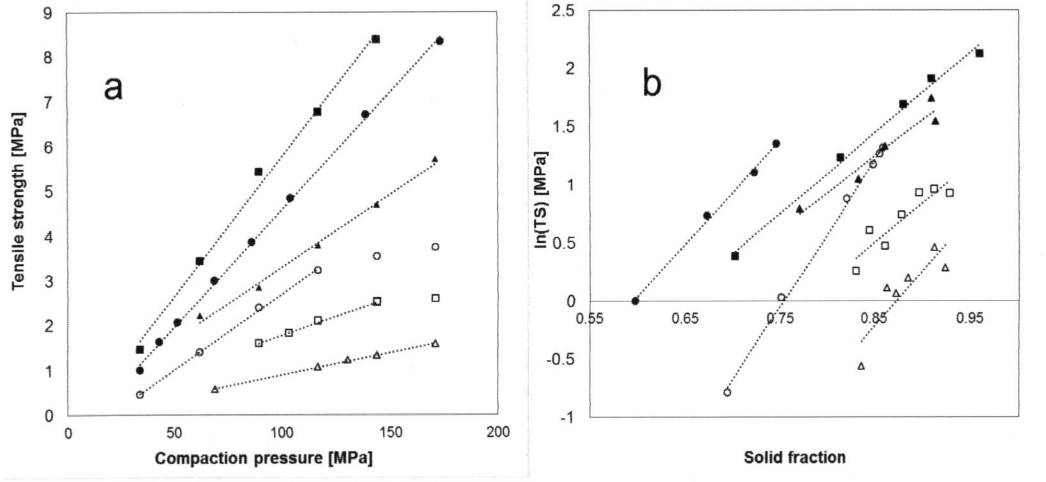

Figure 8: Tabletability (a) and Compactability (b) plots of spray dried (filled circles) and milled extrudate (open circles) & the effect of including Av200 (squares) or ManG (triangles).

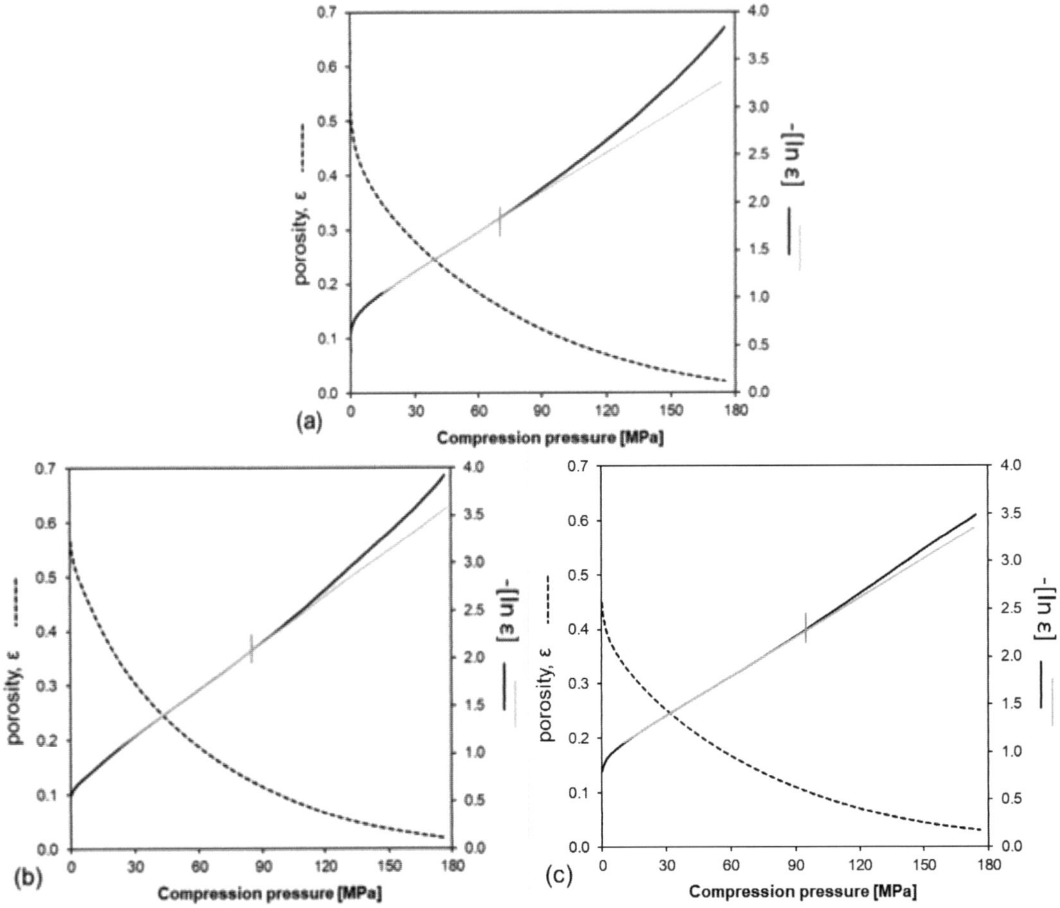

Figure 9: The change in porosity of the powder bed during compression (--- dashed line) of (a) extruded powders, (b) Formulation-3 and (c) Formulation-4 and the resulting Heckel plots, illustrating the logarithmic change in porosity (full line –). The linear portion of the Heckel plot, indicating the region where the powder bed is undergoing plastic deformation, is indicated by a grey line of best fit.

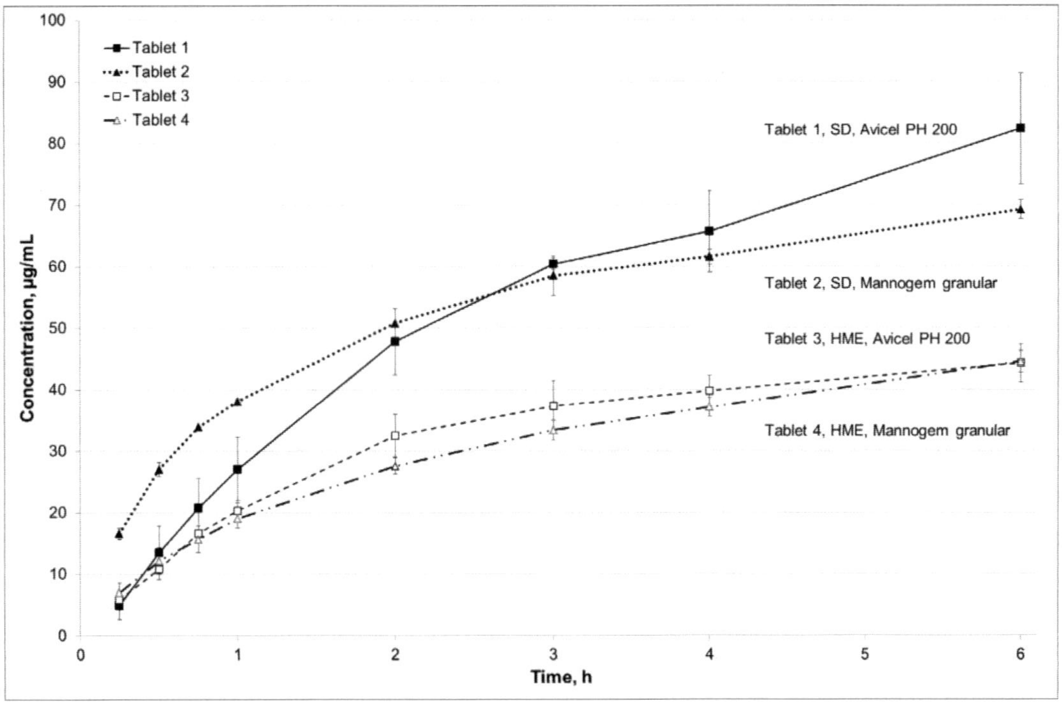

Figure 10: Dissolution profiles of 500 mg prototype tablets, containing a 100 mg dose of amorphous itraconazole prepared by spray drying (tablets 1 & 2) or hot melt extrusion (tablets 3 & 4). ASDs were blended with excipients (Table 4) before compression into spherical disks and dissolution (n = 3) in HCl (pH 1.2), with HPLC analysis at λ = 263 nm.

532310 *Intrusive Probes in Riser Applications*

Abstract:

Many of the probes used to understand hydrodynamics in circulating fluidized bedrisers intrude into the environment they are measuring, although assumptions are typically asserted that the intrusive probes do not affect the data collected. This could be a poor assumption in some cases and conditions. We found that intrusive fiber optic probe measurements consistently mis-predicted the solids concentration compared to the non-intrusive pressure drop measurements outside the fully developed flow region of a riser containing fluid catalytic cracking catalyst or glass bead particles. The discrepancy was sensitive to superficial gas velocity, solid circulation rate, probe position and flow direction. Barracuda VRâ„¢ computational fluid dynamics simulations confirmed this, and indicated that particle momentum was lost at the leading edge of the probe and particles were spilling over to the probe tip. Accordingly, new probe designs were proposed to mitigate the intrusiveness of a fiber optical probe for more accurate characterization.

537127 Non-Intrusive Measurement and Imaging of Circulating Fluidized Beds Using Electrical Capacitance Volume Tomography

Abstract:

Real time flow rate measurements and characterization of gas-solid flows is desirable in many process applications utilizing solids transport. To date, limited methods for non-intrusive measurement of solid circulation rate exist and even fewer exist for the imaging of these flows, particularly at high temperatures. The Ohio State University (OSU) has been developing electrical capacitance volume tomography (ECVT) sensors, measurement techniques, and processing algorithms for imaging multi-phase flows. ECVT is a noninvasive 3-D imaging technology capable of providing real time images of multi-phase flow systems. The ECVT sensor consists of an array of capacitance sensor plates designed to externally fit around the sensing domain. Capacitance measurements from all possible plate pairings are taken by sequentially applying a voltage between each electrode pair. The permittivity distribution of dielectric material within the sensing domain can be derived from the measured capacitance values through the Poisson equation. The solution can be quickly estimated using one of many reconstruction techniques including but not limited to: iterative linear back projection, multicriteria optimization image reconstruction technique, and the Hopfield neural network model. Thus, ECVT is able to provide real time, high fidelity imaging of the internal multiphase flow.

In this study, ECVT technology was utilized and developed to study flow characteristics of gas-solid systems found in circulating fluidized bed systems including packed moving bed and fluidized bed regimes. Algorithms and techniques were developed to extract solid circulation rate, linear velocity of solids, and 3-D imaging of flow patterns. Additionally, the ability of the ECVT sensor to detect small variations in solids packing was studied. Flow rate and velocity measurements were calculated using cross correlation and peak matching techniques. Data measurement and analysis techniques of both the raw sensor data and reconstructed 3-D images will be discussed. Experiments at elevated temperatures were also conducted to examine temperature effects and to confirm the ECVT concept at high temperatures.

518703 Method of Estimating the Solids Mass Flow Rate in a Gas-Solids Riser Using the Integrated Mixture Momentum Equation and the Dynamic Pressure Gradient Distribution

Abstract:

The determination of the solids mass flow rate (G_p) is important for efficient control of conversion processes run in risers. For this reason a significant amount of effort has been dedicated to the development of solids mass flow rate determination methods which often focus on the closure of a one-dimensional, two-fluid model using a differential approach. This requires accurate prediction of the effective drag coefficient (C_d), which is not possible at this time because of the complexity of relating the effects of turbulence in a riser and particle acceleration to C_d. In this work, a novel approach to closing a one-dimensional, two-fluid model is developed. The result is a means of estimating the G_p from riser conditions using three readily measurable and physically relevant quantities; the superficial riser gas velocity (U_r) and the dynamic pressure drop across (ΔP_{ra}) and gradient at the end of ($P'_{ra}(1)$) the acceleration section of a riser.

The method of analysis is made possible by augmenting a one-dimensional, two-fluid model by the inclusion of a theoretically derived expression for the dynamic pressure gradient distribution that was obtained from a solution to an isoperimetric problem of the calculus of variations. The resulting expression for the dynamic pressure gradient requires two parameters to define it; the dynamic pressure drop across (ΔP_{ra}) and gradient ($P'_{ra}(1)$) at the end of the acceleration section. The ratio of these two quantities provides the acceleration parameter $a_p \equiv \Delta P_{ra}/(-P'_{ra}(1))$ which reflects the extent to which particles increase in velocity and reduce in volume fraction in the acceleration section of a riser.

A series of approximations made to the one-dimensional, two-fluid model provide a means of integrating the mixture momentum equation. The resulting expression allows relationships between integrated terms to be quantified providing additional expressions used for closure and prediction of the solids mass flow rate (G_p) from measured values of the gas velocity (U_r) and the defining parameters of the dynamic pressure gradient distribution. The model prediction of G_p is tested using literature data obtained transport experiments using several types of particles transported through risers of varying diameters. The agreement between the predicted and experimentally measured values of G_p is good to excellent.

533152 Influence Parameters and Modeling of Solids Circulation Rate in the High-Density Circulating Fluidized Beds

Abstract:

Circulating fluidized bed (CFB) has been utilized in many industrial processes. Two typical commercial applications are circulating fluidized bed combustor (CFBC) and the fluid catalytic cracking (FCC). The operating parameters are different in these two processes. For example, CFBC is usually operated with the solids circulation rate below 200 kg/m^2s, while FCC process is operated with much higher solids circulation rate of 400-1200 kg/m^2s. To distinguish these two processes, the circulating fluidized beds with high solids circulation of more than 200 kg/m^2s along with high solids holdup of 10 % are defined as the high-density circulating fluidized beds (HDCFB). Those with solids circulation rate of lower than 200 kg/m^2s are called low-density circulating fluidized beds (LDCFB). Right now, studies on CFB always focus on the LDCFB. The reason for less fundamental researches focusing on HDCFB is because it is hardly to obtain the high density operating condition in the laboratory-scale setup. The hydrodynamics of HDCFB are different from that of LDCFB. So itâ€™s necessary to work on high-density circulating fluidized beds, especially on influences and modeling of high density operations.

In this work, experiments are mainly carried out in a riser with 80 mm diameters and 18 m height using the FCC particles with mean diameter of 85 Î¼m and particle density of 1500 kg/m^3. The experimental setup includes a downcomer with 450 mm diameters and a solids storage tank with 660 mm. The downcomer and storage tank can provide enough drive force to supply enough solids particle to the riser. Parameters on the solids circulation rate including superficial gas velocity, solids inventory height in the storage tank, the opening degree of the butterfly valve in the feeding pipe and the aeration air rate in the bottom of storage tank are comprehensively studied. The solids circulation rate can reach as high as 1300 kg/m^2s at superficial gas velocity of 9 m/s with solids holdup nearly up to 20% in the riser indicating the high-density operating condition. The solids circulation rate increases linearly with the increase of superficial gas velocity from 5 to 9 m/s. Similar results are also found with increasing the solids inventory in the storage tank. The sensitivity of influence parameters is also studied in this work. Based on this work, an empirical model is established by fitting the experimental data which is shown as follows:

$$Gs = f(A, U_g, V_A) = (a*exp(A)*U_g^2 + b*V_A*U_g + c*A*V_A)*(A^d*V_A^e)$$

Where

G_s --- solids circulation rate, kg/m^2s

A --- ratio of valve opening area, -

U_g --- superficial gas velocity, m/s

V_A --- aeration air rate in the bottom of the stored tank, m^3/h

a, b, c, d, e --- the empirical coefficients, -

There are 3 independent variables and 5 empirical coefficients in the equation and those coefficients are shown in the Table 1. The solids circulation rate can be predicted by the empirical models according to the value of the solids inventory height in the storage tank, the opening degree of the butterfly valve, the aeration air in the bottom of the tank and the superficial gas velocity. The relative error between the predicting result and the experimental data is lower than 20%.

Table 1. Coefficients for the empirical predicting model

solids inventory height	a	b	c	d	e
4	-2.0035	88.4986	-109.8504	0.8906	-0.9896
6	-7.2611	158.2493	-282.6492	1.1061	-1.05
8	-13.2719	182.8993	-190.4830	0.8496	-1.0698
10	-9.2338	124.6352	158.7743	0.6943	-0.9279

533453 Full-Loop Simulation of Gas-Solids Flow in Circulating Fluidized Beds with Different Sizes

Abstract:

Circulating fluidized beds (CFB) are widely used in petrochemical, coal, metallurgy, and other industrial fields. To obtain detailed information and fully understand gas-solids flow behaviors in CFB, computational fluid dynamics (CFD) had been applied to simulate gas-solid flow behaviors in CFB. Circulating fluidized bed is a complex and integrated process coupling all inter-connected units including the riser, cyclones, downcomer, storage tank and valve. Therefore, the full-loop simulation is a promising way to describe complex gas-solids interaction in CFB under fully consideration of effects of the whole system by treating the whole CFB system as a simulation object. Most studies focused on gas-solid behaviors in a certain CFB device, and there were little comparison between different sizes of different CFB devices such as heights and diameters of riser. In fact, sizes of CFB are the most critical factors in design, which affect the length of region affected by entrance and exit structure, determining length of fully developed region finally. To keep stable operation and ensure the validity of experiment data, enough length of fully developed region is necessary, which closely related to the sizes of CFB. Therefore, effects of sizes of CFB need to be fully discussed.

In this work, effects of CFB sizes including riser heights and riser diameters were investigated by full-loop simulation method. A pilot-scale gas-solid circulating fluidized bed with riser in height of 18 m has already been set up in our group. The simulation condition was chosen according to experiments with different superficial gas velocity, solids inventory, valve opening and fluidizing air flow rates. A reliable model applying to multiple flow regimes was established, and the validity of the model was guaranteed by comparing simulation results and experiment data. Then simulation of different CFB sizes were conducted, and the effects of CFB sizes in different operating conditions on distribution of pressure, solids holdup, velocity and length of fully developed region in system were also investigated. The lengths of fully developed region with different CFB sizes and different operation conditions were analyzed, and optimum ranges of CFB sizes under different operation condition were provided to give a reference to the design.

536372 A Very, Very Small-Scale Experiment of Fluidized Particle Segregation: A Prerequisite for the Uncertainty Quantification of CFD-DEM Simulations

Abstract:

Systems involving fluidized particles are ubiquitous in industrial operations for which thorough mixing of particulate material is often a requirement. Computational fluid dynamics coupled with the discrete element method (CFD-DEM) allows for a deeper physical understanding of the how the fluid-particle and particle-particle interactions within these operations affect the overall behavior. Just as the uncertainty in experimental results require quantification, the predictions obtained by CFD-DEM should also include bounds associated with their uncertainty. Such uncertainty quantification requires thousands of simulations or more. The already high computational cost associated with simulating the large number of particles in a typical lab-scale experiment with CFD-DEM makes performing such uncertainty quantification prohibitive. In this work, a very, very small scale (< 7,000 particles), segregation experiment was designed such that CFD-DEM simulation could be performed in a reasonable time â€" i.e., less than 8 hours with a single CPU. The details of the segregation experiment and system conditions, along with the careful characterization of particle properties, is explained. Additionally, the sensitivity of the experimental results (intruder segregation time) to various system conditions and intruder properties, e.g., intruder particle size, initial position, number of intruders, and superficial velocity, on the segregation rate are discussed. Preliminary CFD-DEM simulations of one of the experimental conditions are overviewed. Finally, the usefulness of a very, very-small-scale device to perform fast experiments with a small number of particles â€" to assess brute-force uncertainty quantification and simplified methods - is discussed.

521837 Modeling the Hydrodynamics of Tapered Gas-Solid Risers

Abstract:

Gas-solid circulating fluidized bed (CFB) systems have been widely used in many industrial processes. In the CFB systems, the riser section is generally the part that the main reactions occur. Altering riser cross-sectional area can greatly enhance the reaction characteristics. For instance, the tapered-out riser can improve the mixing and thus will suit the reactions with high reaction rate, while the tapered-in riser can enhance the particle residence time and thus will fit to the reactions with low reaction rate.

This study will focus on the multiscale modeling of the hydrodynamics of the tapered-out and tapered-in CFB risers, in which the computational complexities mainly lie in addressing the problems related to the continuous variations of superficial gas and solid velocities with height as well as much more significant wall effect in diameter-changing fluidized bed risers than that in constant-diameter ones. By utilizing the energy-minimization multiscale (EMMS) theory, the structure-based modeling of the both types of riser reactors are performed to compute the axial and radial heterogeneous distributions of hydrodynamic parameters such as solids concentration and particle velocity. The calculation results are found to be qualitatively consistent with the experimental data in small- and large-scale gas-solid CFB risers.

In order to validate the calculation results numerically, a computational fluid dynamics (CFD) method is also utilized to simulate the hydrodynamics of the tapered gas-solid risers, in which the axial variation of the EMMS drag in the tapered risers is calculated according to the variations of superficial gas and particle velocities with height and further incorporated into the two fluid model (TFM) simulation by a user definition function (UDF). It is found that a three-dimensional (3D) transient Eulerian approach can facilitate the prediction of the hydrodynamics of the tapered gas-solid risers, which agrees with the calculation results of the EMMS-based modeling of the tapered-out and tapered-in CFB risers.

This study can be expected to further enrich the theory of full-loop modeling of complex gas-solid processes with various geometries and sizes.

522838 Validation Study on an Eurerian-Lagrangian Method in a Circulating Fluidized Bed

Abstract:

A circulating fluidized bed (CFB) is one of typical gas-solid flow systems in chemical engineering. For better understanding of mechanism in the CFB in industries, numerical simulation is indispensable. In the CFB, the flow characteristics is regarded to be extremely complex, since dilute and dense solid particle regions are appeared simultaneously. An Eurerian-Lagrangian approach becomes promising when the complex gas-solid flow is numerically simulated, where solid and gas phases are modeled by the discrete element method (DEM) and computational fluid dynamics (CFD). Although the DEM-CFD method was well employed in the gas-solid flows, it had critical problems from a viewpoint of restriction of calculated particles and arbitrary shaped boundary. In order to solve these problems, we applied a coarse grain model of the DEM, sign distance functions (SDF) for wall boundary model in the DEM and immersed boundary method (IBM) in the CFD in the existing DEM-CFD model. This numerical approach makes it possible to simulate a complex shaped system such as the CFB with a simple algorithm. In this study, validation tests were performed to show adequacy of the new computation technology. The validation system was quite large, namely, number of the solid particles were about 4 x 10^8. Through the validation tests, particle spatial location and pressure drop were shown to agree between the simulation and experimental results. Consequently, we conclude that our method could simulate large-scale circulating fluidized bed accurately.

Acknowledgement: This study was financially supported by JSPS KAKENHI Grant Number JP17H03451.

527578 Carrier Based Dry Powder Inhaler Formulation – a Particle Engineering Perspective

Abstract:

Introduction

Dry powder inhalers (DPI) are the preferred devices among orally inhaled drug products due to their ease of handling and administration as well as superior product stability[1]. They deliver a dry powder of the active pharmaceutical ingredient (API) either in the presence or absence of larger size excipient particles (usually a sugar) as the carrier. Carrier-based DPIs are manufactured via adhesive dry mixing of the fine drug particles (1-5 µm) with the coarse excipient particles. A delicate balance of the inter-particle forces is necessary to guarantee a stable homogeneous mixture during processability (e.g. capsule filling, transport, and storage) and eventually to allow API detachment from the carrier during delivery. Rational engineering of particles can offer an invaluable opportunity to tailor API and carrier with the desired physicochemical characteristics and thereby enable fine-tuning the inter-particle cohesive-adhesive balance (CAB). Thus, the present work intends to explore diverse particles engineering approaches for API and carrier and evaluate the impacts on particle properties and product performance.

API particle engineering

Salbutamol sulphate (SS) (Selectchemie) was chosen as a model API and engineered to the inhalable size using spray drying (Nano Spray Dryer B-90, Büchi Labortechnick AG) and jet-milling (AS50 Air Spiral jet mill, Hosokawa Alpine AG) The samples were stored and characterized immediately, 7, 14 and 21 days after production with respect to their solid-state, micromeritics and surface properties. Spray drying resulted into rugged spherical amorphous SS particles ($Dv_{0.5}$= 3.75 ± 0.08 µm) that were able to form strong agglomerates through "amorphous bridging". On the other hand, jet-milling produced smaller ($Dv_{0.5}$ = 2.06 ± 0.08 µm), crystalline, irregular shaped particles with a very large surface area (11.04 ± 0.10 m^2/g) that, over time, formed looser particle aggregates of decreasing size.

API particle performance

The SS particles generated by different engineering routes were further formulated into carrier-based DPIs to evaluate their performance. Both spray dried SS (SDSS) and jet-milled SS (JMSS) particles were mixed with α-lactose monohydrate (Lactohale® 100, DFE Pharma) to yield adhesive blends at low (1%) and high (10%) drug loads. By increasing the drug load, propagation of API particle properties effects on blending (Turbula blender TC2, Willy A. Bachofen Maschinenfabrik), capsule filling (Labby, MG2) and *in vitro* aerosolization performance (Next Generation Impactor – NGI, Copley Scientific) could be evaluated.

Adhesive mixing

Notable differences were found between CAB values of JMSS and SDSS particles. SDSS particles present a greater tendency to attach as agglomerates to the "active sites" within the carrier surface compared to JMSS counterpart. Moreover, it was also apparent that smoother lactose particles were seemingly unable to bind the SDSS to their surfaces, reinforcing the hypothesis that cohesive forces of SDSS particles (drug-drug) are higher than their adhesive ones (drug-carrier)[2]. Consequently, it is proposed that under the same formulation conditions, (i.e. API-carrier pair and blending process), when compared to JMSS, SDSS particles adhere more poorly to the carrier being more prone to detachment during powder handling.

Capsule filling

Capsule filling results supported blending observations revealing larger differences in blend uniformity for the SDSS mixtures, being that this was particularly evident at low loads and smaller powder bed heights, where drug content and powder densification effects were minimized. It is hypothesized that during processing, the SDSS particles are able to detach very easily from the carrier surface resulting in rapid segregation of the powder bed. It is well-known that the interaction forces between the drug and excipient host particles are beneficial with respect to handling of the powder blend as they reduce the risk of segregation[3]. Thus, as expected when compared to the JMSS blends API detachment in SDSS mixtures resulted in capsules filled with a highly variable amount of the targeted drug content.

In vitro aerosolization performance

In vitro aerosolization also supports SDSS poor attachment to carrier surface as demonstrated by the extensive deposition in mouth and throat. In opposition the JMSS blends where greater losses were seen in the pre-separator. It is known that the higher fraction of free API particles (not adhered to carrier surface) can lead to larger drug deposition in the mouth and throat than in the pre-separator[4] . Finally, it was also evident that the SDSS agglomerates are aerosolized as such resulting in a significant larger mean median aerodynamic diameter (MMAD) when compared to their JMSS counter-parts.

Carrier particle engineering

Using an integrated approach, lactose was spray dried in the presence of polyethylene glycol (PEG) 200. Anomeric composition of lactose in lactose-PEG 200 feed solutions of variable compositions was analyzed via polarimetry at different temperatures. These results were correlated with the solid state and anomeric composition of the resulting spray dried particles. Characterization of the dried powders revealed that crystalline (in an anomeric ratio 0.8:1 of α to β) spherical particles with a median particle size ($Dv_{0.5}$) of 50.9 ± 0.4 µm could be produced. In addition, the surface of these particles was composed of a plate like structure of PEG 200 and segregated β-lactose.

Carrier particle performance

A blend of lactose-PEG 200 composite particles with JMSS was produced and the role of carrier properties on DPI efficiency assessed. For this, the performance of the composites blend was compared with a series of other adhesive mixtures containing distinct lactose grades with different anomeric compositions and/or physical properties. This way, it was possible to deconvolute the contribution of solid state, surface chemistry and morphology of carrier particles on the dry adhesive mixing (Turbula blender TC2, Willy A. Bachofen Maschinenfabrik) and *in vitro* aerosolization performance (Next Generation Impactor – NGI, Copley Scientific).

Adhesive mixing

Blending of spray dried lactose-PEG 200 composite particles with JMSS resulted in the presence of a notable number of drug agglomerates. For instance, it has been reported that particles with lower shape coefficients (SC), a coefficient representing the surface area, density and aspect ratio of the particle, achieve mixing homogeneity faster, but tend to show higher segregation propensity[5]. Considering the lower density, regular shape (spherical) and smoother surface of the lactose-PEG 200 composites, the aforementioned provided a rationale behind the observed results.

In vitro aerosolization performance

Comparison of the blends containing carriers composed of α-lactose monohydrate evidenced the impact of particle morphology on DPI performance. More precisely, blends containing the spherical α-lactose monohydrate resulted in a poorer *in vitro* aerosolization performance of JMSS when compared to their tomahawk shaped counterparts. It is known that powder bed fluidization behavior can be affected by particle shape; for intance, spherical particles tend to pack more uniformly, resulting in powder beds with higher tensile strengths [6] and more difficult to fluidize *via* airflow. These powder beds are often lifted as plugs or fractures [7] that result in an inferior and variable lung deposition of fine drug particles.

Conclusion

API and carrier particles engineered by different routes yield distinct physicochemical properties. Assessment and understanding of particle properties and process interplay can enable tuning of DPI product performance. Thus, an integrated understanding of particle characteristics in relation to their processability and intended performance is necessary to develop DPI products capable of tailored drug delivery.

References

(1) Telko, M. J.; Hickey, A. J. Dry Powder Inhaler Formulation. *Respir. Care* **2005**, *50* (9), 1209–1227.

(2) Pinto, J. T.; Radivojev, S.; Zellnitz, S.; Roblegg, E.; Paudel, A. How Does Secondary Processing Affect the Physicochemical Properties of Inhalable Salbutamol Sulphate Particles? A Temporal Investigation. *Int. J. Pharm.* **2017**, *528* (1-2), 416–428.

(3) De Boer, A. H.; Chan, H. K.; Price, R. A Critical View on Lactose-Based Drug Formulation and Device Studies for Dry Powder Inhalation: Which Are Relevant and What Interactions to Expect? *Advanced Drug Delivery Reviews*. 2012, pp 257–274.

(4) Kumon, M.; Suzuki, M.; Kusai, A.; Yonemochi, E.; Terada, K. Novel Approach to DPI Carrier Lactose with Mechanofusion Process with Additives and Evaluation by IGC. *Chem. Pharm. Bull. (Tokyo).* **2006**, *54* (11), 1508–1514.

(5) Wong, L. W.; Pilpel, N. Effect of Particle Shape on the Mixing of Powders. *J. Pharm. Pharmacol.* **1990**, *42* (1), 1–6.

(6) Fukunaka, T.; Sawaguchi, K.; Golman, B.; Shinohara, K. Effect of Particle Shape of Active Pharmaceutical Ingredients Prepared by Fluidized-Bed Jet-Milling on Cohesiveness. *J. Pharm. Sci.* **2005**, *94* (5), 1004–1012.

(7) Kaialy, W.; Nokhodchi, A. Freeze-Dried Mannitol for Superior Pulmonary Drug Delivery via Dry Powder Inhaler. *Pharm Res* **2013**, *30* (2), 458–477.

526546 Spray-Dried Nanocomposites and Amorphous Solid Dispersions with Identical Formulation for Comparative Assessment of Drug Dissolution Enhancement

Abstract:

The number of new drug candidates coming out of drug discovery having poor aqueous solubility has increased within the last few decades. Due to the poor aqueous solubility, intestinal absorption of these drugs turns out to be rate-limiting, leading to low bioavailability. Over the years, significant academic and industrial researches have been carried out and directed toward developing various formulation/processing strategies for efficient solubilization/dissolution rate enhancement of poorly water-soluble drugs. Among different approaches, two major platform approaches have gained prevalence both in the scientific literature and marketed products: nanoparticle-based dosage forms (nanocomposites) and amorphous solid dispersions (ASDs) [1]. Both nanoparticle-based dosage forms and amorphous solid dispersions (ASDs) can improve the dissolution rate and bioavailability due to very high surface area and higher saturation solubility, respectively, compared to the micron-sized crystalline drug particles. Although both drug nanocomposites and ASDs have been studied for decades, no head-to-head comparison of these two forms of drugs having identical formulation is available in the literature. Using a novel nanoextrusion process, a comparative assessment of relative dissolution enhancement capability of nanocomposites vs. ASDs was performed [2]. However, the nanoextrusion process requires the use of different polymers to produce nanocomposites or ASDs of a given drug via differing drug–polymer miscibility/interactions; thus, it is usually infeasible to produce nanocomposites and ASDs having identical formulation. Therefore, a direct and scientifically fair head-to-head comparison of drug dissolution performance of nanocomposites vs. ASDs having identical formulation is still lacking in the literature.

In this study, we aim to assess the dissolution enhancement of griseofulvin (GF, a poorly water-soluble drug) from nanocomposites vs. ASDs, both having identical formulations. GF nanocomposite was prepared by wet media milling (nanomilling) an aqueous GF suspension followed by spray drying. Two polymers, hydroxypropyl cellulose (HPC) and Soluplus®, with an anionic surfactant, sodium dodecyl sulfate (SDS), were used for the stabilization of drug nanoparticles. ASDs were prepared by spray-drying a solution of the respective drug–polymer in acetone–deionized water. Laser diffraction, XRPD, Raman spectroscopy, polarized light microscopy, and dissolution testing were used for characterization. XRPD confirmed that for both polymers, spray-drying of the aqueous nanosuspension led to drug nanocrystals dispersed in the polymeric matrix (nanocomposite), whereas that of the drug solution led to ASD. At supersaturating dissolution conditions (100 mg GF dose), both nanocomposites and ASDs significantly improved the dissolution performance as compared with the as-received GF and physical mixtures. For GF–Soluplus®, ASDs exhibited higher supersaturation generation than nanocomposites, whereas for GF–HPC, both dosage forms of the drug exhibited only slight supersaturation. Interestingly, regardless of the solid-state of the drug, higher extent of supersaturation was observed in the dissolution medium for GF–Soluplus® than GF–HPC. Overall, this study demonstrates how to produce both nanocomposites and ASD of the same drug with identical formulation using spray dryer as a tool and to perform a comparative assessment of their dissolution performance.

References:

[1] Brough, C., Williams, R., 2013. Amorphous solid dispersions and nano-crystal technologies for poorly water-soluble drug delivery. Int. J. Pharm. 453(1), 157–166.

[2] Li, M., Ioannidis, N., Gogos, C., Bilgili, E., 2017. A comparative assessment of nanocomposites vs. amorphous solid dispersions prepared via nanoextrusion for drug dissolution enhancement. Eur. J. Pharm. Biopharm. 119, 68–80.

537497 Fabrication of Biodegradable Rod-Shaped Drug Carriers with Modified Two-Step Emulsion Solvent Evaporation Technique

Abstract:

There has been growing interest in recent years towards utilization of non-spherical particles for drug delivery applications. Rod-shaped particles have demonstrated reduced uptake by immune cells, increased circulation time and enhanced margination and binding to the vascular wall making them favorable compared to their spherical counterparts. Oil-in-water emulsion solvent evaporation, a simple and scalable technique for fabricating particulate carriers, has recently been applied to fabricate poly(lactic-co-glycolic) acid (PLGA) spheroids. However, this technique has been unable to generate small-sized rods applicable for usage in drug-delivery.

In this study, we have developed a two-step fabrication protocol enabling stretching of smaller sized droplets by separating the droplet formation and stretching steps and inducing a sudden increase in the viscosity and capillary number of the system. Evaluating different surface-active molecule and oil phase solvents demonstrated that sodium metaphosphate and chloroform are respectively the optimum choices for maximizing the fabrication yield of PLGA rods. With our optimized two-step fabrication protocol, we were able to generate PLGA rods with major and minor axis sizes as small as 3.2 µm and 700 nm. Nile Red loaded rods conjugated with targeting antibodies fabricated with our modified protocol were able to successfully bind to inflamed endothelium under shear flow. The *in vitro* and *in vivo* binding capability of the rods were evaluated and compared to spheres to demonstrate the effect of shape on the performance of biodegradable drug carriers.

532919 Surface Engineering of Lactose Particles By Atomic Layer Deposition for Modified Release

Abstract:

Atomic layer deposition (ALD) is an established technique for the synthesis of thin films for various applications ranging from semiconductors to energy storage devices. Recently, it has been gaining attention in the pharmaceutical field to modify the particle properties, for example the drug release. Compared to the conventional methods of drug particle coating, ALD has a number of advantages: control over the amount of deposited material, conformality, and its solventless nature. A few studies about ALD of metal oxides, including Al_2O_3, TiO_2 and ZnO, on acetaminophen, lactose and budesonide particles showed that the deposition of nanoshells can effectively slow down the drug release [1, 2, 3]. However, the dependence of the film growth on pharmaceutical particles on the ALD operating conditions has not yet been investigated. Furthermore, the relationship between the properties of the coating (e.g, uniformity, conformality) and the dissolution behaviour of the resulting core-shell structures is not clear. Therefore, as the dissolution profile strongly depends on the coating quality, understanding how process parameters (e.g., number of cycles, co-reactants and substrate surface) affect the coating uniformity and conformality is crucial.

In this work, we study the effect of co-reactants, namely H_2O and O_3, on the Al_2O_3 growth on two kinds of lactose particles (i.e., fully crystalline lactose and crystalline lactose with micronization-induced amorphous surfaces) and evaluate the dissolution behaviour. The ALD process is carried out at ambient conditions in a fluidized bed reactor for a low number of cycles (i.e., from 4 to 14) using trimethylaluminum (TMA) as a metal precursor. Time-of-flight secondary ion mass spectrometry and Transmission Electron Microscopy (TEM) show that TMA/O_3 ALD on crystalline lactose particles offers greatly improved control over the coating uniformity and conformality compared to TMA/H_2O ALD. In fact, by causing severe agglomeration of the particles, water deteriorates the fluidization quality and thus the homogeneity of the coating process. In-vitro dissolution tests reveal more sustained release for the O_3-based process than for the H_2O-based one, thus underlining the benefit of O_3 in providing uniform and conformal coatings. However, little to no difference between TMA/O_3 and TMA/H_2O ALD is observed on lactose particles with amorphous surfaces. In fact, such amorphous surfaces firstly are inherently inhomogenous across the particles and secondly present micropores, as suggested by BET measurements. This inevitably leads to non-conformal films regardless of the co-reactant and moreover results in both surface and subsurface growth.

[1] Tommi O. Kääriäinen, Marianna Kemell, Marko Vehkamäki, Marja-Leena Kääriäinen, Alexandra Correia, Hélder A. Santos, Luis M. Bimbo, Jouni Hirvonen, Pekka Hoppu, Steven M. George, David C. Cameron, Mikko Ritala, Markku LeskelÃ¤, Surface modification of acetaminophen particles by atomic layer deposition, International Journal of Pharmaceutics, 525 (1), 2017, 160-174

[2] Joel Hellrup, MÃ¥rten Rooth, Anders Johansson, Denny Mahlin, Production and characterization of aluminium oxide nanoshells on spray dried lactose, International Journal of Pharmaceutics, 529 (1â€"2), 2017, 116-122

[3] Di Zhang, Mike J. Quayle, Gunilla Petersson, J. Ruud van Ommen and Staffan Folestad, Atomic scale surface engineering of micro- to nano-sized pharmaceutical particles for drug delivery applications, Nanoscale, 2017, 9, 11410-11417

531907 Improving Blend and Tablet Properties of Binary Mixtures Containing Cohesive and Poorly-Compactable APIs Using Surface Engineered MMC Based Fine Excipients

Abstract:

Improving blend and tablet properties of binary mixtures containing cohesive and poorly-compactable APIs using surface engineered MMC based fine excipients

*Liang Chen, Zizhou He, Siqi Fan, Xiaoyi Ding, Rajesh N. Davé**

Abstract

High functional excipients with good flow, high packing density, and excellent compaction have been developed via dry particle coating in recent work [1]. Even though these new fine grades of excipients demonstrate high functional properties in placebo tablets compared to other silicified excipients such Prosolv 50, and Prosolv 90, it is questionable that these high functional properties will preserve in blends with cohesive and poorly-compactable active pharmaceutical ingredients (APIs). The objective of this presentation is to study the effect of MCC-based, fine engineered excipients on improving the properties of blends and tablets in binary mixture with cohesive and poorly compactable APIs. Ibuprofen 50, and two grades of Acetaminophen were selected as model APIs and Avicel PH-105 dry coated with 1wt% hydrophilic silica A200 was used as the engineered excipient in preparing binary mixtures at three different drug loadings (10%, 30%, 60%). The functional properties of the blends were compared with the blends made with as received Avicel PH-105, Prosolv 50 and Prosolv 90, respectively. The results indicated that as expected, the bulk density and flow of binary mixtures of engineered excipient, dry coated Avicel PH 105, were significantly improved as compared with as received Avicel PH-105 while there was a slightly decreased tablet strength at lower drug loadings. Most interestingly, at higher drug loading, there was an improvement in tablet strength. When compared with commercial high functional excipients, Prosolv 50 and Prosolv 90, at low drug loadings, the surface engineered excipients still exhibited a better or comparable bulk density, flow properties, and tablet strength. However, as drug loading increased commercial silicified excipients showed adverse effect on blends and tablet properties. In contrast, the surface engineered excipients demonstrated significant improvement in all three desired properties, suggesting improved manufacturability. Overall, it is hoped that these new very fine surface modified engineered excipients will promote high-speed direct compaction tableting of cohesive APIs at relatively high drug loadings.

1. Chen, L., et al., *Surface engineered excipients: I. improved functional properties of fine grade microcrystalline cellulose.* International Journal of Pharmaceutics, 2018. **536**(1): p. 127-137.

548153 Fast, Immediate Drug Release from High Drug-Loaded Spray-Dried Nanocomposites: Criticality of Polymer Molecular Weight

Abstract:

Owing to their large surface area and supersaturation capability, drug nanoparticles prepared via wet stirred media milling in the form of nanosuspensions have been widely used to address inadequate bioavailability of poorly water-soluble drugs [1,2]. Particle size growth and aggregation occur in drug suspensions during wet media milling, storage, and downstream processing, and the extent of such phenomena must be reduced as they reduce drug surface area available for dissolution [1]. To prevent gross physical instability, drug nanosuspensions are usually dried into nanocomposites that are then incorporated into standard solid dosage forms such as tablets and capsules [2]. This approach also helps to achieve patient compliance because solid oral dosage forms incorporating drug nanoparticles are preferred over parenteral/injectable–oral drug nanosuspensions.

In this study, we examine the impact of various polymers/MW on the stabilization of wet-milled ITZ suspensions and ITZ dissolution from spray-dried suspensions (nanocomposites). To this end, aqueous ITZ suspensions containing HPC with SSL, SL, and L grades having MW of 40, 100, and 140 kg/mol, respectively, HPMC (E3 grade with 10 kg/mol MW), PVP (K30 grade with 50 kg/mol MW), SDS, and HPC SL–SDS were wet-milled in a high-energy stirred mill. The milled suspensions were spray-dried to prepare ITZ-laden nanocomposites. Laser diffraction, viscometry, and scanning electron microscopy were used to characterize the milled suspensions, while optical microscopy, differential scanning calorimetry (DSC), X-ray powder diffraction (XRPD), and UV-spectroscopy were used to characterize the nanocomposite powders. Using the modified Washburn method, drug wettability enhancement by the polymers was investigated. Redispersion of the nanocomposites was examined by off-line analysis via laser diffraction and optical microscopic imaging. Dissolution of the nanocomposites was performed in a USP II paddle apparatus. By calculating the difference and similarity factors and fitting the dissolution data via the Korsmeyer Peppas model, initial ITZ release kinetics was examined.

This study has demonstrated that wet-milled stable 10% ITZ nanosuspensions showing near-Newtonian flow behavior can be prepared with 4.5% HPC SL/L (100 and 140 kg/mol MW, respectively). At 4.5% concentration, HPC SSL (40 kg/mol), HPMC E3 (10 kg/mol), and PVP K30 (50 kg/mol) could not suppress ITZ nanoparticle aggregation, leading to significant pseudoplastic behavior. Spray drying of the ITZ suspensions yielded nanocomposites with 60–78% mean ITZ loading, which is higher than that in the ITZ nanocomposites produced in prior studies, and acceptable content uniformity. Severe aggregation occurred during the milling/drying when 4.5% polymers with MW £ 50 kg/mol were used. Their nanocomposites did not redisperse into ITZ nanoparticles/aggregates due to negligible/slow matrix erosion in the redispersion test; thus, they did not exhibit immediate release during the dissolution test. While the use of higher MW HPC (100 and 140 kg/mol) was more favorable from both nanosuspension stabilization and ITZ release perspectives, there exists an optimal MW; the fastest ITZ release among the nanocomposites with 4.5% polymer was achieved by HPC SL (100 kg/mol).

References:

1. Li, M. Azad, R. Davé, E. Bilgili, Nanomilling of drugs for bioavailability enhancement: a holistic formulation–process perspective, Pharmaceutics, 8 (2016) 27.
2. Bhakay, M. Rahman, R. N. Dave, E. Bilgili, Bioavailability enhancement of poorly water-soluble drugs via nanocomposites: Formulation processing aspects and challenges, Pharmaceutics, 10 (2018) 86.

533910 Formulation of Microcarriers for Levodopa Delivery Via the Pulmonary System

Abstract:

It is estimated that as many as 10 million people worldwide are suffering from Parkinsonâ€™s Diseases (PD). Levodopa is currently one of the most effective drug to treat PD patients. While oral delivery of Levodopa is the most common administration method, a significant portion of Levodopa is metabolized by the gastrointestinal enzyme and less than 30% of the drug is retained in the body. This study aims to develop Levodopa in inhalable solid dosage forms by incorporating into polymeric microparticles. Inhalation delivery of Levodopa can help reduce the tendency of choking among elder PD patients. As-prepared particles will be characterized by X-ray diffraction spectra (XRD), differential scanning calorimetry (DSC), scanning electron microscopy (SEM), thermal gravimetric analysis (TGA) and Andersen Cascade Impactor (ACI).

534631 Surface Modification to Improve Drug-Excipient Mixing for DPI Formulation

Abstract:

Recent advances in inhaled pulmonary and anti-microbial therapies using dry powder inhaler (DPI) has the limitation of blend homogeneity and uniform dose due to the cohesive nature of very fine drug particles. This study emphasizes the influence of API-excipient interaction on physical mixing for DPI formulation. Aceclofenac and lactose monohydrate are used for preparing blends by varying drug loading (1, 5, 10, 14, 20, 33 wt. %). Some of the blends are further modulated by surface modification of lactose with 1 wt. % nano-silica (R972 type). The quality of mixing is predicted by surface energy approach using work of adhesion/cohesion and Mixing Index (M.I.) is calculated using Reverse Phase-High Performance Liquid Chromatography (RP-HPLC). The morphology of drug, excipient and their blends were observed by Field Emission Scanning Electron Microscopy (FESEM). The flow property of the blend was determined using FT4 Rheometer. While the crystal level changes were observed using Powder X-Ray Diffraction (P-XRD), the physicochemical interaction between aceclofenac and lactose was confirmed by Fourier Transform Infrared Spectroscopy (FTIR), Differential Scanning Calorimetry (DSC). The results show that M.I. is high for blends with high work of adhesion. Interestingly for all blends without any surface modification shows that there is a dip in M.I.when drug loading is in the range of 10 to14 wt. %. Also, Liquid Chromatography-Mass Spectroscopy (LC-MS) results confirm the presence of formation of the condensed compound. The surface coating over the lactose changes the inter-particle interactions. The results of coated blend showed improved mixing along with high flow function coefficient which is effective for pulmonary drug delivery. Moreover, there is an indication that the blends with surface modified lactose are more stable due to partial attachments of nano-particles on some of the most active sites on the lactose surface.

537256 Impregnation of Catalysts with Viscous Metal Solutions Using Experiments and DEM Simulations

Abstract:

Dry catalyst impregnation of active metals onto a porous catalyst support is an important step in the preparation of heterogeneous catalyst. In a typical dry impregnation process, metal solutions are sprayed over a particulate bed in a mixing vessel until the pore volume is reached. The inter-particle variability of the impregnated liquids inside the particles and metal content significantly affects the activity and selectivity of the resulting catalyst. In this process, metal salts or complexes are dissolved in a typically aqueous solution and added to porous catalyst support particles. The aim of this work is to understand how the viscosity of the metal solution affects the dry impregnation process.

We studied impregnation of Alumina particles with viscous solutions consisting of different concentrations of Nickel Nitrate and Polyvinyl alcohol (PVA, 80% hydrolyzed). PVA solution was used as viscous liquid in our experiments. The viscosities of the PVA solution with different concentrations were measured by a Brookfield DV3T viscometer. Results show that the viscosity of PVA solution is increasing with the increasing concentration.

Langmuir isotherm can be applied to illustrate the adsorption of metal in a viscous solution, where K_L is the equilibrium constant and q_{max} is the maximum adsorption capacity. The two parameters used in the Langmuir model (K_L and q_{max}) need to be determined for the adsorption of a viscous solution on alumina. A series of adsorption experiments were performed for different PVA concentrations. The metal concentration at equilibrium (C_e) was measured by UV-Vis Spectroscopy, and the amount of metal adsorbed (q_e) was calculated using a mass balance. Plot $1/C_e$ vs. $1/q_e$, a straight fitting line was obtained and the two Langmuir equilibrium parameters were derived from the intercept and slope for different viscosities. Experimental results show that the viscosity slows down the impregnation process. The higher the viscosity the longer it takes to reach the equilibration state. The metal absorbed in the pores depends strongly on the viscosity of the solution. The higher the viscosity the smaller the amount of metal is absorbed. We also found the viscosity of the solution affects the Langmuir equilibrium parameters K_L and q_{max}, which decrease with increasing viscosity.

In addition DEM simulations incorporating these parameters for a viscous solution were considered to reveal the amount of metal absorbed and the uniformity of the particle bed. Using the Langmuir model coupled with the mass balance, the total amount of metal absorbed on solid and the amount metal in the liquid phase can be calculated. Our simulation results show good agreement with the experiments.

531644 Understanding the Pressure Loss in Wyes for Dust Collection and Vacuum Cleaning Systems

Abstract:

The wye is a common gas/solids flow element in both dust collection systems and vacuum cleaning systems. They allow the branching of systems which permits multiple pick-up points. This is especially important in dust collection systems where there can be dozens of pick up points. The literature is reasonably complete for tees and a number of particular cases of wyes. These include the cases of the following: a) a simple 90Â° degree tee or a symmetric wye where all branches are of equal cross section; b) one of the branches is the same diameter as the straight downstream section or â€œrunâ€; c) the sum of the cross sectional areas of the branches is equal to the run cross sectional area; and d) the sum of the cross sectional area of the branch is less than that of the run. The situation of where the sum of the areas of the branches is greater than the run has not been explored adequately. In this work, we examine the latter using a combined experimental, theoretical and computational approach to calculate the pressure loss across these types of flow elements.

539016 Limiting Flow Rate of Fine Powders through Hoppers â€“ Investigation of Methods to Increase the Flow Rates

Abstract:

The discharge rate of fine powders through a bin or hopper tend to be much smaller compared with the famous Beverloo equation (which is applicable for coarse bulk solids). This is mainly due to the high compressibility and low permeability associated with fine cohesive bulk solids. Based on our previous work to model the two-phase flow effects of powder flow in cylindrical and conical hoppers, methods to increase the flow rate are investigated in the present work. These include air injection systems within the hoppers or installing standpipes below the hopper. Theoretical predictions of gas pressure profiles within the standpipes as a function of flow rate will be presented in this paper.

538581 Prediction of Loss-in-Weight Screw Feeder Performance and Quantification of Failure Modes from Attribute Measurements of Pharmaceutical Materials

Abstract:

Many pharmaceutical processes such as continuous blending of active ingredients with excipients rely on highly accurate, robust, and tunable loss-in-weight feeders for control of composition and associated quality attributes. The ability of such feeders to accommodate the wide variety of material properties inherent in formulation components is therefore an important process design constraint. It is desirable to predict throughput accuracy and performance robustness and identify potential feeder failure modes in the absence of long-duration experimental runs, which may consume excessive quantities (10s to 100s of kg) of potentially high value ingredients.

To this end, we have conducted empirical studies to assess feeder throughput accuracy for commercially available loss-in-weight feeder configurations. A variety of components spanning a range of cohesivities, compressabilities and yield stresses have been assessed. Feed failure modes indicative of process robustness have been categorized and statistically quantified to develop predictive relationships with lab measurements of materials properties performed with gram-scale quantities. These include particle size distribution, bulk and tapped density; permeability, morphology and flowability (as assessed by both direct arching tests and by shear cell measurements of flow function and angle of internal friction). Along with equipment geometric descriptors, performance envelopes are defined. Results are evaluated against theoretical considerations and guidance around practical strategies for predictive assessment of feeder performance is outlined.

528270 The Use of Fine Excipients to Improve the Manufacturability of Pharmaceutical Tablets

Abstract:

Tablets comprise the majority of pharmaceutical products for oral solid dosage forms in part due to their low cost of manufacturing, convenience of dosing to patients, and high stability. Tablets are typically produced by compacting powders or granules in a die using a rotary press. The formulation intended for tabletting must have adequate flow properties to consistently fill a tabletting die from a hopper and must also have sufficient compression properties to form a tablet with the mechanical strength to undergo subsequent coating, packaging, and handling. Because the drug substance is often poorly flowing and poorly compactable, it is necessary to add fillers, glidants, lubricants, or other excipients to obtain a manufacturable formulation.

The recent work of Capece et al. [1] showed that the tabletability of formulations can be significantly improved by using fine sized excipients (20 Î¼m) while also maintaining adequate flowability. This novel approach is contrasted by the common and conventional formulation strategy of using coarse sized excipients which usually have a mean size of about 100 Î¼m. Due to the greater number of high strength bonds provided by contacts between fine excipients, tablet tensile strength is increased even at high drug loadings (>50%). In addition, the use of fine excipients can also enable the use a direct compaction process.

While adequate flowability and tabletability are critical to attain during development for all formulations intended for tableting, some drug substances may also have a propensity to adhere to the tableting punch. So-called sticking can result in cosmetic defects or can prevent manufacturing altogether for an otherwise acceptable formulation. This work further investigates the use of fine excipients to improve the manufacturability of tablets with a focus on sticking. The effect of various sizes of a common excipient, microcrystalline cellulose, on the sticking propensity of a proprietary pharmaceutical active is investigated. To quantify sticking propensity, a removable-tip punch is used to quantify the mass of material adhered to the punch surface. The work shows that fine sized excipients can reduce and prevent sticking while also improving tabletability. This is due to the reduction of surface area of the sticky drug substance relative to the non-sticky excipient as well as the increase in the number of high strength bonds in the formulation. The reduction in sticking and the increase in tabletability give further credence to a formulation strategy that uses fine sized excipients in place of more commonly used coarse sized excipients.

[1] M. Capece, Z. Huang, R. Davé. Insight into a novel strategy for the design of tablet formulations intended for direct compression. J. Pharm. Sci. 106 (2017) 1608-1617.

All authors are employees of AbbVie. The design, study conduct, and financial support for this research was provided by AbbVie. AbbVie participated in the interpretation of data, review, and approval of the publication.

537944 DedusterÂ® : The Leading Dust and Streamer Removal System

Abstract:

During dilute phase conveying of granulated plastic material, high velocity and friction leads to the generation of dust and streamers. Dust and streamers if not separated prior to downstream processing leads to defects in final products. A Pelletron DeDuster® is designed specifically for plastics granulated material and can remove short and long streamers and electrostatically charged dust from the pellets in a single step. In this work LDPE material containing dust and streamer was used to evaluate the separation performance of DeDuster®.

539633 Improved Combustible Dust Minimum Ignition Energy (MIE) Test Method and Prediction Using CFD Simulation

Abstract:

Dust explosions continue to be a persistent problem in the process industries due to recurring incidents. The combustible dust Minimum Ignition Energy (MIE) is an important dust hazard parameter that guides the elimination of possible ignition sources in solids handling facilities. Partial inerting is an important but underutilized mitigation technique in which the dust cloud MIE is increased through inerting, reducing the risk of an accidental explosion or more accurately, a dust deflagration. The MIKE3 is the predominant apparatus for combustible dust MIE measurement worldwide. The current version of the MIKE3 device is not specifically designed to measure partial inerting MIE accurately. Therefore, a primary objective of this work is to demonstrate that a properly designed add-on device and technique can accurately produce partial inerting MIE results with the MIKE3 or any other similar device. In the process industries, combustible dust MIE prediction could result in considerable time and cost savings during process development. The existing MIE prediction models for dusts are known to deviate significantly from experimental findings. Current MIE-O_2 prediction models in literature consider critical equation terms to be a constant.[1,2] This study demonstrates these critical terms are not a constant, but are linked to the type of combustible dust.

The add-on device developed in this work ensures complete purging of the MIE apparatus with the desired gas concentration before experimentation. The same gas is then used to make the dust cloud dispersions for ignition testing. This approach results in a very uniform gas concentration which is essential for producing proper measurements. Therefore, an important finding of this work is that purging before partial inerting MIE testing results in a proper characterization of the relationship between the MIE and oxygen for the dust. This work has also explored the influence of purge time on partial inerting MIE measurements. Oxygen sensor measurements were used to determine the purge time required to achieve the desired composition. It was observed that the apparatus should be purged for > 40 seconds. Additionally, an ANSYS Fluent CFD model was developed that supports the experimentally determined purge time. The CFD model also revealed ways to further improve the initial add-on device. These findings demonstrate the need to amend existing or develop new standards for this type of dust testing.

The performance of the add-on device and validity of the techniques are demonstrated through the experimental determination of partial inerting curves for Niacin (CaRo15), Anthraquinone, Lycopodium clavatum and Calcium Stearate using the MIKE3 apparatus. A mathematical model for the MIE-O_2 relationship was proposed and compared with the existing models in literature. The proposed model led to more accurate prediction of the MIE-O_2 experimental relationship. Therefore, through both experimentation and modeling, this study aimed at providing a scientific foundation for a partial inerting MIE test method to supplement existing testing standards such as ASTM E2019-03.

542620 The Impact of Shale Gas and Oil on the Chemical Industry

Abstract:

Lower alkanes and aromatics are very useful as both fuels and chemical raw materials. Occasionally supply disruptions constrained availability but also encouraged investigation of alternative feedstocks and chemistries. Recent new technologies for the production of unconventional gas and oil sources from tight shale formations significantly increased availability of both methane and condensate, radically altered process economics, resulted in the construction of much new intermediates capacity, and multiplied the ways traditional and alternative chemistries might be exploited. This presentation will outline historically how natural gas components became useful as chemical feedstocks, how government regulations occasionally led to perturbations in availability and price, how supply crises led to the development of alternative chemistries, catalysts, and ultimately the exploitation of shale resources. More specifically, will be discussed how excess coproducts, especially condensate, from increasing shale gas and oil production for energy are having significant unexpected impacts on the chemical process industries and opportunities for new process chemistry research.

541186 Sustainable Energy and Chemicals: Past, Present, and Future

Abstract:

The last 40 years has seen the U.S. energy situation vacillate from gasoline rationing and concerns over Peak Oil, to projected energy and feedstock prices that were noncompetitive in global markets for chemicals production, to the economic revolution of frack gas and light tight oil. The importance of low cost energy and feedstocks in driving economic opportunity will be discussed relative to this history, along with the importance of sustainability in decision making to gain stakeholder buy-in and acceptance. Strategies for managing career paths and competencies despite changing targets and high uncertainty will be addressed.

542618 Disruptions: What the Future May Hold

Abstract:

Disruptive forces are all around us. They change the way we communicate and work. They change the way we develop goods and services. They also change the types of products that are manufactured. Examples such as the way in which we produce and consume energy or the feedstocks we use to produce product have seen many changes over the course of the last few decades, and these disruptions are expected to continue. The chemical engineering discipline and the expectations for a chemical engineerâ€™s skills has also changed and adapted as a result of these disruptive forces. A discussion of the broad themes associated with these forces and the implications that they may have to chemical engineers in the future will be discussed.

542621 Geopolitical Factors Influencing the Evolution of the Chemical Industry

Abstract:

The basic chemicals and plastics industry is built on a simple principle: find a cheap feedstock and convert it into something more valuable. However, the right feedstock depends on place and time. Certain policies and actions of various governments have influenced the evolution and profitability of the chemical industry during the past 50 years, sometimes in surprising ways. For example, high oil prices and US government funding encouraged the development and commercialization of coal gasification technology during the late 1970s and early 1980s. This enabled China to develop a competitive basic chemical industry using coal as a feedstock during the last 25 years. Similarly, an abrupt increase in oil prices during the first decade of the 21st century encouraged the present shale gas revolution, causing a major shift in the economy of producing basic chemicals and polymers in the US. If history is our guide, then what governments choose to fund and regulate may be as important as market forces in determining the shape of our industry and profession in the future.

542622 Agility & Resilience: How to Maintain Career Competitiveness in the Changing Chemical Industry

Abstract:

The world around is changing. The shift to protectionism, global instability and price volatility of raw materials and feedstocks has the potential to wreak havoc on the chemical industry. With the revolution of shale gas in North America and the sudden drop in feedstock costs, the chemical sector in the US has reasserted its historic dominance. Learn how the role of a chemical engineer can contribute to industry growth and resilience, while adapting to allow for personal agility throughout the career life cycle.

526549 Griseofulvin-Laden Extrudates Prepared Via Nanoextrusion: Impact of Dry-Milling on Dissolution Enhancement

Abstract:

Extrusion processes, including the traditional hot melt extrusion (HME) and a relatively novel nanoextrusion [1–4], have been used to enhance the dissolution rate and bioavailability of poorly water-soluble drugs. Nanoextrusion has been used as a continuous dryer to convert a wet-milled drug nanosuspension along with an extrusion polymer into nanocomposites [1,2]. During the process, the drug nanoparticles are dispersed in the polymeric matrix while water evaporates, which yields extrudates in the form of nanocomposites. Being a continuous process and having the capability to handle viscous drug suspensions, the nanoextrusion process is capable of preparing extrudates with good content uniformity even at very low drug loading [3]. In a recent investigation, Li et al. [4] used the same nanoextrusion process as a platform to produce extrudates either in the form of nanocomposites or amorphous solid dispersions (ASDs), thus enabling comparative assessment of nanocomposites vs. ASDs in a scientific manner.

The traditional HME process or the nanoextrusion process is usually followed by milling of the extrudate threads into powders for further downstream processing during the manufacture of final solid dosages forms. Modulating the particle (matrix) size via dry-milling of extrudates potentially allows for manipulating the drug dissolution performance because matrix size could affect drug release from the extrudates significantly [4]. In the majority of previous investigations where extrudates in the form of ASDs were produced via HME, the extrudates were dry milled and sieved into a *single size fraction*. For example, Fule et al. [5] prepared ASD of artesunate in the matrices of both Soluplus® and Kollidon® VA64, and the extrudates were milled and passed through a 200 µm sieve for various characterizations. Similarly, Juluri et al. [6] produced ASDs of griseofulvin and caffeine anhydrous in the matrix of kleptose linecaps DE17, and the milled extrudates between #40 and #50 sieves (297–420 µm) were collected and characterized. Pudlas et al. [7] produced the ASD of ibuprofen in the matrices of Soluplus® and copovidone, and the milled extrudates were sieved to exclude particles larger than 250 µm. Javeer and Amin [8] produced the ASD of carbamazepine in the matrix of hydroxypropyl methyl cellulose, and the extrudates were milled and passed through a #40 sieve with opening size of 420 µm. Also, Nagy et al. [9] produced the ASD of spironolactone in the matrix of Soluplus® and the milled extrudate was passed through a 300 µm sieve. Only in a limited number of studies, the milled extrudate particle size was actually measured. For example, Martinez-Marcos et al. [10] produced the ASD of albendazole in the matrix of polyvinylpyrrolidone. After dry milling and sieving the D_{10}, D_{50}, and D_{90} of the extrudate particles were reported in the range of 8.6 33.9 µm, 88.9 197.4 µm, and 425.1 463.6 µm, respectively. As exemplified above, in the majority of the published literature, the actual extrudate particle (matrix) sizes in the sieved fraction were not measured or reported. More importantly, the impact of matrix size from various sieved size fractions has not been examined and related to the specific surface area of the respective size fractions. Moreover, the impact of the extrudate matrix size on drug dissolution has not been studied. Similarly, when nanocomposites were produced by nanoextrusion [1–3], the impact of matrix size has not been thoroughly investigated. Hence, it is fair to assert that the impact of particle (matrix) size of the milled extrudates on drug dissolution has been studied thoroughly neither for nanocomposites nor for ASDs produced by an extrusion process, especially by the nanoextrusion process, which is the focus of this work.

The objective of current study is to examine the impact of matrix size, drug loading, and polymer–drug interactions, which modulate the solid state of the drug, on *in vitro* drug release from extrudate powders prepared via nanoextrusion. Soluplus® (Sol), Kolliphor P407 (Kol), and Hydroxypropyl cellulose (HPC) along with sodium dodecyl sulfate (SDS) were used to stabilize wet-milled griseofulvin (GF: drug) suspensions and form matrices of the extrudates. The wet-milled drug suspensions along with additional polymer (Sol/Kol/HPC) were fed to a co-rotating twin-screw extruder, which dried the suspensions and formed various extrudates. The extrudates with various polymer formulations and two target drug loadings (2% and 10% w/w) were dry-milled and sieved into three size fractions to examine the impact of matrix size. The drug nanosuspensions were characterized via laser diffraction, while the extrudates and their milled powders were characterized by laser diffraction, thermogravimetric analysis (TGA), scanning electron microscopy (SEM), and X-ray powder diffraction (XRPD). Digital microscopy was used to visualize the changes of different polymeric matrices when exposed to water. Drug wettability enhancement by Sol, Kol, and HPC solutions was studied using the modified Washburn method. Two drug doses, i.e., 8.9 mg and 100 mg, were used to investigate the dissolution response under non-supersaturating and supersaturating conditions in the bulk dissolution medium. It is hypothesized that extrudate matrices with smaller sizes can dissolve faster and that ASDs show a markedly and qualitatively different matrix size dependence of dissolution rate than nanocomposites under both dissolution conditions. Moreover, we assess the drug dissolution from ASDs vs. nanocomposites, which was modulated by various drug–polymer interactions. The impact of matrix size on GF dissolution from extrudate powders was significant especially for ASDs. We have found that the matrix size dependence of ASDs vs. nanocomposites is markedly and qualitatively different. Matrix size of ASDs can be modulated to control drug dissolution rate from ASDs; however, such modulation may not be very effective for nanocomposites that exhibit a weaker dependence of drug dissolution on matrix size. Overall, this study will provide significant insights to formulators about the impact of matrix size and differing dissolution response from nanocomposites vs. ASDs.

References:

1. R. Baumgartner, A. Eitzlmayr, N. Matsko, C. Tetyczka, J. Khinast, E. Roblegg, Nano-extrusion: A promising tool for continuous manufacturing of solid nano-formulations, Int. J. Pharm., 477 (2014) 1 11.
2. R. Baumgartner, J. Matić, S. Schrank, S. Laske, J. Khinast, E. Roblegg, NANEX: Process design and optimization, Int. J. Pharm., 506 (2016) 35 45.
3. X. Ye, H. Patil, X. Feng, R.V. Tiwari, J. Lu, A. Gryczke, K. Kolter, N. Langley, S. Majumdar, D. Neupane, Conjugation of hot-melt extrusion with high-pressure homogenization: a novel method of continuously preparing nanocrystal solid dispersions, AAPS PharmSciTech, 17 (2015) 78 88.
4. M. Li, N. Ioannidis, C. Gogos, E. Bilgili, A comparative assessment of nanocomposites vs. amorphous solid dispersions prepared via nanoextrusion for drug dissolution enhancement, Eur. J. Pharm. Biopharm., 119 (2017) 68 80.
5. R. Fule, V. Paithankar, P. Amin, Hot melt extrusion based solid solution approach: Exploring polymer comparison, physicochemical characterization and in-vivo evaluation, Int. J. Pharm., 499 (2016) 280 294.
6. A. Juluri, C. Popescu, L. Zhou, R.N. Murthy, V.K. Gowda, C.K. P, M.B. Pimparade, M.A. Repka, S.N. Murthy, Taste Masking of Griseofulvin and Caffeine Anhydrous Using Kleptose Linecaps DE17 by Hot Melt Extrusion, AAPS PharmSciTech, 17 (2016) 99 105.
7. M. Pudlas, S.O. Kyeremateng, L.A.M. Williams, J.A. Kimber, H. van Lishaut, S.G. Kazarian, G.H. Woehrle, Analyzing the impact of different excipients on drug release behavior in hot-melt extrusion formulations using FTIR spectroscopic imaging, Eur. J. Pharm. Sci., 67 (2015) 21 31.
8. S. Javeer, P. Amin, Solubility and dissolution enhancement of HPMC-based solid dispersions of carbamazepine by hot-melt extrusion technique, Asian Journal of Pharmaceutics, (2014) 119.
9. Z.K. Nagy, A. Balogh, B. Vajna, A. Farkas, G. Patyi, Á. Kramarics, G. Marosi, Comparison of

electrospun and extruded soluplus®-based solid dosage forms of improved dissolution, J. Pharm. Sci., 101 (2012) 322 332.

10. L. Martinez-Marcos, D.A. Lamprou, R.T. McBurney, G.W. Halbert, A novel hot-melt extrusion formulation of albendazole for increasing dissolution properties, Int. J. Pharm., 499 (2016) 175 185.

530395 the Effect of Inorganic Salt on Disintegration of Tablets with High Loadings of Kollidon® VA64-Based Amorphous Solid Dispersion

Abstract:

Amorphous solid dispersions (ASDs) can improve dissolution of many poorly soluble drugs, and potentially enhance bioavailability in oral formulations. While the inclusion of ASDs in tablet formulations is increasingly common, tablets with high ASD loadings are often associated with slow disintegration, which hinders the development of immediate-release formulations with high drug loading. This slowdown of disintegration is believed to come from gel formation of the polymer, a major constituent of the ASD, upon exposure to aqueous media. In this study, we use a model ASD, composed of a hydrophobic drug with Kollidon® VA64 (a polyvinylpyrrolidone/ vinyl acetate copolymer) and a non-ionic surfactant, to explore formulation options that prevent such disintegration slowdown. We found that the pH, the ASD loading, and quite interestingly, the inclusion of inorganic salts in the tablet, all play important roles in the dissolution of VA64, and hence the tablet disintegration dynamics. We demonstrated that certain kosmotropic salts, when added in the formulation outside of the ASD, can significantly accelerate tablet disintegration. Contrary to expectation, we showed that the rank order of these salts in their effectiveness to help disintegration does not exactly follow the Hofmeister series; in addition to ionic effects, it also depends on other factors like particle size and dissolution rate of the salts. This highlights the dynamic nature of the tablet disintegration process, which is not to be fully predicted by steady-state properties. From our results, we provided a mechanistic explanation of the disintegration process: fast-dissolving kosmotropic salt builds high local concentration inside the restrained tablet matrix, thus lowering the cloud point of the polymer and preventing it from quickly dissolving and gelling, thereby helping the tablet break apart. This fundamental understanding has enabled design of a new platform formulation that accommodates higher ASD loadings for immediate release of poorly-soluble drug compounds.

527812 Miniaturized Formulation and Processability Screening for the Rational Design of Ethylene Vinyl Acetate Based Co-Extrudates

Abstract:

Intra-vaginal rings (IVRs) present an alternative strategy for the sustained release of estradiol (E2) in the vaginal tract. The currently marketed IVRs comprising E2 are reservoir-type silicon-based systems. One of the disadvantages of silicon elastomers is the requirement of additional curing and post-curing steps during the manufacturing process. Ethylene vinyl-acetate (EVA) copolymers represent a versatile polymeric carrier and they have been successfully used in a marketed IVR for contraception (NuvaRing®). EVA-based reservoir-type IVRs do not require additional curing steps and are manufactured via co-extrusion. A rapid, material-sparing and predictive screening methodology for IVR product development is, therefore, desirable for the rational (pre) formulation assessment and early feasibility studies.

The first aim of this study is to implement and evaluate a recently developed screening procedure (i.e., vacuum compression molding, VCM) to facilitate formulation development of EVA-based co-extrudates. The VCM screening procedure enables the manufacturing of small-scale reservoir systems (referred here as VCM systems) which can predict the rate and mechanism of release of full-scale IVRs. The second aim is to investigate the feasibility of EVA-based reservoir-type IVRs with broad daily in-vitro release rates of E2, considering the effect of the EVA copolymer type and drug loading. The equilibrium solubility of E2 in various EVA copolymers at low temperatures (i.e., 25 and 37 °C) was determined using polymer films immersed in an E2-saturated aqueous solution until equilibrium was reached. The drug-saturated polymer films were subsequently analyzed via HPLC. Additionally, the metastable solubility of E2 in the same copolymers was estimated via hot-stage microscopy (HSM) in samples produced with the VCM tool. Small-scale reservoir systems (VCM systems) with different drug loadings and different core-skin EVA copolymer combinations were produced. To evaluate the predictability of the screening procedure, full-scale IVRs were prepared with hot-melt co-extrusion to elucidate the impact of process parameters (e.g., shear forces) on the E2 drug release. The average daily E2 amount release (per cm^2) from the IVR was higher by a factor of 1.154 ± 0.160 compared to the corresponding VCM systems, dependent on the drug core loading. This fact can be explained by the enhanced E2 solubility in the core polymer, due to the high shear forces during co-extrusion. Still, the average release rates per cm^2 were in the same order of magnitude. All formulations showed near zero-order release ($R^2 > 0.99$) over 28 days. It is well known that drug release from a reservoir-type system is a function of the core and skin polymer type, the drug loading and the solubility of the drug in the core polymer. Adaptation of these parameters, based on the outcome of the rational screening approach, resulted in average E2 release rates between 1.87 ± 0.65 and 21.01 ± 3.84 $\mu g/day/cm^2$.

The generated data show that the applied screening method can be used as a time- and cost saving procedure to predict and facilitate formulation design of IVRs prior to co-extrusion. Moreover, it was shown that a broad and constant E2 release from EVA-based reservoir-type systems can be achieved by optimizing formulation parameters, such as the polymer type and drug loading. IVR formulations with broad release rates of E2 could present a versatile strategy for sustained vaginal drug delivery expanding from topical and systematic menopausal symptoms treatment to contraception purposes.

535109 *Advances in Dissolution Modeling for Oral Dosage Forms with Amorphous Solid Dispersions*

Abstract:

We present further developments in a mathematical model of disintegration and dissolution of amorphous solid dispersion-based oral dosage forms. Amorphous Solid Dispersions (ASDs) of poorly water soluble active pharmaceutical ingredients (APIs) in hydrophilic polymeric matrices are a commonly used strategy to increase the solubility and dissolution rate of oral-dosage forms [1]. The dissolution kinetics of ASD-based dosage forms is substantially different from crystalline based dosage forms due to API supersaturation capability, the gelling and swelling of the polymer matrices during dissolution and the nucleation, precipitation and micellization kinetics.

The proposed models allows the streamlining of oral dosage formulation and process development in achieving an optimal tradeoff between mechanical strength, disintegration time and dissolution profile. This is achieved by an early assessment of dissolution profile with a first principles mathematical model which takes into account intrinsic material properties of the formulation such as drug diffusivity in the polymer matrix [2], the polymer swelling/erosion rate [3] and whether the API is crystalline or is in an amorphous solid dispersion. The disintegration dynamics are handled with a Population Balance Equation [4] and are coupled to the dissolution model. Whenever direct assessment of intrinsic material properties are not possible, these parameters are obtained via curve-fitting the model to a reference experimental dissolution data using a particle swarm optimization algorithm.

We present a case-study on the application of the methodology to the downstream formulation and process development of an amorphous solid dispersion which illustrates the advantages of guiding the development with a computational tool in terms of process and drug performance.

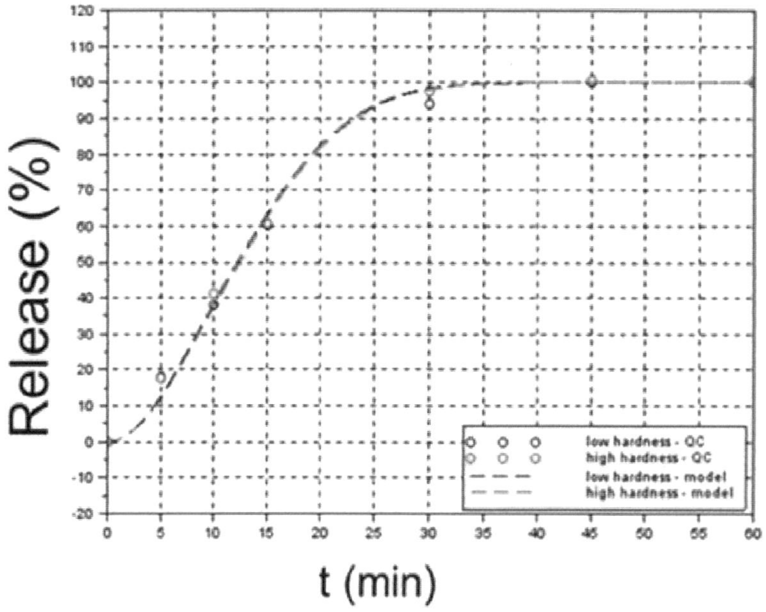

Figure 1: Dissolution profile of a tablet formulated from an amorphous solid dispersion intermediate obtained both experimentally and from the presented mathematical model.

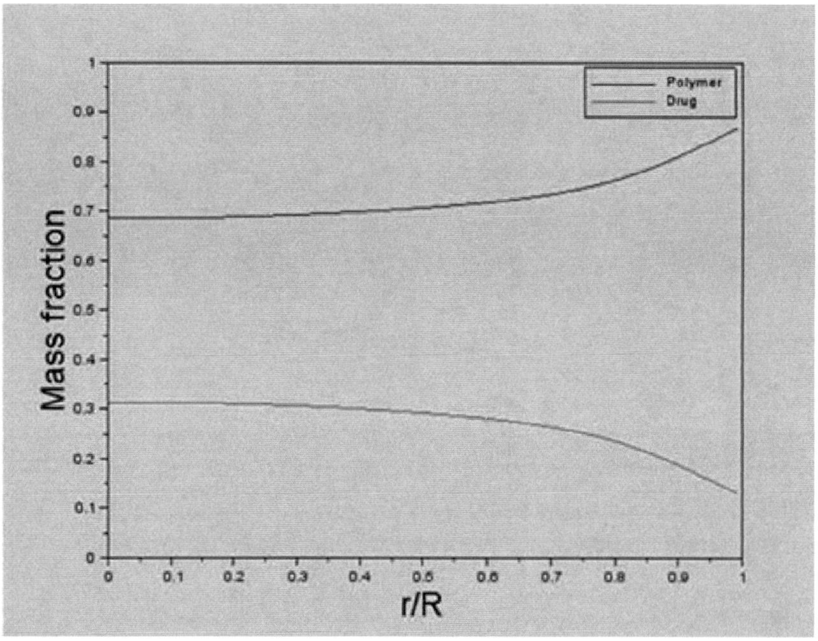

Figure 2: Snapshot of the predicted drug and polymer distribution during dissolution as a function of the radial position within a dissolving particle showing the polymer-rich/drug-poor particle surface caused by the flux of drug dissolved. The abscissae is normalized by the instantaneous radius of the particle which evolves over time as the polymer swells and erodes

References

[1] Perrie Y., Rades T. "Pharmaceutics - Drug Delivery and Targeting", Pharmaceutical Press, 2010.

[2] Wang, Y., Abrahamsson, B., Lindfors, L. & Brasseur, J.G. "Comparison and Analysis of Theoretical Models for Diffusion-Controlled Dissolution", Mol. Pharmaceutics, 9, 1052-1066, 2012.

[3] Siepmann, J., Kranz, H., Peppas, N. A., & Bodmeier, R. "Calculation of the required size and shape of hydroxypropyl methylcellulose matrices to achieve desired drug release profiles", Int. J. of Pharm., 201(2), 151–164, 2000.

[4] Wilson, D., Wren, S. & Reynolds G., "Linking Dissolution to Disintegration in Immediate Release

Tablets Using Image Analysis and a Population Balance Modelling Approach", Pharm. Res., 29:198-208, 2012.

531867 Modeling the Bead-Dissolution Kinetics for Composite Melt-Spray Congealed (MSC) Multiparticulates

Abstract:

Melt-spray congealed (MSC) beads are a promising drug-delivery platform that offers tunable dissolution profiles and drug-release characteristics, taste masking, and ease of dosing for pediatric and geriatric populations. These small, composite, multiparticulate beads (~50-500 mm) are comprised of a mixture of wax and pore former. This wax and pore-former matrix is embedded with active-pharmaceutical ingredient (API) particles, i.e., the drug powder. Once submerged, the pore former rapidly dissolves away leaving behind a porous network. The API then dissolves into the fluid-filled pores, diffuses through the pores, and is finally released at the bead surface into the dissolution media.

Theoretically, the pore-former level, API particle size, bead size, and other adjustable parameters can be engineered to obtain the desired dissolution profile. In reality, this presents a challenge as the exact quantitative dependence of the dissolution rates on the adjustable formulation parameters (pore-former level, drug level, API size, bead size) and unchangeable API material properties (solubility, diffusion rate) are not well understood. To address this issue, this present work describes a model, entitled BeaDiPUB (an acronym for Bead Dissolution Predictions Using BeaDiPUB), that is developed to predict the dissolution of MSC beads as a function of these formulation and material parameters. The current model version is focused on an immediate-release-type formulation without any protective coating.

A calibration/validation procedure is developed for BeaDiPUB which allows application of the model to real compounds. Subsequently, parametric studies are performed to analyze the sensitivity of the dissolution kinetics to the formulation and material properties. Finally, the various regimes for drug release with different rate-limiting steps are examined. The intent of the model is to allow for virtual exploration of the formulation-design space, and thereby reduce the number of experimental iterations needed to manufacture multiparticulates with the desired dissolution characteristics.

535336 Linking Process, Product and Performance By Raman Imaging Analysis

Abstract:

Linking process, product and performance by Raman Imaging analysis

We employ Raman imaging to assess the spatial distribution of the components of a pharmaceutical tablet and its impact on performance measures such as *in vitro* dissolution and mechanical strength. We focus on tablets containing amorphous solid dispersion powders produced by spray drying and we vary the degree of heterogeneity by altering process parameters during sieving and blending steps.

Tablets containing Amorphous Solid Dispersions (ASDs) of poorly water soluble active pharmaceutical ingredients (APIs) in hydrophilic polymeric matrices are a commonly used strategy to increase the solubility and dissolution rate of oral-dosage forms [1]. Furthermore, many of the constituents of the amorphous dispersion/tablet are functional excipients selected to further enhance the bioavailability of the API [2].

Raman Imaging is a versatile analytical tool to provide chemical and spatially resolved information on dosage forms. Mapping the chemical composition throughout a section of a tablet provides a quantitative measure of the heterogeneity of each of the components within that section as shown in Figure 1. Additionally, it allows the detection of clusters which may arise from agglomerates not broken down during the blending step.

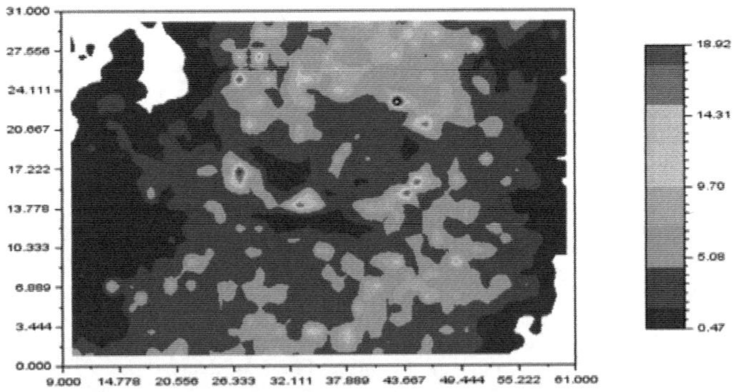

Figure 1: Raman imaging of the centre section of a pharmaceutical tablet indicating the spatial distribution of one of the components.

Herein a case-study is presented for a model drug where the same ASDs spray dried powder is formulated into a tablet with different process parameters following a design of experiments (DoE). In the DoE we vary the sieve mesh size and blending shear rates, known to have impact in individual component distribution within the tablet. Advanced analytical characterization by Raman imaging will be employed to evaluate tablet homogeneity. The influence of those parameters on the tablet tensile strength, friability, disintegration time and dissolution will also be studied, allowing an integration between process and product performance.

[1] Jermain, S. V., Brough, C., & Williams, R. O. / Amorphous solid dispersions and nanocrystal technologies for poorly water-soluble drug delivery – An update. International Journal of Pharmaceutics, 535 (1–2) (2018) 379–392

[2] S. Poozesh et al. / Understanding the process-product-performance interplay of spray dried drug-polymer systems through complete structural and chemical characterization of single spray dried particles. Powder Technology 320 (2017) 685–695

537663 Repurposing Pollen Grains for Oral Delivery of Biologics

Abstract:

Background:

In the recent years, use of biologics as therapeutic agents has significantly increased. However, oral delivery of biologics is challenging, especially due to the need to protect them from the harsh environment of the stomach. Other challenges include limited permeability of biologics through the intestinal epithelium. To overcome some of these challenges we have repurposed pollen grains (PGs). PGs are naturally occurring microcapsules with a tough outer shell. Their natural function is to safely transport the male gamete contained within the pollens to the flower for fertilization. To perform this function, the PG wall is naturally tough and can withstand the assault of low pH and enzymes in the stomach. Furthermore, due to microscopic ornamental features on their surface PGs are mucoadhesive, which can increase their residence time over the use of polymeric particles. Thus, we postulated that PGs could be used as â€˜Trojan Horsesâ€™ to safely carry biologics through the stomach. However, before PGs can be used for oral administration they need to be cleaned to remove intrinsic biomolecules and allergens. Herein, we have developed a new treatment method that uses sequential steps of an organic solvent, acid and an alkali to treat PGs. To augment the properties of PGs we have also created a matrix of an enteric polymer in the PGs to offer protection to the drug and a controlled release property.

Methods: PGs were treated with different protocols to help remove biomolecules from the inner cavity of the PGs, and to obtain clean empty shells. Elemental analysis, scanning electron microscopy (SEM), and Fourier-transform infrared (FTIR) spectroscopy techniques were used to characterize the PGs after treatment. Ragweed (*Ambrosia elatior*) pollen were then chosen as the model pollen species to evaluate their ability to formulate bovine serum albumin (BSA) as a model biologic. Eudragit L100-55 was used as the enteric polymer and glycofurol was used as the solvent. Formulation was optimized with respect to the amounts of Eudragit L100-55 and BSA that were required to achieve high encapsulation efficiency without causing clumping of PGs (to have them free flowing). The formulation was tested for its ability to protect BSA by incubation in simulated gastric fluid. The release profiles of the formulation were also evaluated in the simulated gastric and intestinal fluids.

Results: SEM and elemental analysis showed that our new chemical treatment method can produce clean PGs from multiple species. FTIR spectra confirmed loss of lipid peaks. Video analysis provided insight into the mechanism of the cleaning steps. After cleaning, apertures, which are naturally weak spots in the PG wall got opened producing orifices in the pollen wall measuring about 1 Î¼m in diameter. To ensure that encapsulated proteins do not leak out from these orifices we successfully produced a matrix of enteric Eudragit L100-55 in the PGs. BSA was encapsulated at about 12% efficiency in this matrix. When PGs with Eudragit+BSA were incubated in simulated gastric fluid, the matrix offered protection against denaturation as determined through measurement of intrinsic fluorescence of tyrosine and tryptophan residues. Lastly, we measured the release kinetics of the PGs with Eudragit+BSA in simulated gastric and intestinal fluids. As expected, the presence of PGs provided controlled release, and this effect was accentuated with the use of Eudragit L100-55. The formulation had a slower release as compared to just the Eudragit polymer. We also confirmed the mucoadhesive nature of ragweed PGs by their ability to stick to ex vivo pig intestine, offering PGs a significant edge over smooth polymeric particles.

Conclusions and comparison of PGs to synthetic polymeric particles: These results demonstrate the ability of our new treatment protocol to clean a variety of PGs, and pave the way to investigate various applications of these naturally-occurring pollen particles. Furthermore, the protection of proteins and peptides in the PGs using the enteric matrix creates a new oral delivery system that could substitute parenterally administered biologics and vaccines. The natural mucoadhesive property of PGs can significantly increase their residence time, which makes them very attractive over synthetic polymeric particles. Synthetic polymeric particles also often require the use of organic solvents for encapsulation, and this can harm the proteins to be encapsulated. In contrast glycofurol is protein-friendly. From a cost and manufacturing perspective, PGs are already commercially harvested in large amounts to cater to the need of the allergy industry. In allergen field, PGs are incubated with buffers to extract allergens for immunotherapy and allergy testing. The PG shells are a waste product of this industry. Cost wise PGs are relatively inexpensive. Overall, use of PGs for oral delivery of biologics is an interesting and a viable option that should receive more attention.

538346 Estrogen Receptor-Targeted Multiplexing Photoacoustic Polymeric Nanoparticles for Diagnostic and Treatment of Breast Cancer

Abstract:

Estrogen Receptor-Targeted Multiplexing Photoacoustic Polymeric Nanoparticles for Diagnostic and Treatment of Breast Cancer

C.Nino-Vargas[1], E.Blatchford-Rodriguez[1], Z.Begnell[1], R.Khalid[1], G.Petruncio[2], M.Paige[2], C.Salvador-Morales[1]

[1]Bioengineering Department, George Mason University, 4400 University Drive, Fairfax, Virginia, 22030, [2]Chemistry Department, George Mason University, 4400 University Drive, Fairfax, Virginia, 22030.

Introduction. Breast cancer is the second cause of death in women after lung cancer, with approximately 41,000 breast cancer deaths expected this year [1]. Systemic chemotherapy is one of the most common approaches to therapy, but its systemic toxicity is a major drawback that limits chemotherapeutic utility and effectiveness [2]. Another significant problem is the generation of the multidrug resistance phenomenon, which, among other factors, arises from inhomogeneous distribution of the drug within the tumor [3]. A recent focus of cancer research is on developments of theranostic systems capable of providing diagnostic imaging and therapeutic functions simultaneously. An emerging scientific and clinical consensus is that theranostic devices may provide several medical benefits including: a) molecular imaging functionalities to help physicians track drug distribution within the solid tumor, b) valuable guidance on selection of a safe and efficacious dose, and c) real-time objective monitoring of therapeutic response. The multiplexing photoacoustic capability of these nanovehicles will render substantial anatomical, physiological and functional information of the breast cancer tumor.

Here, we report the design and synthesis of a novel targeted cancer theranostics device for breast cancer treatment that leverages the highly tunable internal properties of Patchy Polymeric Nanoparticles (PPNPs). Estrogen Receptor-Targeted Patchy Polymeric Nanoparticles loaded with Indocyanine Green (ICG) and IR-26 consist of poly (lactic-co-glycolic acid) (PLGA) and lipid-PEGylated-Estrone. PLGA and ICG are biocompatible and FDA-approved biomaterials. These nanoparticles have unique patch-core-shell structural features: hollow core and patchy surface. The particle's patchy surface can be utilized to enhance tumor targeting because of the high concentration of targeting moieties in one region of the particle's surface.

Materials and Methods. The surface of our nanoparticles is functionalized with Lipid-PEGylated-Estrone molecules that will bind specifically to estrogen receptors that are over-expressed in breast cancer cells. We performed a comprehensive characterization of the targeting moiety using Nuclear Magnetic Resonance (NMR), Matrix Assisted Laser Desorption/Ionization (MALDI) and High Performance Liquid Chromatography (HPLC). Our results show that we successfully synthesized the targeting moiety. Next, we synthesized the targeted nanoparticles using a modified single emulsion method previously published [4]. Moreover, we characterized these nanoparticles with different experimental techniques including Dynamic Light Scattering (DLS), Transmission Electron Microscopy (TEM), Absorption and Near-IR Spectroscopy. Our results show that these particles have an average hydrodynamic diameter of 165 nm ± 5 nm. Also, the particles' size is stable for months in frozen and lyophilized storage conditions. Furthermore, we assessed the absorption ability of these nanoparticles in the Near-IR region of the electromagnetic spectrum. These nanoparticles present an absorption peak at 786 nm and 895 nm wavelength. TEM shows the spherical morphology of these nanoparticles. Additionally, we evaluated the cellular toxicity of the engineered nanoparticles at different concentrations. We observed that these nanoparticles are not toxic as expected since PLGA and ICG are biocompatible, and IR is enveloped by the polymeric matrix. Moreover, we evaluated the cell uptake of untargeted and targeted nanoparticles at 2 hrs. and 24 hrs. time points using T-47D breast cancer cells. We observed that targeted nanoparticles are avidly taken by breast cancer cells at both 2hrs and 24 hrs. However, there cell uptake at 24 hrs. is higher than that at 2hrs. The assessment of the photoacoustic performance and ***in vitro*** cellular uptake of these nanoparticles is underway. Future experiments will be carried out to evaluate their in ***vivo*** photoacoustic performance. Equally important, we have carried out drug release studies to determine the drug release profile of these nanoparticles. Our most recent studies demonstrate that these nanoparticles release 70% of the drug over a month.

Conclusions: These preliminary results show the feasibility of engineering multiplexing photoacoustic theranostic carriers for breast cancer treatment. Thus, these theranostics have the potential to provide meaningful information about the tumor vasculature because of their ability to penetrate deep into tissue while killing at the same time tumor cells selectively.

537430 Engineering of Charge Transfer Complex Nanocrystals By Electrocrystallization

Abstract:

Crystal structure and morphology of nanoparticles determine their interfacial atomic arrangements and thus control their properties. To date, controlling the morphology of the nanoparticles is still a major barrier that greatly limits their applications. This work discusses the experimental and theoretical methods to study and control the morphology of the nanocrystals. Guided by the thermodynamics and kinetic transition theory, we used electrocrystallization to make K(def)TCP nanocrystals and tuned the electrochemical parameters to alter the nanocrystal morphologies. Our results suggest that the electrochemical parameters influence the K(def)TCP nanocrystal morphology. We obtained nanoparticle shapes ranging from nanorods to rhombohedral shape depending on the solution concentration, the applied potential and the current. We modeled the observed crystal growth behavior with a Monte-Carlo based computer simulation. The simulation results qualitatively match the experimental findings. This work contributes to the understanding of the crystal growth behavior and the thermodynamic/kinetic morphology transition in electrocrystallization. We will show the application of the engineered K(def)TCP nanocrystals in gas sensing.

539030 Silica-Coated, Near-UV Activated $YVO_4:Eu^{3+},Bi^{3+}$ Nanophosphors for Dynamic Cell Imaging

Abstract:

Luminescent rare-earth-based inorganic nanoparticles (nanophosphors) are promising bioimaging agents due to their high photostability, sharp emission bands and relatively low toxicity [1]. Flame aerosol technology provides a scalable and highly reproducible process for production of such nanophosphors (***e.g.*** $Y_2O_3:Eu^{3+},Tb^{3+}$) with precise control of their composition and properties [2,3]. Nanophosphors that can be excited in the near-ultraviolet and visible region, such as $YVO_4:Eu^{3+},Bi^{3+}$, provide a useful tool for bioimaging and ***in vitro*** dosimetry studies using conventional fluorescence microscopes.

Here, $YVO_4:Eu^{3+},Bi^{3+}$ nanophosphors are made by flame spray pyrolysis and in-situ coated by a nanothin amorphous silica coating. The optimal Bi content for maximum red-shift of their excitation band edge towards the visible region is identified through systematic experiments. The nanophosphors with the optimal composition are highly crystalline and appear bright red under a conventional fluorescence microscope. Their photostability during dynamic imaging of HeLa cells ***in vitro*** is confirmed, contrary to commercial fluorescent (organic-dye labeled) SiO_2 nanoparticles that exhibit 50% photobleaching within 3.5 h [4]

The potential of these nanophosphors for dynamic cell imaging is demonstrated and compared to conventional organic dyes, demonstrating the nanophosphor superior performance/ We employed cell-bacterial co-cultures to identify how GFP-expressing ***S. pneumoniae*** bacteria interact with human epithelial lung cancer cells (A549). Furthermore, the nanoparticle interaction with these cells is further examined by employing upright and inverted cell culture orientations, and by the surface biofunctionalization of the nanoparticles with targeting antibodies compared with non-targeting bovine serum albumin. The amorphous silica coating here facilitates the biofunctionalization of the nanoparticle surface. The developed silica coated nanoprobes here will open up several avenues on dynamic nanoparticle-cell imaging.

REFERENCES

1. A. Escudero, C. Carrillo-Carrion, M. V. Zyuzin & W. J. Parak. Luminescent rare-earth-based nanoparticles: A summarized overview of their synthesis, functionalization, and applications. ***Topics Curr. Chem.*** **374**, (2016).
2. G. A. Sotiriou, M. Schneider & S. E. Pratsinis. Color-tunable nanophosphors by codoping flame-made Y_2O_3 with Tb and Eu. ***J. Phys. Chem. C*** **115**, 1084-1089 (2011).
3. G. A. Sotiriou, D. Franco, D. Poulikakos & A. Ferrari. Optically stable biocompatible flame-made SiO_2-coated Y_2O_3: Tb^{3+} nanophosphors for cell imaging. ***ACS Nano*** **6**, 3888-3897 (2012).
4. A. Spyrogianni, P. Tiefenboeck, F. H. L. Starsich, K. Keevend, F. Krumeich, I. K. Herrmann, J. C. Leroux & G. A. Sotiriou. "Near UV Activated, Photostable Nanophosphors for ***in vitro***Dosimetry and Dynamic Bioimaging" ***AIChE J.*****in press**, DOI: 10.1002/aic.16166 (2018).

539060 Stimuli Responsive Nano-Agents: From Drug Delivery to Oil and Gas Industry

Abstract:

Several concepts that have been successfully used for targeted drug delivery for medical diagnostics and treatment can lend themselves to some key oil and gas applications, despite the clear differences in the systems and environments involved. An example of such concepts is the utilization of stimuli responsive materials to deliver treatment or image contrast agents to specific targets in the human body, or the oil reservoir, without the side effects and unnecessary costs associated with excess release or delivery of agents to where they are not needed.

The Responsive Reservoir Nano-agents Program was initiated at Saudi Aramcoâ€™s upstream Advanced Research Center in 2016. The main objective of this program is to develop nano-agents that can travel deep into the reservoir to collect reservoir information or deliver chemical treatments to *hard-to-reach* regions, autonomously by exploiting the induced or natural chemical and physical conditions in the reservoir. These regions are hardly accessed by conventional water or gas flooding techniques.

In this talk, a proof-of-concept on the effect of salt and surfactant gradients on the migration of reservoir nano-agents will be presented. The experiments use novel microfluidic devices and direct microscopic visualization to quantify the migration of solid colloidal particles (<1μm) under different gradients and to understand the underpinning electrokinetic and interfacial phenomena. The results demonstrate the ability of salinity and surfactant gradients to transport reservoir nano-agents up or down the imposed gradients in a predictable way. In this way, natural or induced gradients in the reservoir can be exploited to direct the reservoir nano-agents to certain regions in the reservoir for discovering or recovering the trapped resources.

533393 Global System Analysis of Twin Screw Granulation Using Population Balance Modelling in gPROMS

Abstract:

Twin screw granulation (TSG) becomes increasingly popular as a method of wet granulation in the integrated part of continuous manufacturing for pharmaceutical industry. Despite many studies of TSG through the population balance model (PBM), the systematic analysis of the input parameters in PBM is still scarce. As a consequence, the influential parameters in PBM are yet fully identified and the holistic understanding of the granule growth in TSG remains elusive. This paper presents a global system analysis of the PBM operation for a ConsiGmaâ„¢ 25/1 continuous twin screw granulator. As a precursor to understand the influence of PBM parameters in the granule attributes, the modelling kernel formation in PBM is first defined in a compartmental way. The modelling formation consists of four rate process kernels, i.e. nucleation, layering, breakage and consolidation. Particularly, the recently developed breakage kernel is implemented in the software of gPROMS, which significantly improves the serviceability of breakage kernel for TSG. The feature of the breakage kernel is to take into account the powder feed number, dynamic yield strength and critical breakage size. The input parameters in PBM are identified and categorized whereas the reference values of all the parameters are specified. With the defined input value of PBM parameters, the comparative values of the testing parameters are also defined. With varying the comparative values of input parameters in a univariate way, the influence of input parameters on the granule size distribution (GSD) is investigated.

Apart from the rate process kernels, another two important process kernels i.e. liquid to solid ratio (LSR) and the average residence time in each compartment are also deployed for sensitivity study. Through the global sensitivity analysis of the input parameters in PBM kernels, the leverage of each parameter on the granule product attributes could be identified, which shed more insights on the validation of PBM model and model-based design of TSG products.

536761 Effect of Crystal Size on the Breakage of High Aspect Ratio Crystals in Stirred Slurries

Abstract:

A number of processes produce high aspect ratio crystals in stirred suspensions. Because breakage of these crystals can produce fine particles that can block filters in downstream operations, determining the quantity of fines is of industrial interest. Particle breakage in stirred suspensions depends on a variety of factors including the agitation rate, material properties, and the particle size and shape. In this work, high aspect ratio crystals are grown to produce sets of paracetamol crystals in different size ranges. These sets of crystals are then stirred and the resulting crystals analyzed to determine the effect of the crystal length on the rate of breakage and on the child particle distributions. This work presents breakage distribution functions based on the experimental results. Both the size and shape distributions of the crystals are reported. Although paracetamol is studied, the approach is applicable to other systems.

530689 Fundamental Prediction of Agglomeration and Entrainment Rates for Cohesive Powders in a Riser Flow

Abstract:

Predictions of entrainment rates of non-cohesive powders in fluidized bed systems given by empirical models can vary by several orders of magnitude for a given system (Chew et al 2015). For flows of cohesive particles, agglomerates can form and break, leading to further uncertainty in the entrainment rate. In this work, a fundamental approach is used to predict the agglomerate fraction and entrainment rate. Specifically, the population balance is coupled with kinetic theory balances. This new continuum framework utilizes fundamental closures for the birth and death rates due to aggregation and breakage in the population balance (Kellogg et al 2017). These novel closures take into account the effect of impact velocity (granular temperature) on the outcome of a collision as agglomeration, rebound, or breakage. Here, the agglomerate fractions and entrainment rates predicted by the new continuum framework are compared to discrete element method (DEM) simulations. The effect of the critical velocity of aggregation (cohesive energy) and the solid volume fraction on the agglomerate fraction and entrainment rate are explored. Preliminary comparisons with experiments will also be discussed.

Chew, J.W., Cahyadi, A., Hrenya, C.M., Karri, R. and Cocco, R.A., 2015. Review of entrainment correlations in gasâ €"solid fluidization. Chemical Engineering Journal, 260, pp.152-171.

Kellogg, K.M., Liu, P., LaMarche, C.Q. and Hrenya, C.M., 2017. Continuum theory for rapid cohesive-particle flows: general balance equations and discrete-element-method-based closure of cohesion-specific quantities. Journal of Fluid Mechanics, 832, pp.345-382.

529428 Molecules As Building Blocks in a Novel Population Balance Model for Flash Nano-Precipitation: Investigation of the Different Good Solvents Effect on Nanoparticle Formation

Abstract:

Polymeric nanoparticles (NP) production is becoming more and more popular in the recent years due to the wide range of applications and research efforts are being focused on the understanding of the phenomena involved in one of the most common preparation techniques: Flash Nano-Precipitation (FNP). More specifically, the effect of fluid dynamics on NP formation and its role in controlling the final NP size is of paramount importance (Lince, at al., 2008). The standard modeling approach used so far considered three different steps separately in the population balance equations with the characteristic NP length as internal coordinate. In this way, nucleation, molecular growth and aggregation are treated separately in the model. An alternative approach that we here propose and validate is to overcome the concepts of nucleation, molecular growth and aggregation, proposing a novel purely-aggregative population balance model (PBM) by using as internal coordinate the number of molecules (building blocks) within each NP, coupled with computational fluid dynamics (CFD).

Molecules are treated as objects aggregating between themselves (self-assembly), once the local solubility limit is reached and overcome (supersaturation ratio greater than one). This is justified by the fact that particle formation is a supersaturation-driven process. The test case shown here is the NP formation via solvent displacement (acetone and water in the role of good solvent and anti-solvent respectively), with a polymer (poly-e-caprolactone, PCL) as solute, in a confined impinging jets mixer (CIJM, Fig. 1). The rate with which PCL molecules aggregate is expressed by the aggregation kernel, expressed in terms of number of molecules which form a particle (say *n*), referred to as aggregation number, and considered here as the internal coordinate of the PBM. When a cluster of *n* molecules collides with a cluster of *n'* molecules, then a cluster of *n+n'* molecules (or molecular building blocks, as shown in Fig. 2) is formed, assuming the freely jointed chain hypothesis (Flory, 1953). The aggregation kernel considers both the aggregation due to Brownian movements and turbulent fluctuations and, for the first time, it is built upon molecular dynamics (MD) (Di Pasquale, et al, 2014); furthermore, the PBM and CFD are coupled together in a solely tool, thanks to a suitable user define function (UDF) implemented in Ansys Fluent 15.

The effect of different good solvents (acetonitrile and THF) is also investigated via the PBM-CFD code, in order to achieve a better insight into both the physical-chemical and transport phenomena (e.g. the role of the aggregation kernel) that lead to different experimental mean NP size (Ferri, et al, 2017). The model results are satisfactory and promising in terms of validation against experiments, showing a good agreement with them (Fig. 3), and in terms of identification of the key parameters able to control the mean NP size when different good solvents are employed in FNP processes.

Future work is recommended to understand what kind of phenomena plays a key role at low solute concentration, and at the same time, also the investigation and validation of this purely-aggregative multiscale model on different geometries (e.g. multi inlet vortex mixers).

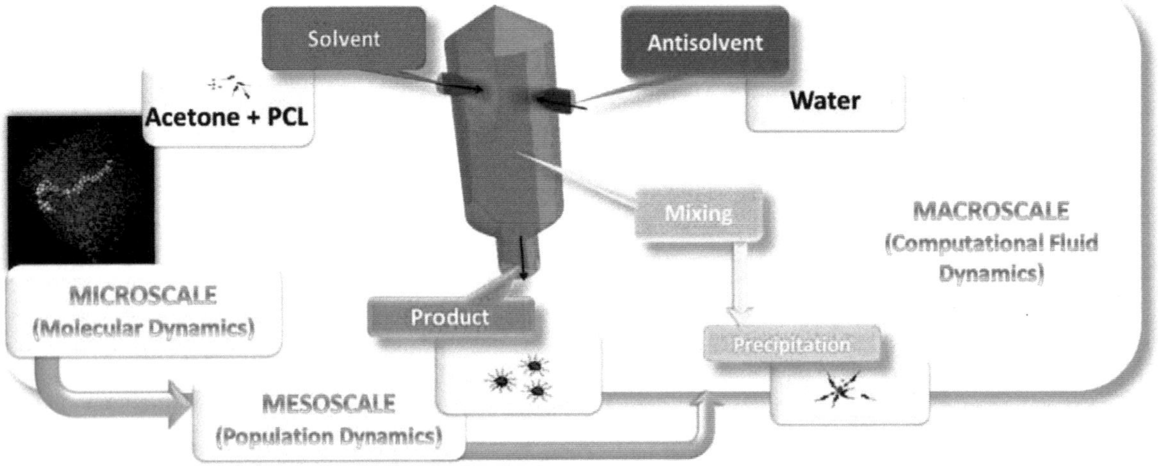

Figure 1. Sketch of the nanoparticle formation process in a CIJM. On the left side of the mixer there is the acetone and PCL inlet (solvent, red); on the other side, the water (antisolvent, blue). The multiscale approach is represented: the molecular level (molecular dynamics scale), the mesoscale (represented by the PBM), and the macroscale (CFD).

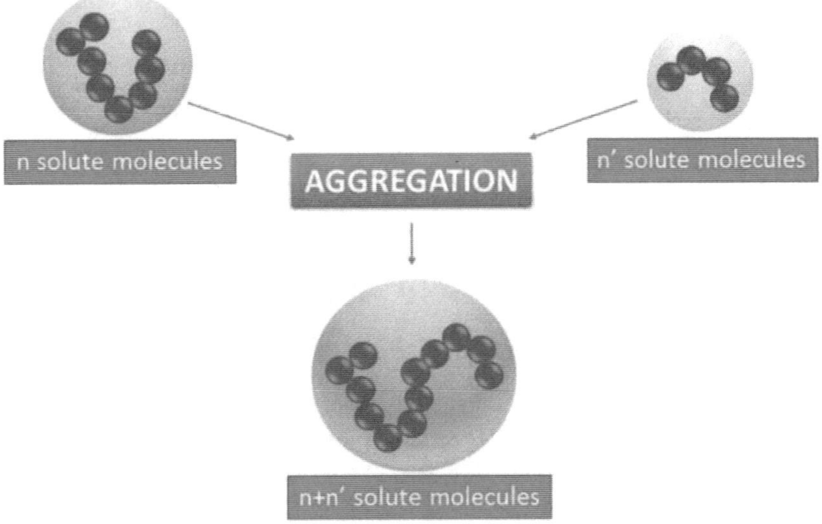

Figure 2. Two nanoparticles, formed by n and n' solute molecules (building blocks) respectively, collide leading to n+n' molecules nanoparticle.

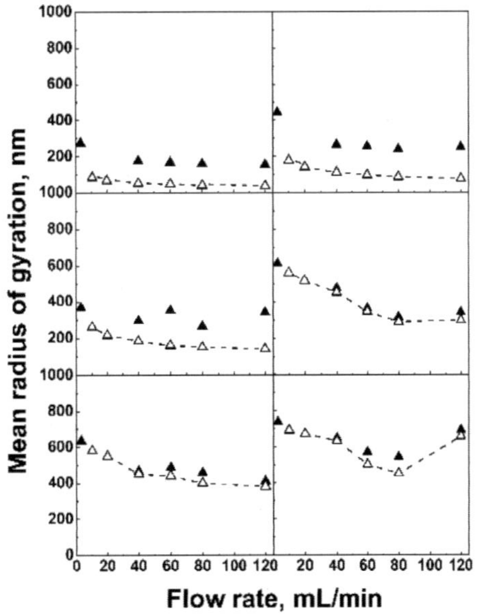

Figure 3. Model predictions (empty symbols) against experimental data (measured by Dynamic Light Scattering, black triangles). From top to bottom and left to right the initial PCL concentration corresponds to 0.5, 2.5, 5.0, 10.0, 15.0, and 25.0 mg/mL.

References

Di Pasquale, N., Marchisio, D.L., Barresi, A.A., Carbone, P., 2014. Solvent structuring and its effect on the polymer structure and processability: the case of water-acetone poly-e-caprolactone mixtures. *J. Phys. Chem. B*, **118**, pp. 13258–13267.

Ferri, A. , Kumari, N. , Peila, R., Barresi, A. A., 2017. Production of menthol loaded nanoparticles by solvent displacement. *Can. J. Chem. Eng.*, **95**, 1690-1706.

Flory, P., 1953. Principles of Polymer Chemistry. Cornell Univ. Press.

Lavino, A. D., Di Pasquale, N., Carbone, P., Marchisio, D. L., 2017. A novel multiscale model for the simulation of polymer flash nano-precipitation. *Chem. Eng. Sci.*, **171**, pp. 485-494.

Lince, F., Marchisio, D. L., Barresi, A. A., 2008. Strategies to control the particle size distribution of poly-ε-caprolactone nanoparticles for pharmaceutical applications, *J. Colloid Interface Sci.*, **322**, pp. 505–515.

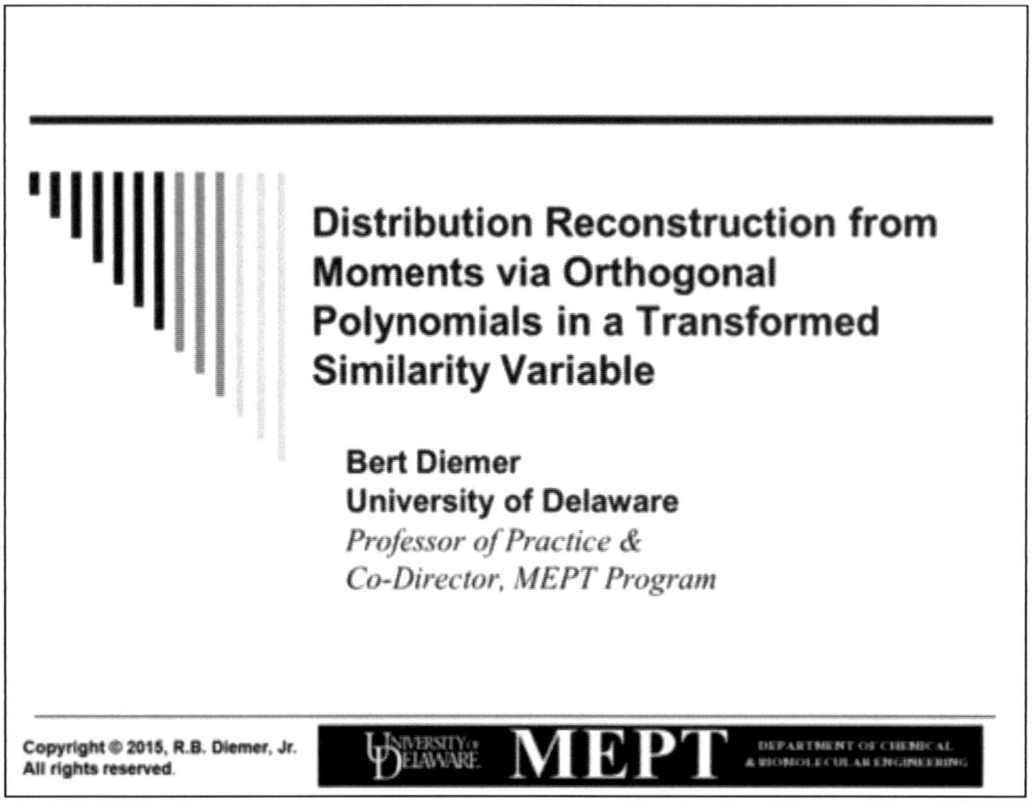

This presentation is a follow up to a talk given at the 2017 Annual Meeting in Minneapolis, and shows additional developments since then in this topic area.

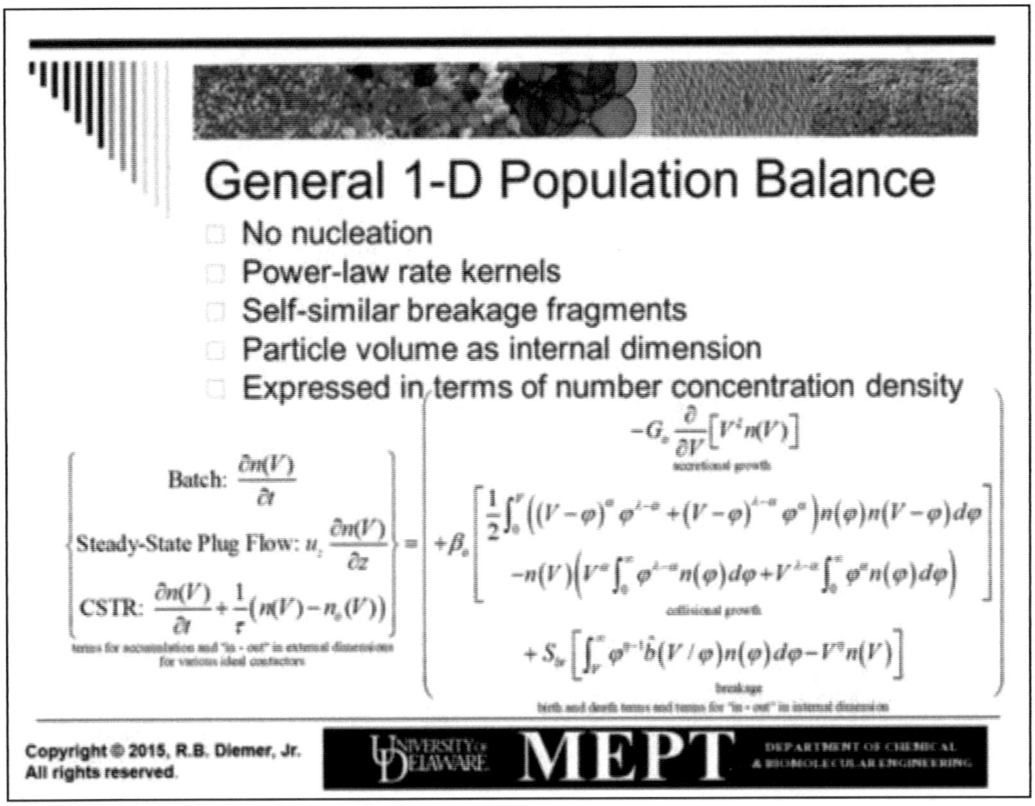

Let's start the discussion by looking at a general population balance for a population with one internal dimension, which will be the particle volume. We express the balance in terms of the number concentration density n(V). We are interested in systems capable of producing either stationary states or similarity solutions, so we will not consider processes that include nucleation. (Such systems are either seeded, or occur downstream of a short nucleation burst.)

We treat general rate kernels for accretional growth, collisional growth and breakage that are power-law in the internal dimension V. In the case of collisional growth, we are treating homogeneous kernels of order λ with a structure parameter α. Such kernels are representative of physical processes typically encountered for example:

- Accretional growth:
  - Reaction-limited, $\xi = 2/3$
  - Mass transfer-limited, $\xi = 1/3$ at low Sherwood numbers
- Vogel-Peukert breakage kernel, $\eta = 1/3$
- Collisional growth:
  - Brownian, 2 terms, one with $\lambda = \alpha = 0$, one with $\lambda = 0$, $\alpha = -1/3$
  - Saffman-Turner low Stokes turbulence, 2 terms, one with $\lambda = 1$, $\alpha = 0$, one with $\lambda = 1$, $\alpha = 1/3$

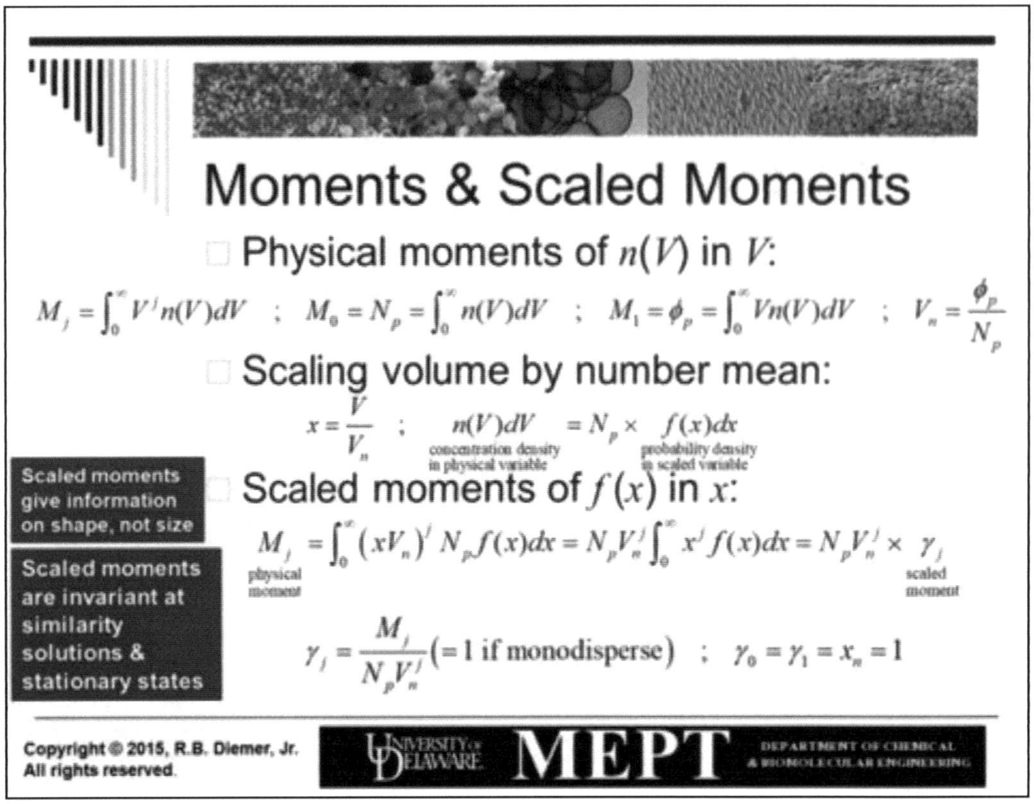

We can form moments about the origin for the number concentration density n(V) which have units and are therefore "physical". The 0th such moment is the total number concentration. The 1st such moment is the particle volume fraction. The ratio of these two gives the number mean particle volume.

It is sometimes that case that the problems reach a similarity solution in a scaled variable in which the particle volume is divided by its time or space dependent number mean. We can define a probability density in this scaled variable that is connected to the physical number concentration density n(V).

We can further show that by substitution into the physical moment equation, we arrive at an expression for a set of scaled moments of the probability density f(x) in the scale variable x.

Since x is a scaled particle volume, its number mean is fixed at 1. Furthermore, since it is a normalized probability density the 0th scaled moment is also 1 and this means that the 1st scaled moment is 1 as well.

The implication is that these scaled moments reflect what is happening to the distribution shape alone since they are pinned on the size axis to have a number mean of 1. When a similarity solution or stationary state is reached, these scaled moments become invariant.

One can show too that if the distribution is monodisperse, all of these scaled moments are equal to 1.

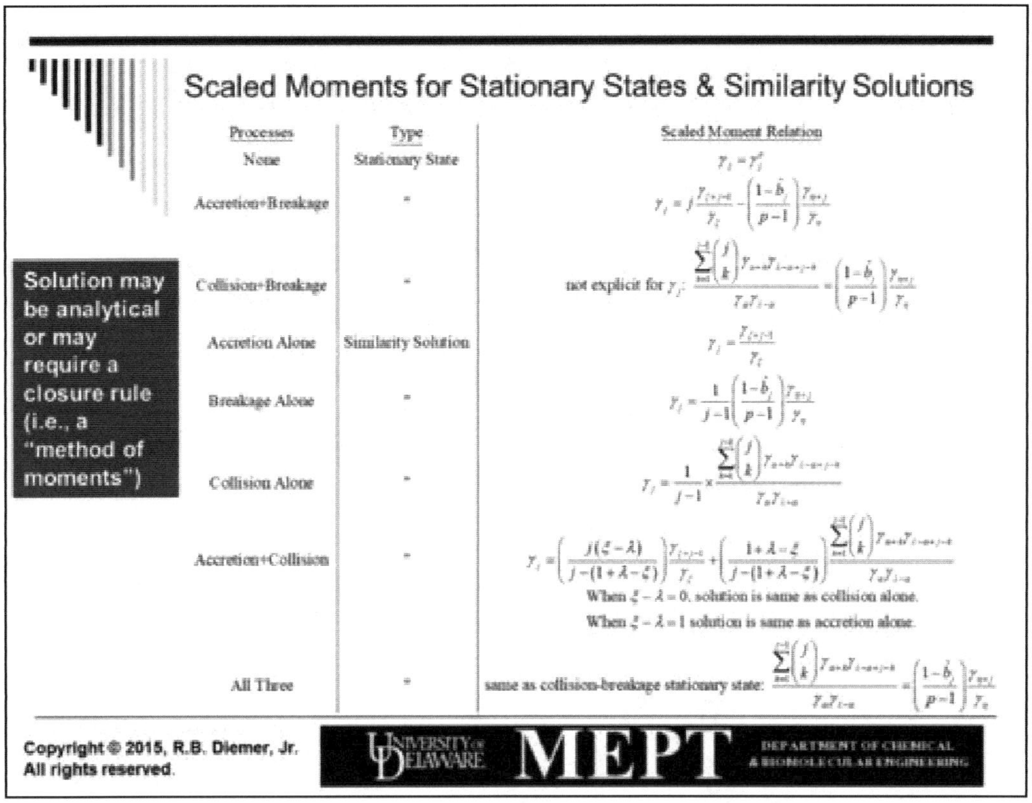

Though not developed here, but it is possible to start from the population balance of the earlier slide and by transforming to the scaled variable, derive equations for the scaled moments when they have become invariant.

There are 8 possible combinations of the accretional growth, collisional growth and breakage processes being considered, and one can show that:

- The 3 combinations of (1) accretion & breakage, (2) collision & breakage and (3) the null case of no processes all result in stationary states defined as those in which the number mean volume and the distribution shape both become constant.
    - Note that in the case of accretion & breakage, both the number of particles and the total particle volume both are increasing but in a way that maintains a constant number mean particle volume
- The other 5 combinations result in similarity solutions in which the number mean particle volume continues to evolve but the distribution shape becomes invariant (*aka* self-preserving distributions).
- The equations for the scaled moments at either stationary states or similarity solutions are as given on the slide.

One can therefore solve for these moments, in some cases analytically, but in the rest, under a "method of moments" that dictates how to close the equation set.

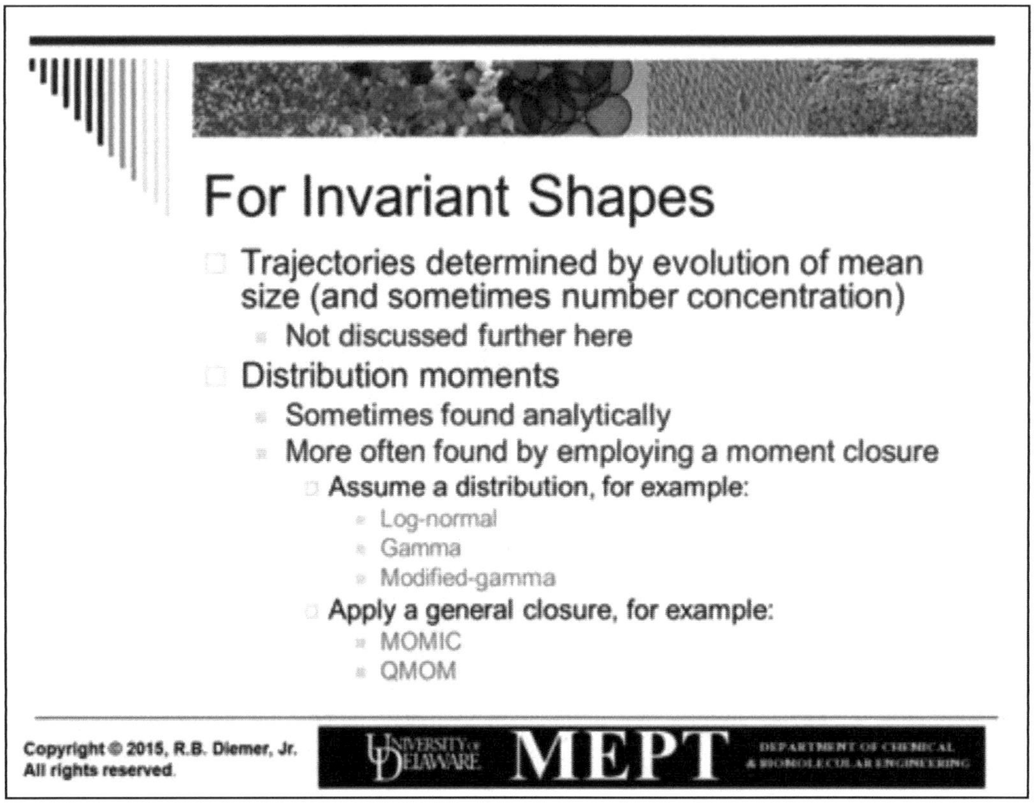

When the shape of the distribution has become invariant, the trajectory of the system is fully described by the evolution of the number mean size (and sometimes the number concentration) once the shape is known.

As mentioned on the previous slide, the finding of the distribution moments usually requires employing a closure rule.

The simplest of such rules involve assuming a distribution shape. Typical choices are log-normal, gamma or modified-gamma distributions. In this approach, the distribution parameters are estimated from a few moments and any other moments needed to calculated the full moment set are filled in by using the moment expression attached to the assumed distribution. Such an approach also offers a crude reconstruction of the distribution since a form has already been assumed and its parameters calculated.

More general closures do not require assuming an particular distribution shape and are in that sense more desirable, offering the possibility of coming closer to the true solution. Examples are MOMIC (method of moments with interpolative closure) and QMOM (quadrature method of moments). If one uses one of these more general closures, one needs to look elsewhere for what the distribution shape may be.

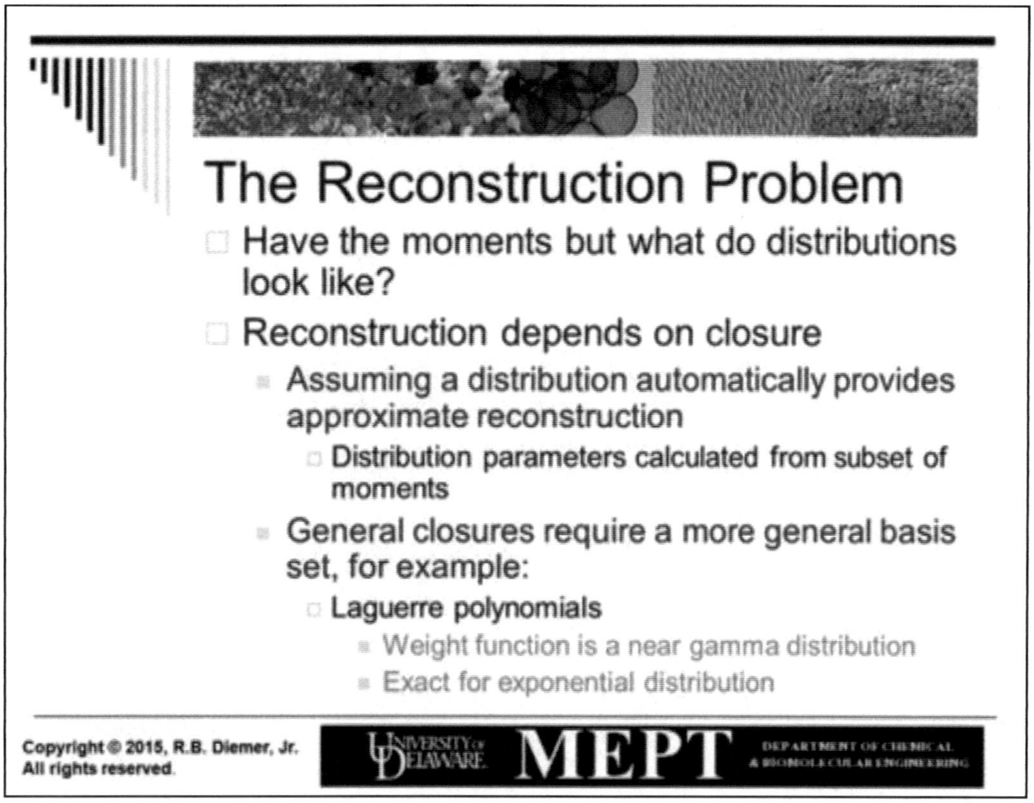

So, we can find stationary state or similarity solutions, but we'd like to know the shape of these distributions, especially in the case of the general closures. In this case, one is looking for a basis set over which to expand the distribution function and from which to develop moment expressions.

The Laguerre polynomials are an orthogonal basis set for the right half-plane ($x \in [0,\infty)$), and so are a natural basis to adopt for this. The Laguerre polynomials are multiplied by a weight function $x^{\alpha}e^{-x}$ which is very nearly a gamma probability density. (To be a true gamma probability density in the similarity variable $x = V/V_n$, there would be a normalization constant that depends on the parameter α, and α would appear not only in the exponent of x but also in the exponent of e.) Note that when $\alpha = 0$, there is an exact match between a gamma probability density and the Laguerre weight function because one recovers the exponential distribution, a member of the gamma (and modified-gamma) family.

.

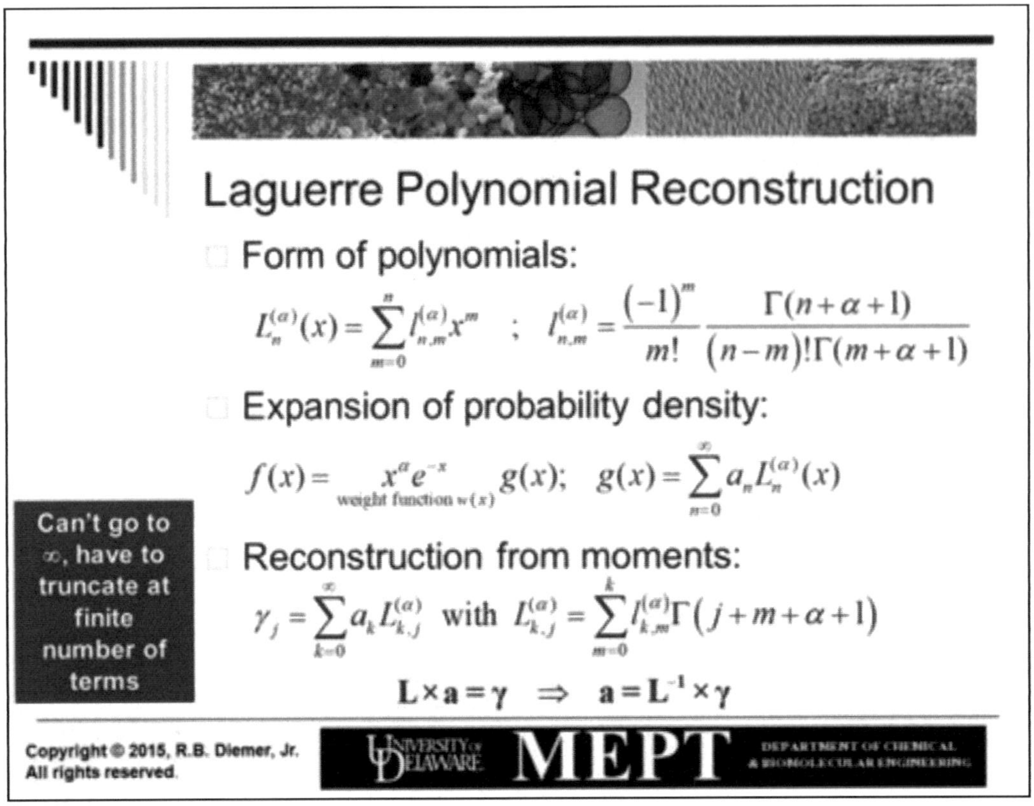

The coefficients of the Laguerre polynomials are shown on the top equation of this slide. One can expand the probability density f(x) in terms of the product of the weight function and an unknown function g(x), which is represented by a linear combination of the Laguerre polynomials.

If we apply the moment expression to f(x) in this form, we can show that the scaled moments are equal to a linear combination of terms calculated as sums of the products of Laguerre coefficients and gamma functions. Of course we can't include an infinite number of terms so we have to truncate the expansion somewhere.

Once we do that, we can populate the matrix $\mathbf{L}$, invert it, and obtain the vector of Laguerre expansion coefficients from an equal dimension vector of scaled moments.

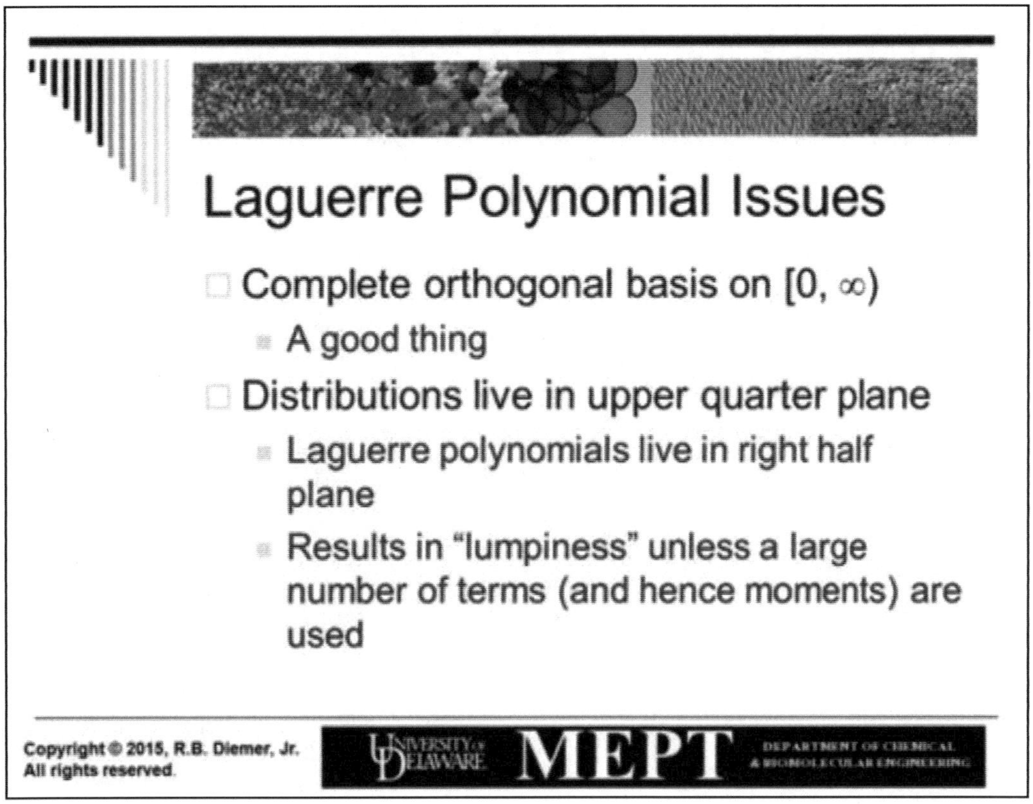

So why don't we just do this?

The Laguerre polynomials are a complete orthogonal basis for all functions that live in the right half-plane. On the other hand, our distributions live in the upper right quarter-plane. The polynomials have to be capable of representing functions that take on negative values while the distributions we wish to reconstruct are fully non-negative.

The consequence is that we need lots of terms in order for the negative parts of the expansion to fully cancel out. If we don't, then we get "lumpy" representations with spurious modes.

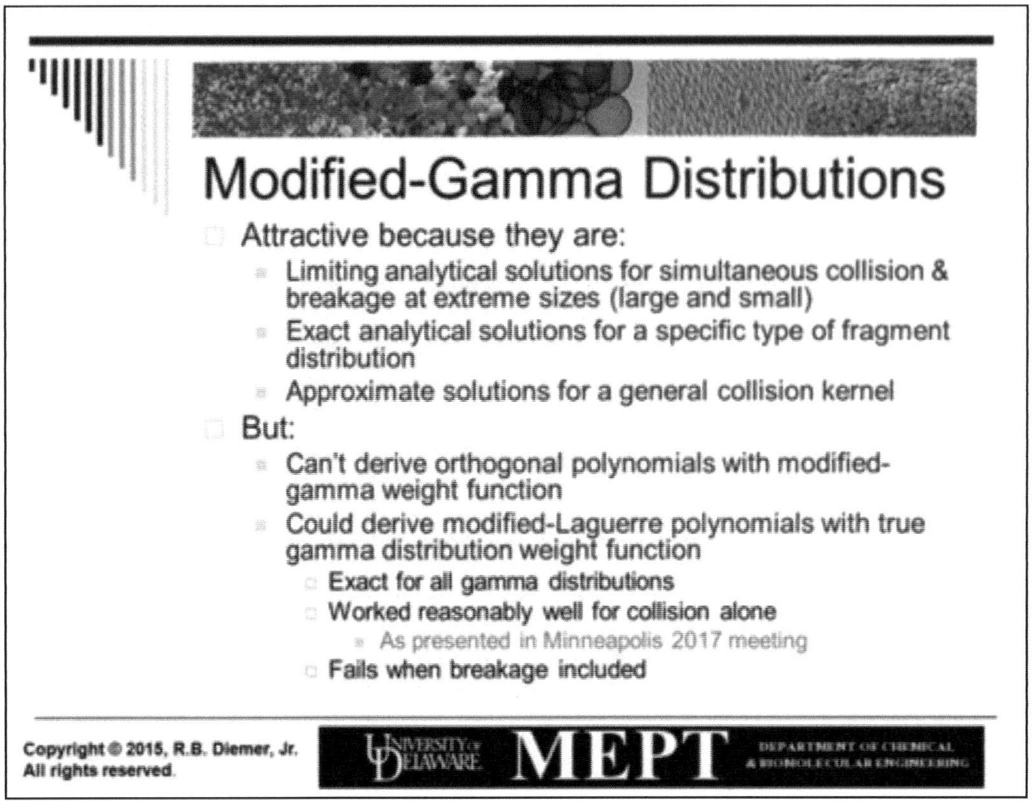

So how can we fix this? It would be nice if the weight function were a modified-gamma distribution so that the polynomials only have to capture perturbations of the solution from it. Why? Because modified-gamma distributions (1) are the analytical solution to the breakage problem for a specific class of fragment distribution, (2) are approximate solutions for a general collision kernel, and (3) are limiting analytical solutions for simultaneous collision & breakage at the large and small ends of the distribution.

1. T. W. Peterson (1986) Similarity Solutions for the Population Balance Equation Describing Fragmentation, Aerosol Science and Technology 5:1,93-101.
2. M. M. R. Williams (1984) On the Modified Gamma Distribution for Representing the Size Spectra of Coagulating Aerosol Particles, Journal of Colloid and Interface Science, 103:2, 516-527.
3. R. B. Diemer and J. H. Olson (2002) A Moment Methodology for Coagulation and Breakage Problems: Part I – Analytical Solution of the Steady-State Population Balance, Chemical Engineering Science, 57:12, 2193-2209.

Unfortunately, we can't derive orthogonal polynomials with such a weight function. A year ago we showed that we could derive a set of modified-Laguerre polynomials for which the weight function was a true gamma distribution. This broadened the ability to reconstruct distributions for the collision alone as was presented at the Minneapolis meeting. But, since then, we found that application to breakage problems did not give good results.

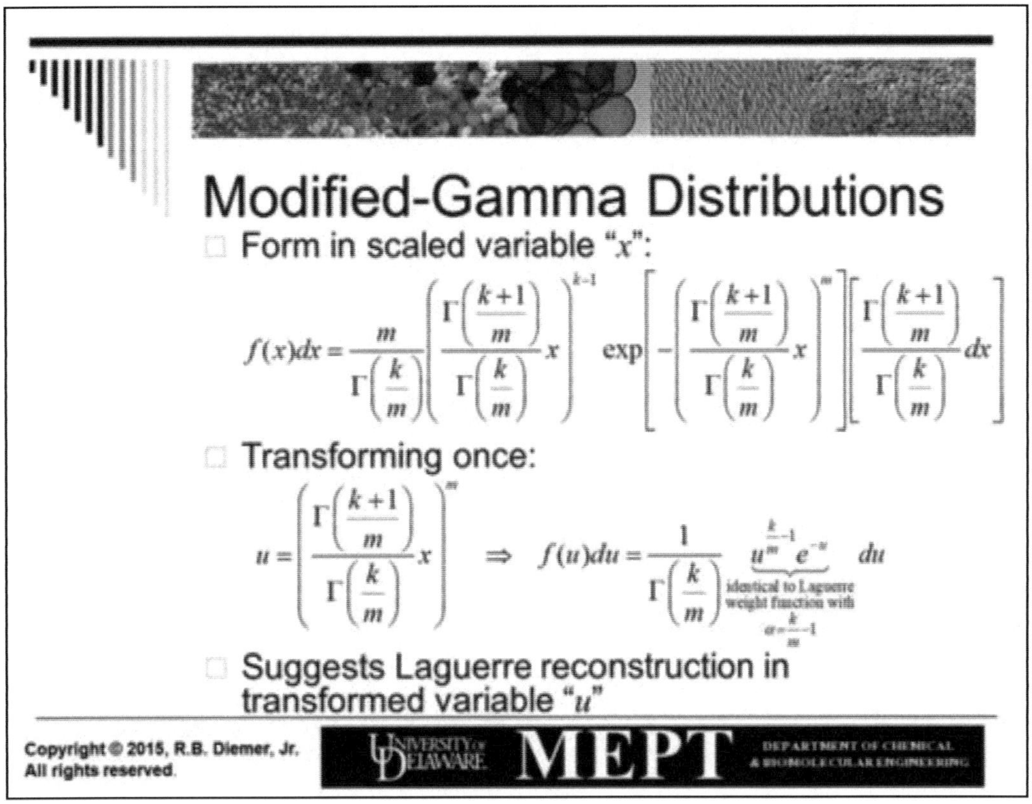

The modified-gamma distribution as a probability density in the scaled variable x, such that its 0th and 1st scaled moments are both unity, is in the form given on the slide. What distinguishes this from the gamma distribution is the exponent m in the argument of the exponential. Note that if m=1, then this becomes identical to the gamma distribution:

$$f_{\mathrm{mod}-\Gamma}(x;m=1) = \frac{k^k}{\Gamma(k)}x^{k-1}e^{-kx}$$

Furthermore, if both k=m=1, then it becomes the exponential distribution:

$$f_{\mathrm{mod}-\Gamma}(x;k=m=1) = e^{-x}$$

But... if we do a change of variable as shown on this slide, where the new variable is given by:

$$u = \left[\frac{\Gamma\left(\dfrac{k+1}{m}\right)}{\Gamma\left(\dfrac{k}{m}\right)}x\right]^m$$

The result is a distribution that contains a factor that is identical to that of the Laguerre polynomials with the parameter α replaced by k/m-1, and also includes a normalization factor absent from the Laguerre weight function.

This suggests that if we expand our distribution function in terms of Laguerre polynomials in this new variable u, then we will only have to use the polynomials to capture perturbations from modified-gamma distributions which have been shown to connect strongly to the expected solution forms which are more general than either gamma or exponential distributions.

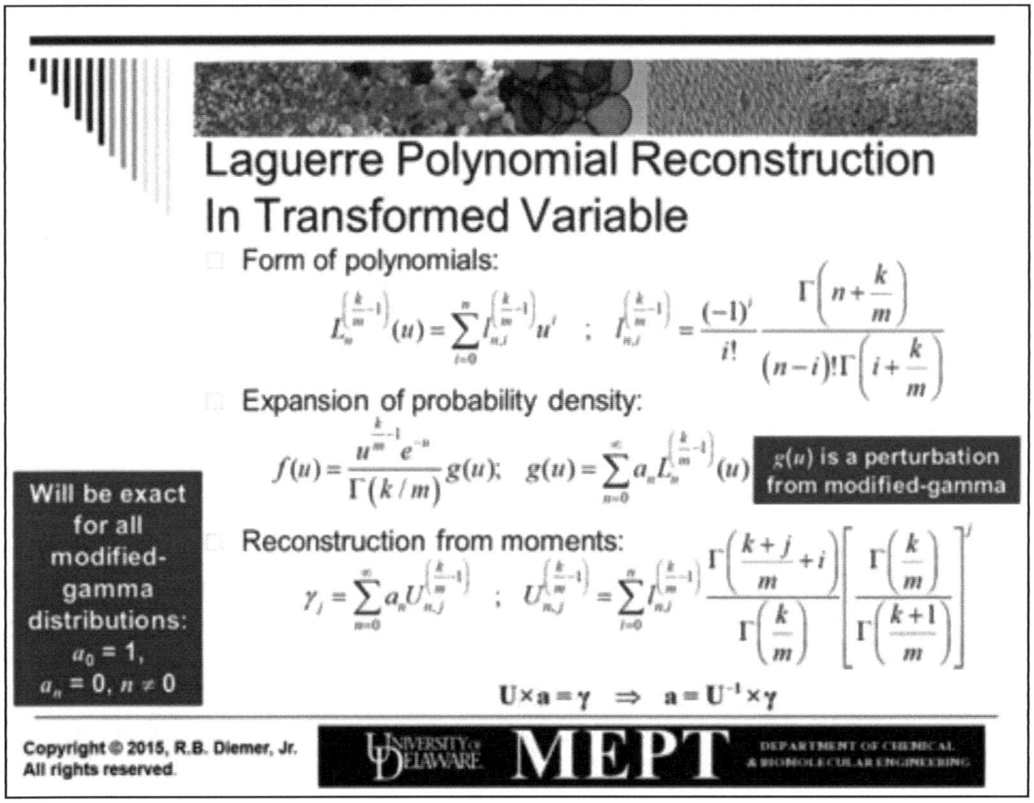

The top equation on this slide shows the Laguerre polynomials now expressed in the new variable u with the parameter α replaced by k/m-1, and also modified so that the weight function includes the normalization factor $1/\Gamma(k./m)$.

Expanding our solutions in Laguerre polynomials with this modified weight factor allows the portion of the solution capture by polynomials of degree greater than zero to be simply perturbations from the modified-gamma distribution.

We can again apply the moment operator to this expansion, and find that the moments are weight sums of slightly more complex terms, but still computable from a combination of Laguerre coefficients and gamma functions.

This means we can populate the matrix $\mathbf{U}$, invert it, and calculate the vector of expansion coefficients from a vector of known moments and thus reconstruct our distribution function. Such a reconstruction will be exact whenever the true solution is a modified-gamma distribution, in which case the coefficient of the 0th degree polynomial will be 1 and all the rest will be zero.

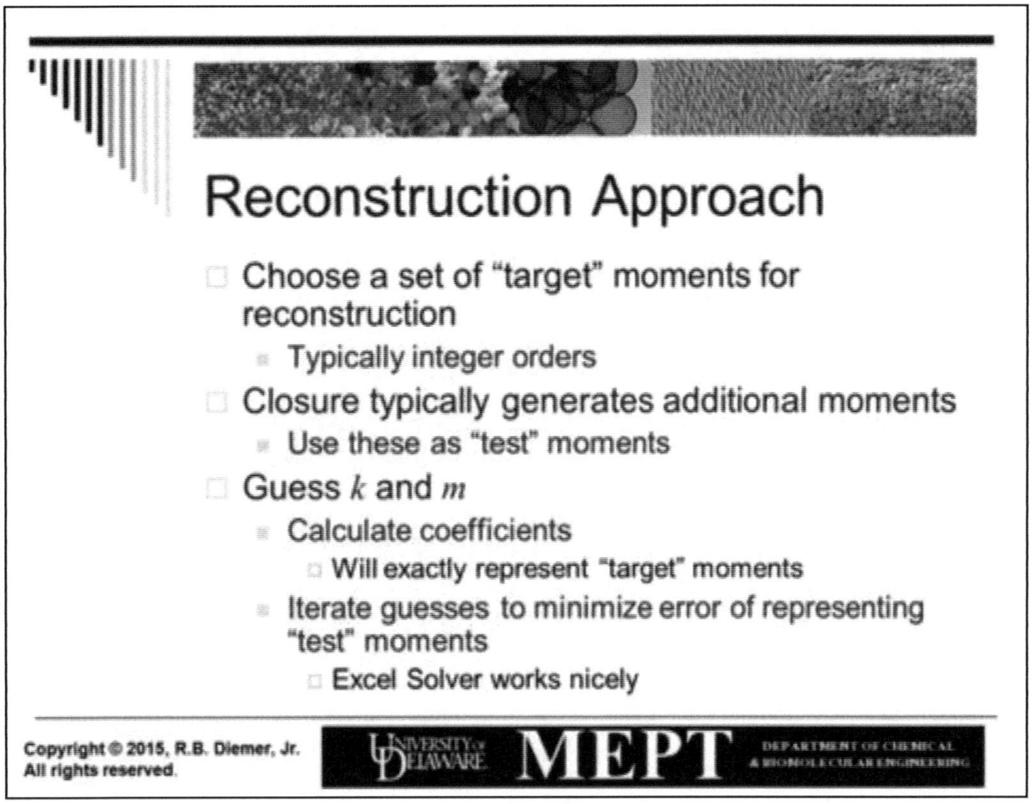

To actually do this, one needs to first solve the moment model. Typically one will solve this for a set of "target" moments, often those of integer order. Closure will typically require estimation of moments of orders outside the target set, hence the need for closure. We can use these additional moments as a "test" set.

If we wish to reconstruct using N Laguerre terms (up to degree N-1), then one needs N target moments. The vector-matrix multiplication of the previous slide (inverse of $\mathbf{U}$ matrix by vector of target moments) will exactly represent the target moment set for any choice of k and m. So, we begin by guessing values of k and m.

We can then calculate the error of representing the "test" moment set for this choice of k and m (typically a root-mean-square percentage error). Using an iterative solver (the one in Excel works nicely), one can then find values of k and m that not only exactly represent the "target" moment set but also minize the error of representing the "test" moment set.

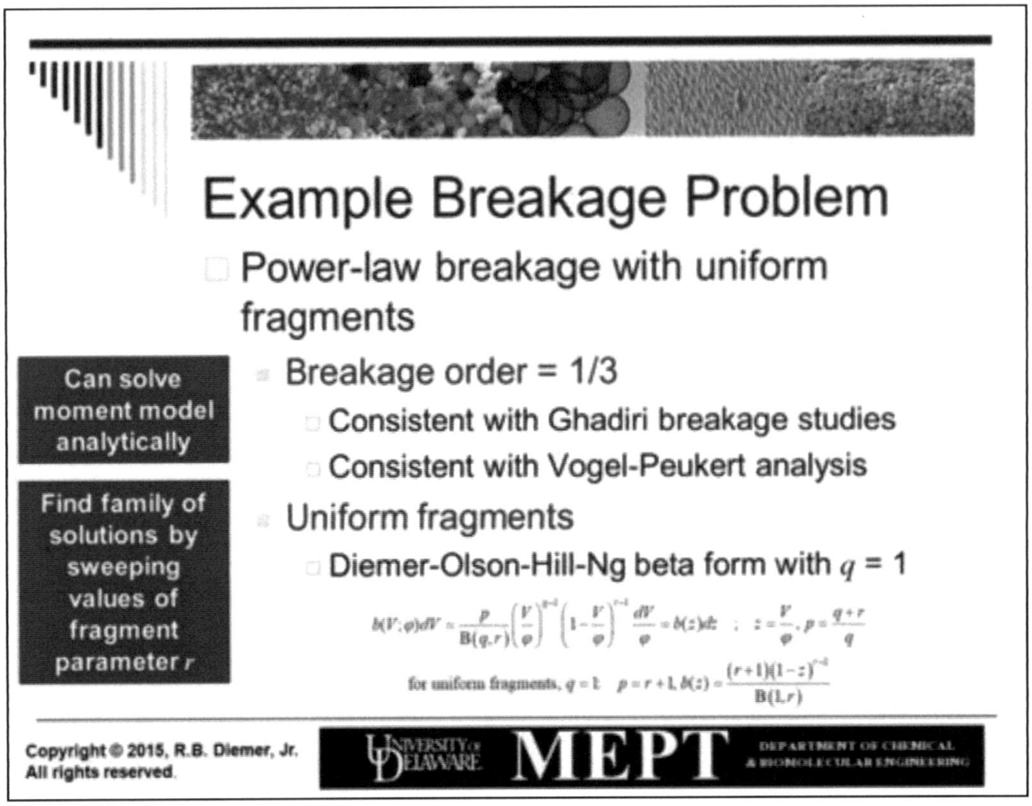

Let's see how this work in an example. Since our lack of success with a breakage problem motivated this work, we'll do a breakage example. We have chose the order of the rate kernel to be 1/3, consistent with both measurements and theoretical arguments concerning the breakage rate of single particles. We'll use a fragment distribution in which the formation of any size particle is equally likely, and we'll look at a family of solutions in which the number of fragments varies.

The moment model for this problem can be solved analytically for moments that are integer multiples of 1/3. We'll choose the target moment set to be those of integer orders 0-9, so that all other moments of fractional order (ranging from 1/3 to 26/3) become the test set.

Multifragment uniform breakage is described by the Diemer-Olson-Hill-Ng beta form in which the parameter q is unity (R. B. Diemer and J. H. Olson (2002) A Moment Methodology for Coagulation and Breakage Problems: Part III – Generalized Daughter Distribution Functions, Chemical Engineering Science, 57:19, 4187-4198.). As shown on the slide, the number of fragments and the distribution function are both described in terms of a single parameter r. We will solve the moment model and reconstruct distributions for various values of this parameter.

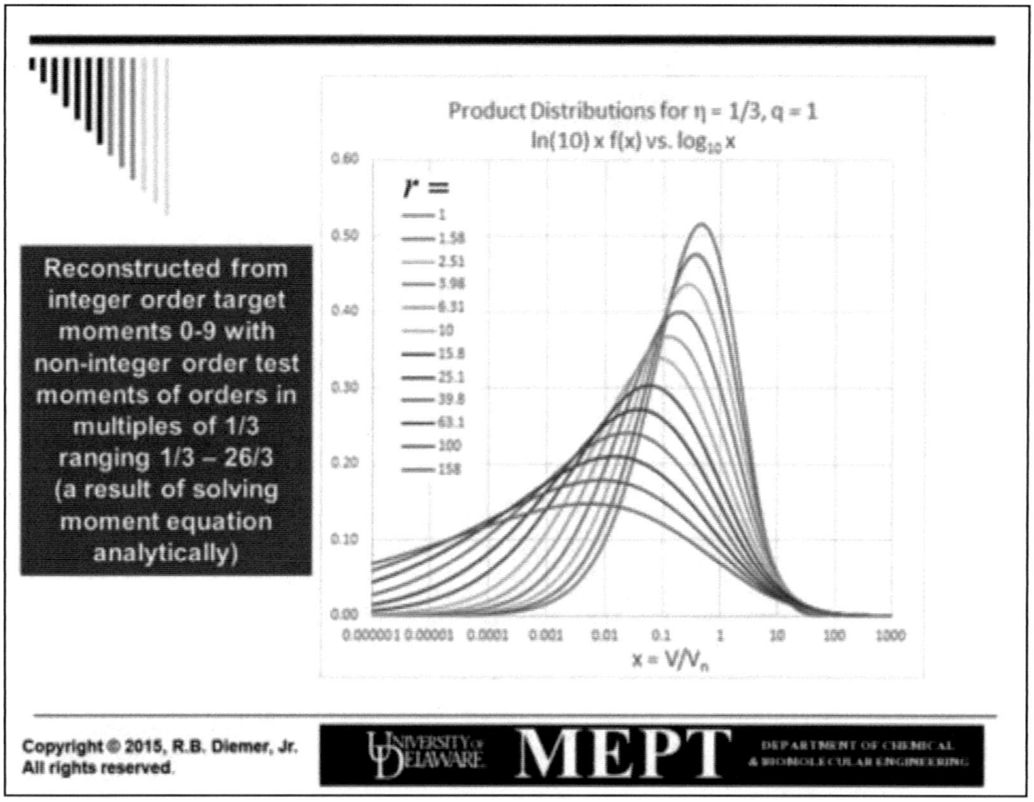

The result is shown on this slide. We can see clearly that as r increases (and the number of fragments increases), the distribution broadens in a smooth way.

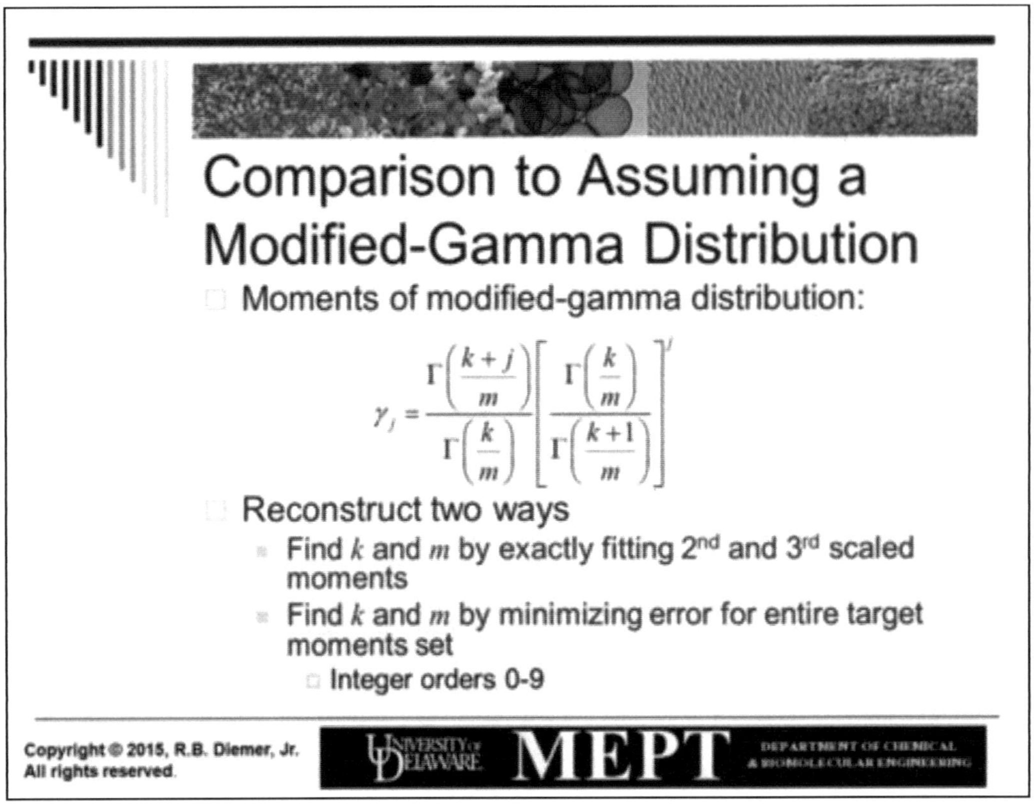

We can compare this reconstruction to what one would find if we simply assumed the modified-gamma form by fitting the parameters k and m in two different ways.

1. We find k and m such that the 2nd and 3rd moments are represented exactly and look at how much error this creates in representing the rest of the target moment set as well as the test moment set.
2. We find k and m so as to minimize the error of representing the entire target moment set (integer orders 0-9) and calculate the test moment set error to go with this solution.

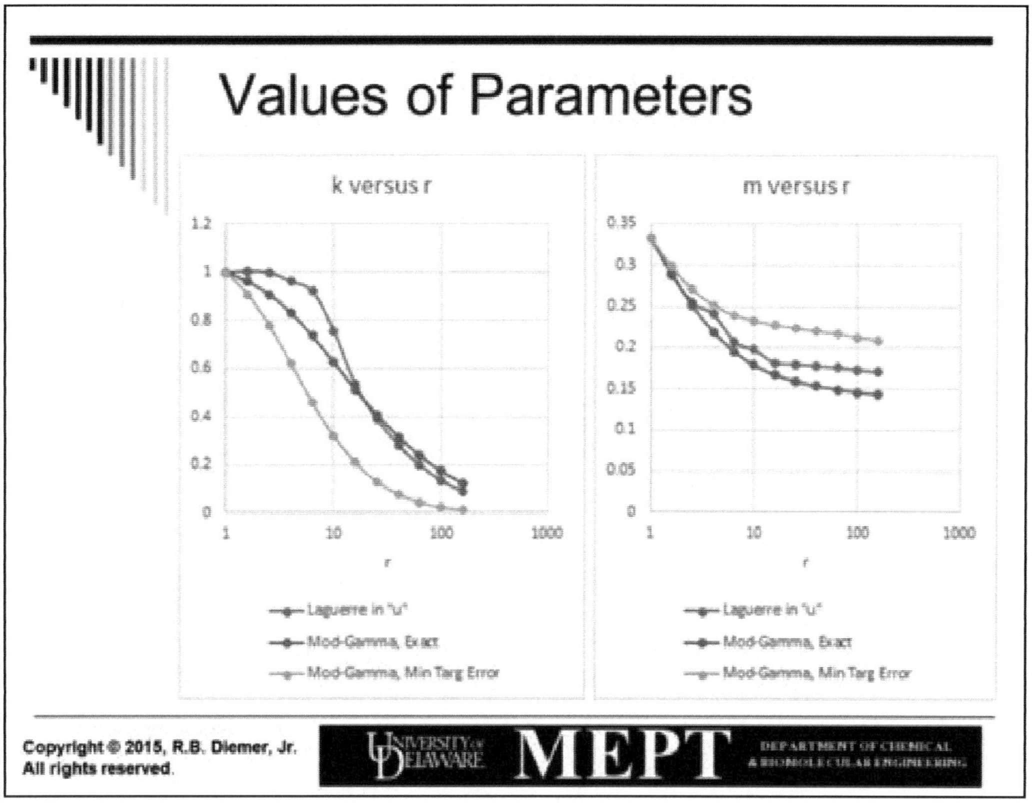

This chart shows the values of k and m found by (1) the Laguerre expansion in the transformed variable u, (2) the exact fitting of a modified-gamma distribution to the 2nd and 3rd moments, and (3) the fitting of a modified-gamma distribution by minimizing the target moment error.

While the values do not agree, the trends in both parameters are the same in the fragment distribution parameter r regardless of the reconstruction technique.

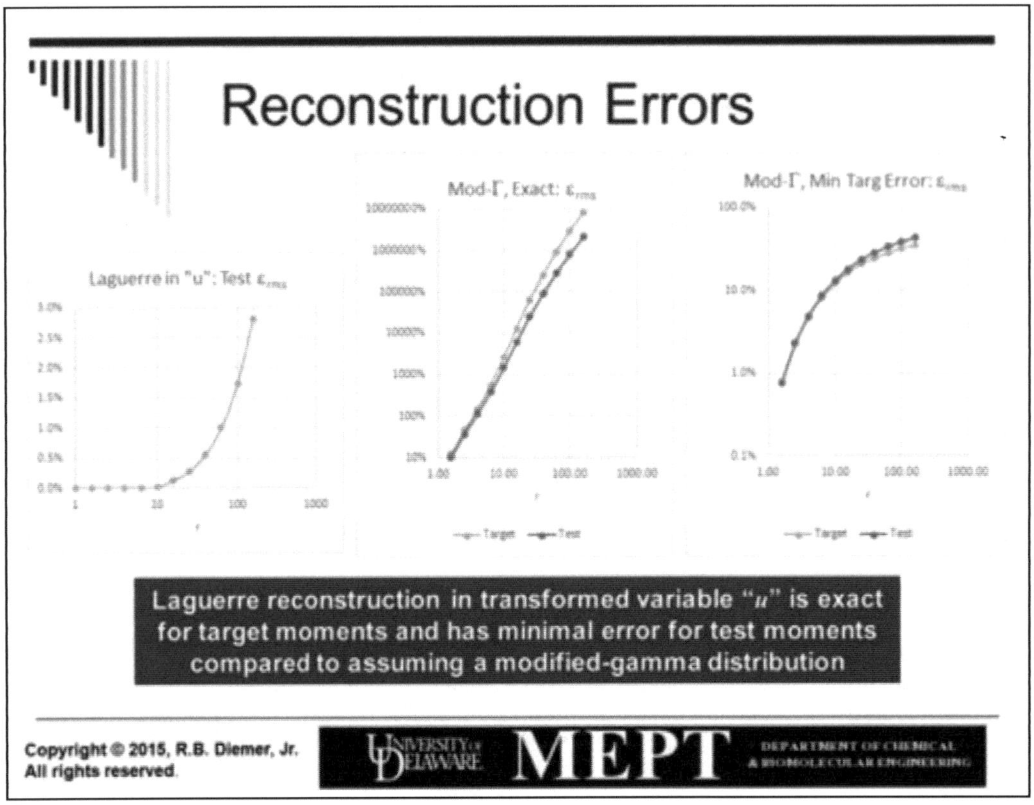

However, the error of moment representation for the Laguerre expansion is dramatically smaller than for either approach simply assuming a modified-gamma distribution. Clearly, the ability of the Laguerre expansion to capture perturbations from modified-gamma is necessary to get a truer reconstruction.

At worst, the test moment error for the Laguerre expansion is a few % and there is no error in representing the target moments. By contrast, simply fitting a modified-gamma to 2 specific moment results in orders of magnitude larger errors for both target and test moments, which are reduced dramatically by minimizing the target moment error. Even so, this second approach to fitting a modified-gamma solution involves much larger errors than incurred by the Laguerre expansion.

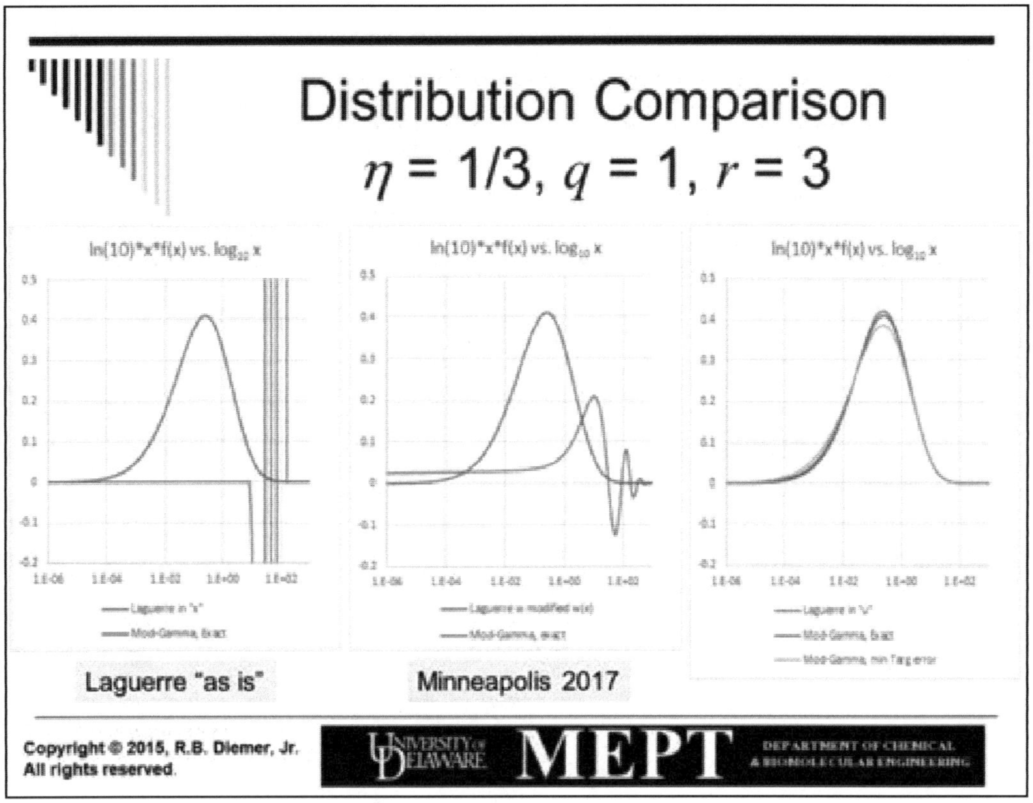

This slide shows a specific case in which r = 3 to show what the evolution to the current state has accomplished. We attempted a reconstruction using the Laguerre polynomials in their standard handbook form and were completely unsuccessful. We compare the result in this graph to what we find if we simply fit a modified-gamma distribution using the 2nd & 3rd scaled moments.

Using the modified-Laguerre polynomials based on a gamma distribution weight function presented last year in Minneapolis, we get a better but still highly unsatisfactory result. This is why we say that last year's approach failed when we tried to do a breakage problem.

On the far right, the same problem is shown using the current approach. Now the Laguerre representation is successfully, and judging by the much lower errors it incurs, is probably closer to true than either of the fits assuming a modified-gamma solution.

This case shows close agreement between the Laguerre reconstruction and assuming a modified gamma distribution. For other case, especially as the number of fragments increases, these three curves begin to agree far less closely. In all cases, the Laguerre reconstruction achieves the smallest error of moment representation is among the three approaches is likely the "gold standard".

Evolution Summary

	Laguerre	Modified Laguerre (Minneapolis 2017)	Transformed Laguerre
Variable:	$x = V/V_n$	x	$u = \left(\dfrac{\Gamma\left(\frac{k+1}{m}\right)}{\Gamma\left(\frac{k}{m}\right)} x \right)^m$
Weight Function:	$x^\alpha e^{-x}$	$\dfrac{(\alpha+1)^{\alpha+1} x^\alpha e^{-(\alpha+1)x}}{\Gamma(\alpha+1)}$ results in modified polynomials	$\dfrac{u^{\frac{k}{m}-1} e^{-u}}{\Gamma(k/m)}$ uses Laguerre polynomials
Parameter(s):	$\alpha > -1$	$\alpha > -1$	$k, m > 0$
Exact for:	exponential distribution	all gamma distributions	all modified-gamma distributions

This table shows how the reconstruction approach has evolved. The "as is" Laguerre polynomials are capable of representing an exponential exactly and are expanded in the scaled variable x = V/V_n. The weight function is shown along with its constrained parameter.

The result presented a year ago is shown in the next column. The weight function is a now a properly normalized gamma distribution, and a new set of modified-Laguerre polynomials were derived from a Sturm-Liouville system employing this new weight function. These polynomials are still in the variable x, and now can exactly represent all gamma distributions, not just the exponential distribution. One the other hand, it was not possible to derive an orthogonal polynomials set for a modified-gamma distribution weight function.

The final column shows the change of variable applied to the modified-gamma distribution resulting in a normalized weight function that is of the standard Laguerre form. This enables use of Laguerre polynomials in the transformed variable so that they are capable of exactly representing all modified-gamma distributions (which include the family of gamma distributions when m = 1 and specifically the exponential distribution when k = m = 1). This dramatically broadens the range of distributions that can be represented. The single Laguerre (and modified-Laguerre) parameter α is replaced 2 parameters such that α = k/m-1, and these can be found such that test moment error is minimized.

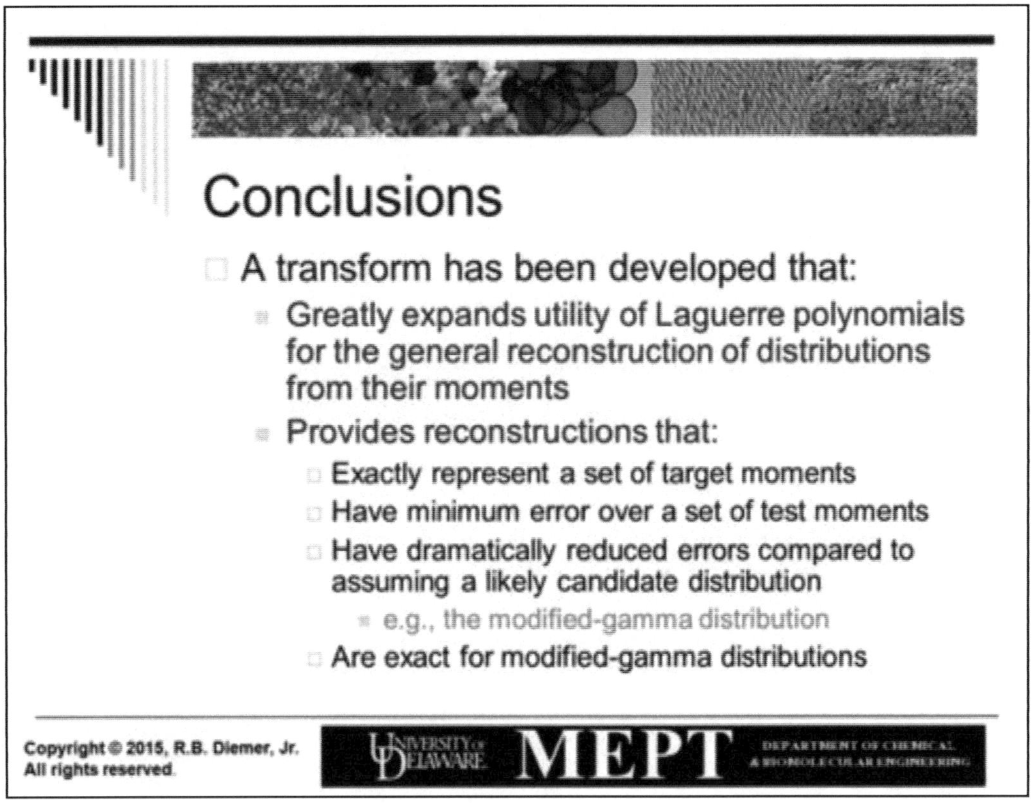

So, to conclude, we have found a way to transform the modified-gamma distribution in the similarity variable x through a change of variable such that form looks like the weight function for the standard Laguerre polynomials if expressed in the new variable (u). This greatly expands the utility of Laguerre polynomials for general reconstruction of distributions from their moments since modified-gamma distribution closely approximate many solutions to population balance problems.

These reconstructions:

- Exactly represent a set of target moments
- Have minimum error with respect to a set of test moments obtained through moment model closure
- Have dramatically reduce error compared to assuming a likely candidate distribution, for example the modified-gamma distribution
- Are exact for all modified-gamma distributions

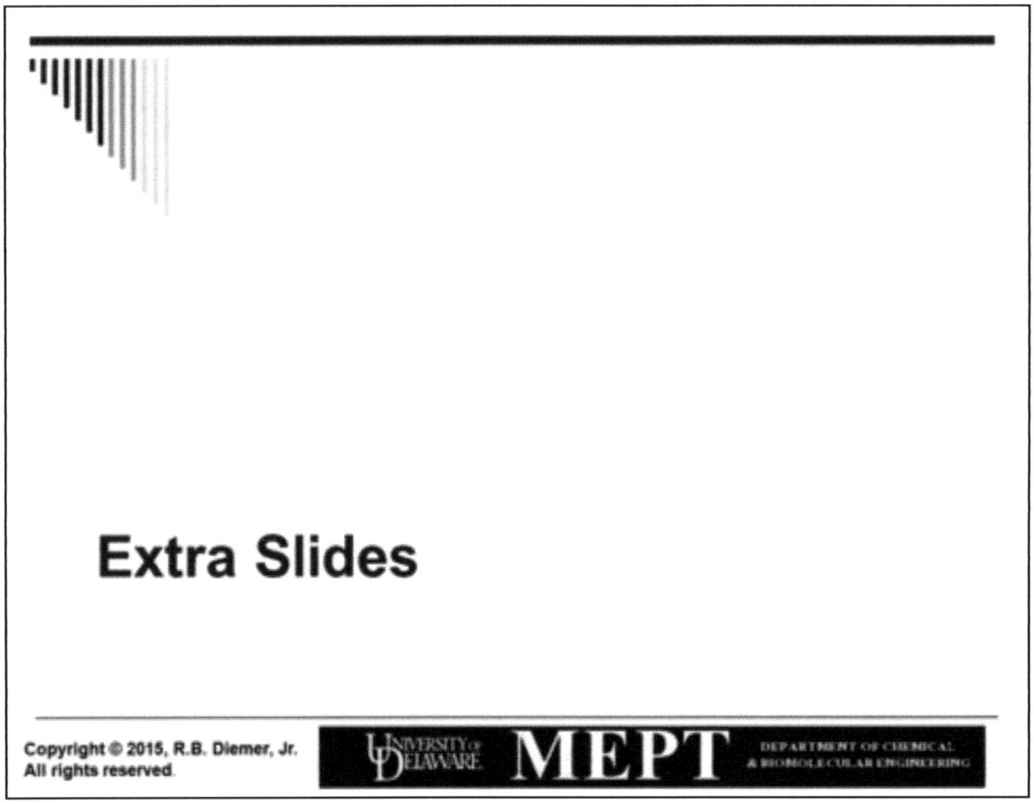

These slides were not used in the presentation. They show the utility for treating other types of problems.

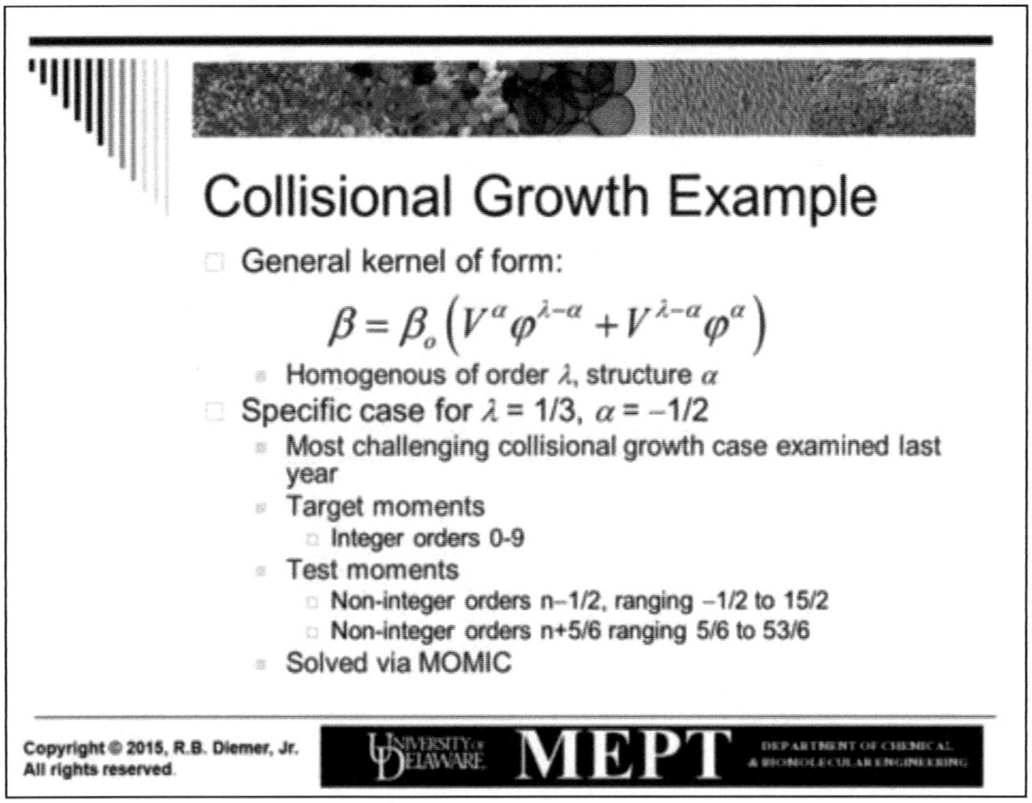

We now look at collisional growth under the general homogeneous kernel of order λ and structure α. The specific case we examine is for a kernel of 1/3 order and -1/2 structure. In Minneapolis last year, we showed that this was expected to admit similarity solutions. We did not show this exact case last year. Instead we showed a less challenging case of the same order but structure of -1/12. This is the most challenging of the cases we looked at then.

Again, the target moments are integer orders 0-9. This time we needed a general closure to find the target moments since no analytical solution could be found. We applied the MOMIC method, and the test moments arising from this closure were of orders n-1/2 and n+5/6.

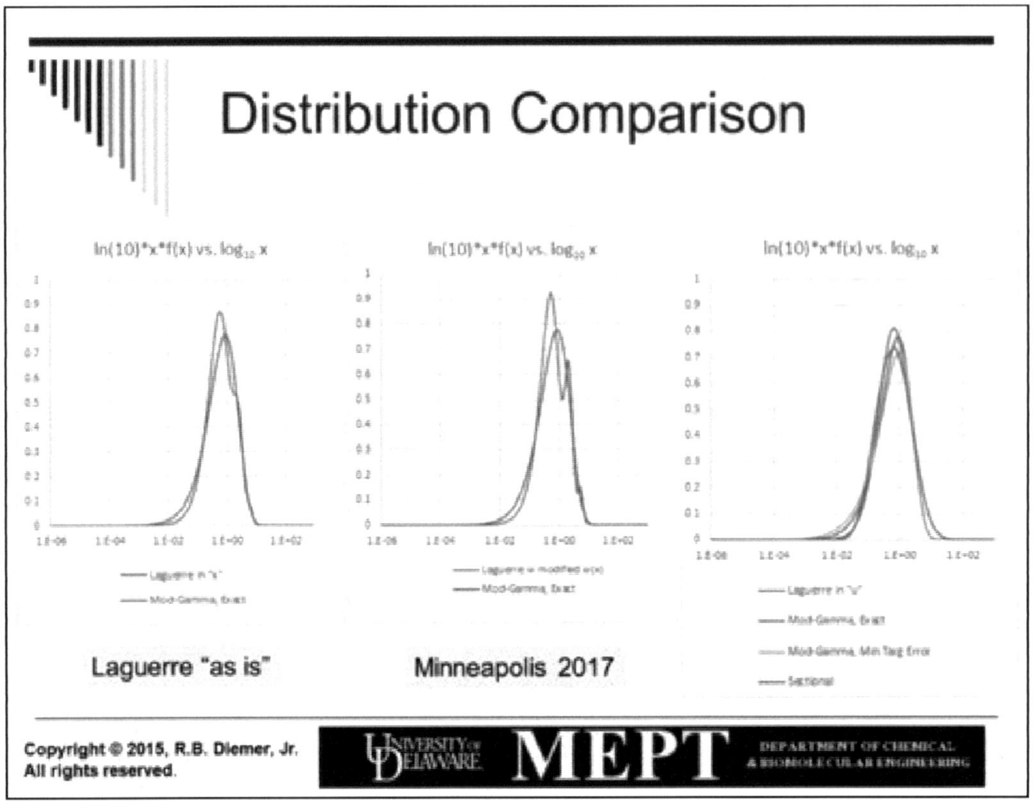

These three graphs show the evolutionary reconstruction sequence from Laguerre polynomials "as is" in the scaled variable (x) to the modified-Laguerre polynomials of a year ago (also in x) to the current Laguerre polynomials in the transformed variable (u).

It is clear that reconstruction for this problem is not nearly as difficult as for breakage. Even the Laguerre "as is" reconstruction (leftmost) at least looks like a distribution function and is fairly close to exactly fitting a modified-gamma distribution to the 2nd & 3rd scaled moments. There is some spurious multimodality, but the width of distribution between the two compare well.

In this case, the reconstruction from modified-Laguerre polynomials (middle) actually looks poorer than the original case on the left.

Using the current approach in the transformed variable (u) eliminates the spurious multimodality. On the right most graph, we show not only this reconstruction, and the exact fit to the modified-gamma distribution, but also what one finds by fitting a modified-gamma at minimum target moment error and also the result of a sectional model. The agreement among all of these is striking. By far, the one representing both target and test moments with the smallest error is the Laguerre reconstruction. The sectional model only captures the 0th and 1st moments exactly.

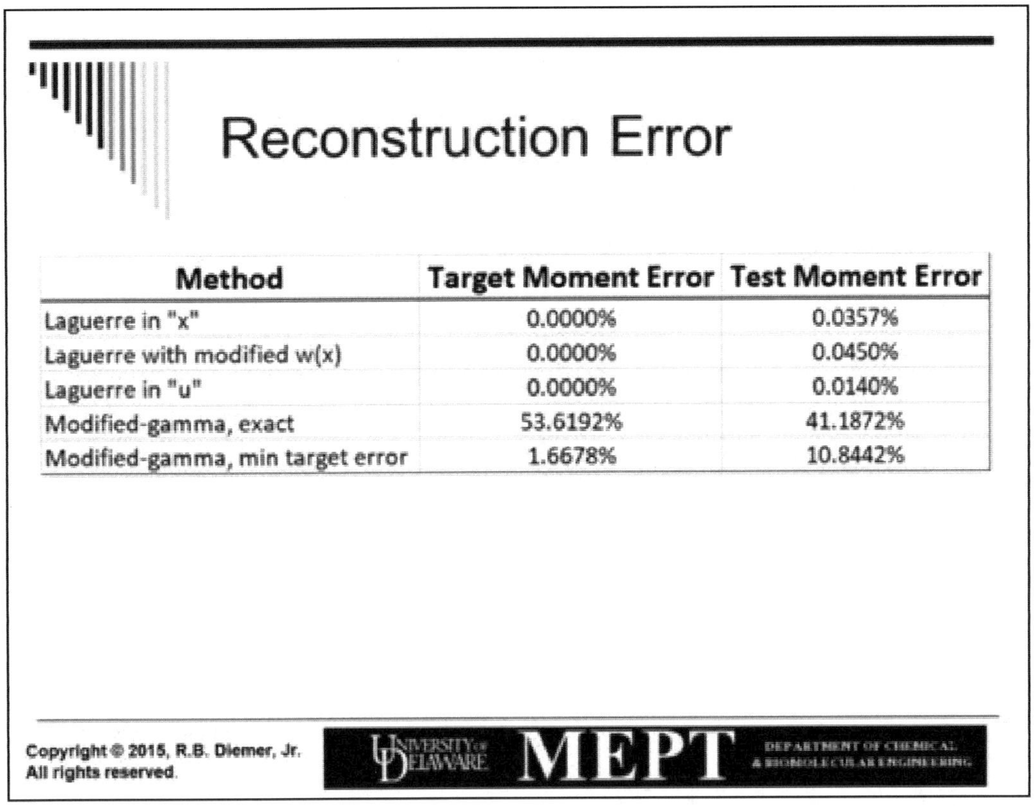

Method	Target Moment Error	Test Moment Error
Laguerre in "x"	0.0000%	0.0357%
Laguerre with modified w(x)	0.0000%	0.0450%
Laguerre in "u"	0.0000%	0.0140%
Modified-gamma, exact	53.6192%	41.1872%
Modified-gamma, min target error	1.6678%	10.8442%

This final slide shows the errors for the reconstructions from moments. All of the polynomial expansions give exact representation of the target moments. By contrast, assuming a modified-gamma distribution results in significant error even when minimized through choice of distribution parameters.

The error of test moment representation is small for all 3 polynomial methods, actually increases slightly for the polynomials with a gamma weight function, but is smallest for the current approach using the transformed variable (u) as the polynomial argument.

This example demonstrates that the approach works well for both collision and breakage and also that breakage is the more difficult problem of the two.

524357 Modeling of Breakage Process Using Monte Carlo Simulations in Spray Fluidized Bed Granulator

Abstract:

Breakage process is an important, yet mostly unexplored phenomenon in a spray fluidized bed granulation process. In this work, the selection function for breakage process is modeled based on the results of constant number Monte Carlo simulation. Here Monte Carlo simulation results are used as the alternative to lab-experimental results. It is found that, the selection function is directly correlated with the overall success factor, collision frequency per particle, total number of particles in the system and the moments of the number density function. Furthermore, another discretized population balance equation is introduced with the discrete form of the breakage distribution function to compute the total number of particles and size distribution of particles numerically. Also, the main properties of a particulate process, such as total number of particles, degree of breakage and size distribution of particles are validated successfully.

522986 Differentiated Research Strategy

Abstract:

The sub-title of the presentation is â€œlooking back over my research in 50 yearsâ€. I will try to speak about the subject of main-title based on my experiences of experiment and DEM simulation. If you are a professional, you must win just like a professional player of sport. To win the game, the important points are selection of the event and strategy. If you or your team has a big budget, it might be natural for you to do research which is gathering the most attention currently. However if you are not in such a good research environment, my story would be something of a useful reference. In the presentation, I will talk about what I have done in my work to seek something different from competitors. Through my experience, I have finally reached the following conclusions. (1) Research with original idea does not need much money. When you scale up the research, you need money. (2) Secure the time to think. (3) Afford to look at other research fields. Not be afraid to enter other fields. Innovation is often done by a new comer. (4) Lucky and unlucky repeat.

529968 DEM Simulation with Scaled-up Particles

Abstract:

After the pioneering work by Professor Tsuji, DEM and DEM-CFD coupling model became a major modelling method for particulate systems. There are many advantages to use DEM over the continuum model; individual particle information can be accessed, particle shape effects can be considered and inter-particle interaction forces can be included, to name a few. On the other hand, one of the existing challenges in DEM is the high computational cost. The massive advances in computational power over the past decades made it possible to simulate millions to billions of particles within a reasonable time. However, this can only be less than a spoonful of particles if using small particles and it is still practically impossible to perform industrial scale simulations. To overcome this problem, several researchers have proposed scaled-up particle models in literature, which is also referred to as coarse grain or similarity models, where the particle size is increased artificially to reduce the number of particles which can significantly speed-up the computation. The present work discusses the recent developments of scaled-up particle models and their effectiveness.

524355 Multi-Scale Modeling of Reactive Dense Flows

Abstract:

Large-scale industrial processes involving coal/biomass gasification and combustion, reactive dense particulate flows coupled with mass, momentum, and heat transfer are frequently encountered. Both inter-phase and intra-particle interactions need to be considered, because these interactions govern the key features of dense particulate flows, such as solid mixing, cluster, and regime transitions. Therefore, these interactions prevail at different length scales and consequently a multi-scale approach is adopted to achieve a quantitative description of these complex flows. Specifically, the relevant details of particle-resolved direct numerical simulation (PR-DNS) and discrete element model (DEM) are used to develop closure laws to feed two-fluid model (TFM) which can be used to simulate dense flows on a much larger (industrial) scale. In recent years, such research has been rapidly developed worldwide. This talk will present an overview of the work in this direction at Zhejiang University. Parallel immersed boundary method has been developed and applied to explore a simplified experimental shallow fluidized bed and a laboratory bubbling fluidized bed coupled with a soft-sphere model. It is demonstrated that the present PR-DNS predicted results show a better agreement with experimental measurements compared with coarser approaches. Large-eddy simulation (LES)-discrete element model (DEM) approach with capability of high performance parallel computation has been developed and validated to explore the dense gas-particle flows in 3-D full-loop circulating fluidized beds (CFBs). Besides, this approach is extended to combine heat transfer and chemical reaction sub-models, used to simulate biomass pyrolysis and gasification in fluidized bed reactors. Moreover, the coarser approaches (coarse grained CFD-DEM, MPPIC, and TFM) are also developed to incorporate chemical reactions for industrial applications.

517364 Real-Time Magnetic Resonance Imaging of Dynamic 3D Granular Systems

Abstract:

Experimental measurements of granular dynamics are vital to test theoretical models and validate numerical simulations. Magnetic resonance imaging (MRI) can provide non-intrusive measurements of particle position and particle velocity of opaque 3D granular systems. However, one major limitation has been the low temporal resolution of particle velocity measurements limiting the application of MRI to stationary systems.

In the present work we combine imaging hardware developments, granular materialsâ€™ engineering and accelerated MRI sequence protocols to increase drastically the temporal resolution of MRI based measurements of granular materials. We apply the new methodology to a variety of dynamic granular systems such as bubbling fluidized beds, granular flow around gas bubbles and the impact of intruders into granular systems. These measurements provide fresh insight into hitherto undiscovered facets of granular dynamics, such as the non-negligible vorticity of granular flow around gas bubbles in fluidized beds. The methodology developed could be used in the future to study other dynamic granular phenomena, such as the jamming transition, granular segregation or wave propagation.

536643 Insights and Model Development Enabled By DEM and CFD-DEM Simulations

Abstract:

Over the past decade, we have applied DEM and CFD-DEM simulations to probe a variety of particulate flows at small scale. They range from rheology of dense granular assemblies (via DEM) to fluidization of particles (via CFD-DEM). Through the DEM studies, we have gained insights on granular shear flow regimes and how inter-particle forces modify this rheology, and formulated continuum rheological models. Through CFD-DEM studies, we have learned how one can incorporate the effects of meso-scale structures into coarse constitutive models of Geldart group B particles, and studied how van der Waals interaction between particles, liquid bridge force between wet particles and electrostatic interactions resulting from tribocharging affect flow. This talk, dedicated to Professor Yutaka Tsuji will highlight some of the results from these studies.

533576 Application of a Modified CFD-PBM Method to the Simulation of a Slurry Bed Reactor

Abstract:

Introduction

Slurry bed reactors, which own many advantages such as high gas-liquid interfacial area and good mass/heat transfer performance between gas and liquid phase, are widely adopted to oxidation and hydrogenation reactions, fermentation, especially Fischer–Tropsch synthesis and so on. In this work, a CFD-PBM method was used to study the hydrodynamics in a slurry bed reactor. In previous work, numerous studies on PBM had been performed to achieve better prediction of bubble characteristics. However, those works were all based on air-water system and had not taken the solid phase into account. In previous experiments, it was found that small amounts of particles can enhance bubble coalescence process, and the average bubble size generally increased with the increase of solid concentration within a certain range, which resulted in the decrease of gas holdup. Therefore, it was proposed that the previous PBM could not be extended to the simulation of slurry bed reactor. However, there are few reports on the modification of PBM for considering the effect of the solid phase. In fact, the main influence of solid concentration lies in the change of liquid apparent properties, including viscosity, density, etc.. There are many reports about the effect of particles on the apparent liquid viscosity and density in slurry bed reactors However, few reports have focused on the apparent liquid surface tension which also has significant effect on bubbles breakup and coalescence. Therefore, the aim of this work is to develop a modified PBM by considering the effect of particles on the liquid surface tension to the application in the simulation of slurry bed reactors.

Methods

Due to the small size (<100 μm) and homogeneous dispersibility of particles in the reactor, the liquid phase and the solid phase was simplified to a pseudo-homogenous phase. Then the two-fluid model (TFM) method based on the Eulerian-Eulerian scheme was applied. It was reported that the addition of small amounts of particles can lead to a higher liquid apparent viscosity and surface tension. So by taking the influence of particles into account, a modified apparent surface tension of the liquid phase was used to modify the PBM. Then Ojima's and Prashka's experiment dates was used to validate the model, and the simulation study was following their experimental equipment and operating conditions.

Results and Discussion

Simulation results that employed the modified and non-modified surface tension were compared to the experiment data, respectively. From the results, it was demonstrated that the modified PBM was suitable to describe the bubbles behaviors in the slurry bed reactor. In addition, it was found that the effect of particles had significant difference under low and higher gas superficial velocity, which may lie on the significant turbulent characteristics between cases under low and higher gas superficial velocity in the slurry bed reactor.

References

[1] Tyagi P., Buwa V. V. Dense gas–liquid–solid flow in a slurry bubble column: Measurements of dynamic characteristics, gas volume fraction and bubble size distribution[J]. Chemical Engineering Science, 2017, 173(7): 346-362.

[2] Wang T., Wang J., Jin Y. Theoretical prediction of flow regime transition in bubble columns by the population balance model[J]. Chemical Engineering Science, 2005, 60(22): 6199-6209.

[3] Wang T., Wang J., Jin Y. A novel theoretical breakup kernel function for bubbles/droplets in a turbulent flow[J]. Chemical Engineering Science, 2003, 58(20): 4629-4637.

[4] Krishna R., De Swarcunt J. W. A., Ellenberger J. et al. Gas holdup in slurry bubble columns: effect of column diameter and slurry concentrations[J]. AIChE Journal, 1997, 43(2): 311-316.

[5] Ojima S., Hayashi K., Tomiyama A. Effects of hydrophilic particles on bubbly flow in slurry bubble column[J]. International Journal of Multiphase Flow, 2014, 58(1): 154-167.

[6] Li H., Prakash A. Influence of slurry concentrations on bubble population and their rise velocities in a three-phase slurry bubble column[J]. Powder Technology, 2000, 113(1-2): 158-167.

540298 Direct Numerical Simulations of Hydrodynamic Forces on Assemblies of Non-Spherical Particles

Abstract:

Historically, the non-spherical particle fluidization simulations are simplified with spher- ical particle assumption. However, there is a large mismatch between the simulation results and experiments. Our work focuses on building a multi-scale modelling procedure for fluidization of non-spherical particles. This work particularly focuses on the direct numerical simulation aspect of the multi-scale modelling. In this work, particle resolved flows are performed around static assemblies of random orientations of non-spherical particles using a multi-relaxation-time lattice Boltzmann method.

The parameter space for assemblies of spherical particles are relatively simpler. They are just limited to different Reynolds numbers (Re) and solids-volume fractions (oÌ¨s). However for non-spherical particles, it is also needed to know the mutual orientation of the particles in the simulation domain and also their mean orientation with respect to the flow direction. Therefore, we propose two parameters S1 and S2 to measure the mutual orientations of the particles and also the incident angle Ï† with respect to the incoming flow. Additional to the drag, we also investigate the dependencies of lift and torques as function of Re, oÌ¨s, S1, S2 and Ï†. From the data obtained, we propose accordingly the correlations for drag, lift and torque coefficients, which can be used to simulate larger scale simulations of non-spherical particles fluidization.

521215 Reactive Crystallization of Metal-Amino Acid Chelates and their Nucleation Kinetics

Abstract:

Zinc(II)-amino acid chelates have been attracted with their anti-inflammatory and even anti-tumoral activities. In the present work, zinc(II) chelate of L-methionine was produced by reactive crystallization. Since it is difficult to measure solubility of chelates sparingly soluble for water, supersaturation ratio (S) was estimated by measuring conductivity of the solution which was correlated in advance with solubility product. To evaluate nucleation kinetics, induction time defined as the time gap between the creation of supersaturation and the appearance of crystals was measured using focused beam reflectance measurement (FBRM) as a function of supersaturation. In an assumption that the formation time of a stable nucleus, t_n, is much larger than the necessary time for the nucleus to grow to a detectable size, t_g, the induction time, t_{ind} ($t_n + t_g$), can be considered as inversely proportional to the nucleation rate. From the linear regression with ln t_{ind} and $\ln^{-2} S$, it was found that the dominant nucleation mechanism is strongly affected by degree of supersaturation.

528674 Investigation of Particle-Size Dependent Charging

Abstract:

Investigation of Particle-Size Dependent Charging

<>X. Liu[1], I. Nwogbaga[1], P. Saba[2], J. Kolehmainen[1], A. Ozel[3], T. Shinbrot[4], S. Sundaresan[1]

[1]Department of Chemical and Biological Engineering, Princeton University, Princeton, New Jersey 08540, USA

[2]Department of Mechanical and Aerospace Engineering, Rutgers University, Piscataway, New Jersey 08854, USA

[3]School of Engineering and Physical Sciences, Heriot-Watt University, Edinburgh EH14 4AS, UK

[4]Department of Biomedical Engineering, Rutgers University, Piscataway, New Jersey 08854, USA

Granular particles tend to develop static charges due to contacts with walls and other particles in transport lines and fluidized beds [1]. These systems often operate with various degrees of particle size distributions. Particles of different sizes of otherwise identical material have been found to charge differently [2]. Different trends were observed in literature, with some studies reporting bigger particles tend to charge positively while smaller particles charge negatively [2,3], as others found the opposite trend [4], but the phenomenon remains poorly understood [1]. In this study, we seek to investigate this effect of particle size on tribocharging.

We first conduct vibrated bed experiments with soda lime glass particles of different sizes in polycarbonate and acrylic containers of different sizes. In each experiment, particles of one chosen narrow size range and one mass loading were shaken for a long time in a container in a dry nitrogen environment and the particle charge was then measured using a Faraday cup. A Discrete-element-method (DEM) simulation [6] complemented each experiment in order to determine the effective work function difference between the particles and container wall that would yield the same level of charge. In this manner, we probe whether the effective function of a particle relative to the wall manifested any systematic dependence on particle size.

Additional vibrated bed experiments are performed using containers whose walls are coated with particles of one size range. Complementary DEM simulations allowed us to probe directly the differences in effective work function difference between particles of different sizes.

The results of these experiments and analyses will be described in this presentation.

[1] Lacks, D., Mohan, S. Contact electrification of insulating materials. J. *Appl. Phys.* 2011; 44(45).

[2] Forward, M., Lacks, D., Mohan, S. Charge Segregation Depends on Particle Size in Triboelectrically Charged Granular Materials. *Phys. Rev. Lett* 2009; 102(2): 028001.

[3] Lee, V., Waitukaitis, S., Miskin, M., Jaeger, H. Direct observation of particle interactions and clustering in charged granular streams. *Nat. Phys.* 2015;11:733â€“737.

[4] Schella, A., Weis, S., SchrÃ¶ter, M. Charging changes contact composition in binary sphere packings. Â *Phys. Rev. E* 2017; 95(6): 062903.

[5] Kolehmainen, J., Ozel, A., Boyce, C. M., Sundaresan, S.. Triboelectric charging of monodisperse particles in fluidized beds. *AIChE J* 2017; 63(6): 1872â€“1891.

[6] Kolehmainen, J., Sippola, P., Raitanen, O., Ozel, A., Boyce, C. M., Saarenrinne, P., Sundaresan, S. Effect of humidity on triboelectric charging in a vertically vibrated granular bed: Experiments and modeling. *Chem. Eng. Sci.* 2017; 173: 363-373.

529078 Effect of Particle Friction on Binary Granular Shear Flows of Inelastic Grains

Abstract:

In this study, shear flows of binary granular systems with various particle sizes and particle friction are modeled using discrete element method in order to explore these effects on the stress behavior. It is found that the stress increases with increasing inter-particle friction until it reaches an asymptote when particles rotate freely. The rotation of particles in shear flows can be restrained in the simulations by incorporating rolling friction or by specifying a large moment of inertia for each particle, in order to account for the effect of an irregular particle shape. Stresses for rotational particles are found to be lower than those for non-rotational particles, as translational energy is partially converted into rotational energy. In addition, for different solid fraction ratios and size ratios, different types of particle-particle contacts (e.g. large-large particle contact, large-small particle contact, and small-small particle contact) play the dominated role. Therefore, the effect of inter-particle friction on shear stress is related to the number and type of particle contacts which is in turn determined by the solid fraction ratio and particle size ratio.

530027 Economic Analysis of Alternative Continuous Crystallization Technologies for Mass Production

Abstract:

Crystallization is an ancient unit operation that remains vital for the chemical process industry [1–3]. Traditional single-effect evaporation consumes a great deal of energy and various alternatives to this method have been proposed. In this work the total cost of producing a fixed quantity of different solid chemicals by crystallization from water is determined for several different technologies: evaporative crystallization (EC), membrane distillation (MD) with porous hydrophobic membranes [4–6], reverse osmosis membrane assisted crystallization [7–8], and eutectic freeze crystallization (EFC) [9–12]. Among the solute properties, the solubility has the greatest effect on the cost of the process since it determines the amount of water that must be removed per unit product produced. If waste heat is available at a unit price lower than that of low-pressure steam produced using coal or natural gas, then the assumed price of waste heat also has a significant effect on the economics. The results indicate that EFC has the lowest cost unless energy (waste heat) is available at a very low cost (less than 1–2 US$/GJ). If EFC is not feasible or not preferred, then a conventional single effect or multiple effect evaporative crystallization or reverse osmosis membrane crystallization process has the lowest cost among the remaining options, depending primarily on the solubility of the solvent.

References

[1] Tung H.-H. Crystallization of organic compounds: an industrial perspective. Hoboken, N.J., Wiley; 2009.

[2] Mersmann A. Crystallization technology handbook. New York, Marcel Dekker; 2001.

[3] Myerson A. S. Handbook of industrial crystallization. Boston, Butterworth-Heinemann; 2002.

[4] Cuellar M. C., S. N. Herreilers, A. J. J. Straathof, J. J. Heijnen and L. A. M. van der Wielen. Limits of Operation for the Integration of Water Removal by Membranes and Crystallization of L-Phenylalanine. Industrial & Engineering Chemistry Research 2009;48(3): 1566-1573.

[5] Curcio E., G. Di Profio and E. Drioli. Recovery of fumaric acid by membrane crystallization in the production of L-malic acid. Separation and Purification Technology 2003;33(1): 63-73.

[6] Curcio E., G. Di Profio and E. Drioli. A new membrane-based crystallization technique: tests on lysozyme. Journal of Crystal Growth 2003;247(1-2): 166-176.

[7] Fountoukidis E., Z. B. Maroulis and D. Marinoskouris. Crystallization of Calcium-Sulfate on Reverse-Osmosis Membranes. Desalination 1990;79(1): 47-63.

[8] Lakerveld R., J. Kuhn, H. J. M. Kramer, P. J. Jansens and J. Grievink. Membrane assisted crystallization using reverse osmosis: Influence of solubility characteristics on experimental application and energy saving potential. Chemical Engineering Science 2010;65(9): 2689-2699.

[9] Randall D. G., J. Nathoo and A. E. Lewis. A case study for treating a reverse osmosis brine using Eutectic Freeze Crystallization-Approaching a zero waste process. Desalination 2011;266(1-3): 256-262.

[10] Fernandez-Torres M. J., D. G. Randall, R. Melamu and H. von Blottnitz. A comparative life cycle assessment of eutectic freeze crystallisation and evaporative crystallisation for the treatment of saline wastewater. Desalination 2012;306: 17-23.

[11] van der Ham F., M. M. Seckler and G. J. Witkamp. Eutectic freeze crystallization in a new apparatus: the cooled disk column crystallizer. Chemical Engineering and Processing 2004;43(2): 161-167.

[12] van der Ham F., G. J. Witkamp, J. de Graauw and G. M. van Rosmalen. Eutectic freeze crystallization: Application to process streams and waste water purification. Chemical Engineering and Processing 1998;37(2): 207-213.

[13] van der Ham F., G. J. Witkamp, J. de Graauw and G. M. van Rosmalen. Eutectic freeze crystallization simultaneous formation and separation of two solid phases. Journal of Crystal Growth 1999;198: 744-748.

530695 Bimetallic Atomic Layer Deposition for Extended Surface Electrocatalysts

Abstract:

Extended surface electrocatalysts based on nanowire templates have emerged as promising candidates for advanced catalysts for the oxygen reduction reaction (ORR) in fuel cell applications. When synthesized by spontaneous galvanic displacement (SGD), these extended surface electrocatalysts have demonstrated exceptional performance (specific activity > 6000 Î¼A/cm^2) and superior durability to benchmark carbon-supported Pt catalysts. However, challenges in scale-up and reproducibility of this fabrication process have prompted an investigation into atomic layer deposition (ALD) as a facile, scalable synthesis alternative for the production of extended thin film electrocatalyst structures. The use of a bimetallic ALD process may also allow for tunable control of multiple deposited metals, as opposed to monometallic deposition with SGD. An ALD analog of the SGD synthesis process was developed by combining oxidative Pt and Ni ALD chemistries into various â€œsupercycleâ€ configurations and depositing these metals onto a cobalt nanowire substrate in a viscous flow reactor. In situ mass spectrometry was used to study the deposition reactions, while elemental analysis, transmission electron microscopy, elemental mapping and X-ray diffraction were used to characterize Pt growth and effects of post-treatment procedures, such as H_2 annealing.

530905 An Experimental Study of Cylindrical Particle's Effective Size in a Rotating Tumbler

Abstract:

There is a dearth of studies that focus on shape segregation. This is due, in part, to the complex nature of the flow of differently shaped particles within a dense flow. In this paper, we explain the segregation of cylindrical tracer particles inside a rotating tumbler that is half filled with a background media comprised of spherical particles. We evaluate all existing methods proposed in the literature to calculate the cylindrical particles' effective size and compare these methodologies to results under a wide range of particle conditions. We find that a newly proposed method not only more accurately predicts the cylindrical particles' effective size, but also shows consistent results for both disk and cylindrical particles. This work represents a positive first step toward merging shape segregation into the more mature and well-studied size segregation ecosystem.

538543 Laser Pyrolysis Synthesis of Novel Nanoparticles Using Spray-Based Precursor Delivery

Abstract:

Nanoparticle synthesis, the production of particles from 1 to 100 nm in characteristic size, is a key component of nanotechnology. Vapor-phase nanoparticle synthesis has advantages including high throughput continuous operation, scalability, and product purity that have given it a dominant role in the commercial production of nanomaterials. Laser pyrolysis is one of the important vapor phase nanoparticle synthesis methods. In laser pyrolysis, a laser beam is used to selectively heat a gas stream containing nanoparticle precursors, such that they decompose, inducing nucleation of nanoparticles. One of the most important advantages of laser pyrolysis is its flexibility to synthesize nanoparticles of diverse materials using appropriate precursors. A key advantage of laser pyrolysis compared to furnace-based methods is that only the gas, which has small heat capacity, is heated and, moreover, only that part of the gas that absorbs the laser energy is heated. As a result, the gas benefits from very rapid heating and rapid cooling by mixing of the heated gas with unheated gas. Gaseous, liquid, or solid precursors can be used in this method. Although gaseous precursors have most often been used in laser pyrolysis synthesis, solid or liquid precursors would be preferable in many cases, due to issues of safety and precursor cost and availability. These precursors can be delivered to the reaction zone as small droplets that rapidly evaporate upon laser heating. Precursors with sufficiently high vapor pressure could also be delivered as vapors from a bubbler. However, for some metals, no sufficiently volatile precursor is available. Accordingly, spray-based precursor delivery is the most general approach to delivering precursors for the synthesis of nanoparticles by laser pyrolysis. This poster will describe our recent progress in using ultrasonic spray delivery of precursors to our laser pyrolysis reactor to prepare nanoparticles from nonvolatile precursors.

533598 Particle Size Techniques/Capabilities Used in the Coatings Industry

Abstract:

Particle size distribution plays an important role in the coatings industry as it affects many paint properties including rheology and optical properties. Three techniques that PPG commonly utilizes for particle size characterization are dynamic light scattering (DLS), laser diffraction, and centrifugation sedimentation. This presentation will focus on each particle size technique, as well as demonstrate the importance of the optical parameters needed for each technique and how to choose the appropriate analysis for an application. Several case studies involving latex particle characterization with DLS and pigment analysis with centrifugation sedimentation will also be highlighted.

535256 Modeling Granular Material Segregation Using a Multi-Scale Model

Abstract:

A multi-scale model is presented for predicting the magnitude and rate of segregation of a binary mixture of particles in a rotating drum and conical hopper. The model combines diffusion and segregation correlations from the literature with advective flow field information from finite element method simulations. Predictions of the segregation rate from the multi-scale model compare well quantitatively to published DEM simulation results and experimental data. A standard method is introduced to calibrate the diffusion and segregation correlations for different materials. The multi-scale model, since it does not directly model individual particles, is expected to scale well to systems of industrial interest.

536193 Oxidation of Fractal-like Soot Agglomerates

Abstract:

Oxidation is the last soot formation stage and determines the mass, mobility, d_m, and primary particle diameter, d_p, of particulate emissions. The fractal-like soot agglomerates produced by nucleation,[1] surface growth[2] and agglomeration[3] react with molecular oxygen and shrink producing gaseous CO and CO_2 pollutants. Accurate soot oxidation kinetics are essential to optimize combustion and aftertreatment devices.

Here, an analytical model for the evolution of soot d_m and d_p during internal and external oxidation is derived. External oxidation takes place on soot primary particle surface reducing d_p and d_m, while internal burning decreases the soot bulk density, Ï. The unoxidized soot $Ï_o$ and $d_{p,o}$ is related to Ï and d_p by $Ï=Ï_o(d_p/d_{p,î¿})^3$, finding excellent agreement with thermogravimetric and microscopic data of carbon black, diesel and ethylene soot. The common assumption that soot nanoparticles are spheres (i.e. $d_m = d_p$) results in 35 % overprediction of the d_m measured during oxidation. Thus, a relationship for the d_m of agglomerates formed by coagulation and surface growth[3] is used to account for the ramified soot morphology. The evolution of soot mobility size distribution during internal and external oxidation dynamics of fractal-like soot agglomerates is in excellent agreement with oxidation measurements of ethylene soot. The analytical model developed here from first principles can be coupled with mobility size measurements to probe accurately soot oxidation kinetics or with population balance models to design efficient and clean combustion processes.

References:

[1] Kholghy, M. R.; Kelesidis, G.A.; Pratsinis, S.E., Reactive polycyclic aromatic hydrocarbon dimerization drives soot nucleation. *Phys Chem Chem Phys* **2018**, *in press*.

[2] Kelesidis, G. A.; Goudeli, E.; Pratsinis, S. E., Flame synthesis of functional nanostructured materials and devices: Surface growth and aggregation. *Proc Combust Inst* **2017,** *36*, 29-50.

[3] Kelesidis, G. A.; Goudeli, E.; Pratsinis, S. E., Morphology and mobility diameter of carbonaceous aerosols during agglomeration and surface growth. *Carbon* **2017,** *121*, 527-535.

536805 Experimentally Validated Computational Models to Predict the Impact of Humidity on the Flow of Pharmaceutical Mixtures

Abstract:

Moisture-induced flow variability in pharmaceutical mixtures lead to multiple impediments during manufacturing of solid dosage formulations as the processing and storage humidity conditions can bear significant impact on the final product quality. Experimentally validated Discrete Element Method (DEM) based computational models has been developed to predict the effects of humidity on pharmaceutical powder flow by altering the cohesive forces based on granular bond numbers in simple geometries. Statistical formalism (Simplex Centroid Design) has been performed to understand the moisture induced cohesion in binary and tertiary mixtures at 20%, 40% and 60% RH. V-blending was applied to prepare the pharmaceutical blends, and mixing characterization was performed using a Raman PhAT probe. Optimum fill volume was established for the mixing conditions to minimize static charging due to blender wall interactions on the pharmaceutical powders. Statistical analysis predicted and quantified the non-linearity of the moisture-induced cohesion between the pharmaceutical powders within the blends, based on their systematic hopper discharge studies (experiments and simulations). A methodical implementation of these quantification tools were hence performed to validate a design space that enables an approach to the appropriate selection of blend concentrations that achieve minimum mixture flow variability across different humidity conditions.

537624 Conduction and Convection Heat Transfer in a Rotary Drum Using an Integrated PIV/IR Technique

Abstract:

A new measuring technique for studying heat transfer in a multiphase rotary drum is proposed using infrared (IR) thermography. An infrared camera is used to record images to give instantaneous thermal and hydrodynamic data of the solid bed inside the drum. The established technique particle image velocimetry (PIV) is performed on the IR images to obtain combined quantitative (i.e., hydrodynamic and thermal) data sets. The combined technique provides insightful information on the conduction and convection heat transfer mechanisms in a rotary drum.

This work also aims to investigate the effects of rotation speed, fill level, particle size and their interaction effects on the conduction and convection heat transfer mechanisms in rotary drums. For this, the discrete element method (DEM) (using MFIX-DEM, an open-source computational fluid dynamics solver suite) simulation technique is used to analyze the thermal behavior of spherical silica beads, which are the most common catalyst support material.

537867 Light Alkane Valorization to Ethylene Via Chemical Looping Oxidative Dehydrogenation

Abstract:

Ethylene is commercially produced from light alkanes, like ethane and naphtha, via high-temperature pyrolysis in the presence of diluting steam. The process requires a high temperature (>1100 K) and a large steam generation load, making it highly energy intensive (17-30 GJ/tonne ethylene). The process also emits significant amount of NO_x and CO_2 and requires periodic shutdowns for coke burnout. Moreover, single-pass ethylene yield is limited by thermodynamic equilibrium. Oxidative dehydrogenation of ethane (ODH) has been investigated as a potentially more efficient approach. In ODH, the feedstock is partially oxidized into ethylene and water. The oxidation of hydrogen makes the ODH reaction exothermic, thereby reducing fuel needs while overcoming the equilibrium limitations for cracking reactions. However, to allow economical product separation, conventional ODH schemes require the use of pure oxygen. The oxygen generation, which uses cryogenic air separation systems, is capital and energy intensive. We propose a chemical-looping ODH (CL-ODH) approach that can address these issues. In this scheme, a metal oxide based redox catalyst provides oxygen for the ODH reaction from its lattice. The oxygen depleted redox catalyst is subsequently oxidized in air, regenerating the catalyst and releasing the heat needed for the process. This allows built-in air separation with minimal parasitic energy loss and better temperature control.

We have identified, in our previous work, a family of mixed oxides which can function as excellent redox catalysts for ODH. They are capable of supplying lattice oxygen at rates comparable to the rate of hydrogen formation via thermal cracking of the alkane feed. The bulk and near-surface properties of the oxide system can be improved by modifying the redox catalyst to produce ethylene with exceptional selectivity by suppressing deep oxidation of ethylene. The facile combustion of hydrogen favored by the promoted redox catalyst leads to high ethylene yield providing the heat required for the endothermic dehydrogenation reaction. CL-ODH of these light alkanes is modeled using ASPEN Plus and is compared with a conventional stream cracking process. Results show that CL-ODH with 85% ethane conversion provides over 80% reduction in the overall energy demand and CO_2 emissions, with over 50% savings, for the naphtha feed. The exothermic nature of the regenerator and elimination of the steam requirement lead to major reductions in the upstream energy consumption.

527882 Optimising Granulate Formulation through Uniaxial Powder Testing

Abstract:

Granulation is often employed during the production of pharmaceutical formulations for oral solid dose delivery. Due to the potency of most Active Pharmaceutical Ingredients (APIs) the quantity required per capsule or tablet is extremely small, therefore a granulation step is often required to ensure content uniformity; this is especially important when other components of the formulation have poor flow properties.

The quality of the final dried granules is dependent on both the formulation and process variables, for example, shear rate and mixing time. Inferior quality granules can reduce process efficiency and result in poor final product quality. Furthermore, as reworking the dried granulate is not possible, this may result in the loss of expensive and scarce APIs. As such optimising, the formulation and process variables is essential for efficient process performance and final product quality.

This study demonstrates the application of an Advance Uniaxial Powder Tester as a tool for the rapid evaluation of dried granulate materials produced using high shear wet granulation. It consisted of two stages, in the initial stage the water content of the wet mass was varied, increasing from 10% to 40% wt/wt whereas the second stage investigated the influence of the premix formulation on the dried granulate properties by adjusting the ratio of granulation grade lactose monohydrate to microcrystalline cellulose (MCC).

In order to prepare the granules, predetermined quantities of lactose and MCC were blended to provide 750mL of premix, the raw materials were mixed for 3 minutes using a low shear Tubula mixer to ensure a uniform mixing. The premix was then transfer to a high shear wet granulator, mixed for a further 30 seconds and then granulated for 270 seconds with water addition over the first 90 seconds of granulation. The wet mass was then screened through an 850 micron sieve before drying at 60°C for 7 hours. Once dried the granules were sieved using a 2 mm sieve to remove large agglomerates. The granules were then evaluated using a Uniaxial Powder Tester (UPT; Freeman Technology, UK) and digital microscopy.

The results from stage 1 showed a clear correlation between the granulate properties and the water content of the wet mass, with lower uniaxial Unconfined Yield Strength (uUYS) values at higher water contents. Lower uUYS values indicate weaker interparticle bonding and improved flowability. A similar trend was also observed for compressibility with 40% water generating the least compressible granules, suggesting more efficient particle packing. Microscopic analysis of the samples showed that granules start to form at 30% water content and dominate at 40% explaining the improved flowability. Below 30% water the improvement in flowability (compared with the dry powder) may be due to a reduction in the small particle fraction and /or changes in the surface properties.

In stage 2, both the premix and dried granules were quantified with the UPT, allowing comparison of the flow properties before and after granulation, with the granulates demonstrated to be much more free-flowing. Both the premix and dried granules exhibit a similar response to increasing the lactose content, generating lower uUYS values at higher lactose contents, however the response was more pronounced for the dried granules. The compressibility results were markedly different; whilst the granules generate lower compressibility values at higher lactose contents, the premix exhibits an increase in the compressibility, demonstrating that compressibility alone is not necessarily an indicator of flowability. Again, the change in the granulate flow properties can be further rationalised by microscopic analysis. As the ratio of lactose to MCC increases, the percentage of granules also increases with a step change at 70% lactose exhibited by a significant reduction in the small particle fraction and more elongated granules. This corresponded with an observed change in the granulation behaviour with both the 70% and 85% wet masses blocking the 850 micron screen, significantly reducing yields.

Overall the results demonstrate that uniaxial testing is a quick, simple and repeatable technique for quantifying the flowability of powders and granules, enabling the influence of formulation and process changes to quantified. The results also demonstrate the granulation can result in a marked improvement in flow behaviour compared with the corresponding dry mix and that there are clear trends between water content and formulation, and flow properties of the granules.

527887 Optimising Powder Properties for DPI Capsule Filling Performance

Abstract:

Most single dose and multi-unit dose Dry Powder Inhalers (DPI) dispense the Active Pharmaceutical Ingredients (API) alongside a suitable carrier from a pre-metered capsule or blister pack. Due to the small particle size of the API (typically less than 5 microns) they tend to be highly cohesive so an excipient, for example lactose monohydrate, is required as a transport medium for the API. As the drug formulation is ejected from the inhaler the carrier is stripped away as the API continues to the deep lung. The selection of the carrier excipient will not only influence the ejection of the powder from the inhaler, and subsequent detachment of the API, but also the filling of capsule and blister packs during the manufacturing process.

The pre-metered dose required for DPIs varies considerably depending on the API and medical condition, from less than 1mg to tens of milligrams. In order to dispense small quantities, dosator technology is routinely used, with modern dosators capable of filling up to 180,00 capsules per hour. The operating principle of dosators is relatively simple, the dosator is inserted into the powder bed, forcing powder into the cavity, the powder is then compacted via the dosator pin, locking the powder into the cavity. The dosator (including the powder plug) is then retracted from the powder bed, and excess powder is then removed during the doctoring stage before the powder plug is ejected into the capsule, blister pack or vial.

Due to the way dosators operate they are more sensitive to the physical properties of the powder compared with other filling mechanisms. As the powder is dispensed by volume rather than mass the bulk density must be uniform throughout the powder bed, the powder should also flow into and fill the voids formed by the dosator, finally the powder should exhibit sufficient cohesion/compressibility to from a stable plug during retraction, doctoring and ejection. Dosators are therefore suited to a limited range of powders necessitating the need to match the powder properties to the dosator.

In order to further investigate this relationship, 5 grades of lactose with varying particle sizes were processed through a lab-scale dosator (3P Innovation, Warwick, UK) using outlets of progressively smaller size. The single arm dosator is capable of delivering up to 450 doses per hour, and in conjunction with a rotating powder bed, conditions the powder prior to and during operation using a blade and pin system, ensuring a uniform bed height. For each dosator-lactose combination, the standard deviation across 30 runs was measured for a target doses of 50 mg. The target relative standard deviation (RSD) was 2%. The powders were also evaluated using an FT4 Powder Rheometer (Freeman Technology Ltd, Tewkesbury, UK) to quantify dynamic flow, bulk and shear properties.

The performance varied significantly depending on the combination of lactose and dosator outlet size. Optimum performance was not related to particle size, and instead correlated with dynamic flow properties. Powders with low Specific Energy (SE) and Aerated Energy (AE) values were demonstrated to have the best performance across all four dosators, however as the dosator outlet size decreased sensitivity to aeration became less influential while the impact of inter-particular friction and inter-locking increased.

With the largest dosator outlet â€" Dosator 1, optimal performance was demonstrated by lactose grades 3 and 4 which both generate low AE and low to mid SE values, whereas powders with high AE values - Lactose 1 and 2 and high SE values - Lactose 5 all presented substandard filling behaviour suggesting that dosator 1 is influenced by both the AE and SE. At the other end of the spectrum, Dosator 4 was less sensitive to the AE, with Lactose 1, 2 and 3 generating acceptable performance despite high AE values, whereas Lactose 4 and 5, both of which presented low AE values but high SE values generated higher than acceptable RSD values indicating that this dosator is much more sensitive to mechanical forces.

The larger dosators allows for a greater interaction with air at the outlet, increasing inter-particle spacing which in turn reduces interparticular interactions, as such there is a stronger correlation with AE. With smaller outlets, bridging/arching across the outlet due to friction and mechanical interlocking, as quantified by SE, dominate.

These results demonstrate that even minor variations in powder properties can result in substandard filling performance, as such itâ€™s important that powders are matched to the dosator. Powder rheology, and dynamic flow testing in particular, can support the development of more easily processed DPI formations and the selection of equipment that will consistently deliver the correct dose.

531429 The Effect of Storage Time on Flow Characteristics of Maic-Modified Compounds

Abstract:

In order to provide more consistent processing and product quality, flow aids, such as fumed silica or magnesium stearate, are typically added to a powder to improve its flow. These flow aids are typically added via a traditional blending method, with the amount of flow aid being the only processing parameter that could be used to control changes in powder flow.

In an ongoing study, Magnetically Assisted Impact Coating (MAIC) has been shown to improve flow properties of 5 Î¼m citric acid (a model organic compound) and 17 Î¼m cornstarch (a model natural polymer) when compared with a traditional blending process, v-blending. A series of samples were created and measured repeatedly in a longitudinal study to look at the effects of storage time on flow properties, especially their basic flow energy (BFE) as measured on an FT4 Powder RheometerÂ®. Samples with and without flow aids are measured.

530262 Bioinspired Silica: A Novel, Green and Biocompatible Drug Delivery System

Abstract:

The development of a drug delivery system (DDS) is essential in many cases to remedy the limitations of free drug molecules. Silica has been of great interest as a DDS, however, their fabrication often involves harsh chemicals and energy intensive laborious methods. This work details the employment of a bioinspired "green" method[1] to develop a one step and one pot method for simultaneous silica synthesis and drug loading.[2]

Using various drugs, we investigated bio-inspired silica (BIS) as a new DDS and compared it to mesoporous silica; the latter have received much attention for drug delivery applications. The BIS synthesis enables controllable *in situ* loading of drugs by carefully designing the DDS formulation conditions. Here we systematically studied these conditions and it was found that the drug loading efficiency, loading and release could be enhanced significantly. By identifying ideal formulation conditions for BIS, we produced a DDS that was able to release fivefold more drug per weight of silica when compared with mesoporous silica. BIS was found to be considerably less cytotoxic[2,3] than their current counterparts. Further it was found that, although 20% of BIS was able to pass through the gut wall into the bloodstream, it was nonhemolytic (2% hemolysis at 500 µg mL^{-1}) when compared to mesoporous silica (10% hemolysis at the same concentration). Overall for DDS, it was clear that BIS has several advantages over mesoporous silica (ease of synthesis, controllability and lack of hazardous chemicals) as well as being less toxic, making BIS a real potentially viable green alternative to DDS.

References:

1. Patwardhan, S.V. *Commun.*, 47, 7567, 2011.
2. Davidson et al. *ACS Biomater. Sci. Eng.*, 2, 1493, 2016.
3. Steven et al. *Mater. Chem. B*, 2, 5028, 2014.

545183 Purdue Universityâ€™s Center for Particulate Products and Processes

Abstract:

This poster presents an overview of Purdueâ€™s Center for Particulate Products and Processes (CP3). The CP3 seeks to address the lack of knowledge related to particulate materials. Our vision is to create new functionality, improve quality, and reduce the costs of particulate-based products. Specifically, our mission is to provide the knowledge, tools, and trained workforce needed to effectively design and manufacture particulate products. The faculty and students in the CP3 are involved in research, consulting, testing, and coursework focused on a range of topics and applications related to particulate materials. We have extensive experience working with pharmaceuticals, agricultural, and energetic materials as well as consumer products, catalysts, food products, and batteries.

541565 25 by 25: Chemical Engineering in the Next 25 Years

Abstract:

In 2008, the centennial year for AIChE, we asked 25 chemical engineers for their visions of the profession from around the US and the world: Nine faculty, eight from industry, and eight post-docs and graduate students. We posed four questions related to each individual's industrial sector or area of research, asking for extrapolation, new impacts on existing sectors, new sectors, and an optional, more open-ended comment on the future of the profession as a whole.

In this 110th year of AIChE, we have revisited these questions with many of the same people. Ten years has seen many developments in our profession, in society, and in our thought-leaders' experiences. Here we report on their perspectives, comparing them to the 2008 visions.

541542 The Future of Chemical Engineering Itself

Abstract:

Chemical engineeringâ€™s future is bright because it is vital to so much of society and industry -- and its future is complicated by its rapidly expanding universe of activities. It must represent depth in core skills centered on the chemical sciences including biology and materials. At the same time, it must be transdisciplinary; it must exemplify the need and value of reaching across disciplines in borderless ways.

One way is to create and disseminate a compelling vision and strategy for chemical engineering, research, and education for the future. AIChE and the National Academies are working together to that end. In 1988, the National Academies published "Frontiers in Chemical Engineering: Research Needs and Opportunities.â€ Its committee of authors outlined a wide-ranging roadmap to turn promising research opportunities into reality, while calling for researchers to embrace new frontiers.

Since 1988, that has happened -- along with many other transformative changes like the Web, fracking, personalized medicine, and social media. A new nationwide effort, aided by international contributions, will address both the identity of chemical engineering and its future directions. Its goal is to ensure that the people in chemical engineering are primed to meet society's needs and rapid changes.

544708 Accelerating Innovation through Academic-Industrial Partnerships

Abstract:

In 2011, The Dow Chemical Company launched the University Partnership Initiative (UPI) to sponsor collaborative research at leading U.S. Universities. This 10-year commitment aims to address fundamental questions in industrial chemistry and chemical engineering with alignment to Dowâ€™s next-generation process technology and product development efforts. To date, this portfolio of projects has generated an impressive array of intellectual property, publications, and numerous examples of the collaborative developments implemented at Dow. This presentation will introduce UPI program and highlight several contributions from our collaborators and leaders in the chemical engineering profession.

540559 Maximizing Uptime, Efficiency, and Safety of Industrial Operations through Early Risk Detection

Abstract:

Achieving zero incidents while maximizing process uptime and efficiency is a key vision for operating companies. An important building block towards this vision is establishment of a proactive risk mitigation culture, supported by effective systems and a strong management. Studies have shown that most of the incidents and unexpected process failures can be avoided if the operating teams get timely information about developing risks and take preventive actions early on. While advances have been made in process monitoring and automation in the past decades, there still remains significant technological and behavioral gaps that prevent plant operations from proactively mitigating process risks.

A modern industrial plant monitors thousands of parameters, generating upwards of 50-100 million data points every day. Thanks to recent breakthroughs in machine learning and artificial intelligence approaches, today there are autonomous systems that can sift through this data and point out meaningful and timely insight. This can help operating teams ascertain process issues that are hidden in the data, long before process variables reach critical levels.

In this presentation, a new autonomous system, Dynamic Risk Analyzer™ (DRA), will be introduced that points out early indicators of process issues using proprietary machine learning. The risk indicators enable operations team to respond and make required process changes, days and sometimes weeks ahead of any alarm, avoiding late stage (expensive) fixes. To realize these objectives fully, a proactive management workflow must also be in place. The presentation will discuss new initiatives and workflows developed by plant operations to implement a proactive culture. Real-life case studies from plants using DRA technology will be presented on how this culture change resulted in increased uptime, efficiency and safety of plant operations.

531309 Gaussian Processes for Hybridizing Analytical & Data-Driven Decision-Making

Abstract:

Surrogate models are widely appreciated in process systems engineering [1]. The typical setting focuses on expensive-to-evaluate, possibly uncertain functions. Examples include: modular process simulators [2], integrated gasification combined cycle processes [3], a carbon capture absorber [4], and many other applications, e.g. [5-8]. Resources are typically limited, so effective decision making requires data-efficient learning.

The data science and statistical machine learning communities typically focus on models learned solely from observed data. But chemical engineering applications may also require explicit, parametric models, e.g. modeling known process constraints, operations constraints, and cost objectives [9]. So, work has integrated semi-algebraic functions with those learned from data [10] or developed semi-physical modeling techniques [11, 12].

This presentation surveys the state-of-the-art in hybridizing analytical and data-driven decision making [13]. We consider three probabilistic modeling applications to these hybrid situations:

Design of experiments for model discrimination [14]. We bridge the gap between classical, analytical methods [15] and Monte Carlo-based approaches [16]. Classical methods may have difficulty managing non-analytical model functions and data-driven Monte Carlo approaches come at a high computational cost. We replace the original, parametric models with probabilistic, non-parametric Gaussian process surrogates learned from model evaluations. The surrogates are flexible regression tools that extend classical analytical results to non-analytical models, while providing us with model prediction confidence bounds and avoiding the computational complexity of Monte-Carlo approaches.

Multi-objective optimization [17, 18]. We make novel extensions to Bayesian multi-objective optimization in the case of one analytical objective function and one black-box, i.e. simulation-based, objective function. The resulting method has been applied to a bone neotissue application [19] and a more general test suite.

Scheduling plant operations under uncertainty. For processes with equipment degradation, we use Gaussian processes to approximate large-scale, mixed-integer optimization problems.

We close by offering a broad outlook on applying probabilistic surrogate models to chemical engineering. Statistical machine learning has recently attracted significant interest in process systems engineering [20]. Here we show that state-of-the-art research in Gaussian processes [21, 22] and probabilistic modeling more generally [23] can have a big impact on chemical engineering.

References

[1] Bhosekar A, Ierapetritou M. Advances in surrogate based modeling, feasibility analysis and and optimization: A review. *Comput Chem Eng*. 108: 250-267, **2018**.

[2] Caballero JA, Grossmann IE. An algorithm for the use of surrogate models in modular flowsheet optimization. *AIChE J*. 54(10):2633-2650, **2008**.

[3] Lang Y, Zitney SE, Biegler LT. Optimization of IGCC processes with reduced order CFD models. *Comput Chem Eng*. 35(9):1705-1717, **2011**.

[4] Cozad A, Sahinidis NV, Miller DC. Learning surrogate models for simulation-based optim-ization. *AIChE J*. 60(6):2211-2227, **2017**.

[5] Boukouvala F, Ierapetritou M. Surrogate-based optimization of expensive flowsheet modeling for continuous pharmaceutical manufacturing. *J Pharm Innov*. 8: 131, **2013**.

[6] Soepyan FB, Cremaschi S, Sarica C, and others. Estimation of percentiles using the Kriging method for uncertainty propagation. *Comput Chem Eng*. 93:143-59, **2016**.

[7] Shokry A, Ardakani MH, Escudero G, Graells M, Espuña A. Dynamic Kriging-based fault detection and diagnosis approach for nonlinear noisy dynamic processes. *Comput Chem Eng*. 106:758-76, **2017**.

[8] Tran AP, Georgakis C. On the estimation of high-dimensional surrogate models of steady-state of plant-wide processes characteristics. *Comput Chem Eng*. DOI 10.1016/j.comp-chemeng.2018.02.014, **2018**.

[9] Boukouvala F, Hasan MF, Floudas CA. Global optimization of general constrained grey-box models: new method and its application to constrained PDEs for pressure swing adsorption. *J Glob Optim*. 67(1-2): 3-42, **2017**.

[10] Boukouvala F, Floudas CA. ARGONAUT: AlgoRithms for Global Optimization of coNstrAined grey-box compUTational problems. *Optim Lett*. 11(5): 895-913, **2017**.

[11] Pearson RK, Pottmann M. Gray-box identification of block-oriented nonlinear models. *J Process Contr*. 10(4):301-315, **2000**.

[12] Cozad A, Sahinidis NV, Miller DC. A combined first-principles and data-driven approach to model building. *Comput Chem Eng*. 73: 116-27, **2015**.

[13] Boukouvala F, Misener R, Floudas CA. Global optimization advances in mixed-integer nonlinear programming, MINLP, and constrained derivative-free optimization, CDFO. *Eur J Oper Res*. 252: 701-727, **2016**.

[14] Olofsson S, Deisenroth MP, Misener R. Design of Experiments for Model Discrimination Hybridising Analytical and Data-Driven Approaches. *arXiv:1802.04170*. **2018**.

[15] Buzzi-Ferraris G, Forzatti P, Emig G, Hofmann H. Sequential experimental design for model discrimination in the case of multiple responses. *Chem Eng Sci*. 39(1): 81-85, **1984**.

[16] Vanlier J, Tiemann CA, Hilbers PAJ, van Riel NAW. Optimal experiment design for model selection in biochemical networks. *BMC Syst Biol*, 8(20), **2014**.

[17] Olofsson S, Mehrian M, Geris L, Calandra R, Deisenroth MP, Misener R. Bayesian Multi-Objective Optimisation of Neotissue Growth in a Perfusion Bioreactor Set-Up. *Computer Aided Chemical Engineering*. 40: 2155-2160, **2017**.

[18] Beykal B, Boukouvala F, Floudas CA, Pistikopoulos EN. Optimal design of energy systems using constrained grey-box multi-objective optimization. *Comput Chem Eng*. DOI 10.1016/j.compchemeng.2018.02.017, **2018**.

[19] Mehrian M, Guyot Y, Papantoniou I, Olofsson S, Sonnaert M, Misener R, Geris L. Maximizing neotissue growth

kinetics in a perfusion bioreactor: An *in silico* strategy using model reduction and Bayesian optimization. *Biotechnol Bioeng*. 115(3):617-29, **2018**.

[20] Lee JH, Shin J, Realff MJ. Machine learning: Overview of the recent progresses and implications for the process systems engineering field. *Comput Chem Eng*. DOI 10.1016/j.compchemeng.2017.10.008, **2017**.

[21] Damianou A, Lawrence N. Deep Gaussian processes. In *Artificial Intelligence and Statistics*. 207-215, **2013**.

[22] Deisenroth MP, Ng JW. Distributed Gaussian processes. In *32nd International Conference on International Conference on Machine Learning*. 37: 1481-1490, **2015**.

[23] Ghahramani Z. Bayesian non-parametrics and the probabilistic approach to modelling. *Phil Trans R Soc A*. 371(1984):20110553, **2013**.

528485 Ceria Nanoparticle Dissolution and Stability in Acidic Aqueous Environments

Abstract:

Ceria (cerium oxide) nanomaterials, or nanoceria, have many applications such as acting as a redox catalysis and in catalyst supports, solid oxide fuel cells, sintering additives, heat resistant alloy coatings, and as UV absorbents and superconducting materials. The cerium atoms on the surface of nanoceria can store or release oxygen, cycling between Ce^{3+} and Ce^{4+}, and can therefore relieve oxidative stress within the body. As an anti-oxidant, nanoceria has to potential to reduce nausea and bacterial reproduction, stimulate the immune system to assist in fighting infection and disease, and prevent cancerous tumor growth. Nanoceria dissolution is present in acidic environments *in vivo*. In order to accurately define the fate of nanoceria *in vivo*, nanoceria dissolution or stabilization is observed *in vitro* using acidic aqueous environments.

Nanoceria stabilization is a known problem even during its synthesis; in fact, a carboxylic acid, citric acid, is used in many synthesis protocols. Citric acid adsorbs onto nanoceria surfaces, capping particle formation and creating stable dispersions with extended shelf lives. For example, citric acid surface coatings prolong nanoceriaâ€™s residence time within the blood stream. Dynamic light scattering (DLS) was utilized to determine the hydrodynamic diameter of nanoceria in aqueous solutions as a measure of dispersion stability. Nanoceria is shown to agglomerate in the presence of some carboxylic acids over a time scale of up to 28 weeks, and degrade in others, at pH of 4.5 (representing that of phagolysosomes). Sixteen carboxylic acids were tested: citric, glutaric, tricarballylic, 2-hydroxybutyric, 3-hydroxybutyric, adipic, malic, acetic, pimelic, succinic, lactic, tartronic, isocitric, tartaric, dihydroxymalonic, and glyceric acid. Each acid is introduced as 0.11M, into pH 4.5 iso-osmotic solutions. Controls such as ammonium nitrate, sodium nitrate, and water are also tested to assess their effects on nanoceria dissolution and stabilization.

To further test stability, nanoceria suspensions are subjected to light versus dark milieu. Light induced nanoceria agglomeration in some, but not all ligands, and is likely to be a result of UV irradiation. Light initiates free radicals generated from the ceria nanoparticles. Some of the ligands completely dissolved the nanoceria when exposed to light. Citric, malic, and lactic acids form coordination complexes with cerium on the surface of the ceria nanoparticle that can inhibit agglomeration. This approach identifies key functional groups required to prevent nanoceria agglomeration. The impact of each ligand on nanoceria are analyzed and will ultimately describe the fate of nanoceria *in vivo*.

Supported by NIH Grant 1R01GM109195.

525648 Probing Peptoid-Carbon Nanotube Coatings for Biological Imaging

Abstract:

Single-walled carbon nanotubes (SWNT) have emerged as ideal candidates for biological imaging due to their tissue-transparent near-infrared fluorescence, and infinite photobleaching lifetimes.[1] However, the hydrophobic surfaces of SWNT require ligand functionalization or polymer encapsulation to promote aqueous colloidal stability for use in biological environments. We introduce peptoids, N-substituted glycine polymers, as easily synthesized programmable polymers with a large monomer space and that are less susceptible to proteolytic degradation than peptide or protein polymers.[2] We show that peptoids can be used as a novel carbon nanotube coating by pinning peptoid polymers to the SWNT surface through pi-pi and electrostatic interactions. Herein, we identify the stabilizing interactions of peptoids on SWNT surfaces. We probe the effect of peptoid polymer length, hydrophilicity, initial dispersion conditions, and ionic buffer conditions on the resulting colloidal stability of peptoid-SWNT conjugates. These findings allow us to stabilize carbon nanotubes with rationally designed peptoid polymers and enable us to effectively use peptoid-based SWNT coatings for biological imaging.

1. Zhang, J. *et al.* Molecular recognition using corona phase complexes made of synthetic polymers adsorbed on carbon nanotubes. *Nat. Nanotechnol.* **8,** 959â€"68 (2013).
2. Nam, K. T. *et al.* Free-floating ultrathin two-dimensional crystals from sequence-specific peptoid polymers. *Nat. Mater.* **9,** 454â€"60 (2010).

529604 Synthesis, Characterization, and Interfacial Properties of Lignin Coated Iron Oxide Magnetic Nanoparticles in Aqueous Solutions

Abstract:

Lignin is an abundant natural raw material on the planet with attractive properties, including high thermal stability, adsorbent properties, and antioxidant characteristics. However, lignin is a waste product with very low-value applications in the paper pulp industry. Ongoing studies are focusing on the development of value-added high-performance materials using lignin, which includes the development of nano- and micron-sized particles. In this study, core-shell nanostructures were synthesized using magnetic nanoparticles as the core functionalized by lignin macromolecules as the shell. In addition to the versatile properties of lignin, the hybrid nanostructure is attractive due to their ease of manipulation under the presence of a magnetic field and their inherent superparamagnetic properties. In this work, magnetic iron oxide-lignin nanoparticles were synthesized using two methods: a co-precipitation technique using iron salts to produce core-shell nanostructures (lignin@Fe_3O_4), and nanoprecipitation using ethanol/water solutions. For the co-precipitation technique, lignin was dissolved in ammonium hydroxide; whereas a lignin solution in ethanol was used for the nanoprecipitation method. Various concentrations of lignin were used to stabilize iron oxide nanoparticles, and the stabilization or the aggregation of the nanoparticles were characterized as a function of the weight ratios between lignin and the magnetic core for both synthetic routes. The nanoparticles were characterized by using a variety of experimental techniques including DLS, TEM, UV-Vis, FTIR, XRD, VSM and zeta potential analyses. The hybrid nanoparticles showed various sizes ranging from 17 -25 nm for single core-shell lignin@Fe_3O_4 nanostructures and from 100-200 nm for the aggregates. TEM images of the particles displayed encapsulation of iron oxide cluster within the lignin polymer matrixes at high concentrations of lignin. The surface tension of lignin@Fe_30_4 nanoparticles in water and ethanol/water mixtures was measured at different nanoparticle concentrations using the pendant drop method. The results of this work show that superparamagnetic iron oxide nanoparticles coated with lignin were successfully synthesized. The outstanding properties of lignin as a renewable material, which could be used as antioxidant, emulsion stabilizer, UV adsorbent, antimicrobial properties, and preparation of biomaterials make the lignin-magnetic nanocomposite a unique substrate for a plethora of future investigations.

530075 Investigation of Interactions between Magnesium Silicate Particles and Diamond-like Carbon Surface By Atomic Force Microscopy

Abstract:

Contamination of nanoparticles on surfaces becomes a growing concern in various industries, especially in microelectronic industries since the size of the components have drastically reduced. Although the atmospheric dust is generally controlled, particles from various sources such as processing components and operating environments are often detected. This work emphasizes on the adhesion of magnesium silicate particles onto diamond-like carbon (DLC) surface. Magnesium silicate particles are applied in many daily uses such as skin-care products, cosmetics, or even papers, while DLC thin film has been considered as a promising material for protection of electronic components because of its high hardness, low friction, high thermal conductivity and chemical inertness. Moreover, due to low hardness of magnesium silicate that leads to strong adhesion to the surface, removal of magnesium silicate particles becomes a challenging problem.

In this work, two types of magnesium silicate, i.e., commercial talcum powder and Mg_2SiO_4 particles synthesized by the hydrothermal method, were studied. The magnesium silicate-DLC interaction was investigated by atomic force microscopy (AFM) operated in force-volume mode, in which DLC coated silicon AFM tip was used. The particles were deposited on a silicon wafer, washed with deionized water, and dried at room temperature before subjected to the AFM analysis.

Preliminary experiments, which compare AFM analyses of talcum particles using a bare silicon tip to that using a DLC coated silicon tip, revealed that the adhesion forces measured are significantly different, although the spring constants of both tips are similar. The average adhesion forces measured using the bare silicon tip and the DLC coated silicon tip are 80 nN and 40 nN, respectively. Hence, tip-particle interaction can be used to represent particle-surface interaction.

In the experiments comparing talcum and Mg_2SiO_4 particles, the normal spring constants of the DLC coated tips were calibrated by thermal noise method (k = 0.02 N/m). The adhesion measurements were conducted several times to achieve statistical representation of the data. According to the measured force-displacement curves, both adhesion force and force acting on both types of particles are significantly different. The average adhesion force on talcum powder tends to be larger than that on Mg_2SiO_4 particles. Although the difference is suggested to be a result from different hardness of the particles (i.e., the hardness of talcum is 1-1.5 in Mohsâ€™ scale, while that of Mg_2SiO_4 is 7), other factors such as chemical composition may also involve.

To verify the effect of chemical functional group on the surface of magnesium silicate particle on the tip-particle interaction, surface modification was conducted. Talcum naturally contains surface hydroxyl, but the content of the hydroxyl group can be varied by a treatment with various kinds of acid. For surface modification, 100 g of talcum powder was immersed in 1 L of concentrated acid at 80°C for 2 h. The acids employed were formic acid, acetic acid and hydrochloric acid. The treated particles were filtered, washed with deionized water and dried in an oven at 70°C for 48 h. According to Fourier transform infrared spectroscopic analyses, the talcum powder treated with hydrochloric acid showed significant increase in the signal corresponding to hydroxyl group, while that of the powders treated with organic acids is roughly unchanged. However, the adhesion force of hydrochloric acid-treated talcum particle is about the same as the untreated talcum. On the other hand, the talcum treated with organic acids showed much increased adhesion forces. It is therefore suggested that hydroxyl group does not contribute to magnesium silicate-DLC interaction. Instead, grafting of the surface of talcum by organic groups, which leads it to be hydrophobic, can enhance the adhesion because DLC is also hydrophobic. Hence, hydrophobicity of the particles plays an important role in adhesion onto the DLC surface in addition to the softness of the particles.

533086 Polymer-Metal Composite Nanoparticles Via Vapor Phase Deposition Processes Onto Liquid Substrates

Abstract:

The tuning of polymer nanoparticles is important for applications in catalysis, optics, and electronics. Metal nanoparticles have already shown promise in these applications because of their high surface area-to-volume ratio and durability. However, methods to create polymer-metal nanocomposites are expensive, time consuming, and require the use of many solvents. In this work, we combine the vapor deposition methods of sputter deposition and initiated chemical vapor deposition (iCVD) to create metal-polymer composite nanoparticles. By combining these two vapor deposition methods, we are able to tune the polymer-metal composite functionality independently. We characterize the resulting structures using dynamic light scattering and transmission electron microscopy.

535945 Controlling Surface Morphology and Spatial Distribution of Active Nanoinclusions in Functional Coatings Via Air-Controlled Electrospray Process

Abstract:

Electrospray process is used to produce electrically charged droplets of sizes in micron and submicron range. This process is used in diverse applications like production of thin and uniform coating, nanodroplet and nanostructure production, biomedical engineering, energy storage and mass spectrometry. In this process, the electrospraying liquid is pumped out of a needle through a syringe and an electric field is applied between the needle and the collecting plate. When the liquid is released out of the needle the electric field causes the charges to separate inside the liquid meniscus and it takes the shape of a cone at the needle, which is called as Taylor cone. When Taylor cone becomes unstable, depending on the process conditions either jets or droplets are ejected from the cone. In the spray process for viscoelastic liquids, the jet extends due to viscoelastic forces and the modulations on the jet radius due to axisymmetric instabilities cause the jet to break-up. Whereas, in non-viscoelastic liquids, the atomization of the meniscus at the nozzle is responsible for the formation of droplets. In the electrospray process, the electric field force is responsible for the growth of the axisymmetric instability and breaking the jet into droplets. In the air spray process, the external air flow is applied axially to the jet and the air drag accelerates the growth of the axisymmetric instability. In this work, we study the spray phenomena with simulations and experiments for the following cases: i) electrospray (only electric field), ii) air spray (only airflow) and iii) air controlled electrospray (airflow and electric field).

In this work, we use a Lagrangian discretized model as it is suitable to model both the jet breakup and droplet dynamics. In the discretized model, we consider the liquid jet to be made up of series of beads attached together with springs. The Newtonâ€™s second law of motion is applied to conserve the momentum. The jet at the nozzle is perturbed by introducing a normal mode disturbance to the radius of the bead at the nozzle. This disturbance results in axisymmetric instability on the jet and with the growth of the instability the jet breaks into droplets. We incorporate equations of motion for the formed droplets to study their dynamics. In this model, we also model the spray process due to external air flow by incorporating the air drag effects for air spray and air controlled electrospray cases.

In addition, we perform experiments to study the air controlled electrospray process of 8 wt% polyvinylidene fluoride in dimethylformamide solution. The air spray, electrospray, and air controlled electrospray processes are visualized with a high-speed camera. The average radius of the droplets obtained from experiments is compared with simulation results. The surface roughness and efficiency of the coating of PVDF/DMF solution on the collector are obtained to study the effect of air flowrate on air controlled electrospray of the solution. We also investigate the effect of air flow in air controlled electrospray process of 3% PVA/H2O solution with 15 vol.% Carbon Black or Carbon Nanotube particles to produce coatings with precisely controllable nanoscale topology, morphology, deposition thickness, and spatial distribution of active nano-inclusions. We further investigated the effect of additional deformation on the deposition homogeneity and the dispersion of carbon NPs (Carbon Black and Carbon Nanotubes) in resulting coatings, as it plays a crucial role in energy-storage applications.

543232 Role of Computational Modeling in Fluid Catalytic Cracking Design

Abstract:

Computational modeling plays an increasingly important role in understanding gas-particle flow dynamics in the Fluid Catalytic Cracking process. TechnipFMC actively uses tools such as computational fluid dynamics (CFD) for FCC design validation and as a troubleshooting tool. CFD provides both the qualitative and quantitative information needed to study, and understand, how hardware modifications and operational changes will impact gas-particle flow behavior and hence the overall performance of the unit.

This paper provides two revamp case studies, one of an FCC regenerator and the other an FCC riser. The FCC regenerator revamp incorporates improvements to the spent catalyst distributor and regenerator catalyst standpipe to mitigate operational problems and improve mechanical reliability. The FCC riser revamp includes design modifications to incorporate larger diameter and longer riser to achieve target residence time at increased capacity. This paper describes how CFD modeling tools were used to confirm the adequacy of the proposed hardware changes, fine tune the design and evaluate the performance to minimize the potential risks on startup. The paper also highlights key pre and post turnaround performance results.

524349 Direct Numerical Simulations of Flow Around Assemblies of Non-Spherical Particles and the Investigation of Voidage Effects

Abstract:

Many industrial applications involve fluidization of different non-spherical particles. Yet there is little understanding about accurate modelling of such fluidization processes. Our work focuses on building a multi-scale modelling procedure for fluidization of non- spherical particles. This work particularly focuses on the direct numerical simulation aspect of the multi-scale modelling. In this work, particle resolved flows are performed around static assemblies of random orientations of non-spherical particles using a multi- relaxation-time lattice Boltzmann method.

For assemblies of spherical particles, the parameter spaces are just limited to different Reynolds numbers (Re) and solids-volume fractions (Îµs). However for non-spherical particles, it is also needed to know the mutual orientation of the particles in the simulation domain and also their mean orientation with respect to the flow direction. Therefore, we propose two parameters S1 and S2 to measure the mutual orientations of the particles and also the incident angle Ï† with respect to the incoming flow. Additional to the drag, we also investigate the dependencies of lift and torques as function of Re, Îµs, S1, S2 and Ï†. From the data obtained, we propose accordingly the correlations for drag, lift and torque coefficients. The proposed correlations can be used to simulate larger scale simulations of non-spherical particles fluidization.

529861 Detailed Analysis of a Large-Scale Wurster Coating Process

Abstract:

The Wurster-coating process is commonly used to layer drug, protective coatings, and other ingredients onto multiparticulate beads in the pharmaceutical industry. Beads can be coated with multiple functional films that control the release rate of drug during dissolution. The coating process takes place inside a fluidized-bed Wurster coater, which achieves a recirculation pattern of the bead flow via a combination of geometric features (Wurster tube insert, air-distributor plate) and operating conditions (air flow rate, spray properties, etc.).

Our past modeling efforts (AIChE 2016:736d, AIChE 2017:776a) have focused on the bead flow and coating in a cold-flow situation; thermodynamics of the process were not included. Similar simplifications were made in other numerical studies (Askarishahi et al. [1]) or, sometimes, the mass transfer effects were completely neglected (Pietsch et al. [2]). However, the thermal phenomena inside the fluid bed are quite complex and influence the final coating attributes. In our present work, we develop a fully integrated CFD-DEM environment including fully-coupled interphase mass and energy exchanges between the solid and fluid phases, in addition to momentum exchange (using the commercial code eXtended Particle System (XPS) coupled with AVL-FIRE^{TM}). The evaporation from the particle surfaces is modeled via constitutive drying models and species balance between the phases. The spray is idealized using a ray-tracing approach which incorporates the droplet size and droplet velocity distribution in the spray region.

We attempt to connect the bead-scale properties (temperature and drying rates) to the process-scale parameters such as spray rate, spray-liquid temperature, and fluidization-air-flow temperature. This understanding allows us to adjust the process parameters to achieve the desired particle-scale properties; e.g., attaining the correct film forming temperature and bead-drying time scales. After careful validation (against Liang et al. [3]) and verification (against Askarishahi et al. [1]) of the newly-implemented physics, we present simulations of the Wurster coating process. We compare our predictions with experimental data from a pilot-scale Glatt GPCG 1.1 Wurster coater; our comparisons are based on process-scale measurements such as the relative humidity, product temperature, and fluid temperature above the bed region. This coupled CFD-DEM model, once validated, will serve as the basis for subsequent production-scale studies and virtual scale-up efforts.

References

[1] M. Askarishahi, S. Mohammadsadegh, S. Radl. Full-Physics Simulations of Spray-Particle Interaction in a Bubbling Fluidized Bed. AIChE Journal, 63(7), 2569-2587, 2017.

[2] S. Pietsch, S. Heinrich, K. Karpinski, M. Müller, M. Schönherr, F. Kleine Jäger. CFD-DEM modeling of a three-dimensional prismatic spouted bed. Powder Technology, 316, 245-255, 2017.

[3] L. Liang, J. Remmelgas, B. G. M. van Wachem. Residence time distributions of different size particles in the spray zone of a Wurster fluid bed studied using DEM-CFD. Powder Technology, 280, 124-134, 2015.

534619 MFIX-Exa: A CFD-DEM Code for Exascale Computer Architectures

Abstract:

The combination of computational fluid dynamics and discrete element method (CFD-DEM) offers an accurate way to model gas-solids reactors. However, CFD-DEM simulations require much computational effort, which currently limits their application to small reactors, typically lab-scale reactors that at most contain tens of millions of particles. To apply CFD-DEM for accelerating fossil energy technology development, the modeling of pilot-scale reactors (~1 MWe) would be required. The modeling of such reactors, which contain hundreds of billions of particles, poses both a computational and data management challenge.

The development of a code called MFIX-Exa, which enables such simulations on exascale computers, began in November 2016. This development project takes place under the Exascale Computing Project (ECP), a collaborative effort of two US Department of Energy organizations – the Office of Science and the National Nuclear Security Administration. ECP was established to accelerate delivery of a capable exascale computing system that integrates hardware and software capability to deliver approximately 50 times more performance than the most powerful supercomputers in use today in the US. ECP's work encompasses applications, system software, hardware technologies and architectures.

The first version of MFIX-Exa has been developed by migrating the core hydrodynamic models from the widely used, open-source code MFIX into AMReX, a new ECP software framework that uses block-structured adaptive mesh refinement algorithms to solve systems of partial differential equations with complex boundary conditions on exascale architectures. The correctness of MFIX-Exa was established by comparing its solutions with MFIX (version 2016-1) solutions for a set of benchmark cases that represent flow regimes encountered in the project's challenge problem – a 1 MWe chemical looping reactor. The parallel performance of the code has been improved by implementing logical tiling, enabling OpenMP threading over tiles, and employing a dual grid approach for fluid and particle load balancing. To handle complex geometries in the simulations, AMReX embedded boundary (EB) data structures and iterators were incorporated and an algorithm to address particle collisions with EB walls was implemented. The first major performance milestone for MFIX Exa has been reached with the 0.025 s simulation of 1 billion particles undergoing homogeneous cooling in 12.9 hours of wall time using 6912 processors of NREL's Peregrine supercomputer.

532084 Simulation As a Tool for Learning from Historical FCC Regenerator Operations

Abstract:

Simulation As A Tool For Learning From Historical FCC Regenerator Operations

John Pendergrass, Peter Blaser, Sam Clark

CPFD, LLC

10899 Montgomery Blvd. NE, Suite A, Albuquerque, NM USA

E-mail: john.pendergrass@cpfd-software.com

Abstract

Simulation of FCCUs is frequently used when anticipating changes to design or process conditions, and is often performed prior to a turnaround. Simulation has been shown to identify root causes of underperformance, test the effects of potential changes prior to implementation, and minimize the risk of unforeseen negative consequences of changes.

Successful utilization of simulation by refiners requires both technology and people. While multiphase gas-particle simulation technology has significantly matured in recent years, other industry trends have been observed which affect the average experience level of personnel working on FCCUs. In many cases retirement has reduced the in-house expertise pool in this area, particularly those with 30+ years of specialty experience with FCC technology. Simultaneously, highly skilled newer staff are rapidly rotated through multiple areas of refinery operations, with seemingly increasing frequency. As a result, the pool of those with deep experience with FCCUs, especially spanning multiple units or multiple turnaround cycles, is declining.

This presentation and discussion is focused on how simulation is used to capture lessons learned from current and historical FCCU operations and convert this knowledge to prediction of future proposed changes. Multiple case studies are presented from both North American and international refiners, where simulations were used to understand and mitigate issues such as erosion, emissions, afterburn, and catalyst losses. The simulation results and historical operational data are analyzed to capture lessons learned, both positive and negative. Finally, examples of simulation to predict the effects of proposed changes are presented.

Evaluation of Filtered Two Fluid Models Against Data from an Industrial Scale Fluidized Bed Reactor

Jan Hendrik Cloete[1], Schalk Cloete[2], Thomas Gurker[3], Günter Gronald[3] & Shahriar Amini[1,2]

[1]Norwegian University of Science and Technology, Trondheim, Norway

[2]SINTEF Materials and Chemistry, Trondheim, Norway

[3]ANDRITZ, Graz, Austria

Abstract

This study presents the validation of two filtered Two Fluid Model (fTFM) formulations against experimental data from an industrial fluidized bed boiler. The results revealed that significant uncertainty exists in applying the fTFM closures at particle/fluid configurations that are very different from the conditions during derivation and at very large filter sizes. It is therefore recommended that dedicated resolved TFM simulations be performed for a specific application, eliminating the need for extrapolation of the closures. Such dedicated simulations will only be feasible in 2D for industrially relevant filter sizes. Fortunately, previous work indicate that closures derived in 2D can reasonably be applied in 3D (Cloete et al., 2018c). Additionally, the simulations show that the fTFM with an anisotropic drag closure predicts qualitatively more realistic results than an fTFM with an isotropic drag closure. This adds to previous research (Cloete et al., 2018a) that suggests that future fTFM drag modelling efforts should move away from the isotropic closures that are standard in the literature.

Introduction

The filtered Two Fluid Model (fTFM) (Igci et al., 2008) provides a possible method to allow accurate simulations of industrial-scale fluidized beds. This is achieved by means of closures that compensate for the effects that gas bubbles and solids clusters, which are not resolved on the chosen grid, would have on the flow behavior. This makes it possible to use very large grid cells and still obtain reasonable predictions of the mean flow behavior. The fTFM closures that are required are identified by performing a spatial average on the governing Two Fluid Model (TFM) equations. The necessary closures can then be derived by spatially-averaging (or filtering), and then analyzing, data from resolved TFM simulations.

Several groups have derived and evaluated different fTFM formulations (Cloete et al., 2018c; Gao et al., 2018b; Igci and Sundaresan, 2011; Ozel et al., 2013; Schneiderbauer, 2017). It is clear from these studies that fTFM closures, of which the closure for the filtered drag force is most important, are necessary to obtain accurate results on grid sizes significantly larger than 10 particle diameters, which is necessary for timely simulations on larger than lab-scale fluidized beds (Cloete et al., 2016). However, evaluation of these fTFM formulations against data from industrial fluidized beds is rare.

The present study will describe the results from such an attempt by evaluating the performance of an fTFM formulation in an industrial fluidized bed boiler from ANDRITZ. The fTFM formulation evaluated uses an anisotropic meso-scale solids stress closure (Cloete et al., 2018b), and both isotropic and anisotropic closures for the filtered drag force (Cloete et al., 2018a) are investigated. The results from this study are then analyzed to make recommendations for future work to further improve fTFMs for simulation of industrial-scale fluidized beds.

Simulations

Filtered governing equations and closures

Previous studies (Cloete et al., 2018a; Cloete et al., 2018b; Cloete et al., 2018c) have thoroughly documented the derivation of the filtered governing equations, the closures that are required and those that can be neglected under certain conditions, and the procedures required to filter and analyze resolved TFM data to derive the necessary closures. These topics will therefore not be repeated here. However, the closures that are evaluated in the present study will be described briefly.

For the meso-scale solids stresses, an anisotropic gradient-based closure (Cloete et al., 2018b) is used for all simulations. It has previously been revealed (Cloete et al., 2018b; Cloete et al., 2018c) that such a closure offers much better predictions on coarse grids, especially considering grid independency of the results, than the Boussinesq approximation-based closures that are common in literature (Gao et al., 2018a; Igci and Sundaresan, 2011; Sarkar et al., 2016; Schneiderbauer, 2017). An anisotropic closure is used for the meso-scale interphase force (Cloete et al., 2018a), which results from sub-grid fluctuations of the gas pressure gradient and which was found to be significant for coarse-grid fTFM simulations of bubbling fluidization (Cloete et al., 2018a).

For the filtered drag force, generally considered as the most important closure for fTFMs, two closures are considered: an isotropic closure and an anisotropic closure. Similar to most of the fTFM literature (Gao et al., 2018b; Igci and Sundaresan, 2011; Milioli et al., 2013; Sarkar et al., 2016; Schneiderbauer, 2017; Schneiderbauer and Pirker, 2014), the isotropic closure multiplies the drag force predicted at the filtered solids volume fraction and the filtered slip velocities by a drag correction factor to account for the effects of sub-grid heterogeneities. The closure for the drag correction factor is derived based on the vertical direction drag, where the drag is considered most important, and then applied to all other directions. Previous work showed that this approach would significantly overpredict the filtered drag force in the directions perpendicular to gravity. Consequently, the anisotropic closure considered in the present study (Cloete et al., 2018a) accounts for this anisotropy in the sub-grid drag correction. This is done by re-writing the filtered drag force based on a drift velocity formulation (Ozel et al., 2013; Parmentier et al., 2012). Contrary to previous drift velocity-based closures, the present formulation also considers an isotropic closure for the non-linearity correction factor, which compensates for assumptions used to derived the drift velocity formulation of the filtered drag force (Cloete et al., 2018a). It was previously shown that the use of such a closure improves the grid independency of coarse-grid simulations of bubbling fluidization (Cloete et al., 2018a). The anisotropy of the sub-grid drag correction is accounted for in the drift velocity closure, which consists of a part that is proportional to the filtered slip velocity in the direction considered, and another which is independent of the filtered slip velocity and only acts in the direction aligned with gravity.

Simulation setup

The geometry of the fluidized bed boiler simulated in the present study is shown in Figure 1. The fluidized bed is over 30 m tall and has a cross-sectional area of about 80 m^2 in the large upper chamber. An attractive feature of the filtered model is that it can be employed on an unstructured grid that allows for convenient and economical meshing of a large and complex geometry. In this case, the complexity arises from the concentrated secondary air injectors in the bottom chamber of the reactor and the heat exchange surfaces in the upper chamber. For the purpose of this validation exercise, the concentrated injections at the bottom are by far the most important, because they have a large influence on the pressure drop profile observed in the reactor. To mesh this geometry, an unstructured polyhedral mesh of about 120 000 cells is used, which allow reasonably fast simulations of this very large domain (about 1 day of computational time on 32 cores).

From the pressure drop measurements provided by ANDRITZ, the amount of solid materials in the reactor could be estimated. To keep the correct mass of solids in the reactor, a condition was imposed that all the solids exiting the reactor at the top is instantaneously reintroduced at the bottom. Following this condition, the two main variables for comparison were the shape of the time-averaged pressure profile and the mean solids circulation rate. These values were obtained by time-averaging over 75 s of simulation time (after a pseudo steady-state was achieved, which typically required about 40 s of simulation).

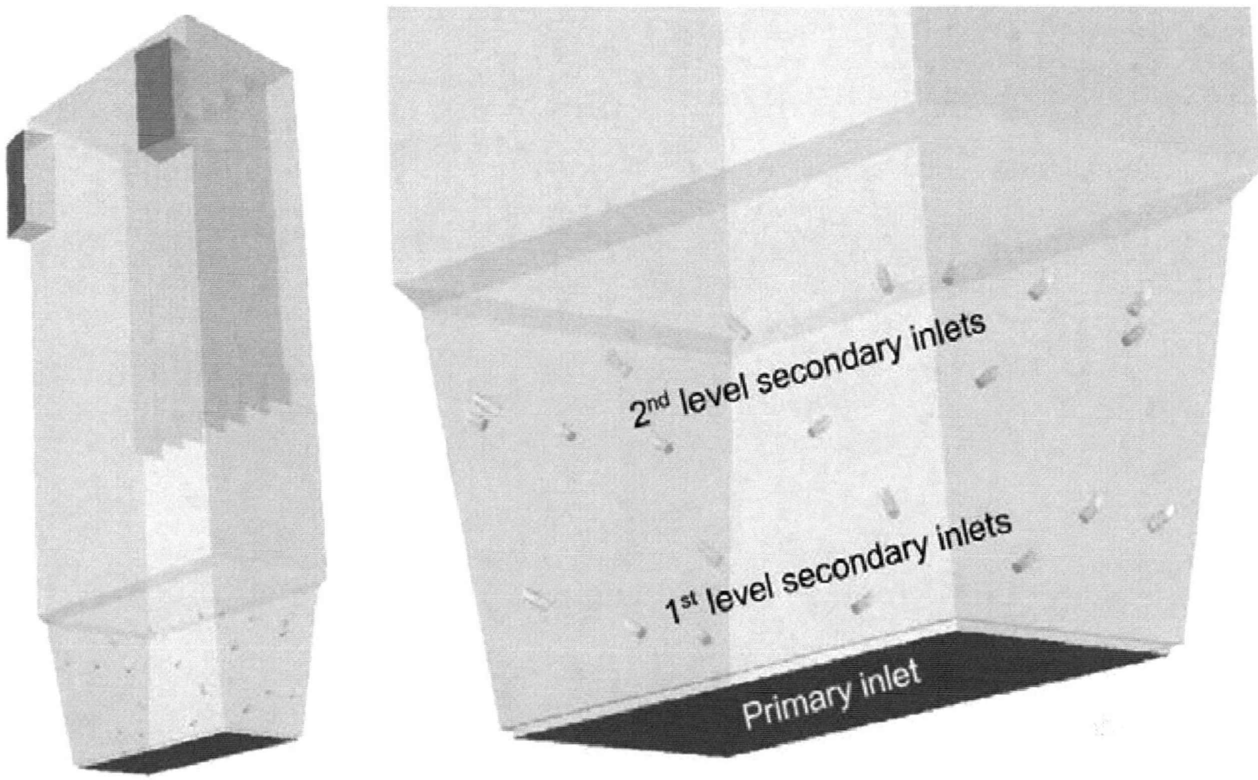

Figure 1 – Illustration of the geometry simulated in the present study. Left: the full geometry. Right: Zoomed view to the bottom region to show the positioning of the primary and secondary inlets. Blue represents inlets and red represents outlets

Significant uncertainty was also involved in the specification of the inflow boundary conditions to the reactor. The total circulation rate is not exactly measurable and additionally, the position where combustion gas products will form is very much dependent on the operational load of the boiler. Furthermore, the reactor simulations are completed under the assumption of isothermal conditions, which may introduce uncertainties, especially at the bottom where cold gases are introduced and most of the reaction takes place. Noting these uncertainties, the inlet flow rates for the gas were specified as: 63.9 kg/s in the primary inlet at the bottom, 38.6 kg/s at the first level of secondary air injection and 25.8 kg/s at the second level of secondary air injection.

The properties of the fluid and solids considered in the fluidized bed boiler are summarized in Table 1. The gas has an outlet temperature of about 855°C and the solids consists of sand, biomass, ash and limestone. Additionally, Figure 2 shows the particle size distribution of the solids.

Table 1 – Summary of the fluid and particle properties considered in the fluidized bed boiler, as well as those considered for the derivation of the fTFM closures.

	Present study	*Closure derivation*
d_p	$\pm 210\ \mu m$	$75\ \mu m$
ρ_s	$2500\ kg/m^3$	$1500\ kg/m^3$
ρ_g	$0.314\ kg/m^3$	$1.3\ kg/m^3$
μ_g	$4.4 \times 10^{-5}\ \frac{kg}{m.s}$	$1.8 \times 10^{-5}\ \frac{kg}{m.s}$

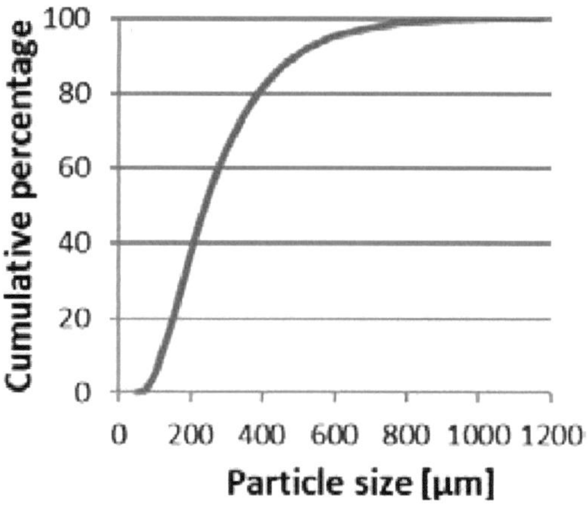

Figure 2 – The particle size distribution of the solids in the fluidized bed boiler.

Coarse-grid fTFM simulation challenges

This simulation exercise pushed the fTFM closures far outside of their comfort zone in three primary ways:

1. Reasonably fast simulations of the extremely large reactor require very large cells sizes (up to 1 m^3). As a result, the filter sizes used in the closures are much larger than the filter sizes used in the model derivation process. This required a large amount of extrapolation of the model, which always introduces significant uncertainty.
2. The particle and fluid properties in the industrial-scale reactor were very different from those used in the derivation of the filtered model (see Table 1). This required a large amount of scaling of the model to the correct conditions, introducing additional uncertainty since it is known (Schneiderbauer and Pirker, 2016) that the scaling methods common in the literature are not highly accurate.
3. The industrial scale reactor uses particles with a fairly wide size distribution (see Figure 2), but the filtered model was derived for monodisperse flows. Therefore, all the solids have to be approximated as a single particle type.

Some additional uncertainty is introduced by the fact that the filtered model was derived from finely resolved TFM simulations on a 2D plane, which will quantitatively affect the closures. However, it has previously been found that closures derived in 2D could perform reasonably well in validation against 3D experimental data (Cloete et al., 2018c), therefore this uncertainty is considered to be of lesser importance.

Results and discussion

Initial evaluation

Initial evaluation of the fTFM with the isotropic drag closure revealed that the fTFM significantly overpredicts the bed expansion. The most likely reason for this behavior is the use of the fTFM closures at particle and fluid properties very different than those used in the derivation, as discussed in the previous section. The extrapolation of the closures to large filter sizes possibly also plays a role.

To reasonably predict the correct solids circulation rate and pressure profile, the scaling with particle and fluid properties were adjusted by non-dimensionalizing the filter size as $\hat{\Delta}_f = \Delta_f / (\frac{v_t^{1.4} v_{t,der}^{0.6}}{g})$, instead of $\hat{\Delta}_f = \Delta_f / (\frac{v_t^2}{g})$. Here $v_{t,der}$ is the terminal velocity of the particles for the setup used in deriving the fTFM closures, Δ_f is the filter size in meters, $\hat{\Delta}_f$ is the dimensionless filter size and g is the gravitational acceleration constant. Since the terminal velocity, v_t, is significantly larger in the fluidized bed boiler than in the derivation case, this adjustment leads to larger dimensionless filter sizes evaluated in the closures. In return, this results in a larger reduction in the modelled drag and therefore less bed expansion. Furthermore, a change was also made to the meso-scale solids stress closure to prevent extremely large stresses returned by the closures from destabilizing the simulation. When dimensionalizing the stress returned by the closure, $v_{t,der}$ was used instead of v_t. This leads to significantly smaller forces due to meso-scale solids stresses. See Cloete et al. (2018b) for more information about the scaling of the meso-scale solids stresses. The effect of the meso-scale stresses is relatively small compared to that of the drag reduction, therefore this modification is considered to be of lesser importance.

It can be noted that these modifications of the scalings, which were determined based on the fTFM with the isotropic drag closure, were similarly applied to the fTFM with anisotropic drag closure. Furthermore, both closures were evaluated with the filter-to-grid-size ratio determined to be optimal in 2D verification studies (Cloete et al., 2018a), i.e., a ratio of 2 for the fTFM with the isotropic drag closure and a ratio of 1 for the fTFM with the anisotropic drag closure.

Sensitivity to particle size

The particle size was varied to investigate the effect of the uncertainty of the particle size specified in the monodispersed simulations. Estimates of 240 μm for the mean diameter, 210 μm for the Sauter mean diameter and 180 μm for the Sauter mean diameter accounting for non-sphericity were considered, as well as two additional points at 150 and 270 μm. Additionally the fTFM was evaluated with both isotropic and anisotropic drag closures. The results of these simulations are summarized in Figure 3 and Figure 4 in terms of the mean pressure profile and solids circulation rate, respectively.

From the pressure profiles in Figure 3, it can be seen that the simulations can reasonably approximate the experimental results with the modifications to the filter size scaling. The main qualitative difference is that the experimental results show a larger pressure gradient at the bottom of the reactor. A possible explanation for this is that the solids injection at the bottom of the reactor was approximated to be uniform across the bottom plane in the geometry, obtained from previous work of the industrial partners. In reality, the solids will enter at the side from a standpipe. This much more concentrated injection will lead to less gas-solids contact, reduced vertical drag, more solids accumulating at the bottom of the reactor and therefore a larger pressure drop. A case with a particle size of 210 μm and using no fTFM closures is also shown for perspective, which clearly demonstrates the need for sub-grid modelling in these simulations employing very large grid sizes.

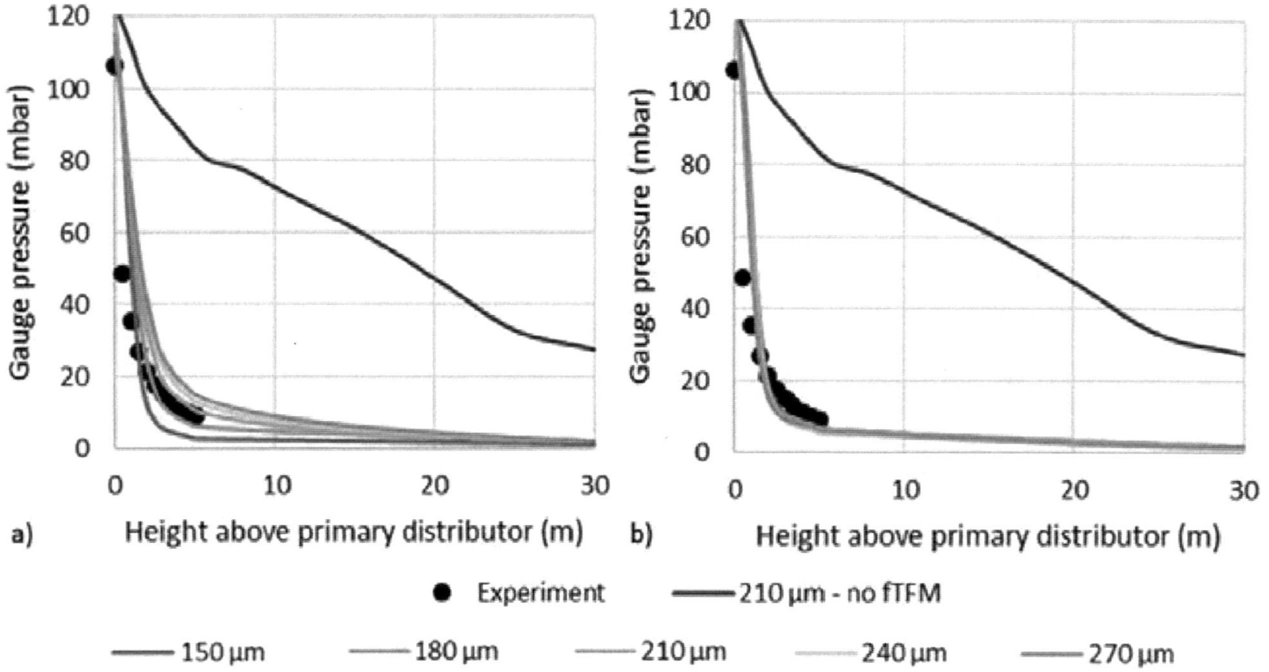

Figure 3 - The time- and horizontally-averaged gauge pressure along the height of the reactor as predicted using a) the fTFM with the isotropic drag closure and b) the fTFM with the anisotropic drag closure, and compared against experimental data.

It is clear from both the pressure profiles and the circulation rates that the fTFMs with isotropic and anisotropic drag closures predict very different behaviours with changing particles size. In Figure 3, the isotropic drag closure leads to a more compact bed with decreasing particle size. This is a counter-intuitive result, since it can generally be expected that finer particles will be easier to fluidize. On the contrary, the anisotropic drag closure predicted negligible dependency of the pressure profile on the particle size. As for the solids circulation in Figure 4, the isotropic drag closure again predicts an unexpected trend of decreasing solids circulation with decreasing particle size. The fTFM with the anisotropic drag closure, on the other hand, predicts a more intuitive increase of solids circulate rate with decreasing particle size. The particle size also has a somewhat smaller effect in this case.

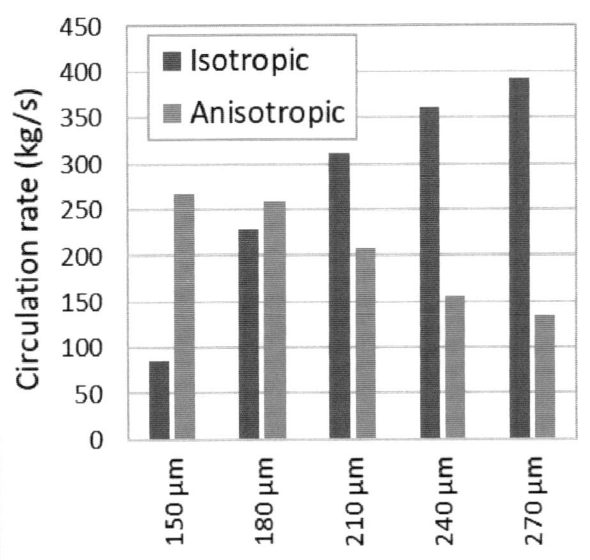

Figure 4 - The mean solids circulation rate as predicted by the fTFMs with isotropic and anisotropic drag closures.

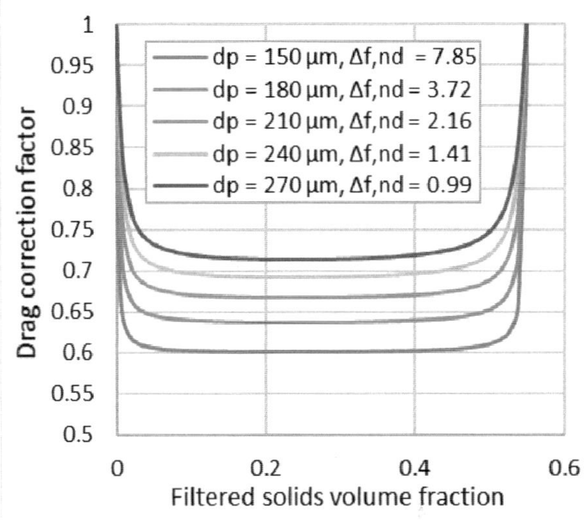

Figure 5 - Effect of the particle size on the drag correction factor in the isotropic drag closure. The cells size is set to 0.28 m (the mean cell size in the simulation) and a scaled filtered slip velocity magnitude of unity is considered. It can further be noted that a drag correction factor of unity corresponds to no sub-grid drag corrections.

The opposite behavior of the two fTFMs with respect to particle size can be explained as follows. The particle size essentially has two effects on the modelled filtered drag force (the interested reader can find the detailed model equation in another study (Cloete et al., 2018a)). Firstly, the modelled filtered drag force is directly proportional to the interphase momentum exchange coefficient of the microscopic drag law evaluated at the filtered conditions. This value will always increase with decreasing particle size.

On the other hand, the filtered drag force is also affected by the sub-grid heterogeneities. In the isotropic drag closure, this effect is modelled by a drag correction factor, to which the filtered drag force is directly proportional. In the anisotropic closure, the sub-grid effects are modelled through the non-linearity correction factor and the drift velocity. These modelled sub-grid effects will increase with increasing dimensionless filter size, since there will be more sub-grid heterogeneity at larger filter sizes. In this study, as is common in literature, the filter size is non-dimensionalized based on the particles' relaxation length, $\frac{v_t^2}{g}$. Since the terminal velocity, v_t, decreases with decreasing particle size, a smaller particle size leads to a larger dimensionless filter size when considering the same grid cell. The physical argument for this behavior is that smaller particles will tend to form finer heterogeneous structures (gas bubbles and solids clusters), leading to more sub-grid heterogeneity compared to larger particles when considering the same grid cell. This means that smaller particles lead to larger sub-grid corrections to the drag (when considering a constant grid size) and therefore to less vertical drag. This behavior is demonstrated in Figure 5 for the drag correction factor in the isotropic drag closure when considering a constant cell size.

It is therefore clear that the particle size has competing effects on the microscopic drag law and the sub-grid drag reduction in the filtered drag force closure. From the results in Figure 3 and Figure 4 it can be deduced that the effect of particle size on the sub-grid drag reduction significantly outweighs the effect on the microscopic drag law in the case of the isotropic closure. For the anisotropic closure the opposite is true: the effect on the microscopic drag law is slightly larger. The reason for this behavior lies in the different filter-to-grid-size ratios used for each of these fTFMs, which were determined from 2D verification studies (Cloete et al., 2018a) to be 2 for the isotropic drag closure and 1 for the anisotropic drag closure. The isotropic drag closure requires larger filter sizes to be evaluated in the closures to compensate for its tendency to overpredict the lateral direction drag, which indirectly leads to overpredicted vertical drag. However, in this case, the larger filter sizes evaluated in the isotropic closure also overemphasizes the effect of the particle size on the sub-grid corrections. On the contrary, the anisotropic drag closure, which uses a filter-to-grid-size ratio of 1, predicts a more realistic behavior of increasing bed expansion and solids circulation rate with decreasing particle size.

From Figure 3 and Figure 4 it can also be seen that, on average, the bed expansion and the solids circulation rate is less for the fTFM with the anisotropic drag closure compared to the case with the isotropic closure. The anisotropic drag closure tends to underpredict the vertical drag compared to the experimental results (a rough estimate of the experimental solids circulation rate was determined to be 350 kg/s). However, it is important to note that this is only because the modification to the filter size scaling was determined based on the fTFM with the isotropic drag closure. The fTFM with the anisotropic closure would actually give a good match against experimental results with less tuning of the filter size scaling common in literature. The tendency of the isotropic drag closure to predict more vertical drag relative to the anisotropic closure (with the same scaling for the filter size and filter-to-grid-size ratios determined to be optimum in 2D verification) can be attributed to the tendency of the isotropic drag closure to increasingly overpredict the lateral drag with increasing filter sizes (Cloete et al., 2018a).

A final observation regarding the simulation results can be made from Figure 5. The drag correction factor in the isotropic drag closure changes rapidly near the dilute and dense limits. It can be noted that the behavior is similar for the drift velocity in the anisotropic drag closure (not shown here).

Furthermore, Figure 5 shows that the drag correction factor also changes more rapidly in the dilute and dense regions with larger filter sizes. The physical interpretation of this behaviour is that a larger filter region will have a good chance of containing at least some clustering even at low average particle loadings. For example, if we consider a small region of 1 cm^3 at an average solids volume fraction of 0.1%, it is highly unlikely that any clustering can take place to reduce the drag. However, if we consider a region of 1 m^3 (a million times more volume), this same 0.1% average solids loading actually represents quite a lot of particles. It may be that these particles are concentrated as clusters in certain regions of this large volume so that they offer much less drag to the resulting upwards flow. Figure 6 shows that large parts of the fluidized bed boiler is very dilute, therefore the simulations of the boiler will be very sensitive to the accuracy of the closure at very low filtered solids volume fractions, especially considering the large filter sizes employed in these simulations.

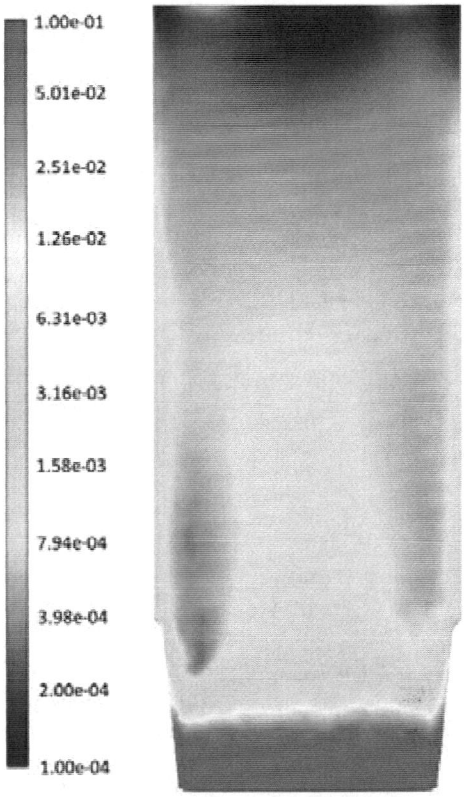

Figure 6 – Contour plot of the time-averaged filtered solids volume fraction on a logarithmic scale on a vertical plane near the center of the bed for the case with 210 μm particles and the anisotropic drag closure

Conclusions and recommendations

In this study, two fTFMs, considering isotropic and anisotropic drag closures, were evaluated against experimental data from a large industrial-scale fluidized bed boiler, with the primary aim to identify areas of improvement that are needed to make fTFMs a reliable tool for investigating industrial-scale problems. The three primary lessons from this experience can be summarized as follows:

1. Using a filtered model at much larger filter sizes and very different flow conditions from those where it was derived introduces significant uncertainties. For optimal results, a filtered model should be derived with a specific application in mind so that it can be applied without the need for great extrapolation or scaling. Alternatively, future work should improve the methods for scaling fTFM closures to different fluid and particle properties.

2. Filtered models derived in 2D appear to function adequately when applied to 3D flows. This is a very important advantage when considering the large filter sizes at which models need to be derived for use in very large equipment such as ANDRITZ's fluidized bed boiler. Deriving models in 2D for the filter sizes required by this geometry will be feasible in about 3 months of computational time, whereas 3D fine-grid simulations at the required scale will be completely infeasible.
3. It is important that preliminary simulations such as those completed in this report are done before a dedicated model is derived for a specific application. Even though such a simulation campaign will involve significant uncertainties from extrapolation and scaling, it will reveal the areas of the parameter space where the model needs to be very accurate (e.g. very dilute flows in this case). This insight can then be used to maximize the accuracy of a dedicated filtered model derived for a specific application.

Additionally, it was concluded that an anisotropic drag closure predicts a more realistic dependency of the fluidization behaviour with changing particle size than an isotropic drag closure. This adds to previous research (Cloete et al., 2018a) showing the benefits of anisotropic closures for the drag. Based on these results and noting that almost all existing sub-grid drag closures are isotropic, it is suggested that future development efforts of fTFM drag closures should focus on anisotropic closures.

Acknowledgements

The authors would like to express their gratitude for the financial support in the following projects: NanoSim (project number: 604656), funded by the European Commission, and GaSTech (project number: 271511), funded by the Research Council of Norway and co-funded by the European Commission under the Horizon 2020 program, ACT Grant Agreement No 691712. We would further like to acknowledge the computational resources provided at NTNU by UNINETT Sigma2 AS, https://www.sigma2.no.

References

Cloete, J.H., Cloete, S., Municchi, F., Radl, S., Amini, S., 2018a. Development and verification of anisotropic drag closures for filtered Two Fluid Models. Chemical Engineering Science 192, 930-954.

Cloete, J.H., Cloete, S., Radl, S., Amini, S., 2018b. Development and verification of anisotropic solids stress closures for filtered Two Fluid Models. Chemical Engineering Science 192, 906-929.

Cloete, S., Cloete, J.H., Amini, S., 2016. Comparison of the Filtered Two Fluid Model and Dense Discrete Phase Model for Large-Scale Fluidized Bed Reactor Simulations, AIChE Annual Meeting San Franciso, USA.

Cloete, S., Cloete, J.H., Amini, S., 2018c. Hydrodynamic validation study of filtered Two Fluid Models. Chemical Engineering Science 182, 93-107.

Gao, X., Li, T., Rogers, W.A., 2018a. Assessment of mesoscale solid stress in coarse-grid TFM simulation of Geldart A particles in all fluidization regimes. AIChE Journal 64, 3565-3581.

Gao, X., Li, T., Sarkar, A., Lu, L., Rogers, W.A., 2018b. Development and validation of an enhanced filtered drag model for simulating gas-solid fluidization of Geldart A particles in all flow regimes. Chemical Engineering Science 184, 33-51.

Igci, Y., Andrews, A.T., Sundaresan, S., Pannala, S., O'Brien, T., 2008. Filtered two-fluid models for fluidized gas-particle suspensions. AIChE Journal 54, 1431-1448.

Igci, Y., Sundaresan, S., 2011. Constitutive Models for Filtered Two-Fluid Models of Fluidized Gas–Particle Flows. Industrial & Engineering Chemistry Research 50, 13190-13201.

Milioli, C.C., Milioli, F.E., Holloway, W., Agrawal, K., Sundaresan, S., 2013. Filtered two-fluid models of fluidized gas-particle flows: New constitutive relations. AIChE Journal 59, 3265-3275.

Ozel, A., Fede, P., Simonin, O., 2013. Development of filtered Euler–Euler two-phase model for circulating fluidised bed: High resolution simulation, formulation and a priori analyses. International Journal of Multiphase Flow 55, 43-63.

Parmentier, J.-F., Simonin, O., Delsart, O., 2012. A functional subgrid drift velocity model for filtered drag prediction in dense fluidized bed. AIChE Journal 58, 1084-1098.

Sarkar, A., Milioli, F.E., Ozarkar, S., Li, T., Sun, X., Sundaresan, S., 2016. Filtered sub-grid constitutive models for fluidized gas-particle flows constructed from 3-D simulations. Chemical Engineering Science 152, 443-456.

Schneiderbauer, S., 2017. A spatially-averaged two-fluid model for dense large-scale gas-solid flows. AIChE Journal 63, 3544-3562.

Schneiderbauer, S., Pirker, S., 2014. Filtered and heterogeneity-based subgrid modifications for gas–solid drag and solid stresses in bubbling fluidized beds. AIChE Journal 60, 839-854.

Schneiderbauer, S., Pirker, S., 2016. The impact of different fine grid simulations on the sub-grid modification for gas–solid drag, 9th International Conference on Multiphase Flow, Firenze.

524389 Finding the Preferred Safe Operating Condition of a Fluidized Bed Incinerator to be Used for the Disposal of Waste Explosives

Abstract:

Explosives are one of the most important components of missiles systems. In general, they have a limited usable period over which performance can be assured, after which they should be disposed of safely. This is often done by incineration of the particulate explosive, though the process is extremely hazardous due to the highly reactive materials such as TNT (2-Methyl-1,3,5-trinitrobenzene), RDX (1,3,5-Trinitro-1,3,5-triazinane) and HMX (1,3,5,7-Tetranitro-1,3,5,7-tetrazocane). Currently rotary kilns are used for this purpose. However, the method has some problems such as incomplete reaction and local hotspots and fluctuating process conditions. Therefore a new and more efficient type of incinerator is required. Since fluidized bed incinerators are well known for effective heat transfer and well mixed conditions, it was proposed that they be considered as an advanced reactor for this incineration process. It is apparent that fluidizing and mixing characteristics are likely to be quite significant in the design of an effective and safe fluidized bed technology. Thus it was proposed to first study the system using a computational fluid dynamics (CFD) model to evaluate the likely bed performance. For the bed fluid dynamics the Eulerian - Lagrangian coupled method is used to rigorously simulate fluidization behavior of the particles in bed, while allowing the effect of various process conditions such as the air injection rate to be considered. In addition, the decomposition reactions of the explosive are modelled to calculate the temperature and pressure conditions in the incinerator. Physically, the bed is designed as a cylinder 2m in diameter and 9m in height, with a target disposal rate of 20,000kg/year, assuming a uniform particle diameter of 3.0mm. For safety, 40% of fluidized bed is filled with sand to buffer against spikes in pressure and temperature. The simulations are performed over a range of air injection velocities and the maximum temperature change with time in the bed for each velocity is calculated. The objective is to find the condition that gives the least spiking of temperature and pressure. The results of the CFD model showed that fluidization begins to be observed at 1.0 m/s of air injection rate, however, fluidization is weak in that the particulate explosive and sand are not mixed well. At 1.5 m/s of air injection velocity, meaningful fluidization is observed and the maximum temperature in the bed decreases. When the air injection rate is 2.0 m/s, maximum temperature is lowest, before increasing again at 2.5m/s. In this case the fluidization is too strong creating is a lot of voids, therefore, the sand cannot buffer well. Given that high temperature will result a less well controlled reaction and a more hazardous operation, 2.0 m/s of air injection rate is recommended as the preferred safe operating condition. This research will be of help to demilitarize old explosives safely and economically and can be useful data to build incineration facilities.

535149 Three Dimensional CFD Simulation of Two-Phase Flow in Pilot Plant Dryer

Abstract:

Drying technology is one of the most important process in industrial applications. Generally, in this technology gas and solid phases are in contact. Depending on the process, dryer regimes may be various. CFD technique is widely used for better understanding of phenomena in different types of apparatus and various scale. For modeling of multiphase flow two approach is be considered, Eulerian-Eulerian and Eulerian-Lagrangian. In this work, applied CFD simulation for investigation of fluid flow in three-dimensional gas-solid pilot plant fluidized bed dryer. Eulerian-Eulerian model combining the kinetic theory of flow (KTGF) for solid phases was applied. KTGF based on continuum assumption and is appropriate for simulating of the large number of solid particles in large scale of fluidized bed. In this theory, some parameters defined for solid phases such as solid pressure, solid temperature, radial distribution and etc. In addition, used Syamlal-Obrien drag function for calculating of the momentum exchange coefficient. The validation is performed by comparing with the experimental results such as bed expansion. Usually, study the effects of operating parameters is very important because the flow field affected by these parameters. The operating parameters gas velocity, initial height of the bed, particle diameters, solid properties and etc. In the end, it was observed that the inlet gas velocity could be significant effects on hydrodynamics and consequently change the heat transfer rate in the dryer. Therefore, by using the validate CFD model can be predicted flow behavior and investigation of efficiency by changing of operating parameters.

529941 Coupling a Continuum Granular Segregation Model with a Flow Model Incorporating Granular Rheology

Abstract:

Segregation in flowing granular materials is ubiquitous and has important implications in industrial systems that handle bulk solids. In recent years, continuum models have been developed to predict segregation based on a modified transport equation that includes advection, diffusion, and segregation. Some of these models have achieved quantitative predictions in simple geometries where the velocity field is known by fitting to data, empirical scaling laws, or simple theory. However, for segregation in more complicated systems or transient flows, the velocity field is often hard to obtain. To resolve this problem, we couple a continuum segregation model with a flow model incorporating the granular Î¼(I) rheology to simultaneously predict the velocity field and segregation. The mathematical formulation consists of the Navier-Stokes equations with the Î¼(I) rheology, a level set convection equation to track the free surface evolution, and the segregation transport equation. The problem is solved by a residual-based variational multi-scale approach in a fully coupled fashion on unstructured meshes. The predictions for the velocity field agree well with experiments and Discrete Element Method (DEM) simulations for highly transient monodisperse granular column collapse and for segregation in bounded heap flow. The coupled model has the potential to capture segregation in industrial-related processes such as the combined hopper filling and discharge problem. Partially funded by NSF grant no. CBET-1511450

528686 Modeling Granular Material Segregation Using a Finite Element Method and Advection-Diffusion-Segregation Equation Multi-Scale Model

Abstract:

A multi-scale model is presented for predicting the magnitude and rate of segregation of a binary mixture of particles in a rotating drum and conical hopper. The model combines diffusion and segregation correlations from the literature with advective flow field information from finite element method simulations. Predictions of the segregation rate from the multi-scale model compare well quantitatively to published DEM simulation results and experimental data. A standard method is introduced to calibrate the diffusion and segregation correlations for different materials. The multi-scale model, since it does not directly model individual particles, is expected to scale well to systems of industrial interest.

520459 Continuous Powder Blending inside Twin Screw Extruder

Abstract:

Purpose: Our study aimed to develop fundamental understanding of powder blending at the early stages of twin screw extrusion as function of material properties (non-cohesive and cohesive powders) and process conditions using Discrete Element Method (DEM) and experimentation. Previous work in the pharmaceutical and food industry has focused on mixing when materials are melted, or homogeneity at the output of the extruder.

Methods: DEM models were developed to study powder flow/mixing inside the extruder. The simulations were performed using EDEM^{TM} and the Hertzâ€"Mindlin contact model and JKR cohesion model. The geometry and dimensions of the extruder were based on the ZSK-18 Coperion Twin Screw Extruder. Experiments were subsequently conducted in the ZSK-18 to characterize powder mixing efficiency. Near-infrared (NIR) spectroscopy was used to evaluate blend uniformity. A clear extruder was built to experimentally visualize powder mixing inside the extruder. The experimental and DEM simulation factorial designs included screw speed and powder flowrate. The screw configuration consists of conveying elements and one reverse kneading element. For the DEM models, the materials properties (inputs to the simulation) and process conditions were defined based on experimental data. In addition, residence time distribution was studied experimentally and using DEM.

Results: Experimental results show that process parameters and materials flowability (non-cohesive vs cohesive) affect the mixing efficiency and mean residence time (MRT). The simulations RTD proï¬les are similar to an ideal continuous stirred tank reactor (CSTR) proï¬le with a sharp peak and a tail.

Conclusion: Feasibility of powder blending inside the extruder using only one reverse kneading element was demonstrated. The developed DEM model was able to elucidate the mixing mechanisms behind the observed phenomena. DEM was also demonstrated to be a useful development and visualization tool that helped answer questions about powder blending inside the extruder.

537581 Investigation of the Effect of Baffles on Axial Mixing and Impregnation in a Double Cone Blender

Abstract:

Discrete Element Method (DEM) simulations coupled with a novel algorithm for fluid transfer were used in combination to model dry catalyst impregnation in a double cone blender with baffles with the goal of improving the mixing and fluid content uniformity in the particle bed. The goals of this work are to understand how baffles affect the mixing process in the presence and absence of fluid impregnation in a rotating vessel and to develop a method to determine the optimum baffle position, number and shape. To increase the extent of axial mixing, baffles that break the line of symmetry along the axis (three along the cylindrical section and two along the conical sections of the double cone) were considered. We also propose and test a new hypothesis centered around the idea that the baffles located in the position of maximum velocity will break the flow and offer best mixing performance.

We locate the baffles in 5 different positions and characterize them by means of the aspect ratio between the fill level and the radius of the double cone. The effect of different properties of the baffles, such as their height, width, angle of inclination and number of baffles with the various process parameters, e.g. fill level, particle size, rotational speed, flow rate on the mixing and impregnation performance were studied. Kramer mixing index based on the distance between the volume centers of an axially segregated system, was calculated to quantify the mixing of the system. Simulation results show that there is an optimal baffle height for low fill levels beyond which the mixing performance, whereas the angle of inclination has little to no effect on the mixing performance. Also, the baffle position affects mixing significantly more at lower fill levels, and an optimum baffle position could be found.

530543 Estimation and Explanation of Adhesive Mixing Efficiency Via Energy-Based Stick/Bounce Model

Abstract:

An energy-based stick/bounce model is presented to explain relationships between particle interactions dominated by *Van-der-Waals* force during mixing of fine and coarse particles at different mixing intensities. The investigation concerns examination of the adhesive mixing efficiency based on relative magnitudes of *Van-der-Waals* energy, dissipated energy, de-agglomeration energy and relative kinetic energy during impacts. indicates good adhesive mixing efficiency, while indicates poor adhesive mixing efficiency. Effect of size, material properties and mixing intensity of fine and coarse particles are investigated to attain a deep understanding of the mechanisms of adhesive mixing. Most interestingly, larger coarse particle size leads to larger agglomerate size of fine particles. Larger fine particle size and high intensity both lead to poor adhesive mixing efficiency. Size of fine particle agglomerates was also considered in the model. De-agglomeration energy was found to be highest amongst other energy terms during processing, indicating the de-agglomeration step was the most energy consuming step of adhesive mixing. The experimental results using a vibrational mixer and two sizes of fine (silica and cornstarch) and several sizes of coarse particles (KCl and cornstarch) validate the modeling as it explains various mechanisms. The guest silica particles tended to attach to the host KCl particles with a good adhesive mixing efficiency, while the guest cornstarch particles tended to detach from the host KCl particles beyond certain mixing intensity. Also, larger KCl particles would lead to the larger agglomerate size of cornstarch, indicating the poorer mixing efficiency. Finally, higher intensity leads to the worse mixing efficiency.

530864 Experimental Study of Particle Density Segregation in Granular Shear Flow

Abstract:

It is well known that as a granular material is sheared, the bed dilates and heavier particles move perpendicular to the mean flow. Our previous computational study established a connection between particle density segregation, solid volume fraction and the so call inertia number. In this work, we perform experiments aimed at physically realizing the predictions of our proposed theory (and computational results). Specifically, we have constructed an experimental annular Couette cell that has a moving lower lid and fixed inner, outer, and top wall. With this device we can control the solid volume fraction by applying either a constant volume or constant pressure condition at the top wall as well as the global shear rate via modification of the bottom lid rotation rate. We perform trials that follow the segregation velocity of heavy steel tracer particles as they move within a background media of either acetate or glass. Similar to our previous computational and theoretical results, we find a saturated segregation velocity regime and a direct connection between segregation velocity and solid volume fraction/inertia number.

531020 Cohesive Particle Segregation and Granular Rheology

Abstract:

Cohesion can play an important role in the behavior of granular materials and is pervasive in industrial practice. Moreover, cohesive interactions can dramatically impact the rate of segregation as well as the overall degree of mixing attained. In this work, the segregation of granular material under wet conditions is investigated in a simulated shear cell. The segregation rate of both dry and wet granular materials is measured and compared. We find that our previous theoretical and computational results for dry density segregation â€“ which leverages granular rheology via the inertia number to predict particle density segregation â€“ can be modified in a simple way to yield similar predictions for wet conditions. By doing so, we extend the utility of our proposed connection between rheology and granular segregation to be applicable for both wet and dry granular flows.

536669 *Solids Metrics for Successful 3D Printing of Energetic Feedstocks*

Abstract:

E&G Associates, Inc.

100 Cherokee Blvd, STE 332

Chattanooga, TN 37405

[1] Presenting Author, brandon.ennis@powdernotes.com, (615) 823-0054

[2] Corresponding Author, benjamin.ennis@powdernotes.com, (423) 544-7298

Abstract:

Traditional manufacturing methods for producing warheads, propellants, and pyrotechnic systems are timely and costly. Additive manufacturing (AM) approaches offer one path of alternative manufacturing technologies. Among various technologies, powder bed fusion (PBF) is one promising approach. A PBF printer is essentially a miniature manufacturing facility in a box, with significant powder handling challenges that must be overcome for robust production performance and part quality [1, 2]. Feedstock handling in a powder bed fusion printer is no different than and faces the same challenges as full-scale solids handling processes.

For a feedstock to function properly in a powder bed fusion 3D printer, it must be designed to meet several material characteristics. Some of the most important material properties include: powder flowability & cohesion, powder and roll friction, permeability, particle size & shape distribution (PSSD), triboelectric charging propensity, wetting characteristics, compaction characteristics and thermal softening point.

This paper presents select studies of developing feedstocks for PBF printing of energetic and other materials. Key attributes benchmarked against included powder flowability, aeration attributes, wetting behavior, spreadability, and fusion, as well as final part performance (bulk density, detonation, sensitivity, and part tolerances.)

[1] Ennis, B.J. *et al.* (2008) Section 21: Solids-Solids Operations & Processing. *Perryâ€™s Chemical Engineersâ€™ Handbook*. Perry, R. H., & Green, D. W. (eds.), New York: McGraw-Hill.

[2] Ennis, B.J. et al. (2017), HPC4Mfg Modeling of Powder Dynamics in Metal Additive Manufacturing, Austin, TX. Link for presentation:

https://hpc4mfg.llnl.gov/pdfs/events/powder2017/Powder_Dynamics_EG_17oct2017_FINAL.pdf

[3] Ennis, B.J. et al. (2017). Navy SBIR Phase I Award. Development of Explosive Feedstock for Commercial-off-the-Shelf (COTS) 3D Printers. Link for award:

https://www.navysbirprogram.com/NavySearch/Award/award.aspx?pk=869CC83B-5945-4E75-A598-AA2527D51045

[4] "*Chattanooga firm to help U.S. Navy figure out how to 3-D print explosives*" (2017), T. Omarzu,

Chattanooga Times Free Press. Story link:

http://www.timesfreepress.com/news/business/aroundregion/story/2017/dec/10/chattanoogfirm-help-us-navy-figure-out-how-3-/458777/

536262 3D Printing of Metal-Polymer Composite Structures Via Fused Deposition Modeling

Abstract:

This work demonstrates the ability to 3D print a fluoropolymer based energetic material that can be used as a structural reactive. The developed filament consists of a fluoropolymer binder (polyvinylidene fluoride, PVDF) with embedded aluminum particles. The filament can be printed with commercial 3D printers to create geometrically complex, three dimensional structures. The printing performance of the energetic filament was found to be comparable with standard 3D printing materials, such as ABS. The reactivity of the 3D printed energetic material was shown to be unaffected by the printing process, as determined by comparing the burning rates, as well as differential scanning calorimetry (DSC) and thermal gravimetric analysis (TGA) of the printed material with the unprinted filament. Quasi-static mechanical testing demonstrates this materialâ€™s strength and ability to function as a structural reactive. The original Al/PVDF material was shown to have a higher mechanical strength than PVDF alone. The metal content was also varied in order to obtain relationships between the energetic performance and mechanical strength based on the Al content in the material.

531868 Polymer Resin Systems for Precision Direct-Ink-Write Printing of Thermite-Loaded Inks

Abstract:

Additively manufactured sample preparation allows control of the material scale and hierarchical structure, and can potentially enable unprecedented tailoring of reaction characteristics in energetic materials. Direct-ink-write 3-dimensional printing offers the simplest test-bed to explore numerous material classes. However, the inks for such printing require specific, predictable properties such as shear thinning to accommodate nozzle extrusion, consistent yield stress point and viscosity for predictable flow, and a rapid transition rate between viscoelastic solid and liquid phases. Most classes of energetic materials such as explosives, propellants, and pyrotechnics are solids or composites comprised of highly-solids loaded polymer systems. Optimizing a desired energetic formulation for print quality while also targeting maximum energy density is problematic due to solid-solid interactions dominating the rheology of the ink. Solid-solid interactions make the flow and rheological properties less ideal for extrusion-based printing methods, and require sufficient rheological tunability to be built into the low weight percent reserved for the resin system. The halting of flow after deposition can be greatly improved by rapid, on demand chemical curing mechanisms initiated by light or heat, but such systems are poorly suited as phlegmatizing agents for sensitive energetic materials. An ideal resin system for energetic materials would have sufficient rheology and yield stress tunability to be precision printed with many types of solids, and cure on demand to expand the print design space.

We report a general, dual-polymer-based composite resin system with dual-curing mechanisms that affords extensive adaptability across multiple applications. This resin system allows for precise, omnidirectional printing of highly solids-loaded inks. Using this resin system, we were able to print single walled filaments, overhangs, spans, and a freeform helix. Our work on this family of inert ink-composites with 68-80% solids loading exhibited independently tunable (i) cure rate, (ii) rheology, (iii) glass transition temperature, (iv) print-ability, and (v) tensile/compression properties. Our inks are demonstrated to print as well as current state-of-the-art inks with little to no solids loading. A reactive ink comprised of Al and CuO solids forming a thermite was prepared and characterized. Burn characteristics of the Al/CuO thermite ink as a function of geometry will be discussed.

This work was performed under the auspices of the U.S. Department of Energy by Lawrence Livermore National Laboratory under contract DE-AC52-07NA27344.

532194 3D Printing of Thermite Mixtures Using Static Mixing

Abstract:

Water-based solids-loaded inks are a general solution for 3D printing of functional materials, including reactive materials. While pre-mixed reactive materials inks are possible, it is desirable from a safety standpoint to formulate single-component inks which each contain only one reactant. These inks are essentially inert while separate and only achieve reactive properties when mixed, minimizing the quantity of reactive material present at any one time. We have been exploring various approaches to on-the-fly mixing of such single-component inks for 3D-printed reactive materials. In this talk, I will focus on Al/CuO thermites printed from single-component inks with on-the-fly "static" mixing. Commonly employed for two-part epoxies or dental materials, static mixers work by repeatedly "folding" two fluids together until they are intimately mixed. This is a useful alternative to dynamic mixing using a rotating paddle because it minimizes the energy input into the materials and thus the risk of premature reaction. The mixing quality is also less sensitive to variations in the printing speed. Our initial work has focused on demonstrating feasibility and identifying how much mixing, as characterized by the number of mixing elements, is required to achieve maximum thermite reactivity. After establishing the critical mixing conditions, we have explored some basic 3D-printed structures and begun to assess the impact of architecture on overall reactivity.

This work was performed under the auspices of the U.S. Department of Energy by Lawrence Livermore National Laboratory under Contract DE-AC52-07NA27344. LLNL-ABS-733101

534724 Additive Manufacturing of Pyrotechnic Ignition Delays

Abstract:

Currently, boron/barium chromate, also known as T-10, is predominantly used in the field of energetics as a pyrotechnic ignition delay. It has shown to be a reliable ignition delay in providing a consistent burning rates at varying effective diameters while also being a gasless reaction. However, barium chromate has been confirmed to be highly carcinogenic. This research focused on the additive manufacturing of five possible replacements for boron/barium chromate as a pyrotechnic ignition delay: zirconium/strontium molybdate, zirconium/barium molybdate, boron/strontium molybdate, boron/barium molybdate, and titanium/silicon. Each formulation utilized a low weight percent methyl cellulose solution in water as a printing medium to provide fluid-like behavior with high viscosity. Viscosity was used to characterize printing formulations, and performance behavior was based on burning rate vs. effective diameter and differential scanning calorimetry (DSC).

539118 The Role of Particle Size on the Combustion of Boron Carbide/Sodium Periodate Biocidal Formulations

Abstract:

The Role of Particle Size on the Combustion of Boron Carbide/Sodium Periodate Biocidal Formulations

Lance N. Kotter[a] and Lori J. Groven[b,*]

[a]South Dakota School of Mines and Technology, Department of Material Science and Engineering, Rapid City, SD, US

[b]South Dakota School of Mines and Technology, Department of Chemical and Biological Engineering, Rapid City, SD, US

***corresponding author email: lori.groven@sdsmt.edu**

Abstract

In our groups previous work, the boron carbide/iodate system has been shown to be an effective biocidal agent. The use of boron carbide mitigates aging issues typically associated with traditional aluminum-based formulations. In this work, boron carbide based biocidal formulations with $NaIO_4$ serving as the primary oxidizer are explored and the role of particle size on the combustion velocities, iodine output, and combustion products are detailed. Formulations are studied with and without the addition of binder and the role of particle size detailed toward developing 3-D printable energetics. It was shown in this study that the effect of changing the boron carbide particle size had negligible effects on both the combustion velocity and iodine recovery. Ignition was however significantly influenced. However, promising results have been established by reducing the particle size of the oxidizer, $NaIO_4$. The $NaIO_4$ was milled to an D_{90} value of 30.30µm resulting in iodine recovery close to 90%, much higher than any formulation examined thus far. By reducing the particle size even further to under 25 µm while maintaining the same fuel content, iodine recovery was close to 100%. Iodine quantification was done using UV-VIS, simultaneous differential scanning calorimetry and thermogravimetric analysis (DSC/TGA), heat of reaction, X-ray diffraction (XRD) and combustion characteristics will be discussed as a function of oxidizer size.

528708 Hydrodechlorination of 1,2-Dichloroethane over Ag-Pd Catalysts Prepared By Controlled Surface Reactions

Abstract:

Hydrodechlorination is an attractive process for the treatment of toxic chlorinated waste, such as 1,2-dichloroethane, to produce valuable hydrocarbon products[1,2]. Previous studies have shown Group VIII metals, such as Pt and Pd to be active for this reaction[1,3–8]. Furthermore, the addition of Group IB metals, such as Cu or Ag has been shown to improve selectivity to produce ethylene over ethane[1,3–11]. Although numerous studies in the literature have examined hydrodechlorination reactions over AgPd catalysts[1,3–6,10–12], traditional catalyst synthesis methods produce bimetallic catalysts with a wide range of structures, making it challenging to elucidate structure activity relationships. In this work, AgPd catalysts have been synthesized by controlled surface reactions to form well defined active site structures.

We present our results on the hydrodechlorination of 1,2-dichloroethane to form ethane, ethylene, and chloroethane over Pd and AgPd catalysts. These Pd and AgPd catalysts are synthesized on carbon, SiO_2, and TiO_2 supports, and we show that the formation of uniform bimetallic structures is most effective on the TiO_2 support. The bimetallic catalysts are synthesized by depositing Pd by controlled surface reactions onto a supported Ag catalyst, as has previously been demonstrated for other metal systems[13–19]. UV visible spectroscopy is used to quantify the extent of interaction between the Pd precursor (cyclopentadienyl Pd allyl) and the support. We observe no interaction between the Pd precursor and the titania support, indicating that selective deposition of Pd onto Ag/TiO_2 structures occurs during the synthesis process. On carbon and silica supports, some interaction between Pd precursor and support is observed, and Pd deposition onto Ag is not entirely selective. It is determined that formation of bimetallic AgPd structures is most effective for AgPd/TiO_2, followed by AgPd/C and then AgPd/SiO_2 catalysts.

FTIR spectra of adsorbed CO on the AgPd catalysts provide additional evidence for variations in the active site structure depending on the degree of interaction between Ag and Pd. On the TiO_2-supported AgPd catalyst, CO adsorbs to Pd primarily in a linear configuration, while CO adsorbs in both linear and multifold configurations on the SiO_2-supported AgPd catalyst. This difference in CO adsorption indicates that Pd is well dispersed on the Ag nanoparticles for the TiO_2-supported catalyst, while a higher fraction of contiguous Pd sites is present on the SiO_2-supported catalyst.

The ethylene selectivity decreases from 97% to 1% on AgPd/TiO_2 catalysts as the Ag:Pd atomic ratio increases from 1:2 to 1:0.35. Similarly, an ethylene selectivity of 82% is achieved on the AgPd/C catalyst. On these catalysts, the selectivity to chloroethane is less than 1% and the balance of the selectivity is to ethane. The $AgPd/SiO_2$ catalyst, however, is 91% selective to ethane and 9% selective to chloroethane. These results indicate that the well dispersed Pd achieved at low Pd loadings on TiO_2 supported catalysts and carbon supported catalysts are selective to ethylene, while the contiguous Pd species formed at higher Pd loadings on TiO_2 and on SiO_2 are selective to ethane.

We have synthesized AgPd catalysts on TiO_2, carbon, and SiO_2 supports with various surface structures by using controlled surface reactions. The range of surface structures formed on these supports at various Pd loadings has allowed us to develop structure-selectivity relationships for the hydrodechlorination of 1,2-dichloroethane, observing high ethylene selectivity when Pd is well isolated because of intimate contact with Ag.

(1) Heinrichs, B.; Delhez, P.; Schoebrechts, J.-P.; Pirard, J.-P. *J. Catal.* **1997**, *172*, 322–335.

(2) Xu, L.; Stangland, E. E.; Mavrikakis, M. *Catal. Sci. Technol.* **2018**, *8*, 1555–1563.

(3) Heinrichs, B.; Noville, F.; Schoebrechts, J.-P.; Pirard, J.-P. *J. Catal.* **2000**, *192*, 108–118.

(4) Heinrichs, B.; Schoebrechts, J.-P.; Pirard, J.-P. *J. Catal.* **2001**, *200*, 309–320.

(5) Heinrichs, B.; Noville, F.; Schoebrechts, J. P.; Pirard, J. P. *J. Catal.* **2003**, *220*, 215–225.

(6) Lambert, S.; Ferauche, F.; Brasseur, A.; Pirard, J. P.; Heinrichs, B. *Catal. Today* **2005**, *100*, 283–289.

(7) Borovkov, V. Y.; Luebke, D. R.; Kovalchuk, V. I.; d'Itri, J. L. *J. Phys. Chem. B* **2003**, *107*, 5568–5574.

(8) Han, Y.; Zhou, J.; Wang, W.; Wan, H.; Xu, Z.; Zheng, S.; Zhu, D. *Appl. Catal. B Environ.* **2012**, *125*, 172–179.

(9) Ito, L. N.; Harley, A. D.; Holbrook, M. T.; Smith, D. D.; Murchison, C. B.; Cisneros, M. D. Processes for converting chlorinated alkane byproducts or waste products to useful, less chlorinated alkenes. EP 0 662 941 B1, December 10, 1997.

(10) Han, Y.; Gu, G.; Sun, J.; Wang, W.; Wan, H.; Xu, Z.; Zheng, S. *Appl. Surf. Sci.* **2015**, *355*, 183–190.

(11) Han, Y.; Sun, J.; Fu, H.; Qu, X.; Wan, H.; Xu, Z.; Zheng, S. *Appl. Catal. A Gen.* **2016**, *519*, 1–6.

(12) Heinrichs, B.; Schoebrechts, J.-P.; Pirard, J.-P. *Stud. Surf. Sci. Catal.* **2000**, *130 C*, 2015–2020.

(13) Aragão, I. B.; Ro, I.; Liu, Y.; Ball, M.; Huber, G. W.; Zanchet, D. *Appl. Catal. B Environ.* **2018**, *222*, 1–2.

(14) Alba-Rubio, A. C.; Sener, C.; Hakim, S. H.; Gostanian, T. M.; Dumesic, J. A. *ChemCatChem* **2015**, *7*, 3881–3886.

(15) Hakim, S. H.; Sener, C.; Alba-rubio, A. C.; Gostanian, T. M.; Neill, B. J. O.; Ribeiro, F. H.; Miller, J. T.; Dumesic, J. A. *J. Catal.* **2015**, *328*, 75–90.

(16) Liu, Y.; Göeltl, F.; Ro, I.; Ball, M. R.; Sener, C.; Aragão, I. B.; Zanchet, D.; Huber, G. W.; Mavrikakis, M.; Dumesic, J. A. *ACS Catal.* **2017**, *7*, 4550–4563.

(17) Ro, I.; Liu, Y.; Ball, M. R.; Jackson, D. H. K.; Chada, J. P.; Sener, C.; Kuech, T. F.; Madon, R. J.; Huber, G. W.; Dumesic, J. A. *ACS Catal.* **2016**, *6*, 7040–7050.

(18) Ro, I.; Sener, C.; Stadelman, T. M.; Ball, M. R.; Venegas, J. M.; Burt, S. P.; Hermans, I.; Dumesic, J. A.; Huber, G. W. *J. Catal.* **2016**, *344*, 784–794.

(19) Sener, C.; Wesley, T. S.; Alba-Rubio, A. C.; Kumbhalkar, M. D.; Hakim, S. H.; Ribeiro, F. H.; Miller, J. T.; Dumesic, J. A. *ACS Catal.* **2016**, *6*, 1334–1344.

534263 Science of Shape-Controlled Synthesis of Metallic Nanoparticles

Abstract:

The hydrophilic polymer, polyvinylpyrrolidone (PVP) is ubiquitously used in the synthesis of shape-controlled colloidal transition metal (Ag, Au, Pd, Pt, etc.) nanoparticles (NPs). Nearly all hypotheses regarding the role of PVP rely on observations from parametric studies of the impact of PVP-related variables on the final shape of the as-synthesized nanoparticles. The dominant hypothesis in the literature is the preferential binding of PVP to Ag(100) facets compared to Ag(111) facets; however, there is no unified agreement on the exact shape control mechanism in Ag NPs growth as both Ag cubes and octahedron are synthesized in the presence of PVP. We experimentally measured adsorption isotherms of PVP on different-shaped Ag NPs to determine the thermodynamics of PVP adsorption to Ag(100) and Ag(111) surfaces and found the preferential binding of PVP is not responsible for thermodynamic shape control in Ag nanoparticles synthesis.

Rather, the addition of Cl^- during synthesis is critical for for the formation of Ag nanocubes. In the absence of Cl-, truncated octahedron are formed in a polyol synthesis, which has been confirmed by recent theoretical investigastions of Wulff constructions on PVP-covered Ag surfaces. We examined the role of Cl- in the synthesis of Ag nanocubes utilizing various ex-situ characterization techniques such as scanning electron microscopy (SEM), x-ray diffraction (XRD) and x-ray photoemission spectroscopy (XPS) to probe the growth mechanisms of Ag nanocubes. The initially formed nanostructures during the early stages of synthesis were AgCl nanocubes which were subsequently etched by incoming Ag^+ flux to form metallic Ag nanocubes containing a half monolayer of Cl^- on the surface. Variation of Cl^- concentration alone at constant H^+ and PVP concentration led to modification of the nanoparticle shape. As the concentration of Cl^- decreased, truncated cubes, cuboctahedron and truncated octahedron formed, which confirms the importance of Cl^- for Ag NPs shape control.

A number of syntheses exploit the observation that the addition of foreign metal ions influences the shape of the final nanostructure. We focused on the ability of Cu2+ to impact the transformation of Ag cubes to octahedron. The conversion is initiated using Ag cube as seeds in the presence of Cu^{2+} to convert cube to a final octahedron product. The evolution of shape from truncated cube, cuboctahedron, truncated octahedron and octahedron was captured by ex-situ microscopy analysis. The existence of Cu^{2+} on the Ag octahedron surface was confirmed by x-ray photoemission spectroscopy (XPS) and the bulk content by inductively coupled plasmon optical emission spectroscopy. Control experiments demonstrate shape transformation only occurs with the addition of Cu^{2+}. Substitution of Cu^{2+} with other metal ions (Ni^{2+} and Co^{2+}) with a lower reduction potential than Ag^+ led to a similar transformation of Ag cubes to octahedron. We hypothesize foreign metal ions adsorb to the (111) facets on the truncated cubes by under potential deposition (UPD), which favors the deposition of incoming Ag on the (100) facets. Shape control can be achieved by the addition of foreign ions during synthesis; the mechanism of the shape control is complex. To the contrary, it does appear organic additives, such as PVP are responsible for shape control in Ag nanocube synthesis. The origin of shape control may be more kinetic in origin than thermodynamic in nature, a concept that has recently infiltrated crystallization theory.

536401 Facile Novel Synthesis and Characterization of Gold-Copper Bimetallic Nanoclusters for Applications in Oxidation Catalysis

Abstract:

Catalysis through active transition metal nanoparticles often allows high reactivity or selectivity in low temperatures. Varying metal selection and amounts in bimetallic nanoclusters offer a new route to tune activity, selectivity, among other optical and structural properties. This work represents a facile approach to the synthesis of gold-copper bimetallic nanoparticles and the properties they exude from this synthesis method. In this synthesis, triphenylphosphine is used as the bulky capping agent to prevent agglomeration and sodium borohydride is used as the reducing agent. HAADF-STEM confirms the average size of the nanoclusters at approximately 2 nm. UV-Vis and fluorescence measurements were also conducted to measure the optical properties of these clusters. DRIFTS studies show the triphenylphosphine ligand interaction varies with the composition of bimetallic clusters. The particles are dispersed onto (<67 um) TiO_2, MgO, and SiO_2 among other metal oxides as a support structures. These supported metal cluster materials were then subsequently used as catalysts for CO oxidation and benzyl alcohol oxidation to determine structure-function relationships relating the composition in the form of the Au:Cu ratio of the catalyst to its activity. These results help to guide future applications and design of engineered materials in oxidation catalysis.

536760 Atomic Layer Deposited Pt-Co Bimetallic Nanoparticles for Selective Hydrogenation

Abstract:

Highly dispersed Pt-Co bimetallic catalysts were deposited on multi-walled carbon nanotubes (MWCNTs) by atomic layer deposition (ALD). Both high-resolution TEM and STEM/EDS analysis verified the formation of Pt-Co bimetallic particles. High selectivity to cinnamic alcohol (> 90%) was achieved with high conversion (> 90%) in hydrogenation of cinnamaldehyde over an optimized Pt-Co/MWCNTs catalyst. The catalyst also showed both high activity and high selectivity in C=O activation (> 90%) in selective hydrogenation reactions of other Î±, Î²â€unsaturated aldehydes, which indicates that the bimetallic catalyst was a universal and highly efficient catalyst for these kinds of reactions. Through a series of experiments and comparisons, the synergistic effects of Pt, Co, and MWCNTs were investigated. The results revealed that the interaction between Pt and Co NPs played a vital role in the high selectivity in C=O activation, and the synergistic effect between Pt-Co and MWCNTs boosted greatly the activity of Pt-Co bimetallic catalysts.

537189 Facile Synthesis of 2D Molybdenum Carbide Nanosheets

Abstract:

With a growing need for development of technologies to fight the effects of climate change, pollution, and create energy security due to the decreasing availability of fossil fuels, renewable electrochemical reactions have become an important global research focus. However, the oxygen reduction reaction (ORR) and oxygen evolution reaction (OER) limit the efficiency of these targeted electrochemical reactions. Material development for these processes has centred around the optimization of cost and stability. Transition metal-based carbides have emerged as non-precious electrocatalysts with noble-metal-like electronic properties that can catalyze the ORR/OER. The activity of a variety of catalyst morphologies has been studied previously; however, the synthesis of nanosheet carbides has proven difficult. This morphology provides a unique vehicle for the study of surface chemistry and electronic structure as well as correlation with theoretical experiments allowing insight into the key parameters that govern their activity and stability. A series of molybdenum oxide, nitride, and carbide nanosheets of varying thicknesses were synthesized through a facile method and their ORR activity and stability were investigated in order to identify trends in material activity and stability. In addition, the use of these materials to investigate the effects of morphology at the nanoscale will be discussed.

537457 Identification of Optimally Stable Nanoparticle Geometries Via Mathematical Optimization and Density-Functional Theory

Abstract:

One of the key research objectives in the area of transition metal nanoparticles is to determine the most stable structure for particles of a given size (mass) and composition. In this work, we propose a mathematical optimization based modeling framework, coupled with density functional theory (DFT) calculations, for determining the proven minimum energy structures of three-dimensional, mono-metallic nanoclusters of a given size. More specifically, we apply this framework to mono-metallic copper (Cu) nanoclusters, as previous work has shown potential uses for mono- and bi-metallic Cu nanoparticles in CO_2 activation and reduction [1],[2],[3].

In our work, we consider the global minimum energy Cu nanocluster of a given size to be the one that attains the maximum average dimensionless cohesive energy, which is the average cohesive energy normalized to the bulk energy. The cohesive energy measures the strength of inter-atomic bonding between atoms and the overall stability. Thus, maximizing the average dimensionless cohesive energy leads to the most stable, ground-state nanocluster geometry. Our proposed model uses a cohesive energy function, first proposed by Tomanek et al., which depends only on the geometric descriptor of coordination number (CN) [4]. Because CN is a local geometric descriptor of the nanoparticle (NP), it can be encoded within a rigorous, "bottom-up" design optimization model. In fact, previous work has successfully utilized CN as a local descriptor to design transition metal surfaces via a mixed-integer linear programming (MILP) [5], a class of mathematical optimization problems that can be solved readily and reliably with appropriate numerical optimization techniques. The current work serves as a first step in extending such tools for targeted nanoparticle design in three dimensions. It should be highlighted that the proposed method deviates from the existing literature, inasmuch as existing methods for determining lowest energy small nanoclusters typically involves heuristic search of the design space and evaluation of incumbent solutions via inter-atomic potentials [6],[7],[8],[9]. However, global minimum energy nanoclusters reported in this way are only considered "putative," as the search methods utilized provide no mathematical proof of optimality. In contrast, the use of MILP based optimization techniques proposed in this work guarantees the optimality of the final designs.

However, the cohesive energy model originally proposed by Tomanek el al. also includes a repulsive potential energy term. Whereas previous work has shown that neglecting the repulsive contribution for sufficiently large nanoclusters compares favorably to the DFT calculated cohesive energy [10], doing so at small NP sizes is not appropriate due to surface contraction and quantum effects. In order to create an optimization model that is accurate for small NPs, we first assemble a large collection of nanoparticles of given size that feature a wide variety of coordination number distributions, and then use first-principles DFT calculations to evaluate their cohesive energy. We then perform linear regression to elucidate a correction term that can be represented in our MILP optimization framework. This way, optimally cohesive NPs at small sizes can be correctly and quickly predicted.

Following the above methodology, a pool of optimal and near-optimal Cu NPs of different small sizes (up to 50 atoms) was collected, and single-point DFT calculations (using the PBE exchange-correlation functional [11]) were performed to determine the expected cohesive energies for these structures. Using this data, a linear regression model was developed in order to identify corrections for the cohesive energy contribution of a single atom in a given Cu NP. These corrective factors were then incorporated into the mixed-integer optimization model, which determined highly cohesive nanoclusters for each given size, establishing morphological trends related to stability in NPs of any size. The results confirm various "magic-number" designs that are well-known to be highly stable, as well as reveal a number of unintuitive designs that were not previously reported in the literature, yet are confirmed to be highly stable after DFT geometry optimizations. These particles can serve as model particles for a multitude of follow-up studies within the realm of computational chemistry to include, among others, adsorption and catalytic activity studies. Finally, by collecting sequences of most cohesive designs for given NP sizes, our results can also be used to determine expected distributions of clusters that arise during experimental synthesis.

References

[1] Tang, W., et al., "The importance of surface morphology in controlling the selectivity of polycrystalline copper for CO_2 electroreduction," *Physical Chemistry Chemical Physics*, 14(1):76-81, 2012.

[2] Austin, N., Ye, J., Mpourmpakis, G., "CO_2 activation on Cu-based Zr-decorated nanoparticles," *Catalysis Science & Technology*, 7(11):2245-2251, 2017.

[3] Kim, D., et al., "Synergistic geometric and electronic effects for electrochemical reduction of carbon dioxide using gold–copper bimetallic nanoparticles." *Nature Communications,* 5:4948, 2014.

[4] Tomanek, D., Mukherjee, S., Bennemann, K. H., "Simple theory for the electronic and atomic structure of small clusters," *Physical Review B*, 28(2):665, 1983.

[5] Hanselman, C. L., Gounaris, C. E., "A mathematical optimization framework for the design of nanopatterned surfaces," *AIChE Journal*, 62(9):3250–3263, 2016.

[6] Darby, S., et al., "Theoretical study of Cu–Au nanoalloy clusters using a genetic algorithm," *The Journal of Chemical Physics*, 116(4):1536-1550, 2002.

[7] Rapallo, A., et al., "Global optimization of bimetallic cluster structures. I. Size-mismatched Ag–Cu, Ag–Ni, and Au–Cu systems," *The Journal of Chemical Physics*, 122(19):194308, 2005.

[8] Lai, X., Xu, R., Huang, W., "Geometry optimization of bimetallic clusters using an efficient heuristic method," *The Journal of Chemical Physics*, 135(16):164109, 2011.

[9] Barcaro, G., Sementa, L., Fortunelli, A., "A grouping approach to homotop global optimization in alloy nanoparticles," *Physical Chemistry Chemical Physics*, 16(44): 24256-24265, 2014.

[10] Yan, Z., et al., "Size, Shape and Composition Dependent Model for Metal Nanoparticle Stability Prediction," *Nano letters*, 18(4):2696–2704, 2018.

[11] Perdew, J.P., Burke, K., Ernzerhof, M, "Generalized gradient approximation made simple," *Physical Review Letters*, 77(18):3865, 1996.

538638 Niau Single Atom Alloys for the Oxidative Coupling of Methacrolein with Methanol

Abstract:

Direct oxidative esterification of methacrolein with methanol is a more environmentally benign route for producing methyl methacrylate in comparison with the traditional acetone cyanohydrin method [1]. Since the early work of Haruta et al., in which, Au showed remarkable activity and selectivity when it was highly dispersed in the form of nanoparticles (<5 nm diam.), it has been studied for many other reactions with high selectivity towards the desired products [2]. Nanoporous Au has been examined in gas phase and UHV studies for the methacrolein â€“ methanol coupling, and an excess of methanol (>90%) was found necessary for the effective cross-coupling [3]. In addition, Au nanoparticles deposited on basic supports has also been tested for the liquid phase coupling reaction, but the reaction mechanism has not been investigated in detail [4].

In the present work, we study Au and NiAu alloy catalysts where Au is either supported in nanoparticle form on an inert support, SiOÂ
methanol. The formation of highly dilute Ni-Au alloys is followed by CO-DRIFTS. Catalytic performance tests in the liquid phase show that these catalysts have almost 100% selectivity towards methyl methacrylate and remarkable stability. Addition of Ni atoms on Au surface greatly improves its reactivity (conversion increases from ~5% on Au to ~25% on NiAu) while selectivity is maintained at the same levels as pure Au catalysts. Both Au and NiAu systems (either in nanoparticle or in nanoporous form) were found to have similar activation energies by kinetic measurements. This is an indication that Au is the active site in both the pure Au and the NiAu single atom alloy systems for this reaction. The promoting effect of Ni is also investigated by operando ATR-IR spectroscopy under reaction conditions. In addition, ATR-IR provides an insight into the reaction mechanism through identification of reaction intermediates-adsorbates on the catalytic surface. The catalyst structure and composition were also followed by ICP, XRD, HRTEM, SEM and XANES â€“ EXAFS.

Acknowledgements: This material is based upon work supported as part of the Integrated Mesoscale Architectures for Sustainable Catalysis (IMASC), an Energy Frontier Research Center funded by the U.S. Department of Energy, Office of Science, Office of Basic Energy Sciences under Award Number DE- SC0012573.

Reference

1. Yamamatsu, T. Yamaguchi, K. Yokota, O. Nagano, M. Chono, A. Aoshima, Catal. Surv. Asia 2010, 14, 124-131
2. Haruta, N. Yamada, T. Kobayashi and S. Iijima, *J. Catal.* 1989, 115, 301-309
3. Karakalos, B. Zugic, K.J. Stowers, M.M Biener, J. Biener, C.M. Friend, R.J. Madix, Surf. Sci 2016, 652, 58-66
4. Suzuki, T. Yamaguchi, K. Matsushita, C. Iitsuka, J. Miura, T. Akaogi and H. Ishida, *ACS Catal.* 2013, 3, 1845-1849

532636 Why and How to Select and Design a Dense Phase Conveying System?

Abstract:

One of the key processing technologies that have been widely utilized in many industries, such as chemicals, food, and pharmaceutics, is pneumatic conveying, regardless of whether it is for unloading raw ingredients from bags/drums/trucks/railcars, transporting ingredients from its storage to their next process step, or moving final products to their packaging area. Many applications can be easily achieved by choosing a dilute phase conveying system. However, there are many situations where the high velocity causes problems of bend wear, reduction in product quality through attrition, or excessive power requirements. Dense phase conveying, due to the use of significantly lower velocity, can address these issues.

Because of the lower gas velocity produced with dense phase conveying, a much slower, gentler conveying action on the material. This gentle action also reduces segregation and attrition issues often experienced with the more aggressive dilute-phase operation. For these reasons, in food operations, dense-phase conveying is often used to convey the pre-blends before going to the next process step as well as the final products going to packaging.

This presentation not only discusses how to determine the dense phase conveyability of bulk solids, but also explains the mechanisms of available dense phase conveying methods. Finally, a case study will be added to describe the details on how to design a blow tank dense phase conveying system.

537199 Assessment of Effects of Flow Enhancers in Pneumatic Conveying of Cohesive Dairy Powders: CFD-DEM Simulation

Abstract:

The aim of this study is to reduce the influence of particle roping on dairy powder deposition and pipeline blockage at pipe sections after a 90° bend. A CFD-DEM simulation of pneumatic conveying of cohesive dairy powder was performed to study the effect of modifying the design of a pipeline downstream of 90° bends. The simulation was performed with a commercial discrete element modelling (DEM) software, EDEM®, coupled with the computational fluid dynamics (CFD) software, FLUENT®. Fat-filled milk powder (FFMP) fines with average particle size of 94μm was used. The simulation was implemented for different flow enhancers (i.e. inward or outward air jet(s), inner/outer offset, converging nozzle etc.) at the downstream of a 90° bend. The simulation results show that the use of the flow enhancers can speed up particle rope dispersion, effect de-agglomeration, reduce agglomerated particle deposition, and eliminate the continuous build-up of cohesive particles at bends, horizontal and vertical pipe walls.

530081 Towards Understanding Fly Ash Transport and Deposition in the Human Respiratory System: Effects of Physiological Conditions and Fly Ash Properties

Abstract:

Fly ash is a common solid residue of incineration plants and poses a great environmental concern because of its toxicity upon inhalation exposure. The inhalation health impacts of fly ash is closely related to its transport and deposition in the human respiratory system which warrants significant research for health guideline setting and inhalation exposure protection. We carried out a series of fly ash transport and deposition experiments in a bifurcation airway model by combining optical aerosol sampling analysis and high-speed photography technique. Five types of coal fly ash samples of different morphologies were tested and their respiratory deposition and transport processes were compared. The deposition efficiencies were calculated and relevant transport dynamics mechanisms were discussed. The influences of physiological conditions such as breathing rate, duration, and orientation [horizontal (lying-down case) and vertical (standing case)] and fly ash physical properties (size, morphology, and specific surface area) were investigated. The fluid mechanisms governing the transport dynamics and deposition patterns of fly ash particles were explored. The study serve as a basis for setting health guidelines and recommending personal respiratory protective equipment for fly ash handlers.

535348 Large Eddy Simulation of Particle Laden Flows

Abstract:

The effects of particle density and diameter on residence time distribution is predicted using large eddy simulations (LES). Unlike RANS approaches which only capture the mean flow, LES simulations capture the effects of mean flow and the flow turbulence on particle transport. This combination provides superior insights into (i) the flow physics driving particle motion, and (ii) the velocity/shear experienced by individual particles as they evolve through the system.

We begin with a brief introduction to the theory and implementation of LES simulation, fluid-particle dynamics, and particle-particle interactions. We then show how the approach can be used to model movement through agitated and non-agitated open systems. For particles interacting with a fully turbulent system, the effects of turbulence on particle dynamics can be linked predictably to the particle settling timescale. For particles interacting with partially turbulent systems, however, the dynamics are more nuanced and strongly informed by local flow conditions.

538318 Strandphase®: The Gentle Pneumatic Conveying Solution

Abstract:

Dilute and Dense phase are the two common conveying methods used for plastic granulated material. Energy and economics plays a major role in both the conveying methods. In dilute phase, high gas velocity disintegrates the particles and leads to the generation of high volumes of dust and streamers. Whereas dense phase conveying systems due to high pressure requirement, demands better structural and pipe support. STRANDPHASE® conveying falls in between dilute and dense phase. Unlike dilute and dense phase, STRANDPHASE® minimizes dust and streamers generation, maintains high product to air ratio and doesn't require special pipe support. In this work LDPE pellets were used for the conveying tests. Dust and streamers were analyzed before and after the conveying to evaluate the performance of STRANDPHASE® and dilute phase conveying.

547351 CFD Study of Drag Model on Simulating Bubbling Fluidized Bed of Geldart A Particles

Abstract:

Fluidized beds are widely used in chemical, petrochemical, metallurgical, and energy industries. Hydrodynamic understanding of gas-solid multiphase flow is critical in the success of fluidized bed design.

A two fluid model with kinetic theory is commonly used in computational fluid dynamics (CFD) models to simulate the gas-solid flows. Interaction between gas and solid phase is described by a drag model, which plays an important role in predicting the hydrodynamic behavior in the fluidized bed. However, few successful simulations were reported on a bubbling fluidized bed of Geldart A particles. Overestimated bed expansion comparing to experimental data was reported by Zimmermann and Taghipour [1], which attributed to overestimated drag from classic drag models of Syamlal–O'Brien [2] and Gidaspow [3]. Most classic drag models were developed based on homogenous flow and weighted by solids volume fraction. Energy minimization multi-scale theory based drag models (EMMS) were reported for better representing the heterogeneous nature in a bubbling bed and having better agreement of bed expansion with experimental measurement [4][5]. The EMMS drag model intentionally resolves the heterogeneous structures by seeking the minimum energy consumption for particles suspension and transport.

In this research, a 3D CFD model is constructed based on PSRI's bubbling bed for the NETL challenge problem III (2010) [6]. A two fluid model incorporating kinetic theory is used in the simulation. A commercial CFD software, Ansys FLUENT 19.0, is chosen for simulation. The effect of different drag models of Gidaspow, Wen-Yu with factor of 0.7 and EMMS on the simulation results are evaluated and compared to experimental data. The EMMS model was implemented into FLUENT by a user defined function (UDF). Two sets of experiments on 3% content of fine particles with different superficial gas velocity are modeled for validating the drag model's wide application on the bubbling regime. The results showed that both Gidaspow and modified Wen-Yu drag model over-predict the bed expansion comparing to the EMMS drag model. The heterogeneous structure in the bottom dense region is better resolved by the EMMS model.

[1] Zimmermann, S. and F. Taghipour, CFD Modeling of the Hydrodynamics and Reaction Kinetics of FCC Fluidized Bed Reactors, Ind Eng Chem J..44, 9818–9827. 2005.

[2] Syamlal, M. and T.J. O'Brien, Computation of flow patterns in circulating fluidized bed. A.I.Ch.E. Symposium Series 85 (270), 22–31, 1989.

[3] Gidaspow, D., 1994. Multiphase Flow and Fluidization: Continuum and Kinetic Theory Description. Academic Press, Boston.

[4] Yang, N, W. Wang and W. Ge, J. Li, CFD simulation of concurrent-up gas-solid flow in circulating fluidized beds with structure-dependent drag coefficient, Chem.Eng J. 96, 71–80. 2003.

[5] Benzarti, S., H. Mhiri, and H. Bournot, Drag models for Simulation Gas-Solid Flow in the Bubbling Fluidized Bed of FCC Particles, International Journal of Chemical and Molecular Engineering 6, 111-116, 2012.

[6] NETL Multiphase Flow Science, Challenge problem III, 2010, https://mfix.netl.doe.gov/challenge-problem-iii-2010/

AIChE Annual Meeting: Advances in Processing and Handling of Energetic Materials

Oral Presentation, October 31, 2018

Porous Silicon Microparticles and Quantum Dots for Energetics

Philip M. Guerieri*, Casey D. Foley, Wayne A. Churaman, Matthew H. Ervin, David Lunking, and Brian Isaacson

ABSTRACT: Energetic porous silicon has emerged in the last fifteen years as a uniquely positioned formulation class featuring the increased energy density of metal-based fuels relative to monomolecular explosives, yet with exceedingly high energy release rates (1-3000 m/s flame speeds). Thus far, porous silicon is "on-chip", etched from a silicon wafer, and is therefore compatible with MEMS fabrication techniques and ideal for augmenting conventional electronics, creating small-scale thrust or actuation, and replacing explosive initiators. However, this same "on-chip" nature also limits potential applications and effectively dilutes the specific energy density when considering its inert substrate base. Efforts to expand the technology off-chip have led to our fabrication of hydrogen-terminated silicon microparticles and quantum dots derived by mechanical attrition from porous silicon commonly used for experimental drug delivery [1]. This work characterizes microparticle ranging in size from 100 microns down to nominally 5 nm Si nanoparticles with specific surface areas over 500 m^2/g. Polydisperse as-sonicated particle mixtures feature open-channel flame speeds similar to on-chip porous silicon propagation rates with $NaClO_4$ oxidizer (exceeding 2 km/s) [2]. Characterization of surface hydrogen termination competing with incipient surface and backbond oxygen passivation increasing with prolonged sonication times by FTIR and DSC is presented in parallel with combustion performance evaluations of constant-volume dynamic pressure and hotwire ignition temperatures. Results suggest replacement of hydrogen termination with oxygen passivation with decreasing particle size is responsible for increasing ignition temperature of Si colloids and decreasing qualitative mechanical ignition sensitivity of smaller particles (together with lower potential of smaller particles to mechanically fracture). However, nanoporous silicon microparticles and nanoparticles exhibit comparable combustion performance to that of on-chip porous silicon, comprising a promising feedstock for energetics additive manufacturing, augmentation of formulations, or assembly of engineered nanoscale energetic composites.

[1] Qin, Z., Joo, J., Gu, L., & Sailor, M. J. (2014). Size control of porous silicon nanoparticles by electrochemical perforation etching. Particle & Particle Systems Characterization, 31(2), 252-256.

[2] Adams, S. K., Piekiel, N. W., Ervin, M. H., & Morris, C. J. (2018). Silicon quantum dots for energetic material applications. Applied Physics Letters, 112(23), 233108.

535798 A Materials Science-Based Approach for the Re-Development of COMP B

Abstract:

Materials science focuses on investigating the relationships that exist between materials processing, composition, structure, property and performance. In this work, we aim to re-develop Composition B (Comp B) using a materials science based approach. Comp B is traditionally produced by a melt-cast process in which a slurry of large RDX crystals dispersed in molten TNT is cast and allowed to solidify into a charge. It is a widely used formulation due to its relatively high performance and low cost. However, Comp B is a relatively shock sensitive formulation, most likely due to the structural defects such as large voids formed during the solidification process of TNT. To eliminate large voids that play a role of sensitizing the explosive, we prepared nanoenergetics-based Comp B (N-Comp B) by compacting a molding powder which was prepared by spray drying and consists of nanoscale RDX and TNT. Reduced shock sensitivity was observed from N-Comp B, however, the decrease seems to have been limited by the presence of a large number of voids. The addition of a small amount of polymeric binder during the spray drying process was found effective in further reducing the size of voids, and more importantly, the number density of voids. This leads to a significant shock sensitivity reduction compared to melt-cast Comp B. This work continues to demonstrate the versatility of the spray drying-based materials processing method for nanoenergetics-based explosive formulations. Furthermore, it suggests that a materials science-based approach can facilitate the re-development of existing formulations or the development of new formulations to fulfill the sensitivity and performance requirements of the next generation explosives.

538640 Replacement of Barium Chromate in the Traditional Tungsten Delay

Abstract:

Replacement of Barium Chromate in the Traditional Tungsten Delay

Barbara A. Hadrava[a] and Lori J. Groven[a]

[a]South Dakota School of Mines and Technology, Department of Chemical and Biological Engineering, Rapid City, SD, USA

ABSTRACT

There is an immediate need to replace Barium Chromate in the current fielded delay due to its hazardous nature. Previous work in this area has explored various alternate compositions such as metal/metal oxide, intermetallic, or metalloid/metal oxide systems. However, in many cases these alternate systems cannot serve as a true replacement due to a lack of tailorable burn times. In this work, we study the fundamental kinetics of the tungsten delay system and the role that barium chromate serves towards the reliability and performance by replacing the barium chromate with barium molybdate. Delay performance was examined at a similar diameter, 6.35 mm, to a common delay housing. The measured combustion velocities ranged from 0.94-18.75 mm/s for 6.35 mm pellets, and from 0.82-23.99 mm/s when pressed into 6.35 mm ID Al housings. For all W/$BaMoO_4$/$KClO_4$ formulations, gas production was measured, and maximum combustion temperatures are measured as a function of mixture stoichiometry. Combustion products were characterized with powder X-ray diffraction and appear benign based upon known compound information. Therefore, the W/$BaMoO_4$/$KClO_4$ system seems suitable as a relatively low combustion velocity, low gas producing, low toxicity delay composition.

537947 Combustion of Multi-Stage Ball Milled Ternary B/Al/PTFE Nano-Sale Composites

Abstract:

Boron has high gravimetric and volumetric combustion enthalpies, almost two times higher than aluminum, and has high energy release potential in a number of applications involving air combustion such as solid fueled ramjet, blast enhancement and pyrotechnic applications. However, it is difficult to extract from boron its full combustion enthalpy due to the surface oxide layer (B_2O_3), especially in micron size boron particles. Thus, the efficiency of boron combustion is low. In contrast to boron, aluminum particles burn more favorably due to the presence in many environments of homogenous flame zones and the differing thermodynamics/physics of combustion-generated oxides. Fluorine has shown the ability to lower the ignition temperature of boron by removal of the oxide layer chemically and to accelerate boron combustion by converting the high boiling temperature oxide product to gaseous BF_3. Specifically, Al/PTFE energetics have shown favorable combustion performance such as high heat/light release and low ignition temperature. While PTFE addition may further reduce fuel content, it may be possible to develop ternary B/Al/PTFE nanoscale composite energetics that exploit the drastically higher combustion enthalpy of boron and ignition/combustion enhancements of aluminum and PTFE in order to produce fuel particles with favorable ignition and combustion energy release properties that have heating values in excess of neat aluminum.

In this work, Al/PTFE was used to improve boron combustion through use of multi-stage ball milling processes in order to explore the effects of staged ball milling on the combustion of the B/Al/PTFE system. The favorable combustion energy properties of the ternary composites are explored through equilibrium calculations and oxygen calorimetry. The ignition performance of ternary composites are measured using T-jump hot wire experiments. The combustion rate and combustion species are explored using optically accessible small volume combustion bomb experiments. In particular, we investigate differences in performance which can occur specifically from variation of ball milling staging rather than overall stoichiometry and further explore reasons for these differences through characterization of the composites and their combustion products. This work provides a fundamental understanding of reactions between B/Al/PTFE ternary compositions in systems with combustion performance that exceeds aluminum and provides insight into how the nanostructure can affect changes in combustion.

539128 Milling of Energetic Crystals with the Labram

Abstract:

Milling of Energetic Crystals with the LabRAM

Lance N. Kotter[a] and Lori J. Groven[b,*]

[a]South Dakota School of Mines and Technology, Department of Material Science and Engineering, Rapid City, SD, US

[b]South Dakota School of Mines and Technology, Department of Chemical and Biological Engineering, Rapid City, SD, US

***corresponding author email: lori.groven@sdsmt.edu**

The LabRAM acoustic mixer has widely been used for mixing and processing of energetic materials from propellant mixes, explosive-binder formulations, to cocrystal synthesis. It was only very recently that it has demonstrated utility in the field of general material processing such as sieving. Size reduction of energetic crystals remains a tremendous processing challenge and is often time and energy intensive. While milling is used extensively to reduce/grind/modify fuel and oxidizer type materials the LabRAM has not been explored specifically for that purpose. The ability for the LabRAM to be controlled remotely while providing low intensity acoustic waves (60 Hz) makes it a suitable candidate for the milling of energetic crystals. In this study, three distinct materials were selected to determine the effective milling of the LabRAM. Ammonium Perchlorate (AP) as a highly sought after energetic with distinct spherical morphology, Sodium Periodate as an irregular crystal that has shown effectiveness in biocidal formulations, and Nitroguanidine for its interest in energetic formulations while having rod-like morphology. In this study, dry milling of all materials was performed on the LabRAM while optimizing parameters such as media size and loading, vibrational intensity, and time duration. The three materials were analyzed by laser particle analysis and SEM to determine the average particle. The D_{50} after 15 minutes of continuous milling for Ammonium Perchlorate, Sodium Periodate, and Nitroguanidine were 7.01, 2.92, and 22.90 μm respectively. In the case of AP this was a reduction of the measured D_{50} from around 250 to 7 μm in as little as 15 minutes. The processing technique has the potential to allow for wet grinding as well or significant particle morphology modification and will be discussed.

540228 The Synthesis and Properties of Cyclic Nitramine Crystals with Metal Particlesinclusions

Abstract:

The main volume of composition of solid propellants is filled with particulate crystals of oxidizer and various energy additives, while high concentration metal particles occupy the rest of intercrystalline space. All these lead to high agglomeration and close contact of metal particles with each other, which is the necessary condition for developing the processes of agglomeration, that are characteristic for majority of modern high-energy materials.

The current research is devoted to creating a principally new approach to the solution to the problems connected with use of metal particles in high-energy materials. As a resulted of our research, we obtained crystals of cyclic nitramine, such as CL-20 and HMX with inclusions of micro- and nanoaluminum particles, submicron-sized boron, micron-sized aluminum and titanium borides and nanosized ferric oxide (III) of different size. This approach lets remove a part of active metal from the direct contact with liquid binding agents and, as a result, reduce the concentration of metal in intercrystalline space of products.

A principally new approach to eliminate the existing problems of the use of different metals in composite energy materials with the synthesis of inclusion complexes was developed.

By the example of CL-20 it was displayed that the inclusion of metal particles into nitramine does not significantly influence its chemical stability, thermal properties, sensitivity characteristics to mechanical exposure as well as specific electrical resistance and minimal ignition energy as compared with parent CL-20.

It was stated that the substitution of CL-20 for its inclusion complexes in propellants leads to the increase of burning rate up to 1.5 â€“ 2 times and the decrease in the processes of slag formation and agglomeration.

The research was performed with the financial support of the Russian Foundation for Basic Research (grant no. 16-29-01060\16 on June 8, 2016).

527161 Synthesis and Characterization of CL-20/Oxidant Crystals

Abstract:

While the discovery of novel energetic compounds has slowed, the desire for energetic materials with improved detonation performance and reduced sensitivity has not. CL-20 is currently the highest performing energetic material available, but has yet to see widespread applications due to sensitivity and production concerns. Cocrystallization has emerged as a valid method of achieving tunable energetic properties without significant reduction in energetic performance and has been in utilized in a variety of energetic combinations. In this work, several CL-20-oxidant crystals are synthesized via slow evaporation. The resulting crystals were characterized via scanning electron microscopy (SEM), Fourier transform infrared spectroscopy (FTIR), Raman spectroscopy, thermogravimetric analysis (TGA), and differential scanning calorimetry (DSC) to assess their physical and chemical properties. The overall applicability, performance and scalability of energetic-oxidant materials is discussed.

525337 Effect of Liquid Hydrocarbon-Based Process Control Agents on Characteristics of Mechanically Alloyed AlÂ·Ti Powders

Abstract:

Aluminum is a common fuel additive in propellants and explosives. Reducing agglomeration of aluminum powders and accelerating their burn rates are important for maximizing the utility of the high aluminum combustion enthalpy in solid propellants. Similarly, higher burn rates are necessary to couple the energy of aluminum combustion to that of a blast in an explosive charge. Previously, mechanical alloying was used to prepare a range of Al·Ti powders. It was observed that such powders ignite at lower temperatures than pure aluminum; particles of such alloys also burn faster than the same size pure Al particles. However, prepared powders were relatively coarse. Previous efforts were not successful in reducing the particle size of the Al·Ti particles because of the ductile nature of both metals and cold welding accompanying mechanical alloying. Recently, it was found that the particle size and surface morphology for Al-based materials can be successfully controlled using a polar hydrocarbon fluid, acetonitrile, as a process control agent, PCA. This work aims to characterize systematically the effect of liquid hydrocarbon-based PCAs on the particle size, surface morphology, and structure of mechanically alloyed Al·Ti powders. Samples of Al-rich Al-Ti alloy were prepared by mechanical alloying using both polar and non-polar hydrocarbon PCAs, acetonitrile and hexane, respectively. Upon recovery of the powders, the surface morphologies and particle sizes are characterized using SEM. It is observed that the particle agglomeration occurs for the powders processed with hexane; conversely, powders processed with acetonitrile contain well-separated, finer particles. Prepared materials are characterized using differential scanning calorimetry; their ignition is explored using a heated filament experiment operated at different heating rates. Particle combustion is studied by introducing the powders into an air-acetylene flame and/or by feeding the powder through a focused beam of a CO_2 laser. Results will be presented and interpreted with the aim of developing the mechanistic understanding of the effect of surface modification of Al·Ti alloys on their ignition and combustion.

526242 Ignition and Combustion Mechanisms ofÂ Mg-Ca(IO_3)$_2$Â Reactive Nanocomposites

Abstract:

Iodine is extensively explored as a potent biocidal additive in bio-agent defeat energetic formulations. Release of iodine into surrounding areas needs to accompany explosion and combustion of energetic charge. In one approach, elemental iodine is added to metal powders serving as high energy density fuels; its release thus occurs during combustion of such iodine-doped aerosolized metal particles. To increase the iodine concentration in the energetic charge and accelerate combustion, a compound containing oxidized iodine can also be added as part of an oxidizer, combined into composite metal-oxidizer reactive material particles. In this work, combustion of one such composite reactive material is explored with Mg used as a fuel and $Ca(IO_3)_2$ as an oxidizer. Recently, a metal rich composite was prepared mechanochemically with four moles of Mg per mole of $Ca(IO_3)_2$, $4Mg{\cdot}Ca(IO_3)_2$. Thermo-gravimetric experiments showed that the material lost 17.5% of its weight when heated to 450 °C, which is about one-half of its theoretical I_2 content. When coated on an electrically heated filament, this material was observed to ignite in two distinct stages, occurring at ca. 400 and 875 °C for the heating rate of 1900 K/s. Powder particles injected into an air-acetylene flame burned producing optical emission pulses with distinct double peaks. The first, sharp and short peak occurred upon ignition. It was followed by a broader peak, somewhat similar to that observed for combustion of similarly sized pure Mg powder particles. When the same composite $4Mg{\cdot}Ca(IO_3)_2$ particles were ignited in air by passing through a CO_2 laser beam, they burned much faster than in the air-acetylene flame and only showed one sharp emission peak. Here, combustion processes leading to the observed unusual emission profiles are explored experimentally. Combustion environments and flow conditions are varied in the laser ignition tests and their effects on the combustion stage durations and temperatures are characterized optically.

538311 Graphene Oxide-Based Microwave Ignitable Energetic Materials with Thermally Switchable Ignition Characteristics

Abstract:

Reduced graphene oxide (rGO) sheets have been previously demonstrated to be a good microwave electromagnetic field absorber while its counterpart, graphene oxide (GO), reflects microwave energy. The differing absorptivities of GO and rGO may enable the capability to thermally switch an energetic from reflection to absorption mode. In this effort, rGO and GO sheets are wrapped around nanoscale fuel and oxidizer particles in order to deposit microwave energy to particle-particle interfaces. The microwave ignitability of composites are thermally switched through heating and subsequent microwave ignition delay times are measured via diode and high-speed video within a resonant microwave cavity.

The rGO/GO wrapped thermite powder is characterized using thermal gravitational analysis (TGA), differential scanning calorimetry (DSC), scanning electron microscopy (SEM), and other techniques to explain the effects of synthesis conditions on ignitability. The development of energetic materials with on-command controllable microwave energy absorption characteristics has potential to lead to the development of insensitive igniters and the ability to dynamically control ignition.

534399 A Novel Semiconductor Nanofiber with Superb Charge Conductivity for Energy and Environmental Applications

Abstract:

A novel nanostructure embedded into a semiconductor nanofiber that results in superb conductivity has been developed. The nanocomposite addresses a key inhibitor to conductivity, with the potential to improve a wide range of applications, from batteries and solar cells, to air purification devices. While semiconductors are widely used, their effectiveness has been limited by the natural process of photo-generated electrons in recombining with 'holes'. This reduces the moving current of electrons generated by light or external power and, as a consequence, reduces the efficiency of the device.

We have designed a composite nanofiber that essentially provides a dedicated superhighway for electron transport once they are generated, eliminating the problem of electron-hole recombination. We have avoided recombination by inserting a highly conductive nanostructure made of carbon nanotubes and graphene into a titanium dioxide (TiO2) composite nanofiber. The electrons and charges can be transported efficiently in the graphene core as soon as they are generated, prior to recombining with the 'holes' in the nanofibre.

We have tested the effectiveness of the nanocomposite in solar cells and air purification photocatalysts. We have embedded the nanocomposite into the TiO2 component of dye-sensitized and perovskite-based solar cells, which are under investigation as alternatives to conventional silicon-based solar cells. The nanocomposite boosted the DSSC' energy conversion rates 40% to 66%.

TiO2 nanoparticles are the most commonly used photocatalyst material in commercially available air-purifying or disinfection devices. However, TiO2 can only be activated by ultraviolet light, which renders it far less effective indoors. It is also ineffective at converting nitric oxide (NO) into nitrogen dioxide (NO2), at a rate of less than 10%. When our nanostructure was embedded into a photocatalyst, it provided a graphene superhighway for electrons to transport more quickly to generate super-anions to oxidize absorbed pollutants, bacteria and viruses. The graphene core also significantly increased the surface exposed for light absorption and trapping harmful molecules. It also harvested more light energy across all wavelengths. The semiconductor nanofibre converted about 70% of NO to NO2, seven times more than plain TiO2 nanoparticles. We have also tested how well their nanostructure breaks down formaldehyde, a nasty volatile organic compound commonly found in new or renovated buildings and new cars. Our embedded graphene photocatalyst again was able to break down three times more formaldehyde than TiO2 nanoparticles without the added nanostructure. The new nanocomposite has a wide range of other potential applications, such as hydrogen generation by water splitting, biological-chemical sensors with enhanced speed and sensitivity, and lithium batteries with lower impedance and increased storage.

Biography of Wallace Woon-Fong LEUNG

Dr. Wallace W-F Leung received his MS and ScD both from MIT. For the past 40 years, he has worked on chemical engineering. For 18 years, he was with Bird/Baker Hughes as Senior Research Scientist and Director of Process Technologies directing centrifugal separation/filtration. Earlier, he has worked for Gulf Oil and Schlumberger on flow in petroleum reservoirs. For the past 13 years, he is Chair Professor of Innovative Products and Technologies at The Hong Kong Polytechnic University. Besides air filtration, he is also interested in applications of nanofiber technologies in renewable energy (photovoltaics, hydrogen production using water spliting), clean air and water (photocatalysis), and health technologies (wound dressing).

Dr. Leung is a fellow of, respectively, American Institute of Chemical Engineers, American Society of Mechanical Engineers, American Filtration and Separations Society, Hong Kong Institute of Engineers, and Hong Kong Academy of Engineering Sciences. He has 49 US patents and two books to his credit.

530236 Bioinspired Nanomaterials for Environmental Remediation

Abstract:

Porous materials such as silica, carbons, metal organic framework materials, porous polymers and others, are used in a variety of environmental remediation applications such as removal of toxic pollutants from air (e.g. volatile organic compounds VOCs) and water (e.g. toxic metal ions). However, these materials are generally prepared under environmentally damaging conditions, leading to secondary pollution â€“ this only shifts the pollution problem. Taking inspiration from organisms and understanding the molecular principles in biomineralisation, we have developed green nanomaterials (GN) synthesis (*Chem. Commun.*, 2011, **47**, 7567). This green method (mild, one-pot and rapid synthesis in water, at room temperature and neutral pH) offers substantial reductions in resources, time and energy usage when compared to traditional routes, yet offers excellent control over the properties and function of the materials. With the examples of VOC removal from air and arsenic removal from water, we demonstrate the applications of GN.

GN exhibited very high extraction efficiencies for a range of VOCs including formaldehyde (*Environ. Sci. Technol.*, 2012, **46**, 13354) and monoaromatic hydrocarbons (e.g. benzene, toluene and cumene). Moreover, GN demonstrated significantly higher capacity per unit surface area. Upon probing their porosity, morphology and pore structure, it was observed that the combination of broad pore size distribution, disordered arrangement of pores and the presence of meso- and micro-porosity in GN contributed to their VOC extraction efficiencies and selectivity. These features were compared with other adsorbents including mesoporous silicas.

For water treatment, bioinspired green synthesis was used to prepare, for the first time, a family of iron supported on GN (*Chem. Sci.*, 2017, **8**, 567). For arsenate ion removal, they exhibited high extraction efficiencies, high adsorption capacities and superior kinetics (threefold higher than the highest removal rates reported to date). Moreover, a method was developed to regenerate GN allowing for full recovery and reuse of the adsorbent in subsequent extractions; strongly highlighting the potential technological benefits of these new green materials.

It is clear from the results presented herein that GN, which can be prepared with a substantial reduction in secondary pollution, form a viable alternative to traditional porous materials in separations. We have also demonstrated that these materials are scalable and can be manufactured cheaply (*Chem. Eng. J.*, 2014, **244**, 483).

537398 Encapsulation of Nanoscale Hybrid Materials for Innovative CO_2Â capture: NOHMs and MOFs

Abstract:

The ever-increasing anthropogenic CO_2 emissions are likely to remain since fossil energy will continue to play an important role in our energy mix in the foreseeable future. Thus, there is an urgent need for efficient carbon capture, utilization and storage (CCUS) technologies. In terms of CO_2 capture, there have been significant advancements in materials science and engineering and a number of novel nanoscale hybrid materials such as liquid-like Nanoparticle Organic Hybrid Materials (NOHMs) and Metal-Organic Frameworks (MOFs) have been developed. While their CO_2 capture capacity and selectivity are promising, the application of NOHMs and MOFs in gas separation is not straightforward due to their high viscosity and nanoscale size, respectively. Recently, a new solvent delivery method via encapsulation has been proposed to provide large interfacial area for CO_2 capture for these viscous fluids. This study showed that NOHMs and other solvents with high viscosity can be successfully encapsulated inside UV-curing polymeric shell which is gas permeable. The CO_2 capture rate of the encapsulated solvents were drastically improved compared to bulk fluid. By tuning the fluid flow rates during encapsulated, the thickness of the shell can be controlled and, in some cases, multi core capsules were produced further increasing the interfacial area. This encapsulation technique was also employed to deliver MOFs by replacing the inner fluid with MOFs precursor solutions and in-situ synthesizing MOFs within the droplet.

536266 Nanostructured Au/Organoclay Materials for Methylmercury Adsorption

Abstract:

In recent decades, the increase in aqueous methylmercury (MeHg) levels due to anthropogenic activity has raised significant public health concerns, and has emphasized the need for an effective, low-cost means of MeHg removal from contaminated wastewater. Highly expandable clays have been investigated as MeHg sorbents; additionally, the affinity of gold nanoparticles (Au NPs) for MeHg has been documented. In several instances, organically modified clays (organoclays) have been successfully doped with Au NPs, but a comprehensive molecular understanding of these materials, which exist at the interface of natural materials and nanoscience, is still lacking.

In the current work, natural montmorillonite (STx-1b) was modified with hexadecyltrimethylammonimum (HDTMA) and doped with Au NPs in the range of 0.5 â€“ 5 wt%, and batch sorption experiments were conducted to test the efficacy of MeHg uptake on these novel Au-organoclays. All synthesized Au-organoclay species were characterized to determine organic composition (total C and N analysis), interlayer spacing (XRD), interlayer crystallinity (FT-IR), Au loading (ICP-MS), and Au NP morphology and size distribution (UV-Vis; HAADF-STEM). In addition, the effect of MeHg sorption on both interlayer crystallinity and Au NP morphology was measured at several MeHg concentrations.

Our results suggest that the presence of Au NPs drastically increases the sorption of MeHg relative to both pure and organomodified clays. MeHg uptake, on a mass-percent basis, was found to consistently exceed 90% for initial MeHg concentrations of between 40 ng/L and 10 Î¼g/L (vs 20-30% uptake for pure clay and 30-40% uptake for organomodified clay). Increased gold loading was found to be roughly correlated with an increase in sorption capacity across all MeHg concentrations. Further, our results suggest that the interlayer environment of the clay is significantly altered by the sorption process, even at contaminant levels as low as 20 ng/L. Aggregation of the gold nanoparticles was observed for all Au-organoclay species following exposure to MeHg; the degree of aggregation appeared to be largely independent of initial MeHg concentration. Additionally, the crystallinity (relative ordering) of HDTMA in the organoclay interlayer was found to decrease following MeHg sorption; but again, crystallinity did not appear to vary significantly with initial MeHg concentration. These findings imply a general rearrangement of the interlayer upon sorption of MeHg, and may suggest that a new stable arrangement is achieved beyond which additional MeHg sorption does not further alter the interlayer environment.

The extremely high sorption capacity and low cost of these novel nanostructured Au-organoclays makes them a highly promising alternative to existing remediation techniques. We aim to further develop insights into material properties and sorption mechanisms in order to maximize MeHg uptake with a minimum of material expense.

536431 Surface Modification of a Mxene with Silane Coupling Agents

Abstract:

MXenes are a family of hydrophilic, electrically conductive, 2D nanomaterials based on early transition metal carbides and carbonitrides with high aspect ratio and few atomic layer thickness [1]. The general formula of MXenes is $M_{n+1}X_nT_x$, where T represents a surface functional group such as OH, F, O or H [2]. The most well-known member of the MXene family is Ti_3C_2. From morphological point of view, MXenes are similar to graphene oxide, as their sheet thicknesses do not exceed 1 nm but their lateral dimensions can be as large as tens of microns [3]. Similar to other 2D materials, MXenes usually appear in the form of stacked sheets and their properties strongly depend on their morphology, stacked bundles versus individual sheets, and the kind of chemical species that exist between their layers. For practical uses, MXenes need to be intercalated. As there are weak interactions between MXene layers, intercalation of MXenes is possible via sonication. MXenes can form freestanding thin films. However, their mechanical properties are not high enough to be used in practical applications [4]. They can be modified with silane coupling agents, enabling them to incorporate into polymeric matrices. Silane coupling agents are chemicals capable of forming durable bonds between organic and inorganic materials. A silane coupling agent usually has a hydrolysable group like alkoxy in its structure, which changes to a silanol group after undergoing a hydrolysis reaction. In addition, it can bear an organo-functional group like amine, diamine, methacrylate, epoxy, vinyl, chloro, or phenyl [5]. A silanol group can react with hydroxyl groups available on the surface of an inorganic material, bonding the inorganic material to an organic one.

This paper presents a study of surface functionalization of $Ti_3C_2T_x$, the most studied MXene, with 3-(trimethoxysilyl) propyl methacrylate (MPS) and 3-aminopropyl (diethoxy) methylsilane (APDEMS). The surface functionalizations were carried out in both anhydrous and hydrous conditions, and advantages and disadvantages of each approach were investigated. The surface functionalizations were verified using characterization techniques including electrophoretic light scattering to measure zeta potential, contact angle goniometry, FTIR spectroscopy, UV-Vis spectroscopy, TGA, XRD, XPS and SEM. The APDEMS- and MPS-modified MXenes, respectively, showed surface charges of +32.77 and -17.98 mv versus -31 mv for the pure MXene. Surface modification reactions caused the disappearance of the hydroxyl group peak in the FTIR spectrum of the MXene and the appearance of new peaks attributed to the silane coupling agents. Water contact angle increased when the surface of the MXene was functionalized with the coupling agents. The UV-Vis spectrum of the MXene was changed due to the surface modification, and TGA results showed a weight loss peak around 370 °C, which is ascribed to the coupling agent decomposition. XRD showed that the surface modifications increase the gallery distance of the MXene sheets, and the extent of this increase depends on the type of the coupling agent used. Finally, XPS results verified the formation of covalent bonds between the MXene and the coupling agents. Due to their gained surface functional groups, the functionalized MXenes can participate in many grafting reactions to form in-situ nanocomposites.

References

1. Lei, J.-C., X. Zhang, and Z. Zhou, *Recent advances in MXene: Preparation, properties, and applications.* Frontiers of Physics, 2015. **10**(3): p. 276-286.
2. Naguib, M., V.N. Mochalin, M.W. Barsoum, and Y. Gogotsi, *25th anniversary article: MXenes: a new family of twoâ€dimensional materials.* Advanced Materials, 2014. **26**(7): p. 992-1005.
3. Anasori, B., Y. Xie, M. Beidaghi, J. Lu, B.C. Hosler, L. Hultman, P.R. Kent, Y. Gogotsi, and M.W. Barsoum, *Two-dimensional, ordered, double transition metals carbides (MXenes).* ACS nano, 2015. **9**(10): p. 9507-9516.
4. Ling, Z., C.E. Ren, M.-Q. Zhao, J. Yang, J.M. Giammarco, J. Qiu, M.W. Barsoum, and Y. Gogotsi, *Flexible and conductive MXene films and nanocomposites with high capacitance.* Proceedings of the National Academy of Sciences, 2014. **111**(47): p. 16676-16681.
5. Padhiyar, S. and R. Desai, *Silane Coupling Agent Work as Crosslinking Agent in Elastomer.*

535103 Flame Synthesis of Crumpled Graphene Nanostructures Decorated with Multicomponent Metal Nanoparticles

Abstract:

Among carbonaceous materials, graphene has become attractive because of its unique properties such as high specific surface area, good electrical conductivity, flexibility, and high mechanical strength. One of the most common methods of producing graphene is based on graphite oxidation to graphene oxide (GO) followed by thermal or chemical reduction of GO to reduced graphene oxide (rGO). In thermal reduction methods, a small amount of GO is typically placed in a high temperature inert or reducing environment for a specific time. In addition, to improve certain properties or to add new features, rGO can be decorated with other compounds which usually requires additional synthesis steps. Here, we demonstrate the continuous single-step synthesis of three-dimensional crumpled graphene (CG) nanostructures decorated with multicomponent metal nanoparticles including CG-cobalt-nickel and CG-iron-cobalt-nickel using the High Temperature Reducing Jet (HTRJ) process. In this process, combustion products of a fuel-rich hydrogen flame pass through a converging-diverging nozzle. An aqueous solution or dispersion of precursors injected at the throat section of the nozzle is atomized by the hot high-velocity gas stream. The resulting droplets evaporate in a reducing environment containing excess H_2. After the reaction zone, products are cooled immediately and collected on a filter. The key advantage of the HTRJ system over common flame-based aerosol synthesis methods is the separation of flame and product formation zones, which allows synthesis of non-oxide nanomaterials that can be reduced by H_2 in the presence of H_2O. We have utilized the capabilities of this system to synthesize CG nanostructures using an aqueous dispersion of GO as the precursor. Moreover, by adding nickel, cobalt, and ferric nitrate salts to the GO precursor solution, we decorated CG nanostructures with iron-cobalt-nickel nanoparticles of less than 10 nm average diameter. The HTRJ process is a potentially scalable, continuous synthesis method of CG and CG-metal nanostructures. The nanostructures made by this process can be used in electrocatalysts for fuel cells, electrodes in batteries and supercapacitors, conductive inks for printed electronics, wastewater treatment, and many other applications where a graphitized carbon-metal nanomaterial is needed.

536197 Modeling of the Flow Dynamics through Incompressible Porous Media in Solid-Liquid Filtration

Abstract:

Solid-liquid filtration is a long-standing engineering practice and has been widely used in the chemical, process and mineral industries. Current models are semi-empirical in nature, thus they require significant experimental and/or computational resources in order to determine the empirical quantities. In contrast, in this work we provide a model to predict the dynamic behavior for both the liquid and solid phase of a filtration process without the requirement of empirical parameters. Instead, our model relies solely on the to-be-captured particle size distribution of contaminants as well as the pore size distribution of the filtration media. Our new algorithm is capable of describing filtration based on both â€œâ€™stericâ€ capture of contaminants as well as capture within dead-end pores in the material. Here we we show the performance of this model in modeling beds comprised of high void fraction materials (diatomaceous earth) that is used for the removal of multi-modal mixtures of contaminant. By formally accounting for the complex pore size distribution, we predict flow dynamics that are much closer to our experimental results than the predictions of the traditional Kozeny-Carmen (KC) model and show that this approach is viable for both statically formed and evolving (dynamic) beds. In an effort to understand the relationship between flow dynamics and pore size distribution more fully, we built a dynamic filter cake model that continuously modifies the pore size distribution as contaminants (polydispere spheres) are deposited. In short, the predicted flow dynamics of this new model match the dynamic experimental results remarkably well, setting the stage for a priori prediction of filtration times, flow-rates, particle capture, and pressure requirements from simple measurements of the size distribution of both the filter media pores as well as the contaminant particles/droplets.

525196 Two-Drop Model of Depth Coalescing Filter Performance

Abstract:

Depth coalescing filters are used to enlarge droplets in liquid aerosols and in liquid-liquid emulsions. The enlarged drops are subsequently separated from the continuous phase by interfacial mechanisms, gravity, or inertial forces. Mathematical models of coalescing filters are useful to predict performance and to guide filter design. Models using CFD are successful at predicting performance at the cost of significant computational effort. For many practical applications, simpler less computationally intensive models are desired.

In this work, a â€œtwo-dropâ€ model is described. The drops are of two sizes. The small drop, carried by the continuous phase, is captured on fibers of the filter. Once captured, the small drops are immobilized. The small drops coalesce to form the second, larger drops. The larger drops move through the filter due to fluid drag force. The model aims to account for significant mechanisms including capture, coalescence, drag, and capillary forces.

A unique feature of the model is the mathematical description of the capillary forces as a gradient of a capillary potential function. This capillary potential function enables prediction of the effects of the capillary forces near the surfaces of the filter medium that can cause liquid accumulation at the surfaces.

536342 Cake Filtration of Catalyst Materials

Abstract:

Cake filtration is widely used in catalyst production. This traditional process is still largely empirical. There is often a lack of knowledge about the required pressure drop or the resultant flow rate or the actual rate of filter cake formation, especially for fine particles. Our approach is to develop a laboratory scale test that can provide insight into how a particular material of interest would behave in a manufacturing scale filtration process. This study will examine the dependence of the Darcy permeability coefficient on the operating parameters during filtration. An additional aim is to map the filter cake microstructure and the variation in particle size distribution and bulk density at different axial locations along the filter cake.

In this work, we have developed tools and protocols for filtration experiments using a benchtop Nutsche filter device, a model material of 5 microns diameter nonporous glass beads and porous catalyst materials. The applied pressure is varied and we aim to relate filtration behavior to particle physical properties, such as bulk density, air permeability, size distribution and zeta potential. Particle size distribution and bulk density within filter cake layers were analyzed. Furthermore, additional recycling and washing steps of the filtration process were included and the displacement of fine particles through the layers of the cake was obtained as well as changes in the bulk density profile.

535856 The Role of Pore Structure and Chemistry on Particle Deposition during Membrane Filtration

Abstract:

Many relevant industrial fluids (e.g. cell culture broths, paints, milk, etc.) can be characterized as colloidal suspensions (i.e. nanometers to micrometers) and are of great interest for microfiltration. This pressure-driven membrane process is one of the oldest, yet, remains not fully understood. A major limitation is membrane fouling. Particle deposition *onto* the microporous membrane, resulting in pore constriction and/or blockage, and eventually in cake formation, has been largely investigated and mitigation strategies have been proposed.

In this study, we focus on particle deposition *into* a microporous membrane, with particular attention for the role of pore structure and chemistry. The membranes used were 0.2 μm and 5 μm pore size poly(ether sulfone) (PES), since PES is widely used due to ease of casting and mechanical stability. Silica nanoparticles, 40 nm and 1 μm, respectively, were selected as model colloidal suspensions.

In vitro studies focused on 2D and 3D characterization of pore distributions, by atomic force microscopy (AFM), scanning electron microscopy (SEM), and focused ion beam SEM (FIB-SEM). In order to understand the forces in play, (i) zeta potential for both membranes was measured using 1-10-100 mM electrolyte solutions in a pH range 4-10; and (ii) AFM in force mode was used to estimate the adhesion force between the membranes and a silica microsphere attached to an AFM probe. Finally filtration experiments were performed using dilute colloidal suspensions (i.e. <1% v/v), and array tomography, combined with SEM imaging, was applied to visualize the distribution of particles into the fouled membranes.

Taking advantage of the experimental characterization, *in silico* modeling using COMSOL Multiphysics software was conducted in parallel to build a better understanding of particle interactions with the membrane walls in the pores. Prototype models will combine an accurate pore geometry representation coupled with experimental information on particle-wall forces to obtain particle trajectories through the structure and deposition rates onto the pore wall. The information will be used to narrow down the key features pointing toward optimal membrane performance.

529376 From Batch to Continuous Reactive Crystallization: A Case Study on Beta-Lactam Antibiotics

Abstract:

Using data from experiments on batch crystallization to design a continuous system is recognized to be a challenge. Additional complications exist when crystallization is coupled with a chemical reaction and reaction and crystallization kinetics are linked through the generation of a supersaturated environment. Should this coupling of kinetics lead to a Class I behavior (i.e. one in which there is significant supersaturation in the product stream), then the product yield and crystal quality, especially as measured by crystal size distribution, become linked [1].

In the present work, penicillin G acylase (PGA) catalyzes the formation of the antibiotic but also catalyzes the degradation of the product into a slightly soluble by-product. In batch, the PGA-catalyzed reactive crystallization process improved selectivity and productivity by 50% and 25%, respectively [2]. Adapting the batch process for continuous operation can improve productivity further while ensuring better quality consistency, as measured by purity and size distribution. However, PGA must be retained within the crystallizer for the process to be economically feasible, complicating product removal. As a stepping-stone that circumvents this issue, a fed-batch process has been used to tune models accounting for nucleation and growth kinetics, reaction kinetics, and catalyst deactivation kinetics.

Achieving high yields while controlling crystal size distribution requires independent control of the residence times of the solutes (reactants) and solids. Issues encountered include washout, where nucleation or reaction or both are insufficient to maintain a stable slurry; enzyme loss, where catalytic activity diminishes due to either PGA escaping the crystallizer or permanently unfolding; and fines-rich size distributions, where crystal residence time is too short for desired growth. Control over the enzyme reaction rate can be used to reach the desired set point within a feasible space defined by the batch operation. The reaction rate of PGA can be tuned by adjusting temperature, pH, and reactant concentration, however all three of these have been found to impact the antibiotic solubility or crystallization kinetics, further complicating the development of a continuous process. A model of an MSMPR with independent crystal and liquor residence times has been used to optimize the continuous process, and an experimental continuous apparatus has been demonstrated to validate the model while operating under more constrained conditions (imposed by issues of scale and equipment). Compared to the previously published batch process, the *in silico* productivity of the continuous process increases by approximately 2-fold at the sacrifice of approximately 10% of the conversion. The *in silico* average crystal size is small, on the order of tens of microns, comparable to that obtained in batch. The *in vitro* experimental results confirm a decrease in conversion compared to batch; however better purity, consistency, and productivity offset this loss. Experimental crystal sizes were also small, however it may be possible to alleviate this issue with fines destruction.

References

[1] Tavare, N. S. (2013) *Industrial crystallization: process simulation analysis and design*, Springer Science & Business Media.

[2] McDonald, M. A., Bommarius, A. S., and Rousseau, R. W. (2017), *Chem. Eng. Sci. 165*, 81-88.

530097 A New Slurry Reactive Crystallization to Improve Process Robustness and Scalability

Abstract:

Several challenges were faced during the development of development of a drug candidate isolated as a salt. These challenges included high solvent levels entrapped in isolated solids, disproportionation and agglomeration. A root-cause analysis pointed to dimethyl sulfoxide (DMSO), a co-solvent utilized in the crystallization as the main cause for solvent entrapment and agglomeration. DMSO was introduced in the procedure because it enhances drastically the solubility of the drug allowing for a clarifying filtration to be performed at the final step. However, because of its low volatility, DMSO is removed at a slower pace during drying causing partial solubilization of crystal surfaces during drying followed by recrystallization. It is believed that this behavior causes solvent entrapment and agglomeration during drying. In order to address these issues, a study was conducted to identify an alternative co-solvent to DMSO. In this presentation, the development of two types of crystallizations, namely an anti-solvent crystallization and slurry-slurry reactive crystallization to address the issues mentioned above is described. In particular, a new slurry-slurry reactive crystallization, based on an approach recently reported (Derdour et al[â€¡]) was found to be superior in terms of improving process robustness and scalability. That procedure performed successfully when executed at industrial scale.

â€¡ Derdour L., Reckamp J.M. and Pink C., Development of a reactive slurry salt crystallization to improve solid properties and process performance and scalability, Chem Eng Res Des, 121, pp 207-2018, 2017

532745 Crystallization Process Development of Large Size TMP Crystals

Abstract:

Abstract

2,2,6,6-Tetramethyl-4-piperidinol (TMP) crystal is a very important intermediate in organic synthesis. To obtain an improved crystallization yield and desirable crystal-size characteristics, several commonly used organic solvents were investigated to determine the solubility of TMP using a gravimetric method. An ethanol solvent and a cooling crystallization method were chosen to obtain an improved theoretical yield. The cooling crystallization process of TMP in the ethanol solution was further optimized to crystallize larger crystals with a narrow crystal-size distribution using an orthogonal array design. Comprehensive analysis showed that water content in ethanol and seed loading are respectively the main influence factors of the average diameter and the diameter span. Final TMP crystals with an average diameter of 1110 Î¼m and a diameter span of 0.715 were obtained under laboratory conditions using the optimized process.

534105 Polymorph Selection in Continuous Crystallizers in the Presence of Wet Milling

Abstract:

The design of controlled processes that ensure the consistent production of a desired polymorph is an important goal in the context of pharmaceutical manufacturing.[1] The most stable crystalline form can be obtained from a batch crystallization process either through seeding or by allowing enough time for metastable forms to convert to the stable form. On the other hand, producing metastable forms in a batch process relies on delicate process control strategies (temperature, solute concentration and batch time control). As shown in the work by Farmer et al.[4] which polymorph is obtained from a continuous crystallizer at steady state can be controlled by selecting appropriate combinations of residence time, operating temperature and inlet concentration. For cases where the crystallization process is dominated by nucleation and growth, they have shown that the polymorphic outcome in such a continuous process can be accurately predicted once two modified Damköhler numbers are known. In essence, these dimensionless groups represent ratios of the characteristic time for nucleation and growth, the process time (residence time of the crystallizer) and the feed concentration.

In this contribution, we are expanding on this previous works and consider the case where wet milling is used to tune the particle size distribution and potentially alter the polymorphic outcome. We thus present a mathematical model of continuous crystallization with wet milling for polymorphic systems. The model relies on one-dimensional population balance equations including breakage terms. Making the process model dimensionless, we identified a surprisingly small number of additional dimensionless groups that describe the process outcome in terms of polymorphic form when breakage is present. The resulting model is solved using a finite volume scheme (FVS)[5]. Operating regions where the metastable/stable polymorph can reliably be obtained can again be identified. In comparison to the work by Farmer et al.[4], we show that the operating region where the metastable/stable polymorph is obtained can be enlarged by tuning key dimensionless numbers in the breakage kinetics. We combine this type of analysis with an investigation of the productivity of such systems and report regions of attainable particle sizes[4,5]

We show that introduction of a wet suspension mill therefore represents an attractive way to obtain the desired polymorph even if the underlying kinetics of nucleation and growth are disadvantageous. Apart from this finding, we show that introducing breakage increases process productivity and allows to tune particle size more extensively.

[1] Hermanto, M. W.; Chiu, M. S.; Woo, X. Y.; Braatz, R. D. Robust optimal control of polymorphic transformation in batch crystallization. *AIChE J.* 2007,53,2643â€"2650.
[2] Farmer, T. C.; Carpenter, C. L.; Doherty, M. F. Polymorph selection by continuous crystallization. *AIChE J.*2016, 62, 3505â^'3514.
[3] Kumar, R.; Kumar, J. Numerical simulation and convergence analysis of a finite volume scheme for solving general breakage population balance equations. *Appl Math Comput.* 2013, 219, 5140-5151.
[4] Vetter, T.; Burcham, C.L.; Doherty, M.F. Regions of attainable particle sizes in continuous and batch crystallization processes, Chem. Eng. Sci., 2014, 106, 167-180.
[5] Vetter, T.; Burcham, C.L; Doherty, M.F. Designing robust crystallization processes in the presence of parameter uncertainty using attainable regions, Ind. Eng. Chem. Res., 2015, 54, 10350-10363.

536947 Polymorph Dynamics of 2-Aminobenzoic Acid System in Continuous Oscillatory Baffled Crystallizer

Abstract:

The polymorph dynamics of 2-Aminobenzoic acid is investigated in a continuous oscillatory baffled crystallizer (COBC). The COBC is a tubular crystallizer containing periodically spaced orifice baffles with oscillatory motion superimposed on the net flow.[1] A Nitech DN15 COBC consisting of eight segments with each segment and elbow jacketed for temperature control is used for the studies. Preliminary experiments show the feasibility of generating the two polymorphs in the COBC via continuous cooling crystallization. However, in order to obtain higher yield combined cooling and antisolvent crystallization (CCAC) is used. In pharmaceutical industry traditionally an antisolvent is added to the solution to increase the yield of the product. However, the presence of the antisolvent can influence the polymorphic outcome.[2] Design of experiments are done to investigate the effects of the parameters – temperature, solvent composition and residence time. The poster will show additional detail studies such as the effect of mixing parameters – piston amplitude and frequency of oscillation on the polymorphic outcome and yield.

[1]Caldeira, R., Ni, X. 2009 Evaluation and Establishment of a Cleaning Protocol for the Production of Vanisal Sodium and Aspirin using a Continuous Oscillatory Baffled Reactor. Org. Process Res. Dev. 13(6), 1080–1087.

[2]Lindenberg, C., Krättli, M., Cornel, J., Mazzotti, M., & Brozio, J. 2008. Design and Optimization of a Combined Cooling/Antisolvent Crystallization Process. Cryst. Growth Des. 9(2), 1124–1136.

535157 In-Situ Optical Imaging and X-Ray Diffraction Techniques to Probe Organic Molecule Thin Film Crystallization

Abstract:

Crystallization of organic materials as thin films is an area of interest in multiple fields, such as in the pharmaceutical industry, organic electronics and for the multiple applications afforded by metal organic frameworks (MOFs). However, many traditional techniques of studying crystal growth and morphology becomes difficult to use for thin film studies, due to the nanometer to micron scale thickness of the films. To overcome this problem, optical and X-ray diffraction techniques can be utilized to probe crystallization of thin films in organic materials. In the crystallization of organic molecules and MOFs, in-situ grazing incidence X-ray diffraction can provide valuable insight in the mechanism of crystal nucleation. In-situ optical studies can then provide complementary information about the crystal growth process in the micron â€“ centimeter scale. Together, in-situ optical and X-ray diffraction technique can provide a clear picture of crystallization spanning multiple orders of magnitude in both temporal and spatial dimensions. This talk will focus on the use of in-situ X-ray diffraction and optical imaging to study crystallization from solution of organic molecules and MOFs. The future of in-situ techniques using novel instrumentation will also be discussed.

523533 Imaging Crystallization Using Deep Learning to Quantitatively Track the Polymorphic Transformation of Carbamazepine

Abstract:

Solvent mediated polymorphic transformation (SMPT) is a crucial phenomenon for a specific polymorph crystalsâ€™ production and stability control in pharmaceutical crystallization processes. In this study, a completely new method based on computer vision was developed using deep learning to quantitatively track crystal individuals. For the model compound carbamazepine (CBZ), two polymorphic forms with different morphologies were identified using deep learning image segmentation. The geometry models of two crystal forms were built to calculate the volume of each crystal in the real-time tracking process. The spatial valid vision volume of a high-speed camera probe was calculated and then verified by glass bead. Furthermore, the mass of two polymorphic crystals in suspension were fitted by f_1 and f_2 through comparing real-time tracking and off-line measurement. At last, the real-time quantitative tracking of polymorphic transformation of CBZ was realized through deep learning-based imaging analysis. The transformation process including dissolution of form II, nucleation and growth of form III, crystal size distribution (CSD) etc. was analyzed and identified by off-line measurement. Our work in this study attempts to bridge the gap between the advanced image tracking technology available today and the specific needs of solution crystallization to track, count and measure the crystal individuals. It is a great opportunity to proceed intelligent manufacturing through visualizing traditional crystallization process with the help of artificial intelligence (AI).

Acknowledgement

The authors acknowledge the financial support provided by the Natural Science and Engineering Research Council (NSERC) of Canada, National Natural Science Foundation of China (NNSFC 21576187, NNSFC 21776203, NNSFC 21621004, NNSFC 81361140344), and the China Scholarship Council.

539027 Mechanoactivation, Initiation and Combustion of Aluminum and Copper Oxide Mixtures

Abstract:

mechanoactivation, initiation and combustion of aluminum and copper oxide mixtures

A. Yu. Dolgoborodov[1,2,3], A. N. Streletskii[2,3], B. D. Yankovskii[1], V. G. Kirilenko[2], S. Yu. Anan'ev[1,3] , I.V. Kolbanev[2], G. A.Vjrob'eva[2]

[1]Join Institute for High Temperature RAS, Izhorskaya St. 13 Bd.2, Moscow, Russia

[1]Semenov Institute of Chemical Physics RAS, Kosygin St. 4, Moscow, Russia

[3] Moscow Institute of Physics and Technology, Dolgoprudny, Russia

e-mail: aldol@ihed.ras.ru

Thermite mixtures based on metals and solid oxidants allow to obtain a significant exothermic effect during combustion (up to 8 kJ/g), while the burning rate of mixtures of micron powders usually does not exceed several tens of mm/s, which limits the field of their application. This is due to the fact that the rate of propagation of chemical reactions in solid mixtures is determined primarily by the effective contact surface of the reactants. To increase this surface, various methods are used: ultrasonic mixing of nanosized powders, electrochemical deposition of submicron metal-oxidizer layers, etc. [1]. One of the relatively new methods for obtaining fast-burning thermite compounds is the preliminary mechanochemical activation of mixtures of micron-sized particles in energy-intensive ball mills. At the same time, the components are shredded and mixed at the submicron and nano levels, and new defects in the crystal structure are formed, which makes it possible to increase the rate of chemical reaction on the surface of the reagents and, accordingly, we get a way to control the processes of energy release during combustion and shock wave action. In Russia, the method of mechanoactivation of solid oxidant-metal fuel mixtures has been actively used since the beginning of the 2000s [2-5], and the resulting materials are called Mechanically Activated Energetic Composites (MAEC).

Among the variety of thermite mixtures, special attention is drawn to the composition of Al + CuO, which makes it possible to obtain one of the highest exothermic effects per unit volume (more than 20 kJ/cc). In this paper new research results on the initiation and burning of MAEC Al/CuO based on micron and nanosized powders using the mechanoactivation are presented.

a)

b)

Fig. 1. SEM images of MAEC Al/CuO, t_a = 8 min (a) and burning products (b) light - Cu, gray - Al_2O_3.

As the initial components, micron and nanosized powders were used: industrial pyrotechnic powder Al PP-2L (flake 50 ÷ 100 μm ´ 2 ÷ 5 μm), CuO 20÷50 μm, nanosized nAl (Alex 100 ÷ 200 nm) and nCuO (50 ÷ 80 nm). Al weight content was from 18 to 25%. Mixing and activation was carried out in the vibration mill of the Aronov design or in the planetary mill "Activator-2sl» with steel drums and balls. A rough estimate of the energy intensity of the two types of mills based on the growth of the specific surface area of the test material (MoO_3) is for "Ativator-2sl" at a total power J = 9.7 W/g, and for the Aronov mill J = 3.7 W/g. Weight load of powders was 10-25 g, the mass of balls was 200-300 g. Hexane was added to reduce frictional heating. The starting powders and MAEC were analyzed by X-ray diffraction, electron microscopy and thermo-gravimetric analysis. Image analysis shows that as a result of activation the obtained material were a polydisperse mixture of fairly large conglomerates of flat fragments of Al particles (~ 1 - 10 μm) with submicron CuO particles (see Fig.1a). The products of high-speed burning of MAEC Al/CuO are shown in Fig. 1b. The bulk of Al_2O_3 is spherical particles of 10 μm, which consist of stuck submicron particles. Cu is in form of large smeared drops on the surface of Al_2O_3 or in the form of nanosized particles condensed from the gas phase.

a)

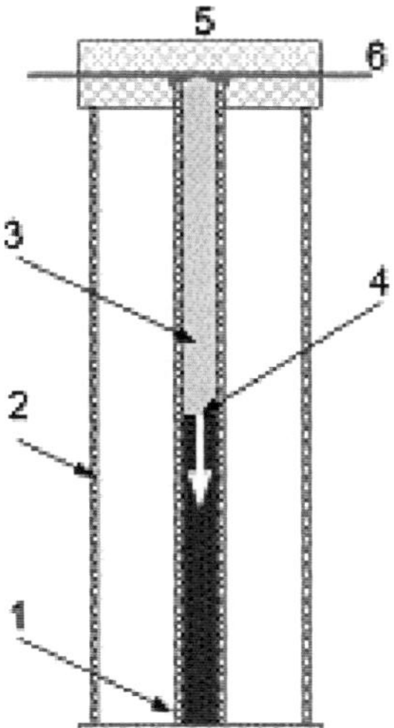

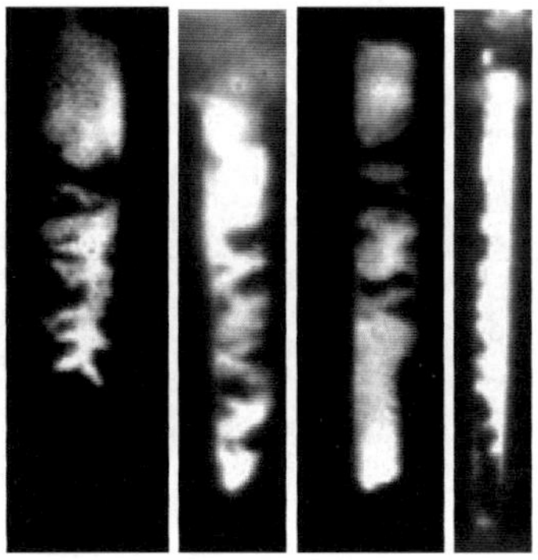

b)

Fig. 2. (a) Experimental scheme: 1–dark part of glass tube (no combustion), 2–glass vessel with water, 3–light part of tube (combustion), 4–a boundary between dark and light parts of tube, 5– spark gap, 6–electrodes. (b) The photos of inhomogeneities of the luminescence at different spark energy

A number of the dependences of the combustion parameters on the activation dose D_a were determined in experiments: ignition temperature by hot surface, brightness temperature of burning products, burning rate in cylindrical channels and electrical resistivity in cloud of products. The dynamics of the expansion of products in free space during electric spark and shock wave initiation was analyzed.

The burning rate was measured in plastic and glass tubes (diameter 4-10 mm) by recording of products emission by light fiber or high speed photography. The porosity of charges was 60-70%. The initiation was carried out by heating the NiCr wire or by an electric spark. The spark energy was regulated by changing the current amplitude in the range of 1.5-50.0 mJ at duration of the current pulse 1.2 μs. Experimental setup with electric spark initiation and individual frames at different sparks energies are shown in Fig. 2. Brightness temperature of products was measured by four - channel pyrometer.

Depending on D_a and the initiation method, the measured burning rate varies from 10 m/s to 700 m/s, and the product temperature is from 2000 K to 3700 K (see Fig.3). In case of low-energy electric spark (<20.0 mJ), combustion has a pronounced heterogeneous character. Photo registration shows that zones of bright glow of hot products alternate with dark zones. As the spark energy increases, the uniformity of the glow increases. The highest reactivity, burning rate and temperature of products for MAEC Al/CuO were obtained at $D_a = 1.8 \div 2$ kJ/g. With increasing D_a there is a partial reaction of the reagents in the activation process.

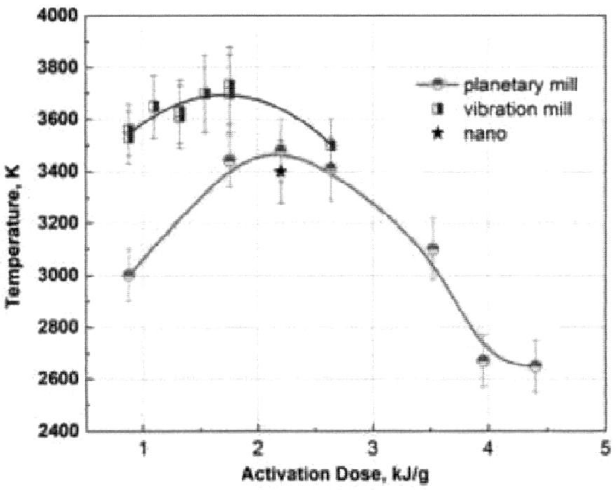

Fig. 3. Dependence of the brightness temperature of products on the activation dose for MAEC Al/CuO 18/82

In the case of shock-wave initiation of compositions in a semi-enclosed volume, the main process of energy release proceeds in the flow of products dispersed in the unloading wave. The initial flow rate of the products is more than 800 m/s. The maximum brightness temperature in the cloud of partially ionized products reaches 3700 K, the resistivity is about 107 Ohm * mm^2/m.

In general, the results of the work have shown the promise of preliminary mechanochemical activation for the production of fast-burning thermite compositions of Al/CuO. The use of the original nanoscale components is inadvisable, since it does not give appreciable advantages. The reactivity of thermite mixtures based on CuO can be controlled by the addition of metals having catalytic action.

1. Energetic Nanomaterials: Synthesis, Characterization, and Application / Ed. by V.E. Zarko and A. Gromov. 2016. Elsevier Inc.

2. A.Yu. Dolgoborodov, M.F. Gogulya, M.N. Makhov, et al. Detonation-like phenomena in Al/S mixture", *Proc. Twenty-Ninth Intern. Pyrotechnics Seminar*, 2002, p. 557-563.

3. A.Yu. Dolgoborodov, M.N. Makhov, I.V. Kolbanev, A.N. Streletskii. Mechanically activated pyrotechnic composition, *RF Patent No. 2235085*, 2004.

4. A.Yu. Dolgoborodov. Mechanically activated oxidizer–fuel energetic composites. *Combustion, Explosion, and Shock Waves*. 2015, V. 51(1), p. 86–99.

5. A.N. Streletskii, M.V. Sivak, A.Yu Dolgoborodov. Nature of high reactivity of metal/solid oxidizer nanocomposites prepared by mechanoactivation: a review. *J. Mater. Sci.* 2017. V. 52(20), p. 11810–11825.

526239 Aluminum-Nickel Fluoride Reactive Materials

Abstract:

Aluminum powders are most common metal fuel additives in diverse energetic formulations. Aluminum as a fuel is attractive due to its high energy density and accessibility. Sustained efforts have focused on modifying aluminum with the goal of accelerating its ignition and combustion and minimizing agglomeration of aluminum particles prior to ignition. One approach is to combine aluminum with a modifier, such as fluorinated oxidizer. The thermite-like reaction yielding aluminum fluoride is expected to boost ignition. Furthermore, in oxygenated environments, metal-rich aluminum-metal fluoride composites are expected to generate gas-phase combustion products, desirable for many applications. Recently, it was observed that aluminum composites with cobalt and bismuth fluorides prepared by arrested reactive milling (ARM) are readily ignited by heating but are insensitive to ignition by electrostatic discharge. This work is exploring nickel fluoride, NiF_2, as a potential additive to improve and tailor combustion of aluminum. Nickel fluoride is one the most stable and easy to handle metal fluorides. Here, it is explored as an oxidizer for a thermite-like composition for the first time. Several Al·NiF_2 composite powders were prepared by ARM with the weight fraction of aluminum varied from 50 to 90 % with respective equivalence ratios for the Al-NiF_2 thermite varied from 5.3 to 47. For each composition, milling conditions were identified leading to preparation of homogeneously mixed, fully dense composite particles. Prepared materials are examined by electron microscopy and x-ray diffraction. Samples of the prepared powders are ignited using an electrically heated filament. It is observed that the ignition temperatures increase significantly, e.g., from 400 to 1200 °C when the concentration of NiF_2 decreases from 50 to 10 %. This suggests an ability to tune the ignition temperature of the Al·NiF_2 composite powders. The effect of heating rate on the ignition temperature for all prepared materials is also quantified. Experiments are presently in progress correlating the ignition of the prepared composites with the exothermic reactions observed in thermo-analytical measurements. Combustion experiments with the prepared composite powders are also in progress. Results of these experiments will be presented and discussed in this talk.

525300 Combustion of Fuel-Rich Boron â€“ Metal Fluoride Composites

Abstract:

The prospective use of boron as a fuel is often hampered by its extended ignition delays and long particle burn times. This research explores ignition and combustion of boron combined with a condensed, fluorine containing oxidizer and exposed to an oxygenated environment. With both fluorine and oxygen available for boron combustion, it is expected that gaseous boron fluorides and oxyfluorides will be the preferred combustion products. Unlike liquid B_2O_3, such products will not cover boron surface delaying its heterogeneous combustion; they are also expected to benefit such applications as solid propellants, where gas-phase combustion products lead to an increased impulse. Composite boron-metal fluoride powders were prepared using arrested reactive milling. The starting material was 95% pure boron powder by SB Boron. In addition to as-received boron powder, a powder subjected to preliminary washing by acetonitrile and toluene, aimed to remove the natural hydrated oxide layer, was used. Both cobalt (II) fluoride and bismuth (III) fluoride served as oxidizers. Prepared composites included 50 wt. % of metal fluoride. These materials are fuel rich with the equivalence ratios of 24.6 for $B{\cdot}BiF_3$ and 13.4 for $B{\cdot}CoF_2$. When coated on an electrically heated filament, the prepared materials using as received boron as a starting material ignite at very low temperatures, ca. 300 °C. The ignition temperatures are about 100 °C higher for the composites prepared using washed boron. It is hypothesized that when present, the nascent hydrated boron oxide layer interacts with the metal fluoride oxidizers at low temperatures triggering ignition and yielding gaseous boron-oxy fluorides as products. Combustion of the prepared composites was studied using a custom-built burner operated with air-hydrogen and air-acetylene flames. The particles were injected into the flame axially. The optical emission pulses produced by burning particles were recorded using filtered photomultiplier tubes and an optical spectrometer. In separate experiments, the composite particles were ignited by passing through a focused CO_2 laser beam and burned in air. Similarly, optical emission pulses generated by the burning particles were recorded and analyzed. Burn times as a function of the particle size were obtained for different oxidizing environments correlating the measured statistical distributions of the emission pulse durations with the particle size distribution for the composite powders. The results for all the combustion experiments shall be presented and discussed for the prepared boron-metal fluoride composites.

538745 *Thermal and Biocidal Inactivation of Di-Isopropyl Methylphosphonate and Tributylphosphate through Static and Transient Heating*

Abstract:

There is currently need to determine the decomposition of chemical warfare agents (CWA) and precursors at high heating rates and in the presence of biocidal agents in order to accurately model CWA decomposition. In this effort, thermal inactivation is studied in thermally static and thermally dynamic environments within a heated cell and by a joule heated hot wire experiment capable of generating heating rates from 100°C/s to 10^{6}°C/s by means of DC mode pulsed capacitive discharge. Inactivation measurements are performed via high-speed IR laser absorption spectroscopy utilizing a tunable IR laser system in the range of 9.6μm to 10.7μm. CWA simulants utilized in this study are di-isopropyl methylphosphonate (DIMP), a Sarin analog, and tributylphosphate (TBP) a VX simulant, in presence of anaerobic, aerobic, and iodine-based biocides generated from thermite decomposition. The effect of biocidal additives on inactivation is investigated to determine effectiveness and necessity in transient thermal decomposition events.

532259 The Characterization of a TNT/Aniline Co-Crystal Solvate: Physicochemical, Explosive Properties and Kinetics of Stability

Abstract:

In the development of crystal engineering and supramolecular chemistry, co-crystallization has been used to develop the novel explosives with desired properties. Herein we present the 1:1 molecular ratio of 2, 4, 6-trinitrotoluene(TNT)/aniline novel co-crystals solvates which has unique physicochemical and explosive properties. The characterization of the crystallographic and physical properties of TNT/aniline co-crystals solvates, compared to TNT, was conducted with X-ray diffractograms, infrared spectra, and thermal analysis. The change in explosive properties of co-crystal solvate was theoretically determined, and the value of detonation parameters indicated that sensitivity of TNT increased in a large extent (drop height > 300cm), without a significant decrease in explosive velocity of detonation (DTNT = 6.93 km/s, Dco-crystal (CHEETAH) = 6.5 km/s). The ruby-color of TNT/aniline co-crystals solvates, which is typically shown in Meisenheimer complexes, gradually changes to yellow-color (which is the characteristics of pure TNT) in the open environment due to the desolvation of aniline. The activation energy (22.1 kJ/mol) of the desolvation process was directly obtained by atomic force microscopy (AFM) analysis.

524826 Transition Metal Catalysts for Boron Ignition and Combustion

Abstract:

Boron is an attractive fuel additive for explosives and propellants due to its high gravimetric and volumetric combustion energy densities. The limitations of using boron in many practical formulations are associated with its relatively slow reaction kinetics leading to long combustion times and ignition delays. Additionally, in hydrogen containing environments, it forms thermodynamically unfavorable but kinetically preferred intermediates such as HOBO. Past studies showed that boron ignition is delayed by its inhibiting oxide layer. It was also observed that boron burns at temperatures substantially below its boiling point and the combustion may be rate-limited by its heterogeneous surface reaction with gaseous oxidizers. Recently, it was shown that the complete or partial removal of the hydrated surface oxide from boron powders by treating them in a polar solvent helps reducing boron particle ignition delays. It was also found that doping boron with a small, less than 5% additive of iron results in reduced particle burn times in oxygen and hydrogen containing oxidizing gases. It was hypothesized that the iron dopant has a catalytic effect on the heterogeneous oxidation of burning boron particles. This work expands these recent efforts to address effect of a broader range of transition metals with high oxygen affinity and multiple oxidation states as potential catalysts improving kinetics of boron ignition and combustion. The candidate transition metals include cobalt, molybdenum, and zirconium. Selected are metals with relatively high boiling points to ensure their lasting effect on heterogeneous combustion of molten boron particles. It is expected that during combustion, metal dopants will be preferentially oxidized by surrounding gaseous oxidizers on surface of the burning particles. It is further expected that boron will readily reduce transition metal oxides. This mechanism is expected to accelerate heterogeneous reactions and thus minimize the chances of generating HOBO as a gas phase intermediate. The boron-metal composite powders are prepared by high-energy mechanical milling using liquid process control agents. Prepared samples are characterized for particle size distributions, elemental distribution of the dopant on the particle surface and interior. Combustion studies are conducted for samples laser ignited in air and in steam created by hydrogen-air diffusion flame. Measured burn times, emission traces, and temperatures will be presented and compared for doped and as received boron powders.

540269 Elutriation Comparison of Particle Separation Systems for Chemical Looping Applications

Abstract:

Fluidized beds are used in many industries for their favorable characteristics of good solids mixing, high rates of heat and mass transfer, and large throughputs. One process of interest is Chemical looping combustion (CLC) reactor systems, which are a recent technology for electrical generation. One process of interest is Chemical looping combustion (CLC) reactor systems, which are a recent technology for electrical generation. One difficulty in the development of this technology is the removal of ash and unburnt carbon (or char) from the oxygen carrier material prior to re-entering the air reactor in coal or other solid fuel applications. Carbon allowed to traverse to the fuel reactor results in lower capture efficiencies – CO2 emissions. To address this issue, several cold flow experimental efforts are being undertaken at the U.S. Department of Energy's National Energy Technology Laboratory (NETL) towards the development of effective particle separation techniques. This work focuses on the study of the hydrodynamics and elutriation characteristics of a mixture consisting of fine (<50μm) particles (serving as a char analog) and larger (~600μm to ~800 μm) particles (serving as the carrier analog). The systems being compared are a 0.1 m meter diameter bubbling bed with a 0.9 m height and a novel riser design called a carbon stripper. The carbon stripper feeds the gas and particle mixture into a conical inlet section with a centralized jet feed and an annular gap between the wall and cone. A sweep gas within the annular gap is present to aid in the separation. The important comparisons are the separation efficiency, the rate of elutriation of fines, and cost of fluidization. The target separation efficiency for the mixture is 80%, with an elutriation flux of 0.5 kg/m^2.

540268 Numerical Investigation of Scale-up and in-Bed Heat Exchanger on the Hydrodynamic Characteristics of the Fluidized Bed Combustor of Coal Direct Chemical Looping System

Abstract:

Coal Direct Chemical Looping (CDCL) for electricity generation is as an energy efficient carbon capture technology with a capability of capturing 99% of the CO2 from fossil fuels. The CDCL system developed by the Ohio State University uniquely uses counter-current moving bed with Geldart group D particles for its reducer. The technology is ready for the scale up to pilot and industrial scales. However, achieving a fundamental understanding of the mechanisms governing the hydrodynamic behavior of the combustor operated under a fluidized bed mode, particularly under industrial operating conditions, still presents a major scientific and engineering challenge. For Geldart Group D particles using in the system, slugging fluidization normally occurs in sub-pilot scaled units whose combustors are with small diameters and large height-to-diameter ratio. While, when the system is scaled-up, the geometry and the height-to-diameter ratio will change, which will change the hydrodynamic characteristics dramatically. The installation of the in-bed heat exchanger tubes inside the combustor may also change the hydrodynamic characteristics. Thus, hydrodynamic characterization on the sub-pilot scaled cold flow model is not enough for understanding the hydrodynamics of pilot and commercial sized combustors. In this study, a two-fluid model (TFM) imbedded in the MFIX model was used to investigate the effect of scale-up and in-bed heat exchanger on the hydrodynamic characteristics of the combustor. Appropriate drag models and wall boundaries were selected for the simulation and validated with experimental data from sub-pilot scale cold flow models. The model and the boundaries are then applied to industrial scaled simulation with the consideration of in-bed heat exchangers.

536997 Chemical Looping Combustion from Biomass Derived Syngas Using a Fluidizable Ni-Co-La/γ-Al_2O_3 Oxygen Carrier: CLC Performance and CPFD Modelling

Abstract:

Biomass steam gasification yields a blend of H_2, CO, CH_4 and CO_2, designated as syngas. Syngas can be further combusted using fluidizable oxygen carriers (OC). This operation is known as Chemical Looping Combustion (CLC). With this end, a Ni-based oxygen carrier is implemented with a Co and La modified γ-Al_2O_3 support. This type of OC limits $NiAl_2O_4$ formation. This novel nickel on g-alumina OC was prepared using an Incipient Wetness Impregnation (IWI) method as described in previous research[1]. This OC also includes Co and La additives. Co and La additives have valuable effects on both $NiAl_2O_4$ and carbon formation reduction, respectively [1]. Syngas is a typical biomass gasification product[2].

In spite of the valuable progress made with OCs by our research team [2,3], the 750°C calcination temperature use was shown to lead to: a) strongly bounded Ni species on alumina and b) undesirable $NiAl_2O_4$ [1]. These species inherently limit the OC oxygen capacity in the 550-650°C operating range. To circumvent this, a new highly performing OC was prepared by reducing all the formed nickel species at 900°C for 2 hours and this prior to 750°C calcination. This methodology ensures the complete decomposition of strongly bounded Ni on alumina, as well as that of $NiAl_2O_4$. This yields a re-dispersing nickel phase which has little interaction with the g-alumina support. This new highly performing oxygen carrier (HPOC) can provide more than 90% of the nominal oxygen lattice at temperatures between 550-650°C. The promising HPOC was characterized using BET, XRF, XRD, H_2-TPR, NH_3-TPD and pulse chemisorption.

CLC runs of syngas and the new OC were performed using the CREC Riser Simulator. This is a mini-fluidized bed reactor with high gas mixing that operates under batch conditions[4]. This unit is designed for oxygen carrier evaluation at expected industrial process conditions. An automatic valve allows the transfer of products to an online gas chromatograph (GC) to detect all product chemical species (CO, CH_4, CO_2 and H_2) accurately. The evaluation of this highly performing OC (HPOC) was developed at 550-650°C, 5-40s and with a ratio of oxygen to be consumed over maximum available oxygen, ψ = 0.5 to 1. A syngas was used in the experimental runs using a 20v% CO, 10v% CH_4, 20v% CO_2 and 50v% H_2 mixture. CLC runs of syngas showed that 60-92% CO_2 yields can be obtained within these operating conditions. The CO_2 yields and the CH_4 conversion steadily increase with reaction time for ψ=1. It is also shown that CO and H_2 conversions stabilize and decrease respectively for ψ =1. These trends were attributed to the influence of reforming reactions under OC starving oxygen conditions. An adjustment of the OC oxygen supply to ψ=0.5, leads to all CO, H_2 and CH_4 species being 78-88% converted.

For the point of view of implementation as an industrial process, a set of twin fluidized bed (oxidizer and reducer) is considered in the present study. Regarding the selected twin fluidized bed configuration, both fluidized beds are interconnected to facilitate oxygen carrier particle circulation. This oxygen carrier particles have a 75 μm mean diameter and a 2500 kg/m^3 particle density. Considering 150KW power generation capacity, and ψ =1, the diameter of oxidizer was designed as 780 mm diameter and with an overall length 4500 mm including transport disengaging height. Similarly, reducer was designed as 800 mm and 3750 overall length. Typical 30 seconds total retention time for reducer and 15 seconds for oxidizer were considered. Reduced particles are collected in a loop-seal pot and furthered transported to the top of oxidizer bed. In a similar manner oxidized particle are transported to reducer bed. Product gas (Flue gas) and exhausted air leave from the reducer and oxidizer top section via cyclone particle separators.

To analyze this twin fluidized bed configuration a multi-scale (Wen-Yu) drag model coupled with computational particle-fluid dynamics (CPFD) and CLC volume averaged kinetics are being considered[5]. This Hybrid Barracuda CPFD model uses the Eulerian-Lagrangian approach called MP-PIC (multi-phase particle-in-cell). The software's numerical methodology considers a direct element method wherein solids are modeled as discrete particles with both size and density distributions, and the fluid is modeled as a continuum. The formation of particle cluster is also considered based on cluster size distribution[6].

On the basis of the data obtained this study allows the following: (a) to establish the twin fluidized bed system performance for CLC; (b) to describe the required oxygen carrier particle circulation; (c) to evaluate the extent of CO_2 gas leaking via the loop-seal.

References

[1] M.R. Quddus, M.M. Hossain, H.I. de Lasa, Catal. Today. 210 (2013) 124–134.

[2] J. Mazumder, H. de Lasa, Appl. Catal. B Environ. 160–161 (2014) 67–79.

[3] M.M. Hossain, K.E. Sedor, H.I. de Lasa, Chem. Eng. Sci. 62 (2007) 5464–5472.

[4] I. Ahmed, S. Rostom, A. Lanza, H. de Lasa, Powder Technol. 316 (2016) 641–649.

[5] M.A. Hamilton, K.J. Whitty, J.S. Lighty, J. Energy Resour. Technol. 138 (2016) 42213.

[6] A. Lanza, H. de Lasa, Comput. Chem. Eng. 101 (2017) 226–242.

540271 Modeling of Circulating Fluidized Bed Reactors for the Selective Oxidation of Alkanes By Chemical Looping

Abstract:

The thermal cracking of alkanes is a major source for olefin production worldwide. The North American shale oil and gas boom has greatly increased the supply of light alkanes, reduced the cost of feedstock for olefin production, and spurred the expansion of worldwide cracking capacity. Unfortunately, conventional cracking processes, which rely upon significant steam dilution and high temperature furnaces, are leading drivers of CO_2 emissions in the chemical industry. We have recently introduced chemical looping systems for Redox Oxidative Cracking (ROC) of natural gas liquids, condensates, and light naphtha. In ROC, a metal oxideâ€™s (or oxygen carrierâ€™s) lattice oxygen is used to partial oxidize the alkane, producing water. Subsequent regeneration of the oxygen carrierâ€™s lattice oxygen in air produces significant process heat. ROC effectively provides heat to endothermic cracking reactions by selectively burning hydrogen byproducts *via* chemical looping combustion. In this work, we present modeling methods for ROC in circulating fluidized bed reactors for evaluation of this emerging technology. The model incorporates experimental measurements of a hexane ROC redox catalyst. We show that lost work in the reactor sections can be reduced compared to both conventional cracker furnaces and oxygen co-feed systems. A significant reduction in lost work is still observed when the effects of different fuels (e.g. H_2 vs. natural gas) are incorporated.

522615 CFD Modeling of a Dual Fluidized-Bed System Using an Eulerian-Lagrangian Approach

Abstract:

A three-dimensional computational fluid dynamics (CFD) model was established to simulate a pilot (6 tons/day, 1 MWth) dual fluidized-bed system. In the CFD model the momentum equation was coupled with the mass and energy transport equations to simulate mass and energy transfer in the dual fluidized-bed system, which includes a gasifier, a combustor, a cyclone separator, and a loop-seal. An Eulerian-Lagrangian approach was applied to describe the interactions between the particulate and gas phases. Heterogeneous particle-gas and homogeneous gas-phase reaction kinetics were developed to simulate char pyrolysis, combustion, and gasification, water gas-shift reaction, steam reforming reaction, and oxidation of CO, H_2, and CH_4 in the dual fluidized-bed system. The CFD model simulated the full-loop of dual fluidized-bed system (a gasifier, a combustor, a cyclone separator, and a loop-seal). It was utilized to predict particle circulation, syngas composition, and temperature distributions in the dual fluidized-bed or gasifier-combustor gasification system.

532349 CFD-DEM with Uncertainty Quantification (UQ) Compared Against Experiments of Horizontal Jets in a Gas-Solid Fluidized Bed

Abstract:

Uncertainty quantification coupled with CFD-DEM generates a distribution of model outcomes and provides valuable insights regarding the uncertain inputs that most influence the model results. In the current effort, a suite of CFD-DEM+UQ simulations are performed and the results compared against existing experimental measurements [Fullmer, W. D., C. Q. LaMarche, A. Issangya, P. Liu, R. Cocco and C. M. Hrenya, "Experimental Data for Code Validation: Horizontal Air Jets in a Semicircular Fluidized Bed of Geldart Group D Particles," *AIChE Journal* (in press)] of opposing horizontal air jets impinging into a semi-circular fluidized bed. The CFD-DEM simulations are implemented using MFIX and are designed to replicate experiments completed at PSRI. Specifically, a cut-cell semi-circular simulation domain is used to replicate the semi-circular cross-section of the experimental apparatus. The uncertainties in the CFD-DEM inputs are identified, tabulated, and ranked using a local sensitivity analysis (LSA). Those input uncertainties which significantly impact the model outputs (as determined by the LSA) are then propagated through the model to estimate the range of uncertainty in the key model outputs. The key outputs for this study are the jet penetration distances (left and right jets) and the bed pressure drop. In addition to the overall methodology, the comparison between the CFD-DEM output uncertainties and experimental results will be discussed, as well as the implications on uncertainty quantification analyses for CFD-DEM simulations.

534513 Validation of the Direct Simulation Monte Carlo (DSMC) Method for Simulating Polydisperse Gas-Solid Flows

Abstract:

Flowing granular materials are ubiquitous in industry and nature, e.g. coffee, snow, coal, and Saturnâ€™s rings, and exhibit complex phenomena not often observed in traditional gaseous or liquid fluids. A better understanding of granular flows can advance industries that use solid particles by providing insight towards the underlying mechanisms that give rise to different flow behaviors and ultimately the development of predictive modeling tools. Small scale experiments can probe the underlying physics of gas-solid systems, but extrapolating small-scale tests to large-scale industrial systems is unreliable. Performing experiments in industrial scale systems is costly and detailed non-intrusive measurements are difficult. Discrete element modeling (DEM) and other deterministic computational models have been used to simulate granular flows, but such methods are often restricted to relatively small-scale systems (~10^7 particles). Continuum modeling approaches can simulate large-scale systems, but there can be challenges with accuracy and inclusion of particle-scale physics. This work assesses the feasibility of the direct simulation Monte Carlo method (DSMC), as an alternative modeling approach that is computationally efficient and accurate compared to discrete element simulations. The DSMC method uses representative particles and models particle collisions in a stochastic manner. The DSMC method has traditionally been used to simulate dilute gases but has been adapted to solve the granular kinetic theory equations (i.e., the revised Enskog equation). We have extended the direct simulation Monte Carlo method to simulate dense to dilute polydisperse gas-solid flows. To test the validity of the DSMC modeling approach for polydisperse flows, we evaluate its accuracy for relatively small scale systems via comparison to high fidelity discrete element simulations. The computational performance of DSMC is compared to discrete element simulations and we assess the suitability of DSMC for use in large-scale gas-solid systems.

535029 Engineered Electrodes for Energy Storage and Battery Safety

Abstract:

The ViPER (Vilas Polâ€™s Energy Research) laboratory at Purdue University focuses research efforts on the development of high capacity electrode materials and their engineering with long cycle life and improved safety. Considering the advantages and limitations of known synthesis techniques, various processing technologies are developed to fabricate a variety of unique anode and cathode materials for lithium-ion, Na-ion, K-ion and Li-S batteries. ViPERâ€™s recent efforts on structural, morphological and electrochemical properties of various electro-chemistries will be demonstrated.

528867 Extended Surface Electrocatalyst Development Via Atomic Layer Deposition

Abstract:

Extended surface electrocatalysts based on nanowire templates have emerged as promising candidates for advanced catalysts for the oxygen reduction reaction (ORR) in fuel cell applications. When synthesized by spontaneous galvanic displacement (SGD) of nickel nanowires (NiNWs), these extended surface electrocatalysts have demonstrated exceptional performance (specific activity > 6000 Î¼A/cm^2) and superior durability to benchmark carbon-supported Pt catalysts. However, challenges in scale-up and reproducibility of this fabrication process have prompted an investigation into atomic layer deposition (ALD) as a facile, scalable synthesis alternative for the production of extended thin film electrocatalyst structures. An ALD analog of the SGD synthesis process was developed using an oxidative Pt ALD chemistry in a viscous flow reactor. In situ mass spectrometry was used to study the deposition reactions, while elemental analysis, transmission electron microscopy, elemental mapping and X-ray diffraction were used to characterize Pt growth and effects of post-treatment procedures, such as H_2 annealing. Electrochemical half-cell testing showed that the highest-performing ALD catalyst was 16 times more active than commercial Pt/C. After integration into membrane electrode assemblies, these materials were also found to exceed the DOE 2020 mass activity target of 440 mA/mg$_{Pt}$. Further work is also underway to leverage the tunable nature of ALD to develop a co-deposition process with Pt and Ni ALD on a cobalt nanowire (CoNW) substrate. Fabrication of these trimetallic extended surface catalysts using ALD â€œsupercyclesâ€ and subsequent characterization experiments will also be discussed.

534820 Thin Oxide Film Coatings for Improved Lithium Ion Battery Cathodes

Abstract:

The application of alumina ALD films to battery cathodes is known to enhance the cycling stability of lithium ion batteries. The true nature and deposition mechanism of these films is not well understood. In this work we will elucidate the nature of low cycle ALD films on NMC cathode materials and discuss the mechanistic causes behind the characteristics of the films. An ultra-thin alumina film is applied to cathode particles by atomic layer deposition (ALD) with a variety of cycle numbers. The films generated within the first 10 cycles of deposition are shown to improve the cycling stability of lithium ion battery cathodes. Low energy ion scattering (LEIS) and computational modeling were used to elucidate the coating mechanism that leads to the observed improved performance.

530256 The Application of Magnesiothermic Reduction of Silica to Produce Porous Silicon for Lithium Ion Batteries

Abstract:

Increasing demand for portable power applications are pushing conventional battery chemistries to their theoretic limit. Silicon has the potential as an anode material to increase lithium-ion cell capacity. The associated volume change of lithiation leads to a decline in capacity during cycling and low lithium diffusion rates within silicon limit high rate discharging. Many silicon morphologies and nanostructures have been studied including porous architectures. Porous morphologies can potentially address the poor cyclability and rate capabilities simultaneously. The template assisted synthesis â€˜magnesiothermic reductionâ€™ of silica to silicon offers a facile and scalable synthesis route to porous silicon structures. However, a review of the recent progress of this method indicates a distinct lack of mechanistic understanding of this phenomenon. To better understand the processes, we have systematically studied the process conditions using a model silica system. Further, we performed a complete characterisation of both the reactant and product structures in order to map how reaction conditions affect the properties of the electrode materials. Finally, we will present the battery performance of these porous silicon structures and compile a processing-structure-property-performance relationship.

539777 Fluidized-Bed CVD of Si@SiC@C-like Nanoparticle and Its Application As Anode Materials

Abstract:

Among the anode materials, silicon (Si) has the poor cycling performance induced by the large volume expansion (300%) which can break the solid electrolyte interface (SEI) during the lithium ion insertion/extraction, although it can provide the highest theoretical capacity as the most promising candidate for the next generation lithium ion batteries. For the purpose of make a stable SEI, a novel Si@SiC@C-like nanomaterial has been designed to gain 800mAh g^{-1} at the current dense of 1 A g^{-1} after 800 cycles. Besides, those nanoparticles can be easily obtained though chemical vapor deposition (CVD) in a fluidized bed at low cost. A reaction mechanism and the rate equations have been proposed based on the understanding of the effect of reaction temperature, partial pressure of reactant gas and time of synthesis on the yield of Si@SiC@C-like nanoparticles.

535397 Functional Nanomaterials for OIL and Gas Discovery and Recovery Applications

Abstract:

Today, the potential of functional nanoparticles and nanostructured materials to transform key upstream applications in the oil industry is evident. Functional nanoparticles can be engineered to traverse the reservoir while reporting their location, recording key reservoir rock and fluid parameters, and actively mobilizing oil. They enable target-delivery of chemical treatments and sensors deep in the reservoir and can be designed to migrate autonomously via harvesting the surrounding natural or induced solute or chemical gradients. Thus, they can substantially improve reservoir characterization and recovery of its target resources without excessive use of water and chemicals, and with minimal pumping energy.

The major challenges to maximizing the benefits of nanoparticle-based technologies in upstream oil applications can be summarized as: 1) lack of long-term stability and mobility of nanoparticles under typical oil reservoir conditions (usually temperatures exceeding 100 °C, salinities between 50,000 and 220,000 ppm, and pressures exceeding 3000 psi), 2) low-cost prerequisite due to the significantly large amounts needed to interrogate pore volumes typical to the reservoirs (millions of cubic meters), and 3) difficulty to precisely characterize the nanomaterials and test their performance under typical reservoir conditions.

In this article, we present three of our successful oil reservoir nanoagents; A-Dots, Nano-Surfactants, and Magnetic Nano-Mappers. The synthesis of these nanoagents and results of experiments to characterize their nature and behavior under typical reservoir conditions are presented.

A-Dots are carbon dots that are persistently stable at high salinity and elevated temperature. They are optically detectable at low concentrations and can migrate efficiently in the reservoir between injectors and producers. They thus allow establishing â€œcommunicationsâ€ between injectors and producers for improved management of the reservoir. Nano-surfactants are 10 to 60 nm nanostructured particles of the seawater-insoluble petroleum sulfonate salt surfactants, stabilized in seawater by by cocamidopropyl hydroxysultaine zwitterionic co-surfactant molecules. They showed resilient colloidal stability in seawater at elevated temperatures (up to 100 °C) for several months and reduced the seawater-crude oil interfacial tension (IFT) by three orders of magnitude. Thus, nano-transformation of the seawater-insoluble, yet highly efficient, abundant, and inexpensive surfactants enabled their utilization in oil recovery applications in high salinity and temperature oil reservoirs. Magnetic Nano-Mappers are coated magnetic nanoparticles that are colloidally and chemically stable under reservoir conditions. They enhance the magnetic permeability contrast between water and oil, thus allow in-situ monitoring of the oil-water interface using EM surveillance for optimal waterflooding operations.

533953 CFD Modeling of a Dry Electrostatic Protein Separation Approach

Abstract:

A lab-scale novel tribo-electrostatic separation approach has been developed as a dry fractionation technique to concurrently produce starch- and protein-rich fractions from legume and cereal flours. This dry separation technology can be easily replaced with the energy-intensive wet fractionation technologies that are currently used in industry for production of plant protein concentrates. A vertical lab-scale tribo-electrostatic separator is consisted of a vibrating bed, tribo-charging device and a separation chamber with two oppositely charged electrode plates. The vibrating fluidized bed is first used to suspend the grain flour particles in the air stream before passing them to the tribo-charger device. In the tribo-charging device, the protein and starch particles, main constituents of cereals and legumes, acquired opposite charges depending on their structures before being separated in the separation chamber with applied electric field. The legume and cereal flours were successfully enriched, almost twice, in proteins using this novel dry fractionation approach. A mathematical model has been also developed and validated using experimental data to describe the motion of charged particles in the vertical tribo-electrostatic separator under both gravitational and electric forces.

ID# 520332: SEPARATION PERFORMANCE OF A COUPLED CYCLONE WITH BUILT-IN CIRCULATING GRANULAR BED FILTER (C-CGBF)

Sihong Gao, Dandan Zhang, Yiping Fan, Chunxi Lu**
State Key Laboratory of Heavy Oil Processing, China University of Petroleum, Beijing, 102249, China

Introduction

Many current industrial processes including the fluidized catalytic cracking (FCC), the biomass gasification (BG), the integrated gasification combined cycle (IGCC) and the pressurized fluidized bed combustion (PFBC), need an effective hazardous particles removal technique for protecting the downstream components from abrasion/erosion/fouling, improving the system efficiency as well as meeting the environmental emission requirements [1–4].

Cyclones [5, 6], electrical precipitators [7] and wet scrubbers [8] are commonly used for particles separation. Recent researches suggest that ceramic filters and granular bed filters maybe the most promising technologies for hot gas clean-up [9]. However, single-stage separating device generally cannot meet the severe demands for high temperature high pressure (HTHP) gas purification. In order to improve the operation stability and the collection efficiency, various separation processes/mechanisms were combined in several new designs [10–15].

In this study, a coupled cyclone with built-in circulating granular bed filter (C-CGBF) was proposed for continuous dry gas purification, combining the centrifugal separation with the filtering separation in one device. Furthermore, it also has the potential for the adsorption separation of gaseous impurities.

Experimental Setup

As shown in Fig. 1, the experimental setup was mainly consisted of three sections: the C-CGBF system, the dust-laden gas generating system and the sampling system.

The C-CGBF System

The C-CGBF system was the main part of the experimental setup, including a cyclone shell, a built-in granular bed and a riser-spouted regenerator. The gas stream through the cyclone shell and the built-in granular bed was introduced by a draught fan. The flow path of the induced-air is marked by the dashed arrows in Fig. 1.

The cyclone shell and the riser-spouted regenerator were made of transparent plexiglass for the purpose of observing conveniently. The inner diameter of the cyclone shell is 380 mm while the wrapping angle of the volute inlet is 180°, as shown in Fig. 2. The built-in granular bed was a coaxial sleeve and the thickness of the interlayer was 50 mm filled with the collector granules. Both the inner and outer walls were made of stainless Johnson screen.

The cyclone shell and two Johnson screen walls were radially spaced about a same vertical axis and thus three distinct spaces were defined. A dust-laden gas was introduced tangentially into the first annular space between the cyclone shell and the outer screen wall, wherein a swirling flow was formed. Then the gas flowed tangentially-radially through the built-in granular bed between the two Johnson screen walls.

Finally, the cleaned gas was obtained in the third space. The dust particles separated by the swirling flow were collected by the dust tank Ⅰ underneath.

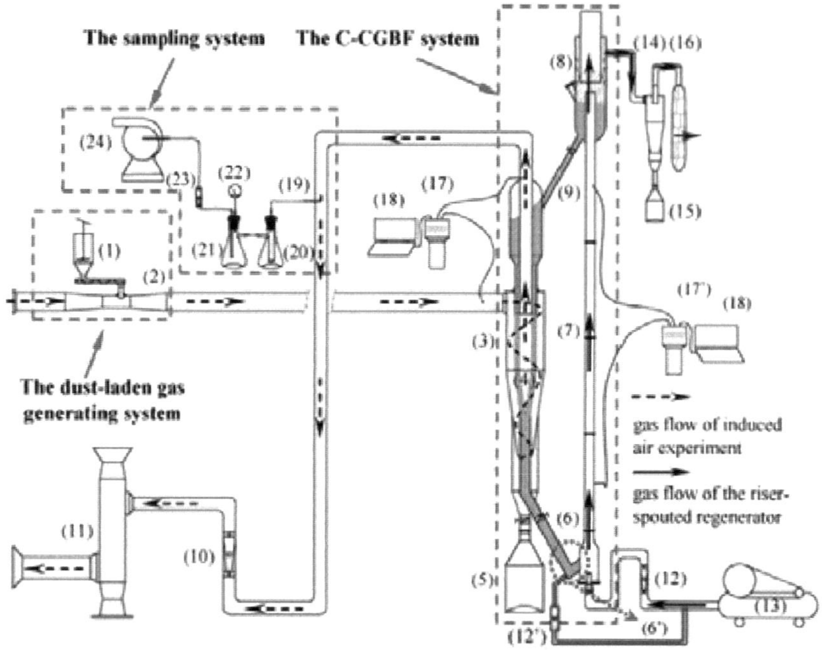

Fig. 1. Schematic diagram of the experimental setup. (1) screw feeder, (2) Venturi throat, (3) cyclone shell, (4) built-in granular bed, (5) dust collecting tank Ⅰ, (6) dirty collector granules valve, (7) riser, (8) spouted bed regenerator, (9) refreshed collector granules valve, (10) (12) and (23) gas rotameters, (11) draught fan, (13) air compressor, (14) small cyclone, (15) dust collecting tank Ⅱ, (16) filter bag, (17) digital manometer, (18) personal computer (PC), (19) sampling probe, (20) filtering flask with a fiber-glass filter cylinder, (21) stabilization flask, (22) pressure gauge, (24) vacuum pump.

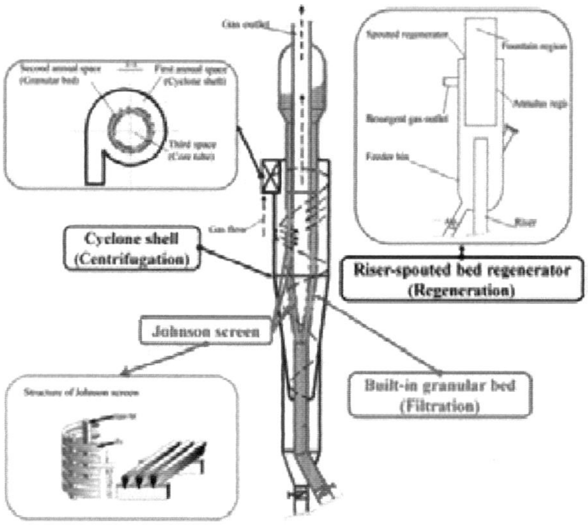

Fig. 2. Detailed structure of the C-CGBF.

Under the moving bed operating regime (MB), the collector granules moved downward from the feeder hopper located at the top of the built-in granular bed to the dipleg. An inclined pipe with a butterfly valve (named by the dirty collector granules valve) connected the dipleg and the riser. The valve was set to control the solids circulation flux in the C-CGBF system. In the riser, the collector granules as well as the captured dust particles were conveyed into the spouted bed regenerator. The refreshed collector granules traveled in an inclined pipe from the regenerator to the feeder hopper, thus the C-CGBF system was operated continuously. The regenerating gas was directed into the following small cyclone and filter bag for collecting the fine particles before expanding to the atmospheric pressure.

The Dust-laden Gas Generating System

The concentration of the dust particles suspended in the inlet gas was controlled by a screw feeder. In order to get a uniform dust-laden gas, the screw feeder was located at the Venturi throat of the inlet duct, as shown in Fig. 1. The inlet dust concentration was determined by the mass flow rate of the screw feeder and the inlet gas flow rate. The experimental conditions for all tests to evaluate the C-CGBF system are summarized in Table 1.

Table 1. Experimental conditions for evaluating the C-CGBF performance

Test number	Inlet gas flow rate m^3/h	Inlet dust concentration g/m^3	Operating regime	Solids circulation flux kg/s
1	800	5.95	Moving Bed (MB)	0.22
2	800	5.95	Fixed Bed (FB)	0
3	800	17.57	MB	0.22
4	800	17.57	FB	0
5	800	30.35	MB	0.22
6	800	30.35	FB	0
7	800	59.13	MB	0.22
8	800	59.13	FB	0
9	1400	17.35	MB	0.22
10	1400	17.35	FB	0

The Sampling System and Collection Efficiency

Eq. (1) is the expression of the collection efficiency η of the C-CGBF system:

$$\eta(\%)=\left[1-\left(C_{\text{out}}/C_{\text{in}}\right)\right]\times 100 \quad (1)$$

Here, C_{out} is the outlet dust concentration while C_{in} represents the inlet dust concentration.

C_{out} was determined by weighing the dust particles isokinetically collected by the sampling system, as illustrated in Fig. 1. A sampling probe with an inner diameter of 8.0 mm was set at the centerline of a vertical outlet duct. A filtering flask that housed a fiber-glass filter cylinder was connected to the sampling probe for capturing the dust particles vented by the C-CGBF system. The sampling rate at the outlet stream was calculated with the requirements of isokinetic sampling and then calibrated by the pressure of the stabilization flask. The filter cylinder containing dust particles was weighed and compared with the mass before sampling.

The Measurements of Pressure Drop and Solids Circulation Flux

A differential pressure gauge was used for transforming the digital signals of the static pressure drop of the C-CGBF system into a PC. The pressure drop was recorded continuously at 1 Hz.

The solids circulation flux W_s was determined by measuring the time for a known volume of collector granules dislodging from the feeder hopper at the top of the built-in granular bed. In this study, W_s was fixed at 0.22 kg/s for the MB and 0 kg/s for the fixed bed operating regime (FB).

Materials

UOP 13X-APG zeolite adsorbents were used as the collector granules in the C-CGBF system. The main physical properties of the collector granules are listed in Table 2. Silica powder was chosen as the dust particles. The particle size distribution and the median particle diameter (D_{50}, 13.2 μm) were measured by a laser particle size analyzer, as shown in Fig. 3.

Table 2. The main physical properties of the collector granules

Granules	Particle density kg/m^3	Bulk density kg/m^3	Fraction （%，mm）		
			>2.36	<1.70	d_p
13X-APG	1049	666	0.4	0.17	2.07

Results and discussion

Pressure Drop of the C-CGBF

Under the fixed bed operating regime (FB), the pressure drop of the C-CGBF system with Q=800 m^3/h increased linearly with time if introducing dust particles, as shown in Fig. 4. The initial average pressure drop before the dusts introduced was about 0.9 kPa. The gradient of the pressure drop with the inlet dust concentration increased evidently. As the dust-laden gas was admitted tangentially into the first annular space, a large part of the dust particles was separated by the swirling flow. The fine particles entrained by the cross-flow gas were then captured by the built-in granular bed. The bed voidage declined continuously as the dust particles depositing in the granular bed; while the pressure drop increased consequentially. As the inlet dust concentration increased from 5.94 to 59.13 g/m^3, the depositing rate was enhanced; hence the gradient also increased. As a result, the FB cannot be operated continuously.

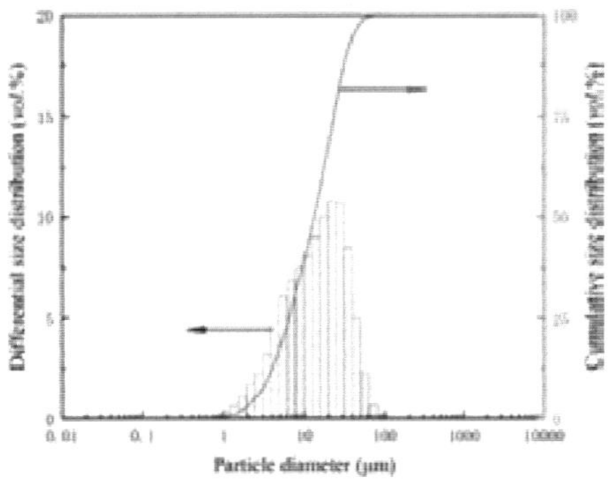

Fig. 3. Size distribution of the silicon powder

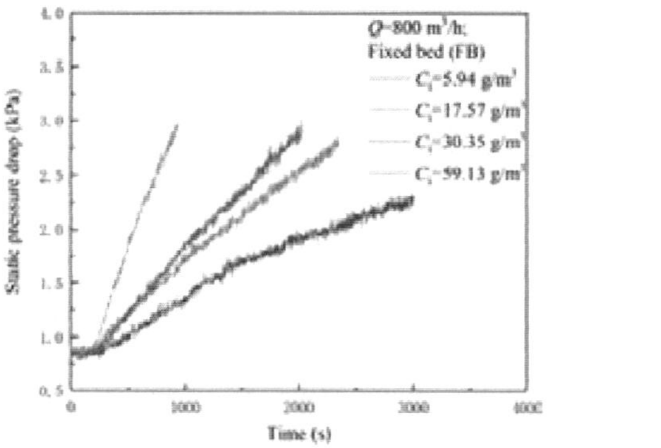

Fig. 4. Static pressure drop vs. time under the FB

The comparisons of the pressure drop sequence under the FB with that under the MB are plotted in Fig. 5. Quite differing from that under the FB, the pressure drop under the MB for all the investigated inlet dust concentrations tended to achieve a steady-state after a period of increase. For instance, as shown in Fig. 5(a), the pressure drop increased as the dust particles were introduced under the MB, but the gradient of the pressure drop was apparently smaller than that under the FB. Besides, the gradient also decreased with time under the MB. In common with the FB, the pressure drop under the MB increased as particles depositing in the granular bed. However, the captured dust particles were gradually carried out by the downward movements of the collector granules under the MB. In other words, the dust hold-up in the built-in granular bed did not keep increase, but present a dynamic equilibrium between depositing and being transported out. For a moving bed granular filter, once steady state is reached, dusts are carried out with the dirty collector granules as quickly as it deposits in the filter; consequentially the dust hold-up in the filter remains almost unvaried and thus the pressure drop assumes a comparatively steady profile. As shown in Fig. 5(a), the pressure drop was quite stable at approximately 1.27 kPa after about 4000 s.

Under the MB, the pressure drop sequences with different inlet dust concentrations are given in Fig. 6. The pressure drop and its increase gradient presented relatively great values under high inlet dust concentration. In such a condition, more dust particles were entrained into the built-in granular bed, leading to a high equilibrium dust hold-up and low voidage therein. The energy consumption of the gas stream inevitably was high compared with that under low inlet dust concentration.

In experiments, the filter cake formed on the outer screen wall of the built-in granular bed was observed. It may explain why the pressure drop still increased slightly after 4000 s. With dust particles accumulating on the outer screen wall, the growth of the filter cake may lead to undesired shut down. Fortunately, a part of the dust agglomerates were removed due to the vortex. The swirling flow, in fact, caused a tangential shear on the filter cake, which restrained the growth of the filter cake. This phenomenon was named as the self-cleaning of the swirling flow in this study.

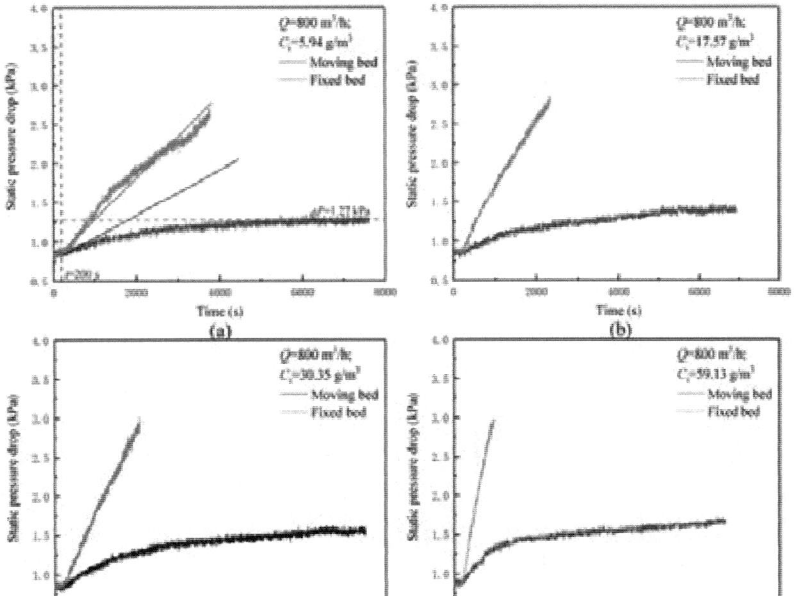

Fig. 5. Static pressure drop vs. time with Q=800 m^3/h, (a) C_i=5.94 g/m^3; (b) C_i=17.57 g/m^3; (c) C_i=30.35 g/m^3; (d) C_i=59.13 g/m^3.

In order to investigate the effects of the swirling flow, the inlet gas flow rate was varied from 800 to 1400 m^3/h with a similar inlet dust concentration (17.57 and 17.35 g/m^3). As shown in Fig. 7, a high inlet gas flow rate led to a comparatively large initial pressure drop. The initial average pressure drop with Q=1400 m^3/h was about 2.6 kPa. The pressure drop sequences with Q=1400 m^3/h, however, assumed the same features with Q=800 m^3/h both under the FB and the MB operating regimes. The difference was that the steady state was achieved more quickly with Q=1400 m^3/h, as illustrated in Fig. 6. Besides, the increase of the pressure drop was lower than that with Q=800 m^3/h. These differences resulted from the considerable shear on the filter cake caused by the high inlet gas flow rate. When the shear increased, the thickness of the filter cake and the pressure drop generated by the filter cake both decreased. High intensity of the swirling flow efficiently suppressed the filter cake; thereby a more stable state was achieved quickly.

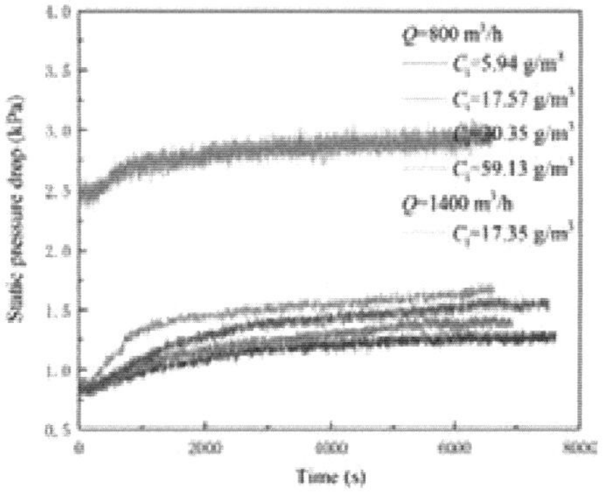

Fig. 6. Static pressure drop vs. time under the MB

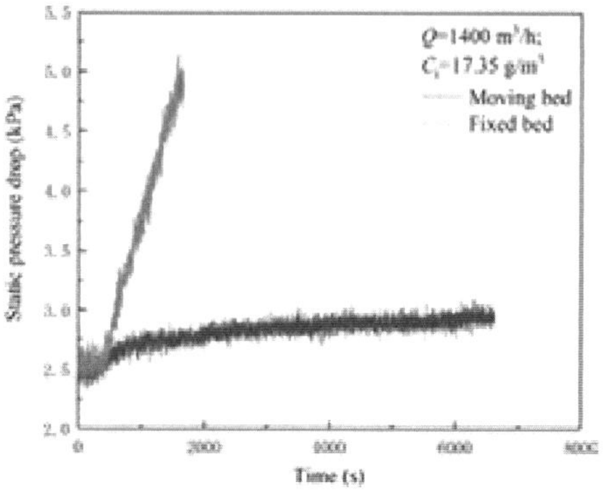

Fig. 7. Static pressure drop vs. time with Q=1400 m^3/h; C_i=17.35 g/m^3

Collection Efficiency of the C-CGBF

Under the MB, the collection efficiency of the C-CGBF system vs. time with Q=800 m^3/h is illustrated in Fig. 8. The collection efficiency was typically higher than 95% with a relatively low pressure drop. As to the FB, the dust hold-up in the granular bed keeps increasing with the filtering operation. As the large voids in the granular bed being filled gradually, the collection efficiency is expected to increase. A similar tendency was also observed under the MB for the C-CGBF. If increasing the inlet dust concentration, the equilibrium dust hold-up in the granular bed raised, while the bed voidage decreased. More dusts were consequentially captured and then the collection efficiency increased. Besides, a high inlet dust concentration was also helpful to the centrifugation. Therefore, the total collection efficiency was increased.

Clearly, it is benefit to operate the C-CGBF system under a high inlet dust concentration to improve the collection efficiency. However, the above high collection efficiency probably leads to some undesired operating conditions such as "clogging". If the diameter of the dust particles is small enough, the adhesiveness between particles goes strong while the flow-ability is then weakened. If the hold-up of the fine particles in the granular bed is high enough, the particles will be difficult to be transported out by the collector granules. With accumulation of the dust particles in the granular bed, the fine particles will clog the collector granules and even transform the moving bed into the fixed bed finally.

The filter cake formed on the outer screen wall also contributes to the high collection efficiency, with the advantage of that it doesn't clog the collector granules. In addition, the filter cake cannot grow continuously because of the self-cleaning of the swirling flow as mentioned above. The filtration performance is anticipated to be improved by paying attention to the control of the filter cake in the future.

The dust collection efficiency vs. time for different inlet gas flow rates is plotted in Fig. 9. The collection efficiency declined when the inlet gas flow rate increased. High inlet gas velocity destroyed the dust agglomerates on the outer screen wall and then the filtration of the filter cake was weakened. In addition, high inlet gas velocity intensified the re-entrainments of the collected dust particles in the granular bed, which also decreased the collection efficiency.

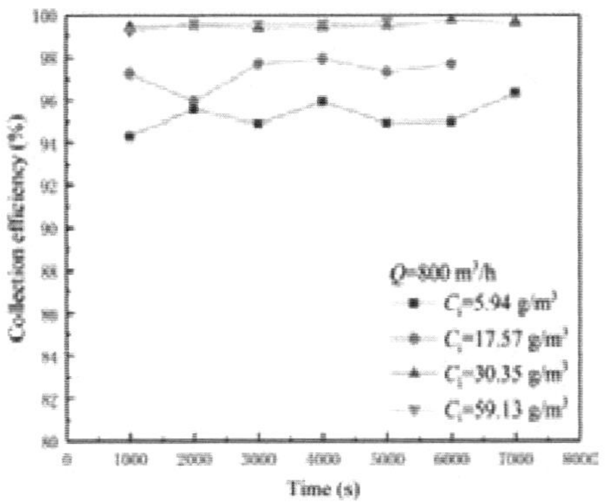

Fig. 8. Collection efficiency vs. time for different inlet dust concentrations

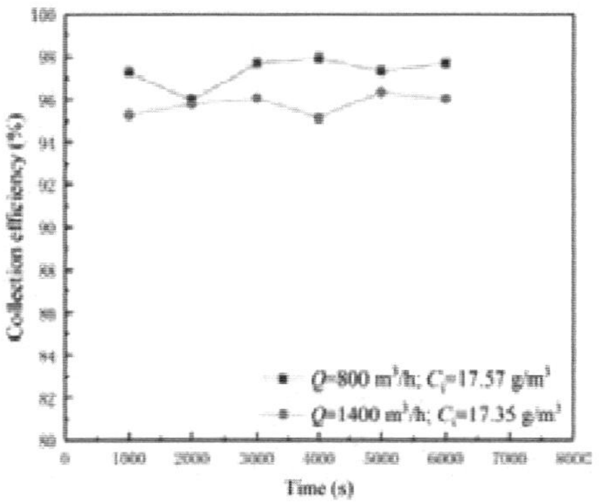

Fig. 9. Collection efficiency vs. time for different inlet gas flow rates

Individual Contributions to the Total Collection Efficiency

As illustrated in Fig. 1, the dust tank Ⅰ was used to collect the dust particles captured by the cyclone shell, while the dust tank Ⅱ and a filter bag for the dust particles entrained by the regenerating gas. The particles carried by the regenerating gas were considered as the dust captured by the built-in granular bed. Due to the formation of the filter cake and the abrasion of the collector granules, the dust particles collected were not exactly equal to all the dust particles introduced. From an engineering viewpoint, however, the individual contribution of the cyclone shell and the built-in granular bed to the total collection efficiency still can be estimated by weighing the particles collected. The contribution ratios of the cyclone shell (e_c) and the built-in granular bed (e_g) are calculated by Eq. (2) and Eq. (3), respectively.

$$e_c = \frac{m_c}{m_c + m_{gc} + m_{gb}} \times 100\% \quad (2)$$

$$e_c = \frac{m_c}{m_c + m_{gc} + m_{gb}} \times 100\% \quad (3)$$

Where, m_c, m_{gc} and m_{gb} are the masses of the dust particles collected by the dust tank Ⅰ, the dust tank Ⅱ and the filter bag, respectively.

As shown in Fig. 10, the contribution ratios of the cyclone shell were around 80% for all 5 test runs under the MB. As the inlet dust concentration increased from 5.95 to 59.13 g/m^3 (test 1, 3, 5 and 7) with Q=800 m^3/h, the contribution ratio of the cyclone shell increased from 78.5% to 84.3%. The dust particles tended to agglomerate under high inlet dust concentration. As a result, the dust particles were more easily separated by the centrifugal force. Besides, the filter cake formed on the outer screen wall was strengthened, thus the separation ability of the cyclone shell was enhanced.

When the inlet gas flow rate was increased from 800 to 1400 m^3/h (test 3 and test 9), the contribution ratio of the cyclone shell increased from 80.1% to 82.5%, as illustrated in Fig. 12. The swirling flow was intensified by the high inlet gas flow rate and then the centrifugal separation was also improved.

Size distributions of the captured particles

The size distributions of the dust particles captured by the cyclone shell and the built-in granular bed were analyzed, respectively. The particle size distributions of all 5 test runs under the MB operating regime assumed the same features, thus the average particle size distributions are illustrated in Fig. 11.

It is known that the centrifugal force provided by the swirling flow is not efficient for separating the small-size particles, especially under a relatively low inlet gas velocity. As shown in Fig. 12, the majority of the particles captured by the cyclone shell were larger than 10 μm; in other words, almost all of the inlet particles smaller than 5 μm escaped. The D_{50} of the dust particles captured by the cyclone shell was 18.0 μm, considerably larger than that of the inlet dust particles (13.2 μm). The escaping fine particles entrained by the cross-flow gas were tangentially-radially introduced into the built-in granular bed. As illustrated in Fig. 11, almost 90% of the dust particles captured by the granular bed was smaller than 10 μm and the D_{50} was merely 4.6 μm. Due to the pre-separation of the cyclone shell, the filtration load of the built-in granular bed was efficiently reduced; while the fine particles were further captured by the granular bed filter. As a result, the gas-solids separation was considerably strengthened.

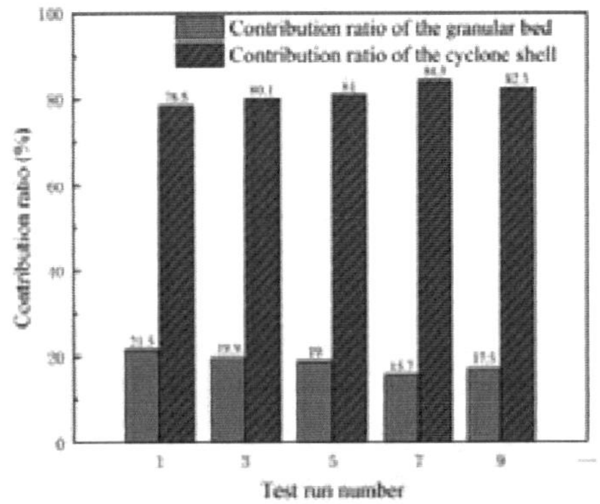

Fig. 10. Contribution ratios of the cyclone shell and the granular bed

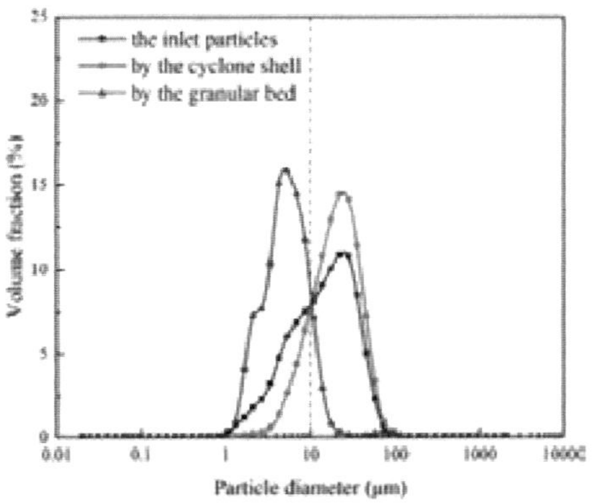

Fig. 11. Size distributions of the captured particles

Conclusion

In this paper, the gas-solids separation performance of the C-CGBF system was experimentally evaluated on a pilot-scale cold-model setup. Experimental results revealed that the pressure drop increased linearly with time under the FB if introducing dust particles. On the other hand, a steady-state pressure drop with high collection efficiency, typically exceeding 95%, was achieved under the MB.

Both the pressure drop and the collection efficiency increased with the inlet dust concentration. The collection efficiency declined when the inlet gas flow rate was increased from 800 to 1400 m^3/h. The dusts hold-up in the granular bed and the filter cake formed on the outer screen wall contributed to the high collection efficiency, as well as increased the pressure drop. The filter cake has the advantage that it doesn't clog the collector granules; while it also can be restrained by the self-cleaning of the swirling flow, especially in a high inlet gas flow rate. The separation performance of the C-CGBF system is anticipated to be improved by paying attention to the control of the filter cake in the future.

The contribution ratio of the cyclone shell to the total collection efficiency was around 80% under the MB. Increasing the inlet dust concentration and the inlet gas flow rate improved the contribution of the cyclone shell. The majority of the particles captured by the cyclone shell were larger than 10 μm. Due to the pre-separation of the cyclone shell, the filtration load of the built-in granular bed was evidently reduced; while the fine particles were further captured by the granular bed. As a result, the gas-solids separation was strengthened considerably.

References

[1] S.C. Saxena, R.F. Henry, W.F. Podolskit, Particulate Removal From High-Temperature, High Pressure Combustion Gases, *Prog. Energy Combust. Sci.*, 11 (1985) 193–251.

[2] P. V. Aravind, W. De Jong, Evaluation of high temperature gas cleaning options for biomass gasification product gas for Solid Oxide Fuel Cells, *Prog. Energy Combust. Sci.*, 38 (2012) 737–764.

[3] P.J. Woolcock, R.C. Brown, A review of cleaning technologies for biomass-derived syngas, *Biomass and Bioenergy*, 52 (2013) 54–84.

[4] Prabhansu, M.K. Karmakar, P. Chandra, P.K. Chatterjee, A review on the fuel gas cleaning technologies in gasification process, *J. Environ. Chem. Eng.*, 3 (2015) 689–702.

[5] C.J. Stairmand, Design and performance of cyclone separators, *Inst. Chem. Eng. -- Trans.*, 29 (1951) 356–373.

[6] A.C. Hoffmann, L.E. Stein, Gas cyclones and swirl tubes: principles, design, and operation., 2 ed., *Springer Berlin Heidelberg*, 2008.

[7] L. Qi, Y. Yuan, Characteristics and the behavior in electrostatic precipitators of high-alumina coal fly ash from the Jungar power plant, Inner Mongolia, China, *J. Hazard. Mater.*, 192 (2011) 222–225.

[8] L. D'Addio, F. Di Natale, C. Carotenuto, W. Balachandran, A. Lancia, A lab-scale system to study submicron particles removal in wet electrostatic scrubbers, *Chem. Eng. Sci.*, 97 (2013) 176–185.

[9] G. Xiao, X. Wang, J. Zhang, M. Ni, X. Gao, Z. Luo, K. Cen, Granular bed filter: A promising technology for hot gas clean-up, *Powder Technol.*, 244 (2013) 93–99.

[10] C.A.P. Zevenhoven, Particle charging and granular bed filtration for high temperature application, *Delft University of Technology*, 1992.

[11] V. Sibanda, R.W. Greenwood, J.P.K. Seville, Particle separation from gases using cross-flow filtration, *Powder Technol.*, 118 (2001) 193–202.

[12] R.C. Brown, H. Shi, G. Colver, S.C. Soo, Similitude study of a moving bed granular filter, *Powder Technol.*, 138 (2003) 201–210.

[13] J.C. Bai, S.Y. Wu, A.S. Lee, C.Y. Chu, Filtration of dust in a circulating granular bed filter with conical louver plates (CGBF-CLPs), *J. Hazard. Mater.*, 142 (2007) 324–331.

[14] S.D. Sharma, K. McLennan, M. Dolan, T. Nguyen, D. Chase, Design and performance evaluation of dry cleaning process for syngas, *Fuel*, 108 (2013) 42–53.

[15] K.S. Lee, J.R. Sohn, Y.O. Park, Filtration performance characteristics of ceramic candle filter based on inlet structure of high-temperature and high-pressure dust collectors, *J. Ind. Eng. Chem.*, 21 (2015) 101–110.

536126 Numerical Simulation of the Capture of Particles By Dead-End Pores

Abstract:

Particle capture by a collecting medium is a common process, which can be applied to industrial applications (e.g., solid/liquid filtration). Ideally, the capture of contaminant particles by a porous filtration medium can be accomplished even when the radius of the particle is smaller than the pore radius. Several studies have been proposed to describe particle clogging within inter-connected (open) pores; however, studies of the capture of particles within dead-end pores are rare, because of the complex pore structure and hydrodynamic conditions. Despite this fact, this mechanism of capture is preferable due to the significantly smaller impact on overall flow behavior. Thus, in this work we describe a simulation model capable of describing the capture of spherical particles within dead-end pores. A 3D discrete element method-lattice Boltzmann method (DEM-LBM) coupling approach is applied to investigate the particle capture under conditions of different particle size and pore structures. The DEM is used to describe the particle dynamics, whereas the LBM is employed to describe the fluid flow through the filter medium. Both the pressure drop and the fluid density are examined to indicate this capture performance. The simulated results illustrate that the capture performance is dependent on the pore geometry and the Stokes number. Finally, a new kinetic model of this process is proposed and verified against experimental data.

525984 In-Line Evaluation of Powder Properties during Mixing Processes Using a Drag Force Flow Sensor

Abstract:

The mixing time required for a dry powder blend to reach homogeneity is highly dependent on many factors, including numerous process parameters and physical properties of the component particles, such as their relative size, shape and density. In order to optimise the efficiency of a mixing operation, it is important for the process to be stopped quickly following the attainment of a homogeneous blend, this can also minimise the effects of over-processing such as particle attrition.

In-line Process Analytical Technology (PAT) tools can generate and display real-time data, enabling the user to accurately monitor and control the physical properties of a blend without the interruptions required to conduct off-line analytical techniques. As a result, important process decisions, such as when the mixing time is sufficient and the process can cease, can be accurately and rapidly made despite novel and modified process conditions and formulations being trialled.

The application of a Drag Force Flow (DFF) sensor overcomes many limitations associated with PAT tools which use spectrographic and particle size analysis techniques. This highly sensitive, in-line measurement tool provides robust and high (temporal) resolution measurements of the forces exerted by the flow of powder within a process. The DFF sensor uses optical technology, with the probe itself containing two Fiber Bragg Grating (FBG) optical strain gauges. As particles impact the probe, the optical strain gauges are stretched or compressed by these minor deflections. The degree of stretching or compression is detected by an optical interrogator which is calibrated to measure DFF and Force Pulse Magnitude (FPM) values are calculated which correspond to changes in the fundamental bulk properties of the particulate flow. The thin, cylindrical shape of a DFF sensor means that it provides minimal intrusion to flow and is therefore appropriate for use across a diverse range of applications, including mixing and blending.

In this study, the suitability of the DFF sensor is evaluated, focusing on assessing its ability to follow the physical changes of powder blends during mixing. The mixing processes of two different formulations within a vertical twin-shaft mixer were monitored using this in-line measurement technique. Repeatable FPM data was generated between repeat measurements of the same formulation and a clear plateau in the FPM data was displayed, associated with the achievement of an even dispersion of components. Agreement between the mixing time required to reach homogeneity was observed between the real-time DFF data and potentiometric titration data conducted on extracted aliquots.

The DFF sensor was demonstrated to be a highly sensitive instrument for routine monitoring and control during mixing, its ability to identify suitable timeframes for a blend to reach homogeneity has vast potential to improve the efficiency of such processes.

536263 *Revisiting the Measurement of Powder Permeability Under Applied Load*

Abstract:

This paper presents an approach to measuring permeability in deep powder beds. Bulk powder permeability is the relationship between interstitial gas velocity and the resulting bed pressure drop as typically governed by Darcy's law [1] at low Reynold's number. The permeability of fine powders can have a significant impact on material handling. For example, a powder's permeability can determine its deaeration time, propensity toward flooding, and rate limiting flow from a hopper.

These analyses often require a knowledge of the permeability as a function of the local bulk density and applied stress, [2] as related by the powder's compressibility relationship. This is easy to determine if low aspect ratio measurements of permeability are appropriately conducted, as the applied column load is approximately equal to the experienced powder stress. However, this is not possible in tall columns or if the wall friction is incorrectly mobilized; the local powder stress may differ significantly from the applied column load due to wall friction and related Janssen phenomena. This is a complex function of wall friction and surcharge due to the applied load as well as powder above the reference measurement. [3]

This paper addresses these challenges of permeability analysis with supporting experimental measurements in tall columns for polymer, food, pharmaceutical excipient, and glass powders.

References:

[1] Darcy, H. (1856). *Les fontaines publiques de la ville de Dijon*. Paris: Dalmont.

[2] Ennis, B.J. *et al.* (2008) Section 21: Solids-Solids Operations & Processing. *Perry's Chemical Engineers' Handbook*. Perry, R. H., & Green, D. W. (eds.), New York: McGraw-Hill.

[3] Nedderman, R. M. (1992). *Statics and kinematics of granular materials*. Cambridge, Eng.: Cambridge University Press. 217-227.

534903 Triboelectrification of Insulator Materials in a Humidified Environment

Abstract:

Triboelectrification is the natural phenomena of the transfer of an electrostatic charge between two materials upon contact. This concept is seemingly simple, yet has left many debating over the factors affecting it, and how this charge is transferred.

The objective of this study was to determine how environmental factors, such as humidity affect different materials and their triboelectric charge. The experimental system is a sealed incubator chamber to control for humidity, and air pressure. Inside the incubator is a downward coiled copper tube (1.5 mm in-diameter), that contains teflon tubing, and is connected to a tin-metal Faraday cup. A single 500-800 micron sized particle of soda-lime glass or polystyrene is dropped into the top end of the coiled tube and the particle is allowed to rolls downward through the tubing accumulating charge before being deposited into the tin metal Faraday cup. The charge was measured and recorded within a period of time as the particle discharges. Measurements taken at a constant low relative humidity are compared to those of higher humidity. This was done by exposing the particles and tubing to higher levels of humidity in a separate chamber and reintroducing the materials at varying wet and dry combinations. It is necessary to determine the humidity affect triboelectrification in insulated materials to better understand the possible benefit and hazards of this process.

535257 Bipolar Charging of Polyethylene Powders: Experimental and Modelling Study

Abstract:

The manufacturing and handling of polyethylene (PE) powders are associated with undesired charging, resulting in agglomeration of charged particles, wall sheeting and eventually leading to plugging of reactors/conveyors. Although some triboelectric theories were developed; these are mostly specialized on one type of charging and are rarely useful for the quantitative prediction of charging. We aim to quantitatively describe the charging of PE powders by charging them in the shaking apparatus, in which we modify the frequency of particle-particle and particle-wall collisions. The key part of our research is the ability to measure the charge distribution, because the particle-particle charging exhibits broadening of charge distribution among the particles of charged sample of PE. To measure charge distribution of our PE samples, we utilized the free-fall electrostatic separator that separated the particles according to their charge densities into several fractions. Afterwards, we measured the charge of each such fraction in the Faradayâ€™s pail connected to an electrometer. Some particles of the PE sample reached significantly larger charges of opposite polarities and therefore they formed agglomeration centres. Such bipolarity in charging due to particle-particle collisions, as well as the formation of the agglomeration centres was also predicted and quantified using our particulate mathematical model based on discrete element method. Apart from agglomeration, fouling can play role in industrial systems, as PE particles of any charge polarity are attracted towards the metal walls of reactors and conveyors due to the effect of electrostatic induction. In our experiments, we observed significant build-up of charge of only one sign in cases where particle-wall collisions were frequent relative to the particle-particle collisions. Particle-wall charging thus doesnâ€™t necessarily leads to the formation of agglomerates, it may, however, lead to the fouling. Our experiments regarding the effect of temperature on charging showed that the saturation charge of PE particles increases with temperature of particles; however, no increase in charging was observed when only the aluminium slide (serving as a counterpart used for charging of PE particles) was heated. This indicates that the increase in charging at elevated temperatures is associated with the properties of PE particles, rather than with properties of the metal. A possible explanation could be the softening of PE particles at elevated temperatures. As for the studies of impact velocity effect on charging, we observed that the saturation charge of PE particles increased with the maximum impact velocity. Such behaviour could be explained by the structure of PE particles that effectively enables the particles to be elastic and also by the surface roughness of the material. Specifically, if force is applied on a PE particle, the asperities on the particle surface bend inside the particle, thus effectively increasing the total particle area accessible for contact with a wall and thus also for charging. This work can be valuable for further work on the understanding of the charging mechanism, as well as it can provide some guidance for the industry.

Konopka, L. and J. Kosek, Discrete element modeling of electrostatic charging of polyethylene powder particles. Journal of Electrostatics, 2017. 87(Supplement C): p. 150-157.

JantaÄ, S., Konopka, L. and J. Kosek Experimental study of triboelectric charging of polyethylene powders: effect of humidity, impact velocity and temperature. Advanced Powder Technology (Submitted)

527911 Evaluating Electrostatic Charging of Powders - the Challenges

Abstract:

Since ancient times, it's been known that materials such as amber attract low mass particles after rubbing, indeed the terms electron and electricity are derived from the Ancient Greek for amber. Whilst this phenomenon is widely recognised and accepted, electrostatic charging and triboelectrification are poorly understood, especially with regards to powder and particulate handling. The propensity of some powders to electrostatically charge can have implications with regards to in-process performance, for example it can lead to adhesion onto equipment surfaces and segregation of different components due to charge differences. Furthermore, there are fire/explosion safety considerations associated with static discharge.

The complex interactions between powder handling, including flowability, and charging make effective measurement difficult, impacting the ability to design processes to control/mitigate charging problems. In most cases the propensity to charge is managed by controlling the relative humidity, earthing equipment allowing the charge to dissipate and/or in-line ionising equipment.

Many measurement methods have been devised for investigating this phenomenon, such as single particle impaction [1]; charging in gas/solids flow – either pneumatic transport/cyclones [2,3] or fluidised beds [4,5]; vibrating plates [6]; or cascades [7], however due to the inability to control the initial state of the powder and the poor reproducibility of the test method, these techniques have not been widely adopted within industry. One reason for poor repeatability is the potential to change the charge state of the powder each time it's handled or comes in to contact with a chemically different surface. For this reason, the test conditions, including the equipment, environment and test protocols must be carefully controlled in order to minimise unwanted powder and equipment charging.

This presentation addresses the challenges of developing an industrially appropriate tool for measuring charge and its effect on powder flowability, investigating the electrostatic behaviour of both powders and contact (equipment) surfaces. This on-going study includes the following areas of investigation:

- Discharging powders and equipment
- Powder handling protocols for reducing unwanted charging
- Propensity of powders to charge – including rate of charge and maximum charge
- Interaction between different powders and equipment surfaces

A strong understanding of the fundamental principles of electrostatic behaviour and triboelectric charge will aid the development of a suitable tool for charging powders and surfaces in a controlled manner and measurement of the resultant dynamic flow properties using powder rheology, allowing process engineers to predict changes in flow behaviour due to charging as well as selecting suitable equipment and equipment settings.

The presentation will highlight the challenges associated with controlling and quantifying electrostatic charging and the steps required to counteract them, followed by discussion of the latest results from the study.

References

[1] Matsuyama, T., Yamamoto, H., 1994. Charge transfer between a polymer particle and a metal plate due to impact. IEEE Transactions on Industry Applications 30, 602–607.

[2] Masuda, H., Komatsu, T., Iinoya, K., 1976. The static electrification of particles in gas–solids pipe flow. AIChE Journal 22,558–564.

[3] Rowley, G., 2001. Quantifying electrostatic interactions in pharmaceutical solid systems. International Journal of Pharmaceutics 227, 47–55.

[4] Iuga, A., Calin, L., Neamtu, V., Mihalcioiu, A., Dascalescu, L., 2005. Tribocharging of plastics granulates in a fluidized bed device. Journal of Electrostatics 63, 937–942.

[5] Zhao, H., Castle, G.S.P., Inculet, I.I. 2002. The measurement of bipolar charge in polydisperse powders using a vertical array of Faraday pail sensors. Journal of Electrostatics 55, 261–278.

[6] Higashiyama, Y., Ujiie, Y., Asano, K., 1997. Triboelectrification of plastic particles on a vibrating feeder laminated with a plastic film. Journal of Electrostatics 42, 63–68.

[7] Oguchi, T., Tamatani, M., 1993. Contact electrification phenomena and powder surface treatments. Wear 168, 91–98

536623 *Powder Permeability As a Measurement Surrogate for Triboelectric Charging*

Abstract:

Powder Permeability as a Measurement Surrogate for Triboelectric Charging

Benjamin Ennis[1], Michael Winn, Naseem Jibrin, Bryan Ennis

E&G Associates, Inc.

100 Cherokee Blvd, STE 332

Chattanooga, TN 37405

[1] Presenting & corresponding author: benjamin.ennis@powdernotes.com, (423) 544-7298

Triboelectric charging is common for many powders in solids handling and other processing applications, sometimes intentional as with laser printing but more often than not unintentional and a processing nuisance. Such particle charging can lead both to wall deposition and increased interparticle agglomeration with substantial increases in minimum bin opening sizes to prevent stable arching and ratholing.

Furthermore, the exact nature of triboelectric charging remains poorly understood, with severe limitations in readily made characterization to provide reliable engineering and process guidance. For example, powders may charge by surface interactions (traditional triboelectric charging) or by particle-particle interactions (self-charge by frictional rubbing). Generation of frictional charge can also lead to (i) a net bulk powder charge, (ii) distributions in particle net charge, or even (iii) surface distributions in charge across single particles. Net bulk charge is readily measured through transfer or Faraday cups; however, the latter two mechanisms involving charge distribution are more likely relevant to powder agglomeration, and far more difficult to assess.

Recent additive manufacturing research has shown that bed permeability is readily impacted by the extent of powder handling and related charging. Bulk powder permeability is the relationship between interstitial gas velocity and the resulting bed pressure drop as typically governed by Darcyâ€™s law [1,2]. This paper presents a novel approach to using powder bed permeability as a rapid surrogate measurement to the extent of powder triboelectric charging. Permeability is shown to track with the extent of drying, humidity conditioning, and time of handling. Limited confirming measurements are also made by Faraday cup for varying processing time and transfer material, as well as by shear cell measurements of minimum opening size and friction.

References:

[1] Darcy, H. (1856). *Les fontaines publiques de la ville de Dijon*. Paris: Dalmont.

[2] Ennis, B.J. *et al.* (2008) Section 21: Solids-Solids Operations & Processing. *Perryâ€™s Chemical Engineersâ€™ Handbook*. Perry, R. H., & Green, D. W. (eds.), New York: McGraw-Hill.

521762 Effect of Intrinsic Materials Properties on the Mechanical and Rheological Behavior of API Agglomeration in Agitated Filter Dryers

Abstract:

The API manufacturing involves a series of processing steps including purification, filtration, drying and milling. Drying is often conducted as one of the last stages of the API production to meet the residual solvent specification, and is considered as a critical step to impact the final product performance. Among various dryer types, the agitated filter dryer (AFD) which integrates filtration and drying process into a single unit operation, is commonly used in potent API production. This single unit operation provides many advantages such as excellent containment. While agitation is required to increase heat transfer and promote homogeneity, it can cause agglomeration of wet API. The nature and material properties of the agglomerates can negatively impact the API performance during drug product processing and hence it can pose a risk to the API physical property control at the commercial scale.

Agglomeration of API during agitated filter drying is a challenging issue in pharmaceutical industry. Surface intermolecular interactions, sintering, mechanical interlocking, and formation of solid bridges have been proposed as potential mechanisms for API agglomeration. In addition, during API drying, the intrinsic API mechanical properties, surface solid-state transformation, surface features and energetics, cake wetness as well as applied external forces such as shear and compaction may lead to the formation of agglomerates and other related effects on API bulk physical properties.

In this presentation, we demonstrate an experimental approach to assess propensity of agglomeration of API particles using a mixer torque rheometer. In particular, the fundamental effect of intrinsic API particle properties on the formation of API agglomerates and resulting agglomerate properties have been investigated. By utilizing a mixer torque rheometer, we define a high risk of agglomeration (i.e. risky zone) of needle shaped API particles as a function of solvent content in a wet cake. This risky zone was demonstrated to be impacted by properties of the API as well as the solvent used. When the wet cake is agitated with a solvent content inside the risky zone, the agglomeration propensity rapidly increased, leading to increased particle size and stronger agglomerates. This work presents opportunities for better process design to avoid API agglomeration during agitated filter drying.

537424 CO_2 Capture and Transport Behaviors of Porous Polymer Beads Containing Metal-Organic Frameworks (MOFs)

Abstract:

A number of novel materials including metal-organic frameworks (MOFs) have been developed to capture CO_2 from various industrial sources. While some recently synthesized MOFs have great CO_2 capture capacity and even water tolerance, their delivery method is still debated. For example, MOFs can be used in the perm-selective layer of gas separation membranes. The direct use of MOFs is not desired due to the safety and process concerns related to fine powder handling. In this study, MOFs-bearing porous polymer beads are prepared and their CO_2 adsorption behaviors are studied in a fixed bed and a fluidized bed reactor. CO_2/N_2 and CO_2/CH_4 gas mixtures are used as the inlet gas and the CO_2 capture conditions are maintained at ambient condition (25 ^oC and 1 atm) while varying the gas velocity to obtain CO_2 breakthrough curves. The experimental findings suggest that the CO_2 adsorption kinetics of MOFs-bearing porous polymer beads are very fast and this delivery method of MOFs can allow for a flexible design of CO_2 capture units ranging from a fixed bed to a fluidized bed. The effect of moisture on the CO_2 adsorption behavior is also investigated while changing the chemical properties of the porous polymer substrate (e.g., hydrophilic vs. hydrophobic). The recyclability of the beads is determined via repetitive long-term CO_2 adsorption and desorption experiments where the partial pressure of CO_2 is controlled as well as humidity.

537437 *Carbon Mineral Sequestration Integrated with the Recovery of Rare Earth Elements from Alkaline Industrial Wastes and Silicate Minerals*

Abstract:

The ever-increasing atmospheric CO_2 concentration has become one of the most difficult environmental issues in recent years. In order to reduce the atmospheric CO_2 concentration, various technologies to capture, utilize and store CO_2 (CCUS) are being developed. One of the CCUS technologies that can permanently immobilize CO_2 and keep CO_2 away from the atmosphere is called mineral carbonation, where CO_2 reacts with Mg or Ca-bearing silicate materials to produce stable, environmentally benign and insoluble carbonates. In a typical pH swing process for mineral carbonation, Ca and Mg in the silicate minerals are extracted by acidic solvents, with silica left in the solid phase. Subsequently, by bubbling CO_2-containing gas streams while adding base into the leachate, solid carbonates are precipitated out.

Serpentine ($(Mg, Fe)_3(OH)_4Si_3O_5$) is a hydrated magnesium silicate mineral which has a large enough CO_2 mineralization capacity to sequester significant amounts of anthropogenic CO_2. Apart from silicate minerals, alkaline industrial solid wastes are also good resources for mineral carbonation. Iron slags and steel slags are particularly interesting because aside from Ca and Mg, iron slags and steel slags can also contain other valuable metals, including Fe, Al, Cr, V, Mn, Ti and rare earth elements (REE). The integrated recovery of these elements during the mineral carbonation process can maximize the exploitation and valorization of the slags, while contributing to the reduction of CO_2 emissions and the sustainable utilization of energy. In this study, the fate of REE in the iron slags and steel slags is investigated during the pH swing process for carbon mineralization. Both dissolution and carbonation kinetics of Ca and Mg in silicate minerals and slags are studied while monitoring the leaching behaviors of REEs. Additionally, the precipitation behaviors of REE are studied under various pH conditions.

531191 Two Dimensional Simulation of Carbon Capture Using Amine-Based Solid Sorbents

Abstract:

Chemical sorption of CO_2 by selective solid sorbents has been considered as promising method with the potential to be economically feasible for carbon dioxide removal from syngas. To obtain better understanding of CO2 capture using solid sorbents, in this study two dimensional CFD simulation of the circulation fluidized bed carbon capture process, based on NETL (National Energy Technology Lab) setup, using amine based solid sorbents has been performed.

The Eulerian-Eulerian two fluid model equations are used to do 2D numerical simulation. Particles interaction is analyzed using kinetic theory based granular flow model along with frictional stress and viscosity for the dense solid flow regimes. Two dimensional CFD simulation (Continuity and momentum conservation equations) were performed without including the carbonation and regeneration reactions. The results of the two-dimensional simulation were compared with experimental data to study the effect of particle stress models, restitution coefficient and specularity coefficient on the flow of the solids at the Circulating Fluidized Bed (CFB) carbon capture unit.

In the second part of the study, carbonation and regeneration heterogenous reactions was added to the CFD model. So, species conservation and energy conservation equations were also solved to calculate the rate and extend of CO_2 sorption and temperature distribution in the system.

528245 Tribocharging of Bidisperse Particles in Fluidized Beds

Abstract:

Tribocharging of Bidisperse Particles in Fluidized Beds

<>X. Liu[1], J. Kolehmainen[1], A. Ozel[2], S. Sundaresan[1]

[1]*Department of Chemical and Biological Engineering, Princeton University, Princeton, New Jersey 08540, USA*

[2]*School of Engineering and Physical Sciences, Heriot-Watt University, Edinburgh EH14 4AS, UK*

Tribocharging refers to the phenomenon that materials pick up static charges upon contact with other materials [1]. In fluidized beds, granular particles undergo tribocharging due to repeated particle-wall and particle-particle collisions, resulting in complex charge distributions. Particle size distribution brings in additional complexities [2-4]. In this study, we seek to investigate the effect of particle size distribution on tribocharging in fluidized beds by using a bidisperse system as a model problem.

The system is simulated using Computational Fluid Dynamics coupled with Discrete Element Method (CFD-DEM) where the tribocharging model is coupled with hydrodynamics [5]. The rate of charge transfer between two surfaces that come into contact depends on the effective work function difference between the surfaces and the local electric field [6]. We first assign different combinations of effective work functions to the two particle sizes and vary their size and volume ratio. The relation between these factors and the resultant charge distribution is examined.

We then simulate the system by assigning the same effective work function to both particle sizes. When the effective work function of the wall is different from that of the particles, a counter-intuitive bipolar charge distribution is found to develop. The source of opposite charges in the system is investigated based on the macroscopic electric field and particle-particle and particle-wall collisions.

[1] Lacks, D., Mohan, S. Contact electrification of insulating materials. J. *Appl. Phys.* 2011; 44(45).

[2] Forward, M., Lacks, D., Mohan, S. Charge Segregation Depends on Particle Size in Triboelectrically Charged Granular Materials. *Phys. Rev. Lett* 2009; 102(2): 028001.

[3] Schella, A., Weis, S., Schröter, M. Charging changes contact composition in binary sphere packings. *Phys. Rev. E* 2017; 95(6): 062903.

[4] Lee, V., Waitukaitis, S., Miskin, M., Jaeger, H. Direct observation of particle interactions and clustering in charged granular streams. *Nat. Phys.* 2015; 11: 733–737.

[5] Kolehmainen, J., Ozel, A., Boyce, C. M., Sundaresan, S. Triboelectric charging of monodisperse particles in fluidized beds. *AIChE J* 2017; 63(6): 1872–1891.

[6] Laurentie, J. C., Traoré, P., Dascalescu, L. Discrete element modeling of triboelectric charging of insulating materials in vibrated granular beds. *J Electrostat* 2013; 71(6): 951-957.

536968 Bubble Hydrodynamic Comparison for Geldart Group a and B Materials at Different Fluidization Regimes

Abstract:

Gas solids contacting in fluidized beds are an important parameter in determining the efficiency of chemical reactions. The bubbles are used to facilitate contact between the particles, and the gas is used to induce mixing of the particles, while increasing the mass and heat transfer efficiency. Studying the bubble hydrodynamics are therefore important in optimizing the reactor design. Scale is an important factor when it comes to fluidized bed operations. Tests were conducted in a 3.0 ft (0.9 m) diameter, 22.0 ft (6.5 m) tall bed and a comparative study was done using Geldart Group A FCC catalyst and Geldart Group B sand material, having a mean particle diameter of 93 m and 162 m respectively. The fluidization regimes ranged from bubbling to turbulent. A proprietary design optical probe was traversed across the column diameter at two different axial locations to determine the cross sectional radial and axial average of various bubble hydrodynamic properties such as gas holdup, bubble rise velocity and the average bubble dimension. The experimental data was then compared with various empirical correlations in literature

538201 Coupled CFD-DEM Simulations forÂ Heat-Exchanger Cleaning

Abstract:

Particle transport in a plate-type heat exchanger having tortuous paths is a ubiquitous topic chemical/environmental engineering. Sub-micron particles experience various types of interactions such as electrostatic interactions, Brownian random forces/torques, gravitational/buoyant forces, and more importantly hydrodynamic forces/torques due to the ambient (often linear shear) flow field. The recent interest of particle science pays more attention to inelastic granules or granular materials, having sizes of at least 100 micrometers. Collision dynamics of these large, inelastic particles in thin flow channels have great importance in process engineering. We use a large-scale, parallel computation dealing with particles of an order of O(10^3) to track all the particles' linear and angular motion at each time step. Fluid fields inside the thin slits are solved using an open-source CFD package, OpenFOAM, in parallel, and simulation results such as velocity, vorticity, and rate of strain are imported to mimic the particle motion. Dynamics of particle-particle and particle-wall collisions are accurately simulated using the dissipative hydrodynamics (DHD) (as a general version of Stokesian dynamics), which can cover frictional interactions and inelastic collisions between two unequal-sized particles. We currently focus on the application of DHD for granular dynamics in confined spaces, including thin or curved flow channels. The fundamental insight of hydrodynamic transport of inelastic granules will provide a deep understanding of the transport of inelastic granules in geometrically confined flow channels.

533065 Cloudy-Zone Modeling of a Gas-Solid Bubbling Fluidized Bed with Liquid Spray

Abstract:

Fluidized bed reactors with liquid spray are widely used in industrial processes especially in the process of olefin polymerization, owing to the excellent heat removal capability and high performance of the polymer products. Liquid spraying from the side wall of a gas-solid fluidized bed results in a gas-particle-droplet interaction zone, or named the â€˜cloudy zoneâ€™, in which the homogeneity of temperature and contact between particles and gas are broken and the hydrodynamics and heat conduction are affected by the conduction and evaporation of liquid. However, such effects are different with the existence forms of liquid. Specifically speaking, the liquid sprayed into the bed usually has two existence formations, droplets and liquid films on particles. The evaporation of droplets absorbs the heat of hot gas whose temperature decreases. However, the heat of gas and particles is absorbed by the evaporation of liquid films. So the temperature of catalyst active center is decreased by liquid films which also provide comonomer. Therefore, droplets and liquid films have different effect on the reaction in the process of olefin polymerization. As a result, the temperature of cloudy zone decreases and there will be a significant dissimilarity between the cloudy zone and non-cloudy (gas-solid) zone. Therefore, it is essential to characterize the ratio of droplets and liquid films and the influence of liquid spray on the fluidized bed.

In this work a mathematical model is setted up to describe cloudy zone and gas-solid zone in the gas-solid bubbling fluidized bed with liquid spray. The model is composed of hydrodynamic equations and heat transfer equations, such as the equations for gas mass balance, bubble phase velocity, bubble phase force balance, emulsion phase heat balance and bubble phase heat balance, etc. The volume ratio of droplets to liquid films is calculated to account for the relative contribution of different liquid existence forms in the bed. For the closure of the equations, an objective function is proposed, containing the energy consumption from the change of surface of liquid phase, the mass-specific suspension energy consumption rates of emulsion and interphases, and the energy consumption from the expansion of bubbles. The model without the consideration of liquid spray is firstly verified through the experimental voidage data from literature [1]. Then a self-designed fluidized bed with back heating and side-wall liquid spray is applied for the verification of predicted voidage and liquid hold-up. According to the modeling results, the volume ratio of droplets to liquid films increases with increasing the velocity of gas because of the increase of voidage and the decrease of the collision frequency between the liquid droplets and particles. The model expresses the effect of liquid spray on the gas-solid fluidized beds and it is of great significance in describing the cloudy zone and the whole reactor more accurately.

Reference

[1] Taghipour F, Ellis N, and Wong C. Experimental and computational study of gasâ€“solid fluidized bed hydrodynamics. *Chemical Engineering Science*, **2005**. 60(24), 6857-6867.

537627 Engineering the Surfaces of Fluorescently-Labeled Polymeric Nanoparticles for Drug Delivery

Abstract:

The use of nanoparticles in drug delivery can address many of the difficulties faced by traditional methods such as systemic toxicity, transport limitations, and drug solubility. By loading drugs into biocompatible polymeric nanoparticles, drugs that were previously unusable due to their solubility or toxicity can now be used. Fluorescently labeling the nanoparticles with a small molecule like 6,13-Bis(triisopropylsilylethynyl)pentacene (TIPS pentacene), the particles can be tracked for biodistribution studies in biological systems. TIPS pentacene is a fluorophore that has been primarily used in semiconductors, LEDs and solar cells.[1, 2] However, its high quantum yield and photostability make it a promising candidate as a fluorescent imaging agent for biomedical applications.

Using the Flash Nanoprecipitation process[3], different loadings of this fluorophore and a filler homopolymer of poly(D,L-lactic acid) (PDLLA) have been incorporated into amphiphilic diblock copolymer nanoparticles of poly(ethylene oxide)-b-poly(D,L-lactic acid) (PEO-PDLLA). We investigated the effects of varying the TIPS pentacene loading as well as the homopolymer to diblock copolymer weight:weight ratio on the characteristics of the resulting nanoparticles such as size, zeta potential, and fluorescent intensity.

The PEO-PDLLA nanoparticles made with varying amounts of fluorophore and homopolymer resulted in particles with sizes ranging from 60-200 nm. The zeta potentials were typically ~-15 to -20 mV. Previous work done by Pansare et al. analyzed a synthesized pentacene derivative Et-TP5 showing promising optical properties.[4] Recent work from our group showed TIPS pentacene loaded nanoparticles readily taken up by macrophages and the TIPS pentacene visible through imaging cytometry even at weight percent loadings in particles as low as 1 wt%.[5] The presentation will delve into the nanoparticle fabrication and engineering of these particles, looking at the fluorescence and quenching properties of the loaded TIPS pentacene. Initial studies showed that even at TIPS pentacene loadings of 1 wt%, self-quenching due to Forster Resonance Energy Transfer (FRET) may be present. Through similar analysis to that of Pansare et al., the work presented shows that the TIPS pentacene may be a comparable fluorophore appropriate for these applications but with the advantage that it is commercially available.

This research also tests the hypothesis that the surface chemistry of the particles can easily be altered by the addition of acid end capped PDLLA homopolymer to increase carboxylate groups on the surface while also changing the size properties. The carboxylate groups provide a useful functional group for coupling streptavidin to the surface. Streptavidin is a glycoprotein from the bacteria *Streptomyces* avidinii and with biotin, form on of the strongest non-covalent interactions with a dissociation constant of K_D â‰ˆ 10^{-15} M. By coating the nanoparticles with streptavidin, it provides a facile method for functionalizing the surface with biotinylated ligands for targeted drug delivery.

Flash nanoprecipitation shows promise in its scalability and efficacy in producing well-defined nanoparticles and this talk explores a potential design and processing framework of engineering polymeric nanoparticles appropriate for biological studies in a variety of different applications within the drug delivery field.

References

1. Anthony, J. E., Silylethyne-Substituted Pentacenes. *Material Matters: Organic and Molecular Electronics* **2009,** *4* (3), 58-60.
2. Platt, A. D.; Day, J.; Subramanian, S.; Anthony, J. E.; Ostroverkhova, O., Optical, Fluorescent, and (Photo)conductive Properties of High-Performance Functionalized Pentacene and Anthradithiophene Derivatives. *Journal of Physical Chemistry C* **2009,** *113* (31), 14006-14014.
3. Gindy, M. E.; Panagiotopoulos, A. Z.; Prud'homme, R. K., Composite block copolymer stabilized nanoparticles: Simultaneous encapsulation of organic actives and inorganic nanostructures. *Langmuir* **2008,** *24* (1), 83-90.
4. Pansare, V. J.; Bruzek, M. J.; Adamson, D. H.; Anthony, J.; Prud'homme, R. K., Composite Fluorescent Nanoparticles for Biomedical Imaging. *Mol. Imaging. Biol.* **2014,** *16* (2), 180-188.
5. McDaniel, D. K.; Jo, A.; Ringel-Scaia, V. M.; Coutermarsh-Ott, S.; Rothschild, D. E.; Powell, M. D.; Zhang, R.; Long, T. E.; Oestreich, K. J.; Riffle, J. S.; Davis, R. M.; Allen, I. C., TIPS pentacene loaded PEO-PDLLA core-shell nanoparticles have similar cellular uptake dynamics in M1 and M2 macrophages and in corresponding in vivo microenvironments. *Nanomedicine: Nanotechnology, Biology and Medicine* **2017,** *13* (3), 1255-1266.

537333 Optimizing the Surface Property-Activity Relationship of Nanoscale Hydrogel Drug Delivery Systems

Abstract:

Ovarian cancer is the deadliest of all gynecological cancers and the fifth leading cause of death due to cancer in women, resulting in more than 15,500 deaths each year [1]. Screening for ovarian cancer in the general population is not recommended by any medical group as many other benign conditions also exhibit expression of the same biomarkers, including endometriosis-induced irritation, pregnancy, and liver disease, leading to a significant rate of false diagnosis [2]. As a result, the development of improved therapies remains a key challenge. Despite broadly increased research into the development of new anti-cancer therapies, there has been no change in the mortality rate for ovarian cancer in the last 30 years [3].

Current treatment plans typically include chemotherapy with radiation or surgical removal of the tumors. However, traditional chemotherapy does not distinguish between healthy and cancerous cells and typically leads to significant side effects. The use of nanoparticles as delivery vehicles has shown promise for enhancing the systemic delivery of chemotherapeutic agents and solving some of the limitations that conventional administration presents [4]. There are three major requirements for nanoparticles to deliver their payload to the tumor. They need to: (i) be stable in the circulation without releasing drug prematurely, (ii) evade opsonization and accumulate in the tumor efficiently, and (iii) intelligently release drug inside the tumor cells. This means the nanogel must have sufficient time in blood circulation to reach intended sites of the body and accumulate in and around the tumors via leaky vasculature. Ultimately, localizing and controlling drug release at the disease site would offer advantages over the current chemotherapeutic regimens because it limits the toxicity to healthy tissues and decreases the side effects for the patient.

In this work, we have developed long-circulating, biocompatible nanoscale hydrogels (nanogels) that respond to the acidic intracellular environment for controlled delivery of chemotherapeutics. The nanogels are comprised of: (i) hydrophilic, cationic 2-(diethylamino)ethyl-methacrylate, (ii) tetraethylene-glycol-dimethacrylate crosslinker, and (iii) surface-grafted poly(ethylene-glycol). First, we show that the nanogel molecular architecture can be tailored to carry a variety of cargos with varying physicochemical properties, and the release kinetics can be tuned to a desired stimuli. The physical properties of the resulting nanogels were compared using dynamic light scattering, zeta potential, titration, pyrene fluorescence, and red blood cell hemolysis to elicit the influence of polymer composition on swelling ratio, surface charge, pKa, hydrophile-hydrophobe phase transition, and erythrocyte membrane disruption capability. The delivery capacity was analyzed using carboplatin and paclitaxel.

Next, we optimized the nanogel surface properties to avoid the body's natural clearance mechanisms and increase circulation time, while still maintaining the necessary characteristics to promote cellular uptake. Stealth coatings shield the particle surface charge and thereby reduce opsonization by blood proteins and uptake by macrophages of the mononuclear phagocyte system. Yet, after localizing at the tumor site, the coating may hinder drug release and target cell interaction and can therefore be an obstacle in the realization of the therapeutic response. To this extent, we developed a novel stimuli-responsive coating on the particle surface that unmasks a ligand that binds to cell surface receptors overexpressed in ovarian cancer cells. Using hyaluronic acid, PEG, and dextran, a library of nanoparticles with varying degrees of surface density was synthesized, and the impact to particle physicochemical properties, protein binding, drug release, and binding affinity was investigated. The influence of the surface properties to cellular cytocompatibility, uptake, degree of nanogel-cell specificity, and therapeutic efficacy was evaluated with OVCAR-3 cells as a function of time and concentration.

We demonstrated that both the surface-grafted and hydrophobic monomer composition can be varied to enable precise control over the nanogel surface and core characteristics to enable effective, controlled delivery. Varying both chain length and steric bulk allowed for precise control over the thermodynamic response (swelling ratio), dynamic behavior (pKa, membrane disruption), and drug-polymer interaction (delivery capacity). Ultimately, we demonstrated that the tunability of our multicomponent nanogel can be exploited to enable long-circulation and effective transport of the drugs to the tumor site.

References:

[1] Ovarian Cancer National Alliance. wwwovariancancerorg. 2015.

[2] Nolen BM, Lokshin AE. Biomarker testing for ovarian cancer: clinical utility of multiplex assays. Molecular diagnosis & therapy. 2013;17(3):139-46. Epub 2013/04/05. doi: 10.1007/s40291-013-0027-6. PubMed PMID: 23552992; PMCID: PMC3670781.

[3] National Cancer Institute. SEER Cancer Statistics Factsheets: Ovary Cancer. http://seercancergov/statfacts/html/ovaryhtml. 2015.

[4] Kirtane AR, Kalscheuer SM, Panyam J. Exploiting nanotechnology to overcome tumor drug resistance: Challenges and opportunities. Advanced drug delivery reviews. 2013;65(13-14):1731-47. Epub 2013/09/17. doi: 10.1016/j.addr.2013.09.001. PubMed PMID: 24036273; PMCID: PMC3849460.

534117 Paramagnetic Cations-Loaded Polydopamine Nanoparticles Cytotoxicity

Abstract:

Melanins are natural occurring biopolymers with different functions in the organisms in which they are present. Among these functions, skin UV protection is perhaps the most recognized. However, they participate in many other biological processes, as the regulation of the cellular concentration of free metallic ions, being able of sequestering them or acting as an ion reservoir [1,2].

Taking into account this fact, polydopamine, a synthetic melanin analogous, has acquired an important role in material science [3]. Since its ability to chelate paramagnetic cations is known, it has been considered to device MRI contrast agents or more advanced theranostic systems. In this line, several studies have demonstrated polydopamine biocompatibility during the first 24 hours of treatment, but its long-time cytotoxicity has not been studied yet [4].

Thus, in the present work, the interaction of polydopamine particles with different cations (Fe^{2+}, Fe^{3+}, Ca^{2+}, Na^{+}, Zn^{2+} and Cu^{2+}) has been carried out, analyzing their metal loading capacity and affinity. A tentative assignation for such metal binding sites has been performed by IR spectroscopy and, finally, polydopamine cytotoxicity has been studied in a cancer (BT474) and a healthy cell lines (NIHT3T3). As result, it has been found that this melanin analogous is toxic for both cell lines, possibly due to a ROS production process. So, in this manner, there would be a concern in the use of these metal-enriched polydopamine particles in some biomedical applications, but they could represent a potential antineoplastic agent [5].

References

[1] Hong L, Liu Y, Simon JD. Binding of metal ions to melanin and their effects on the aerobic reactivity. Photochemistry and Photobiology. 2004;80:477-81.

[2] Hong L, Simon JD. Current understanding on the binding sites, capacity, affinity and biological significance of metals in melanin. The Journal of Physical Chemistry B. 2007;111:7938-47.

[3] Lee H, Dellatore SM, Miller WM, Messersmith PB. Mussel-inspired surface chemistry for multifunctional coatings. Science. 2007;318:426-30.

[4] Cho S, Park W, Kim D-H. Silica-coated metal chelating-melanin nanoparticles as a dual-modal contrast enhancement imaging and therapeutic agent. ACS Applied Materials & Interfaces. 2017;9:101-11.

[5] Vega MA, Nieto C, Marcelo G, Martín del Valle EM. Cytotoxicity of paramagnetic cations-loaded polydopamine nanoparticles. Colloids and Surfaces B: Biointerfaces. 2018. doi: 10.1016/j.colsurfb.2018.04.027

536485 Two-Photon Microscopy for Deep-Tissue Imaging of Dopaminergic Neuromodulation in the Brain

Abstract:

Non-invasively imaging the brain neuronal structure and neuro-chemical activity stands to inform our understanding of psychiatric and neurological disorders and their pharmacological therapies. However, the high optical density of brain tissue makes imaging disease-relevant regions at its center a challenge using conventional fluorophores and microscopy techniques, particularly for intact brains in living animal models. We present a microscopic approach that could enable whole-brain, through-skull, and continuous imaging of neuromodulation. To this end, we have combined two imaging modalities that have emerged in parallel with opposing imaging strengths and weaknesses: infrared microscopy, and two-photon microscopy.[1] Recently, functionalized single walled carbon nanotubes (SWNTs) have emerged as promising near-infrared (NIR) fluorescent nanosensors for the sensitive and rapid detection of dopamine, a neuromodulatory neurotransmitter of clinical importance to brain function and disease.[2] SWNT nanosensors fluorescence in the NIR-II window (1000-1700 nm), where absorption and scattering of light is minimal, making them optimally suited for non-invasive imaging in brain tissue. However, to overcome the challenge of absorption and scattering of excitation photons in deep-tissue imaging, we show that intrinsically-near-infrared fluorescent SWNT nanosensors can be efficiently imaged with two-photon excitation (in lieu of one-photon excitation) through photoexcitation by a NIR-II femtosecond pulsed erbium fiber laser. We measure an effective, average quantum yield of 0.2% and a two-photon cross section of 216,000 GM. Furthermore, we demonstrate that two-photon photoexcitation improves the imaging quality of nanosensors in brain-tissue phantoms at depths approaching 2 mm and confirm that non-linear excitation preserves the molecular recognition of dopamine. Finally, we show that this imaging paradigm can be expanded to a laser-scanning system paired with a NIR photomultiplier tube, enabling layer-by-layer 3D imaging, motivating future efforts to achieve real-time imaging of dopamine neurotransmission in awake and behaving animals.

1 J. T. Del Bonis-Oâ€™Donnell, R. H. Page, A. G. Beyene, E. G. Tindall, I. R. McFarlane and M. P. Landry, *Adv. Funct. Mater.*, 2017, **27**, 1702112.

2 A. G. Beyene, I. R. McFarlane, R. L. Pinals and M. P. Landry, *ACS Chem. Neurosci.*, 2017, **8**, 2275â€“2289.

538019 Ultrafast Post-Formulation Core Radiolabeling of Biodegradable Nanoparticles for PET Contrast Agents

Abstract:

Among almost all cancers cases, diagnostic accuracy and speed remains one of the major factors for patient prognosis. Positron emission tomography (PET) is among the most sensitive of imaging modalities but requires accumulation of radionuclide at the targeted site. ^{18}F-FDG (Fludeoxyglucose) is the current gold standard for PET diagnostics but still has challenges with accessibility, sensitivity and specificity. The short half-life of ^{18}F limits PET usage to only institutions with easy access to a cyclotron. Furthermore, since resolution is dependent on concentration of positron emitters at the imaging site, smaller and low metabolism cancers might not be detectable using FDG. Lastly, non-tumor areas with high metabolism, such as brain tissue or inflammation sites, produce high background signals that limit the specificity of FDG-PET. Nanoparticle-based PET modality has the potential to overcome all these limitations. Targeted particles can both localize PET signal and drastically improve sensitivity by clustering the PET imaging agents in one place. Additionally, nanoparticles have more flexibility in labeling with longer half-life PET isotopes such as Cu-64 (12.7 hrs) and Zr-89 (3.3 days), suitable for long-time imaging and drug biodistribution studies. However, the lack of a clinically relevant formulation and radiolabeling method is the current bottleneck for nanoparticle contrast agent development. Traditional surface labeling methods are not industrially scalable and can lead to off target effects *in vivo*. Alternatively, radiolabeling the nanoparticle core requires careful selection of biocompatible chelators that can rapidly bind to radionuclides to form stable complexes. To this end, we here present the use of Flash Nanoprecipitation (FNP), a continuous and scalable self-assembly process, to form porphyrin encapsulated polymeric nanoparticles (NPs) at various sizes ranging from 50 to 200 nm. The core component, hematoporphyrin derivative, is a biocompatible small molecule that has been FDA approved for photodynamic therapy (Photofrin). Its iron(III) complex precursor is Heme, a major component of hemoglobin. When incubated in a solution of [^{64}Cu]CuCl2, a long half-life positron emitter, the porphyrin core rapidly chelated copper in a stable and irreversible manner, generating PET active nanoparticles. Under very mild temperature and pH conditions (25C and pH 5.5), the porphyrin nanoparticles' chelation to ^{64}Cu was complete in less than 15 minutes with essentially all the initial ^{64}Cu activity (1.1 MBq) localized within the nanoparticle core. This core radiolabeling method is comparable or even faster than currently used nanoparticle surface labeling methods. Additionally, the porphyrin:^{64}Cu complex in the nanoparticle core was found to be stable for at least 24 hours and resistant to leaching from external chelators such as ethylenediaminetetraacetic acid (EDTA). Chelation rate modeling was done based on absorbance changes of the porphyrin nanoparticles when incubated with non-radioactive copper and was found to be consistent with experimental results. Drugs could also be encapsulated into the core of these nanoparticles to facilitate a theranostic approach to cancer diagnostic and treatment. This development of a formulation and rapid radiolabeling method to generate stable and biocompatible PET nanoparticles facilitates more accessible, sensitive and specific cancer diagnostics.

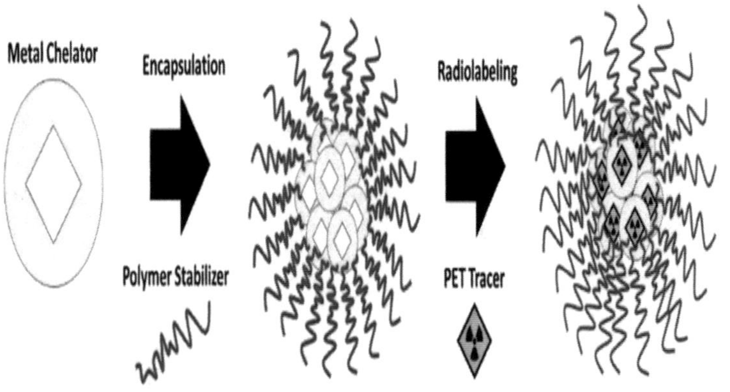

Figure 1-Schematic of nanoparticle core radiolabeling. The biodegradable metal chelator, hematoporphyrin, is encapsulated into nanoparticles using FNP through an amphiphilic block co-polymer stabilizer precipitation process. These nanoparticles are then incubated in a solution of 64Cu. Copper then binds irreversibility to the chelator inside the core, generating PET active nanoparticles.

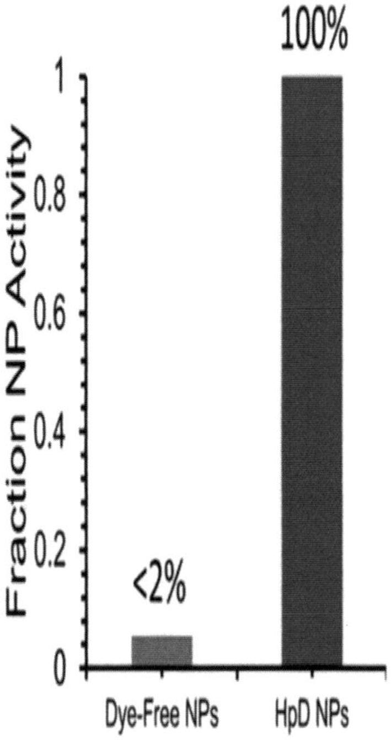

Figure 2-Radiolabeling of hematoporphyrin NPs. Nanoparticles with and without the hematoporphyrin chelator were incubated in a solution of ^{64}Cu. After just 15 minutes, all the initial ^{64}Cu activity was localized inside the hematoporphyrin nanoparticles. This contrasts with the "Dye-Free" NPs which had no hematoporphyrin and thus show minimal uptake of ^{64}Cu.

This page intentionally left blank.

539532 Voltage-Sensitive Ultrasound Enhancing Agent: *In Vitro* and *In Vivo* analysis

Abstract:

We have created a voltage-sensitive ultrasound enhancing agent comprised of perfluorocarbon nano-droplets coated in a monolayer of polymeric surfactant that is then nested inside a negatively charged phospholipid bilayer. The particle, termed Electrast, is responsive to an applied electric field allowing for its internal components to become acoustically activated at ultrasound intensities that are inadequate to cause activation by ultrasound alone. Electrast is novel because activation depends on the presence of an electric field rather than the intensity of ultrasound. This talk will discuss key results from the bench top and the clinic, while commenting on the role of water transport in nesting structures, lipid relocation and reorientation, and lateral phase separation. Calcein and cobalt chloride leakage studies were performed to quantify the amount of water transported across the phospholipid bilayer when exposed to high frequency ultrasound, an applied electric field or a combination of the two. A custom-built, tissue-mimicking gel phantom coupled with a function generator was used to determine the contrast-to-tissue ratio of two different nesting architectures at varying electric field strengths and imaging parameters. Electrast has shown to be selectively activated in coronary circulation in two different animal models (rat and swine) showing an increase in brightness (dB) in the myocardium while leaving the ventricles largely unchanged. This talk will show that Electrast has the potential to be used clinically for myocardial perfusion imaging, assessing heart failure, and evaluating left ventricle wall motion.

537768 Polymorphic Selection of Biominerals By Anionic Polyelectrolytes

Abstract:

In most biomineralization processes many soluble biopolymers exist in minute quantities; these control the nucleation, growth, orientation, and polymorph selection, leading to the formation of minerals with diverse biological functions and distinct properties but also to pathological mineralization, e.g., bone disease, dental calculus, and renal stone formation. In this work, we study the effects of supersaturation, temperature, and additives on the nucleation and growth of calcium oxalate (the major constituent of kidney stones) in aqueous solutions. We have demonstrated that certain anionic polyelectrolytes can be very effective toward kidney stone inhibition by reducing the growth rate, altering the crystal morphology, and favoring the formation of the metastable dihydrate (COD) over the thermodynamically stable monohydrate (COM) which is mainly responsible for kidney, gall, and bladder stones. We compare our results to citrate, a naturally occurring inhibitor in the urinary system and currently the standard pharmacological treatment for stones. Our findings offer a better understanding of the calcium oxalate crystallization kinetics and provide information on the viability of synthetic polymers as potential calcium oxalate stone therapies.

538910 Coupling Experimental Results with Molecular Dynamics Simulations to Describe Polymorphism Obtained Using Solution Shearing

Abstract:

During pharmaceutical drug development, it is important to carry out an extensive search to identify all drug forms. Stabilizing metastable polymorphs is necessary to ensure that the correct drug form is selected for development to balance form stability and drug bioavailability. However, searching for rare forms is a nontrivial process. Here, we show that a combination of fluid dynamics and surface chemistry can be used to overcome some of the challenges associated with searching this extensive landscape of polymorphism. In this study, multiple forms of glycine and acetaminophen were isolated, including metastable forms that are difficult to obtain using conventional methods. The mechanisms through which multiple polymorphs are obtained are described herein. Molecular dynamics simulations and in-situ grazing incidence wide angle X-ray diffraction lend insight to the phase transitions that occur during the crystallization process for fundamental understanding of solution shearing.

533104 Investigation of the Polymorphy Phenomenon of the Carotene Lutein

Abstract:

Carotene is a special group of hydrocarbon pigments characterized with π-π conjugated double bonds in the molecular backbone. This group of compounds is supposed to have great potential in pharmaceutical, healthcare and optical materials because of its excellent properties in antioxidative, light-harvesting, photo-protective, and conductive. Carotene is unstable, due to the conjugated double bonds intend to abstract electron by photo-oxidation. Several methods such as encapsulation have been employed to enhance the stability of carotene. However, seldom studies concerned crystal engineering of carotene to improve its stability has been reported. Nonetheless, H and J aggregation phenomenon of carotene moleculars have been investigated with specific solvents. Recently, β-carotene were reported to have the ability to bind decanuclear homo-and heterometal chain through μ_{10}-briding π-coordination. These findings suggested that the π-π conjugated bond in carotene molecular might have strong plasticity in molecular self-assembling process. As a member of the carotene family, lutein also posseses two hydroxy group (which is easy to occurs polymorph due to the H-bonds between molecular chains) in the molecular structure besides the conjugated double bonds.Thus, cooling crystallization, diffusion zone process and vapor diffusion crystallization methods with selected solvents were tested for lutein, and crystal powders of lutein with needle like, platelet and pyramids morphylogy were obtained. By XRD, TG and DSC analyzing, three crystalline polymorphic forms of lutein were identified. The three crystalline forms of lutein exhibited distinct characters in morphylogy and stability, and the correlation between the stability and the crystal form will be investigated further.

A Kinetic Study of Crystallization Process of Imatinib Mesylate with Polymorphic Transformation Phenomenon

Mengxing Lin, Yuanyi Wu, Sohrab Rohani *
The University of Western Ontario, Department of Chemical and Biochemical Engineering, London, Ontario N6A 5B9, Canada
* Corresponding author: srohani@uwo.ca

Abstract

For pharmaceutical crystallization design and control, the kinetics of nucleation and growth of crystals are significant parameters, especially for the system exhibiting polymorphic transformation. In this study, imatinib mesylate, whose solubility, nucleation and growth kinetics are lacking in the literature, was thoroughly explored in the aspects of characterization, solubility and polymorphic transformation, and evaluation of nucleation and growth rate. Two forms of imatinib mesylate, α and β, were characterized by X-ray powder diffraction, scanning electron microscopy and differential scanning calorimetry. The solubility measurement of the two forms was performed with ultraviolet–visible spectroscopy in three solvents from 278.15 to 333.15 K at atmospheric pressure. The results indicated the β-form is more stable in methanol,1-propanol, and 2-propanol under the experimental condition. The solubility order is methanol >1-propanol >2-propanol. The solvent-mediated polymorphic transformation (SMPT) of imatinib mesylate from α-form to β-form was studied with the in-situ Raman spectroscopy and conductivity meter. It is found that the SMPT process of imatinib mesylate from α-form to β-form in methanol is controlled by nucleation and growth of β-form. Finally, the nucleation and grow rate of β-form of imatinib mesylate were estimated by minimizing the difference between the calculated and experimental solution concentration with MATLAB optimization function.

Keywords: A1. Crystal morphology, A1. Solubility, A2. Growth from solutions, B2. Organic compounds

1. Introduction

Depending on the arrangement of molecules, a solid can be a crystal, a quasicrystal or an amorphous. A phenomenon called polymorphism, exists widely in the crystalline solids, especially in organic compounds[1]. Even though polymorphs have the same molecular formula, they are different in molecular conformations and/or packing arrangements in the solid state, resulting in the distinct physical and chemical properties, such as morphology, melting point, flow properties, mechanical properties, solubility, and Gibbs free energy[2]. In terms of active pharmaceutical ingredients (API), polymorphism would affect the processability, bioavailability and stability of the API[3]. Therefore, the identification and control of polymorphs of a given API are essential for the pharmaceutical industry.

Comparing with the stable polymorph, the metastable polymorph has higher solubility and Gibbs free energy. The divergence in Gibbs energy drives polymorphism to transit from the metastable form to the stable state. Generally, there are two ways to realize the transition: solid-state polymorphic transformation (SST) and solvent-mediated polymorphic transformation (SMPT)[4]. The SMPT process consists of three steps: dissolution of the metastable form, nucleation of stable form, and growth of stable form. The slowest step of the SMPT process limits the whole transition rate. Based on the relative rate of dissolution, nucleation, and growth, O'Mahony et al. rationalized the SMPT process into four possible scenarios: dissolution-controlled, nucleation-dissolution-controlled, growth-controlled, nucleation-growth-controlled [5].

In order to model and control a crystallization process, the kinetic parameters, namely nucleation rate, growth rate, dissolution rate, agglomeration, and breakage have to be determined first through experiments. For the nucleation and growth rate, there are three experimental methods commonly used to determine the nucleation and growth rate. The first one relies on the crystal size distribution of the final product obtained from a Mixed-Suspension, Mixed-Product Removal (MSMPR) crystallizer. The slope and intercept of the "particle number density vs. crystal size" curve provide the information necessary to extract the nucleation and growth rates [2]. The second method which is somewhat approximate, is based on the count and chord length distribution measured by the Focused Beam Reflectance Measurement (FBRM) device [6]. The change in the count of fine crystals presents the nucleation rate, while the change in average chord length refers to the growth rate. The last method uses a rigorous parameter estimation algorithm and works by finding a set of proper constants in the kinetic equations to minimize the difference between the experimental data (measured solute concentration and/or crystal size distribution) and modeling results [7].

The compound studied in this work is imatinib mesylate. Imatinib mesylate (Figure 1), chemically 4-[(4-methylpiperazin-1-yl) methyl]-N-[4-methyl-3-[(4-pyridin-3-ylpyrimidin-2-yl) amino]-phenyl] benzamide methanesulfonate, was developed by Novartis under the trade name "Gleevec/Glivec®" in the late1990s. It was designed as a protein tyrosine kinase inhibitor and shows a remarkable therapeutic effect in Philadelphia chromosome positive chronic myelogenous leukemia (CML) and metastatic malignant gastro intestinal stromal tumors (GISTs) [8]. Many polymorphs of crystalline imatinib mesylate have been observed by precipitating from a solution that consists of an organic solvent, imatinib base and methanesulfonic acid. The different solvents lead to different polymorphs of the final product[9]–[16].

Among the dozens of polymorphs of imatinib mesylate, the α-form and the β-form proposed in WO99/03854 [9] are the most frequent and commercialized. The cubic-shaped β-form has the lower hygroscopic ability, higher flow properties, and higher stability at temperatures below 140° C than the needle-shaped α-form. Therefore, the β-form shows better performance than α-form in terms of processability, manufacturing, pharmaceutical preparation, and storage [17].

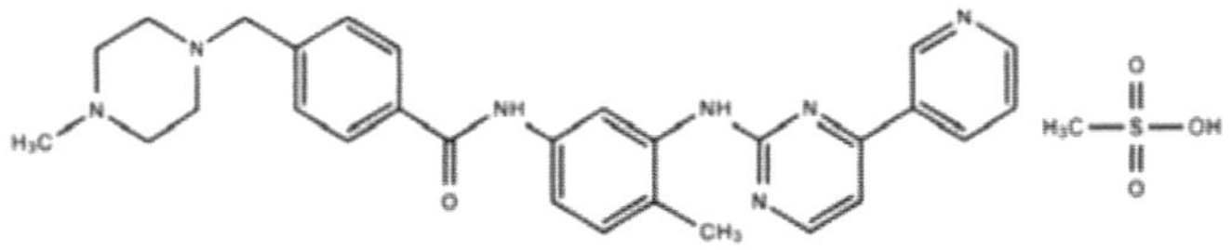

Figure 1. Schematic of imatinib mesylate

Grillo et al. revealed the crystal structure of α-form and β-form imatinib mesylate and found these two polymorphs have significantly different molecular conformations. By grinding them, the amorphous imatinib mesylate was produced that converted to a crystalline phase by either thermal treatment or aging [18]. Veverka et al. modified the morphology of α-form imatinib mesylate with various organic solvent mixtures by precipitation [19]. Srivastava et al. presented a comprehensive study of the vibrational spectroscopies of β-form with the FT-Raman and FT-IR spectra [20]. The polymorphic composition of two polymorphs mixture was determined quantitatively with PXRD, ATR-FTIR, and DSC in the literature [21]. The results show that PXRD technique is more accurate than ATR-FTIR and DSC.

Despite the mentioned studies, the investigation of the imatinib mesylate properties, particularly those related to the crystallization process, was still lacking. The solubility data has not been published up to date. In this study, we measured the solubility of two forms of imatinib mesylate in three solvents (methanol, 1-propanol, and 2-propanol) from 278.15 to 333.15 K at atmospheric pressure to study their relative stability. Subsequently, the solvent-mediated polymorphic transformation (SMPT) from metastable α-form to stable β-form was studied in-situ with Raman spectroscopy and conductivity meter. Finally, the nucleation and grow rate of β-form imatinib mesylate were estimated by minimizing the difference between the calculated and experimental solute concentration with MATLAB optimization function.

2. Material and Experiments

2.1 Material

The α-form imatinib mesylate was supplied by Apotex Pharmachem Inc. (Brantford, ON, Canada) and used without any further processing. The β-form was obtained by the solvent-mediated polymorph transformation from α-form in methanol. The solvents (methanol, ethanol, isopropanol, n-propanol, 1-butanol, acetone, acetonitrile, and tetrahydrofuran) with HPLC grade were purchased from Sigma-Aldrich (Milwaukee, WI,) and no further purification was performed.

2.2 Characterization

2.2.1 Powder X-Ray Diffraction (PXRD)

The polymorphs of solid-state α-form and β-form imatinib mesylate were identified with powder X-ray diffraction (Rigaku, Miniflex, Tokyo, Japan) with the Cu-Kα source ((λ for Kα = 1.54059 Å) at room temperature. The conditions of voltage (30kV), current (15mA), scan angle (2° to 50°) and scan speed (2°/min) were kept the same for all samples.

2.2.2 Differential Scanning Calorimetry (DSC)

Thermal properties of crystals, such as melting point and enthalpy, were measured by differential scanning colorimeter ((DSC, Mettler Toledo, Chicago, United States). The samples (5-10 mg) were placed in a 40-µl aluminum crucible with a pierced lid for gas escaping and then heated from 25°C to 250°C with 10°C/min heating rate under nitrogen purge at a flow rate of 40 ml/min.

2.2.3 Optical microscopy and Scanning Electron Microscopy (SEM)

The optical microscopy (Axioskop 40, Carl Zeiss Light Microscopy, Germany) with a magnification of 500 times was used as an off-line method to give a rough judgement of the polymorphic composition of solid-state imatinib mesylate in suspension. Scanning electron microscope (LEO 1450XB , Zeiss, Oberkochen, Germany) was employed for detailed morphologic information. The samples have been coated with coated with 5 nm of osmium metal in Osmium Plasma Coater (OPC80T, Filgen Inc. Japan) before SEM testing.

2.2.4 Raman spectroscopy

A Raman RXN Analyzer spectroscopy (Kaiser Optical Systems, Inc., Ann Arbor, MI, USA) equipped with a diode laser (784.8-nm) and a fiber optic probe was applied to monitor the polymorphic composition of solid-phase during the SMPT process from α-form to β-form. The characteristic peaks of α-form and β-form of imatinib mesylate are at 1664 cm^{-1} and 1656 cm^{-1}, and the single point baseline is at 1640 cm^{-1}.

2.3 Solubility measurement

To study the stability of two forms, we used Ultraviolet-visible spectroscopy (UV-Vis, Cary Bio 100 spectrophotometer, Varian, Mississauga, ON) to measure the solubility of imatinib mesylate in methanol, 1-propanol, and 2-propanol. The calibration curve of "Absorbance-Concentration" should be determined with a series of known concentration solutions at room temperature at first. The characteristic UV-Vis absorption band at 271-nm and 275-nm was applied for quantification in methanol and propanol, respectively. Then, an excessive sample was added to the selected solvent in a 5-ml glass vial. The vials were immersed in a constant temperature bath with an accuracy of 0.1 °C and agitated for several hours at a specific temperature to reach equilibrium. After 30 mins standing, the supernatant was taken by a syringe and filtered with a 0.45-µm membrane syringe filter (VWR, Mississauga, ON), followed by proper dilution and concentration measurement with UV-Vis. The excess solids in the vials were filtered, dried, and checked with PXRD.

As the α-form imatinib mesylate transformed to β-form in methanol during the solubility measurement, the solubility of α-form imatinib mesylate was determined during the transition process (see section 2.4 for more details).

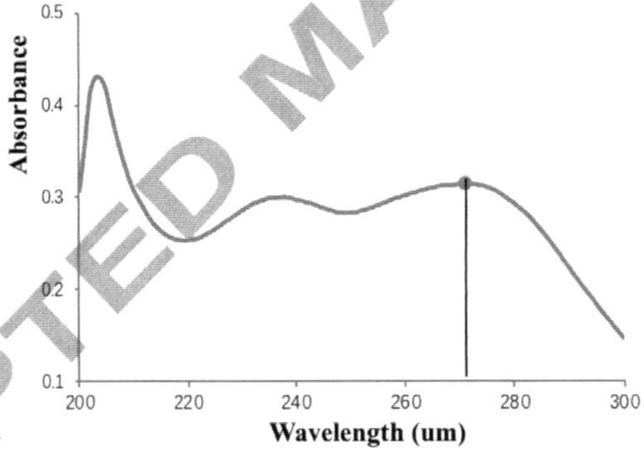

Figure 2. UV-Vis absorption spectra of imatinib mesylate in methanol with a characteristic peak at 271-nm

2.4 SMPT experiments

Eight solvents (methanol, ethanol, isopropanol, n-propanol, 1-butanol, acetone, acetonitrile, and tetrahydrofuran) were screened for mediating the polymorphism transformation. An excessive amount of α-form imatinib mesylate was added to the solvents and agitated in a thermostatic shaker for 2 weeks. The start time of transition was determined by the morphology of the crystals remaining in the solution with the microscope. After filtration and drying, the crystal polymorph was determined by PXRD.

Subsequently, the SMPT process from α-form to β-form imatinib mesylate was investigated in methanol at different temperatures. A carefully excess weighed amount of α-form was added to 35-ml methanol in a 50-mL double-jacketed crystallizer. The crystallizer temperature was kept constant by a water bath circulator (FP-50, Julabo LABORTECHNIK GMBH, Germany). The solute concentration and solid-phase polymorphic composition were monitored by a conductivity meter (Pinnacle Series: M541P) and Raman spectroscopy. The start and end time of the transition process was determined through the morphology of the undissolved crystals with the optical microscope. The solubility of α-form was detected during the SMPT process when the transition did not happen.

3. Theory

3.1 Solubility models

In this work, the solubility of two forms of imatinib mesylate in the selected solvents was fitted with the modified Apelblat equation, as shown in Eq. (1)

$$lnx = A + \frac{B}{T} + ClnT \tag{1}$$

where x is the molar fraction of solute in the solution, T is the absolute temperature in Kelvin, and A, B, and C are dimensionless parameters. The value of three parameters were estimated with MATLAB curve fitting toolbox.

The solubility, in molar fraction, can be calculated using Eq. (2):

$$x = \frac{m/M}{m/M + m_s/M_S} \tag{2}$$

where m and m_s are the mass of solute and solvent, M and M_S are the molecular mass of solute and solvent, respectively.

Solubility reflects the Gibbs energy and stability. The more stable a polymorph is, the more stable a polymorph is, the lower the solubility it is at a given temperature. The difference of Gibbs free energy between two polymorphs relates to the solubility by Eq. (3)

$$\Delta G = RTln\left(\frac{a_1}{a_2}\right) \approx RTln\left(\frac{x_1}{x_2}\right) \tag{3}$$

where ΔG is the difference of Gibbs free energy (J/mol), and a is the solute activity in the solution[22].

3.2 Transition process

The SMPT process consists of three steps: the dissolution of metastable form, nucleation of stable form and growth of stable form. The empirical equations for secondary nucleation and growth rate are expressed as Eq.(4)-(5) [23]:

$$B_{sn} = \frac{dn_0}{dt} = k_{sn} {M_T}^j S^b \tag{4}$$

$$G = \frac{dL}{dt} = k_g S^g \tag{5}$$

where B_{sn} is secondary nucleation rate [#/(m^3· s)], G is growth rate (m/s), S is relative supersaturation, M_T is suspension density (kg solute /kg solvent). k_{sn}, k_g, j, b and g are parameters, representing the nucleation rate coefficient, growth rate coefficient, nucleation order on suspension density, nucleation order on supersaturation, and growth order on supersaturation.

$$S = \frac{C - C^*}{C^*} \tag{6}$$

where C is the solution concentration, while the $C^*$is the equilibrium concentration, namely solubility. In this work, their units are kg solute /kg solvent.

3.3 Parameter estimation

The population balance equation (PBE) coupled with solute mass balance provides a mathematical framework of the crystallization processes. The PBE with the Lagrangian approach[23] describes the change in crystal population density with respect to time and space, as shown in Eq. (7).

$$\frac{\partial n}{\partial t} + \nabla.(v_e n) + \nabla.(v_i n) = B - D \quad (7)$$

where n is population density, v_e is external velocity vector, v_i is internal velocity vector, B is crystal birth rate, and D is crystal death rate. Assuming that a crystallizer has a uniform distribution of suspended solids, a size-independent growth rate, and negligible breakage and agglomeration, Eq. (7) can be simplified to a 1-D PBE as following when $L \neq 0$:

$$\frac{\partial n(t,L)}{\partial t} + G(t)\frac{\partial n(t,L)}{\partial L} = 0 \quad (8)$$

The boundary condition at $L = 0$ is

$$n(t,0) = \frac{B(t)}{G(t)}|_{L=0} \quad (9)$$

Hu et al.[24] proposed a methodology by converting Eq. (7) from a partial differential equation (PDE).to a set of ordinary differential equations (ODE). The method was applied in this study to solve Eq. (7) in MATLAB software. The single objective to minimize the sum of squares of the difference between the calculated and experimental solute concentration of all independent experiments is defined:

$$min\ \ F = \sum_{k=1}^{k}\sum_{i=1}^{N}\left(C_{i,k}^{exp} - C_{i,k}^{cal}\right)^2 \quad (10)$$

where N is the sample number and k is the experiment number. $C_{i,k}^{exp}$and $C_{i,k}^{cal}$ experimental and calculated solution concentration, respectively. In this work, the last step of SMPT process at different temperature was used to estimate the parameters of nucleation and growth rate of β-form imatinib mesylate. The single objective function was solved with the MATLAB optimization function, *fmincon*, based on the sequential quadratic programming (SQP) method.

The solute mass balance is

$$\frac{dC(t)}{dt} = -3k_v\rho_c\int_0^\infty G(t,L)\,n(t,L)L^2dL \quad (11)$$

where k_v is the crystal volume shape factor (k_v=1 for cube), and ρ_c is crystal density.

The average relative deviation (ARD) was introduced to evaluate the accuracy of estimation,

$$ARD = \frac{1}{N}\sum_{i=1}^{N}\left|\frac{C_{i,k}^{cal} - C_{i,k}^{exp}}{C_{i,k}^{exp}}\right| \quad (12)$$

4. Results and discussion

4.1 SEM, PXRD and DSC

α-form imatinib mesylate supplied by Apotex Pharmachem Inc, and β-form transformed from α-form in methanol at 293K were used for SEM, powder XRD, and DSC test.

Figure 3 shows the SEM images of the α-form and β-form imatinib mesylate. Obviously, the α-form is needle-shaped while β-form is cubic-shaped. According to the morphology, we can distinguish the crystal polymorph easily.

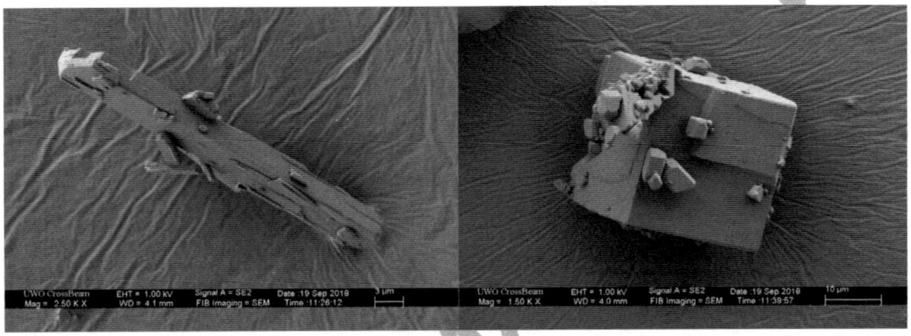

Figure 3. SEM images of imatinib mesylate. Left: α-form (3-μm), Right: β-form (10-μm)

Figure 4 displays the powder XRD patterns of α-form and β-form imatinib mesylate. The results are consistent with the α-form and β-form single-crystal XRD patterns, which were collected at 293K by Grillo et al.[18] and deposited in the Cambridge Crystallographic Data Center with publication numbers CCDC 821869 and 821871. α-form shows strong peaks at scattering angle (2θ) of 4.8°, 10.4°, 14.8°, 19.0°, 24.8° and 28.5°, whereas β-form has characteristic peaks at 9.7°, 13.9°, 18.2°, 20.0° 21.1° and 30.8°. The distinct differences on the XRPD spectrum can be used for polymorph identification of imatinib mesylate.

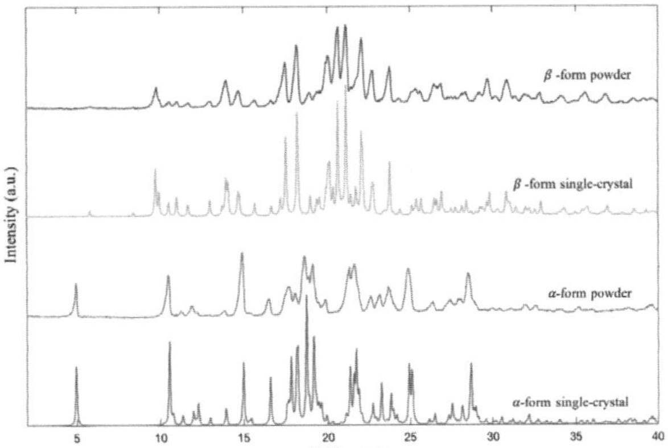

Figure 4. The XRD patterns of α-form and β-form imatinib mesylate

For both of α-form and β-form imatinib mesylate, there is only one endothermic peak observed in the DSC curves (Figure 5). The peaks are related to the melting event, which was confirmed with hot-stage microscopy and thermogravimetric analysis. The onset (start point) of the peak corresponds to the melting point, and the heat of fusion is equal to the integral of heat flow over time divided by sample mass. According to the heat of fusion rule [25], α-form and β-form imatinib mesylate are enantiotropic, as the α-form imatinib mesylate has higher melting point and lower heat of fusion than β-form.

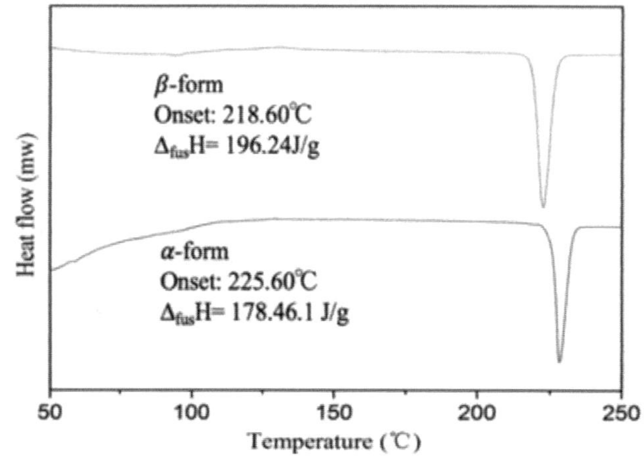

Figure 5. The DSC curves at a heating rate of 10 K/min of α-form and β-form imatinib mesylate

4.2 Solubility

The PXRD results of the residual solids demonstrated that no transformation or conversion occurred during the solubility experiments, except for the α-form in methanol. Figure 6 shows the solubility of α-form and β-form imatinib mesylate in methanol, 1-proponal and 2-proponal. As shown in Figure 6, the α-form has higher solubility than β-form in methanol,1-proponal and 2-proponal. The solubility order for both forms is methanol>1-propanol >2-propanol. The fitting parameters are listed in Table 1 and the results indicate that the modified Apelblat equation matches the solubility well with all R^2 greater than 0.99.

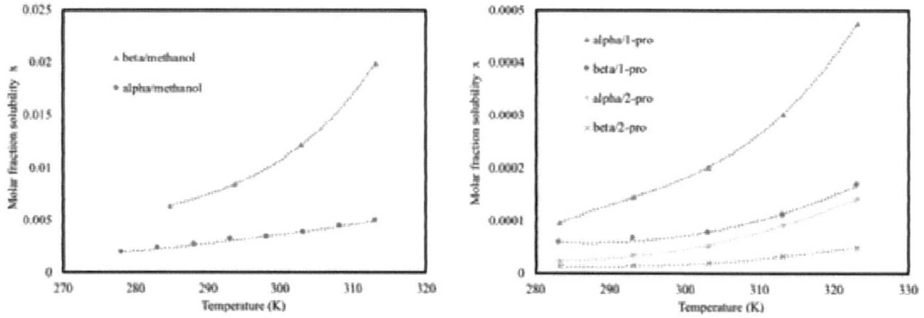

Figure 6. The molar fraction solubility of α-form and β-form imatinib mesylate in methanol, 1-proponal and 2-proponal. Dashed line: calculated curve with modified Apelblat equation

Table 1. Parameters of the modified Apelblat equation for imatinib mesylate in three solvents

		A	B	C	R^2
Methanol	α-form	-685.8	2.73E+04	103.5	1
	β-form	144.2	-8.61E+03	-21.24	0.9938
1-proponal	α-form	-279.2	9116	42.11	0.9983
	beta	-762.2	3.18E+04	113.4	0.9972
2-proponal	α-form	-455.2	1.66E+04	68.38	0.9982
	β-form	-933.4	3.89E+04	139	0.9949

As concluded above, α-form and β-form imatinib mesylate are enantiotropic. Figure 7 presents that the difference of Gibbs free energy between α-form and β-form imatinib mesylate increases with temperature. So, the transition temperature should be lower than 283.15K, which means that the β-form is always more stable than α-form above 283.15K.

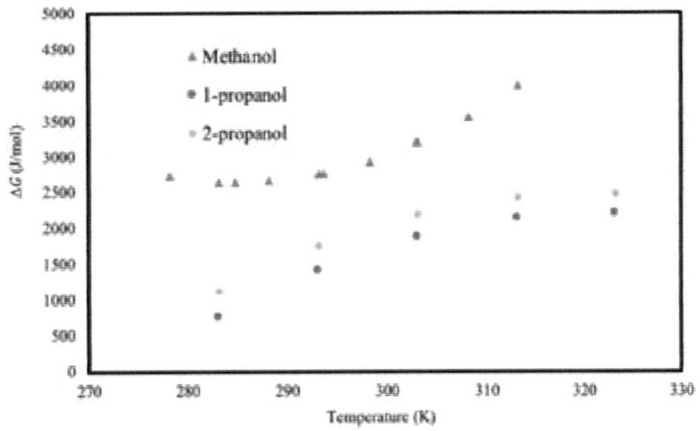

Figure 7. The difference of Gibbs free energy between α-form and β-form imatinib mesylate in methanol, 1-proponal and 2-proponal

4.3 Solvent-mediated polymorphic transformation (SMPT)

Through the solvent screening, we found that the SMPT from α-form to β-form imatinib mesylate could occur in methanol, ethanol, isopropanol, n-propanol, and 1-butanol. The transition time increased with the solvent in the order of methanol, ethanol, 1-propanol, 2-propanol and 1-butanol. Thus, methanol was chosen as the solvent to investigate the SMPT process at different temperatures by in-situ Raman spectroscopy and conductivity meter.

Due to the apparent fluctuations in the baseline, the relative heights of the Raman spectrum were used for quantitative analysis to reduce the effect produced by disturbance variables [26]. Here, the peaks at 1664 cm^{-1} and 1656 cm^{-1} were selected as the characteristic peaks of α-form and β-form imatinib mesylate, and the single point baseline was at 1640 cm^{-1}. Generally, the conductivity of a binary conducting solution is a function of temperature and solute concentration. For an isothermal process, conductivity increase in direct proportion to concentration, so the conductivity meter can be used for real-time measurement of the concentration.

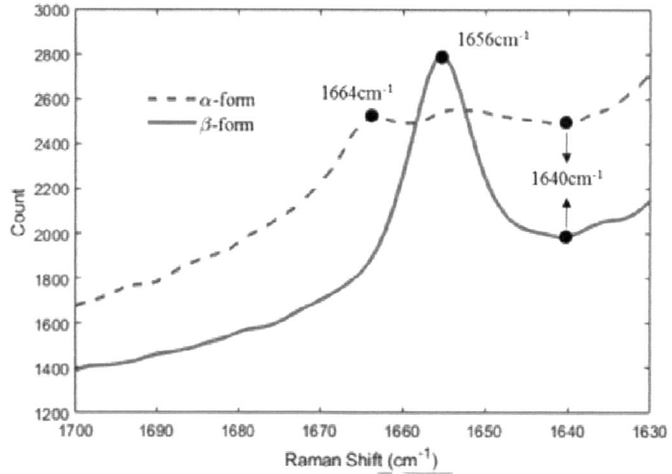

Figure 8. Raman spectra of α-form and β-form imatinib mesylate

Figure 9 displays how the concentration and solid composition change during the SMPT process in methanol at 293.15K. The process can be divided into four steps. At the beginning, the concentration increased rapidly as α-form dissolved, and the undissolved solid was pure α-form. The second step involves the nucleation of β-form. Meanwhile, the conductivity and the Raman intensity for both forms kept constant. The dissolution of α-form and growth of β-form happened during the third step, resulting in a decreased Raman intensity of α-form and an increased intensity of β-form. In the last step, the Raman intensity of α-form did not change, illustrating that all α-form crystals had been dissolved and the solid-phase were pure β-form. The nucleation and growth of β-form led to the drop of conductivity and a gradual increase of Raman intensity of β-form.

During the SMPT process, namely the third step, the concentration was close to the solubility of α-form, which indicated dissolution rate of α-form is much faster than growth rate of β-form. Therefore, it can be concluded that the SMPT from α-form to β-form imatinib mesylate is nucleation-growth controlled of β-form.

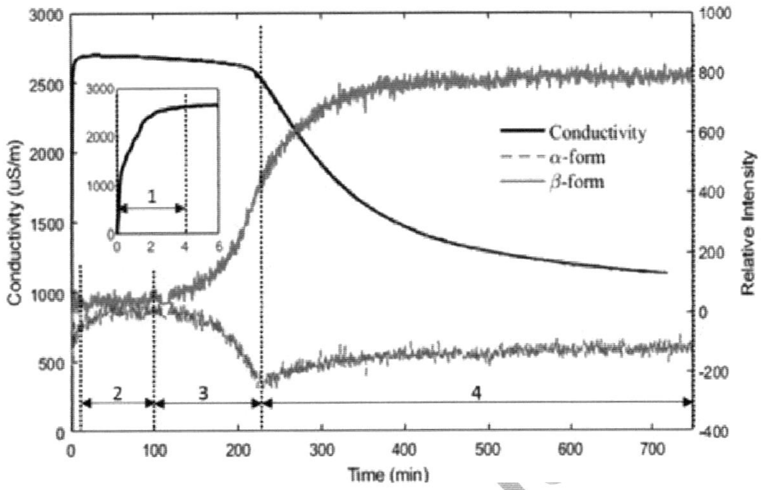

Figure 9. Solvent-mediated polymorphic transformation from α-form to β-form imatinib mesylate at 293k

4.4 Kinetic parameter estimation

As mentioned in section 4.3, all α-form crystals had dissolved and only β-form crystals existed in the solid-phase in the last step (step 4) of the solvent mediated polymorphic transformation. Therefore, the last step of the SMPT process can be regarded as a seeded isothermal batch crystallization. During the process, the solute concentration was measured by the conductivity meter and the suspension density was calculated by solute mass balance. Hence, the nucleation and growth can be estimated through minimizing the difference between the calculated and experimental solute concentration measured in all independent experiments. The estimated parameters are listed in Table 2.

The nucleation order with respect to supersaturation (b) is much higher than the nucleation order with respect to suspension density (j), indicating that supersaturation affects the secondary nucleation rate more significantly than slurry density does. $b > g$ shows that the nucleation rate has a stronger relationship with supersaturation than the growth rate, hence the mean crystal size will decrease with supersaturation. These estimated kinetic parameters would be very useful for optimizing and controlling the cooling crystallization process of β-form imatinib mesylate.

Table 2 Kinetic parameter estimated from optimization

Kinetic parameter	Estimated value	Units
k_b	9.22×10^{11}	$[(\#/m^3 \cdot s)/(kg/kgsolvent)^j]$
j	0.754	$[-]$
b	3.12	$[-]$
k_g	9.12×10^{-11}	$[m/s]$
g	1.23	$[-]$

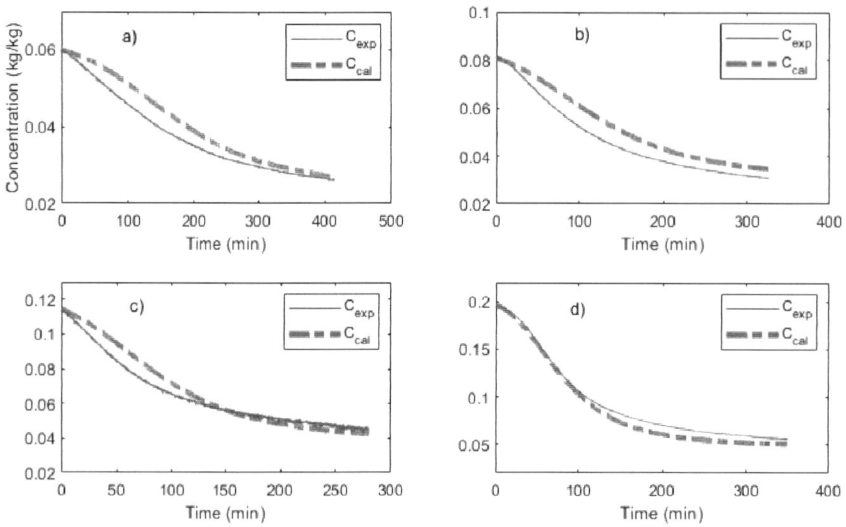

Figure 10. Calculated and experimental solution concentration
a)284.75 K, b)293.65 K, c)302.95 K and d) 313.15K

The calculated and experimental solute concentration at different temperatures are plotted in Figure 10. The ARD of the four experiments are 7.6%, 12.19%, 6.44 %, and 8.62%, respectively, which indicates the empirical equations (Eq. 4, 5, 7) used in this work are suitable for β-form imatinib mesylate. The PBE combined with agglomeration and breakage would give better estimation performance with considerable increase in computation time.

5. Conclusions

In this work, the α-form and β-form imatinib mesylate, were characterized by powder X-ray diffraction, scanning electron microscopy and differential scanning calorimetry firstly. Then, we measured their solubility in methanol,1-propanol, and 2-propanol with UV-Vis. The modified Apelblat equation was used to fit the solubility data. According to the DSC and solubility results, we found that α-form and β-form imatinib mesylate are enantiotropically related and the transition temperature is lower than 283.15K. Above 283.15K, α-form always has a higher solubility and lower stability than β-form in methanol, 1-proponal and 2-proponal. The solubility order in selected solvents is methanol>1-propanol>2-propanol. After solvent screening, the solvent-mediated polymorphic transformation (SMPT) from α-form to β-form imatinib mesylate was investigated in methanol. The solute concentration was monitored by a conductivity meter, while the solid phase was monitored by Raman spectroscopy. As the concentration was close to the solubility of metastable form during the SMPT process, it can be concluded that SMPT process of imatinib mesylate from α-form to β-form in methanol is controlled by nucleation and growth of β-form. Finally, the nucleation and grow rate of

β-form of imatinib mesylate were estimated by minimizing the difference between the calculated and experimental solution concentration with MATLAB optimization function.

6. Nomenclature

a	solute activity	-
b	nucleation order on supersaturation	-
B	crystal birth rate	$\#/m^3 \cdot s$
B_s	secondary nucleation rate	$\#/m^3 \cdot s$
C	solution concentration	kg solute /kg solvent
C^*	equilibrium concentration--solubility	kg solute /kg solvent
C^{cal}	calculated solution concentration	kg solute /kg solvent
C^{exp}	experimental solution concentration	kg solute /kg solvent
D	crystal death rate	$\#/m^3 \cdot s$
F	objective function	
g	growth order on supersaturation	-
G	crystal growth rate	m/s
ΔG	difference of Gibbs free energy	kJ/mol
j	nucleation order on suspension density	-
k	number of experiments included in the objective function F	-
k_g	growth rate coefficient	m/s
k_{sn}	secondary nucleation rate coefficient	$(\#/m^3 \cdot s)/(\text{kg/kg solvent})^j$
k_v	the crystal volume shape factor	
L	characteristic size of crystal	m
m	mass of solute	kg
m_s	mass of solvent	kg
M	molecular mass of solute	kg/mol
M_s	molecular mass of solvent	kg/mol
M_t	suspension density	kg solids /kg solvent
n	population density	$\#/m^3 \cdot m$
N	number of samples in each experiment included in the objective function F	-
R	universal gas constant	kJ/mol·K
S	relative supersaturation	-
t	time	s
x	molar fraction of solute	-
ρ_c	crystal density	kg/m^3

7. Reference

[1] G. P. Stahly, "Diversity in Single- and Multiple-Component Crystals. The Search for and Prevalence of Polymorphs and Cocrystals," *Cryst. Growth Des.*, vol. 7, no. 6, pp. 1007–1026, Jun. 2007.

[2] A. Myerson, *Handbook of Industrial Crystallization*. Butterworth-Heinemann, 2002.

[3] A. Y. Lee, D. Erdemir, and A. S. Myerson, "Crystal Polymorphism in Chemical Process Development," *Annu. Rev. Chem. Biomol. Eng.*, vol. 2, no. 1, pp. 259–280, 2011.

[4] J. W. Mullin, *Crystallization*. Elsevier, 2001.

[5] M. A. O'Mahony, A. Maher, D. M. Croker, Å. C. Rasmuson, and B. K. Hodnett, "Examining Solution and Solid State Composition for the Solution-Mediated Polymorphic Transformation of Carbamazepine and Piracetam," *Cryst. Growth Des.*, vol. 12, no. 4, pp. 1925–1932, Apr. 2012.

[6] M. Trifkovic, S. Rohani, and M. Sheikhzadeh, "Kinetics Estimation and Polymorphic Transformation Modeling of Buspirone Hydrochloride," *J. Cryst. Process Technol.*, vol. 02, no. 02, pp. 31–43, 2012.

[7] G. Morris, G. Power, S. Ferguson, M. Barrett, G. Hou, and B. Glennon, "Estimation of Nucleation and Growth Kinetics of Benzoic Acid by Population Balance Modeling of a Continuous Cooling Mixed Suspension, Mixed Product Removal Crystallizer," *Org. Process Res. Dev.*, vol. 19, no. 12, pp. 1891–1902, Dec. 2015.

[8] B. M. H. Al-Hadiya, A. H. H. Bakheit, and A. A. Abd-Elgalil, "Chapter Six - Imatinib Mesylate," in *Profiles of Drug Substances, Excipients and Related Methodology*, vol. 39, H. G. Brittain, Ed. Academic Press, 2014, pp. 265–297.

[9] J. Zimmermann, B. Sutter, and H. Buerger, "Crystal Modification of a N-Phenyl-2-Pyrimidineamine Derivative, Processes for Its Manufacture and Its Use," WO9903854 (A1), 28-Jan-1999.

[10] R. B. Parthasaradhi, R. K. Rathnakar, R. R. Raji, R. D. Muralidhara, C. R. K. Subash, and H. D. Limited, *Novel polymorphs of imatinib mesylate*. 2003.

[11] K. Amala, R. T. Srinivasa, K. S. B. R. Adibhatla, S. Rachakonda, C. N. Venkaiah, and K. Podili, "Novel polymorphic form of imatinib mesylate and a process for its preparation," WO2005077933A1, 25-Aug-2005.

[12] S. L. Pathi, R. Puppala, R. N. Kankan, D. R. Rao, C. Limited, and C. Wain, *Stable crystal form of imatinib mesylate and process for the preparation thereof*. 2005.

[13] H. Patel, R. Jani, and R. Thennati, "Imatinib Mesylate Crystal Form and Process for Preparation Thereof," WO/2006/048890, 12-May-2006.

[14] A. K. Kompella, K. S. B. R. Adibhatla, K. Podili, and C. N. Venkaiah, "Polymorphic forms of imatinib mesylate," WO2006054314A1, 26-May-2006.

[15] Mutz M., "Delta and Epsilon Crystal Forms of Imatinib Mesylate," WO/2007/023182, 02-Mar-2007.

[16] M. Mutz, "F,g,h,i and K Crystal Forms of Imatinib Mesylate," WO/2007/059963, 01-Jun-2007.

[17] J. Zimmermann, B. Sutter, and H. M. Bürger, "Crystal modification of a N-phenyl-2-pyrimidineamine derivative, processes for its manufacture and its use," US6894051B1, 17-May-2005.
[18] D. Grillo, G. Polla, and D. Vega, "Conformational Polymorphism on Imatinib Mesylate: Grinding Effects," *J. Pharm. Sci.*, vol. 101, no. 2, pp. 541–551, Feb. 2012.
[19] M. Veverka, P. Šimon, J. Lokaj, and E. Veverková, "Crystal habit modifications of imatinib mesylate under various precipitation conditions," *Monatshefte Für Chem. - Chem. Mon.*, vol. 143, no. 1, pp. 65–71, Jan. 2012.
[20] A. Srivastava, B. D. Joshi, P. Tandon, A. P. Ayala, A. K. Bansal, and D. Grillo, "Study of polymorphism in imatinib mesylate: A quantum chemical approach using electronic and vibrational spectra," *Spectrochim. Acta. A. Mol. Biomol. Spectrosc.*, vol. 103, pp. 325–332, Feb. 2013.
[21] E. Bellur Atici and B. Karlığa, "Quantitative determination of two polymorphic forms of imatinib mesylate in a drug substance and tablet formulation by X-ray powder diffraction, differential scanning calorimetry and attenuated total reflectance Fourier transform infrared spectroscopy," *J. Pharm. Biomed. Anal.*, vol. 114, pp. 330–340, Oct. 2015.
[22] A. Grunenberg, J.-O. Henck, and H. W. Siesler, "Theoretical derivation and practical application of energy/temperature diagrams as an instrument in preformulation studies of polymorphic drug substances," *Int. J. Pharm.*, vol. 129, no. 1, pp. 147–158, Mar. 1996.
[23] A. Ranodolph, *Theory of Particulate Processes: Analysis and Techniques of Continuous Crystallization*. Elsevier, 2012.
[24] Q. Hu, S. Rohani, and A. Jutan, "New numerical method for solving the dynamic population balance equations," *AIChE J.*, vol. 51, no. 11, pp. 3000–3006, Nov. 2005.
[25] A. Burger and R. Ramberger, "On the polymorphism of pharmaceuticals and other molecular crystals. I," *Microchim. Acta*, vol. 72, no. 3–4, pp. 259–271, May 1979.
[26] A. Caillet, F. Puel, and G. Fevotte, "In-line monitoring of partial and overall solid concentration during solvent-mediated phase transition using Raman spectroscopy," *Int. J. Pharm.*, vol. 307, no. 2, pp. 201–208, Jan. 2006.

8. Acknowledgment

Thanks to Apotex Pharmachem Inc. (Brantford, ON, Canada) for supplying the raw imatinib mesylate, Dr. Tim Goldhawk from Western University Nanfabrication Facility for help with SEM imaging, and Chenxi Li for her assistance in the solubility measurement. The authors acknowledge the financial support provided by the Natural Science and Engineering Research Council (NSERC) of Canada and the China Scholarship Council.

Highlights

- Two forms of imatinib mesylate, α and β, were thoroughly characterized.
- Solubility of the two forms was measured in methanol, 1-propanol, and 2-propanol.
- Developed kinetics of polymorphic transformation from α-form to β-form in methanol.
- The nucleation and grow rates of β-form of imatinib mesylate were estimated.

532399 Polymorphism of D-Mannitol: Selective Nucleation and Crystal Growth Mechanism

Abstract:

Polymorphism of D-mannitol: Selective Nucleation and Crystal Growth Mechanism

Weiyi Su, Chunli Li, Honghai Wang, Jing Fang

School of Chemical Engineering, Hebei University of Technology, Tianjin 300130, PR China

Polymorphism has gained more and more attention in the field of pharmaceuticals[1]. D-mannitol is a natural hexahydric alditol which has been widely used as a nephropathy treatment medicine and also an excipient in the formulation of various tablets and granulated powders[3]. It has been reported that there are three anhydrous polymorphs[4-6] and a hemihydrate[7] of mannitol. However the nomenclature of these polymorphs is often diverse in different literatures[8, 9] in despite of similar unit cell parameters. Therefore a review of mannitol polymorphism is initially presented along with the crystal structure parameters. Then the influence of supersaturation on the polymorphic nucleation of mannitol is investigated in aqueous solutions and ethanol-water mixtures. Moreover, a method reported by Kuldipkumar[10] was extended to determine the crystal growth of metastable ¦Ä mannitol, which was found to follow the 2D nucleation-mediated mechanism.

1. Polymorphs of D-mannitol

It has been confirmed that there are only three pure anhydrous polymorphs of mannitol even though numerous names have been used[6, 8]. Hereby, for clarity and to avoid any confusion, the unite cell parameters of mannitol polymorphs in various literatures are summarized in Table 1. It is clear in Table 1 that the first six ¦*Â* references have a similar structure, which is named ¦*Â* in our work. While focusing on the polymorphs from the 7th to 9th in Table 1, the parameters were nearly the same disregarding what they were called in the original literature. So this form is referred to as ¦*Á* mannitol. Finally, the last three items are significantly distinct from the others, and they are referred to as the ¦*Ä* form of mannitol in this paper.

Table 1. The review of unit cell parameters and nomenclature of D-mannitol anhydrous polymorphs in literatures

Mark	Literature	Space group	a (Å)	b (Å)	c (Å)	β (°)	V ($Å^3$)	Z	D_x ($kg \cdot m^{-3}$)
β	Marwick, 1931[11]	$P2_12_12_1$	8.65	16.90	5.56	--	203*	4	1.488
β	Rye, 1952ª[12]	$P2_12_12_1$	8.65	16.90	5.56	--	203*	4	1.488
β	Berman, 1968[13]	$P2_12_12_1$	8.672	16.875	5.560	--	--	4	1.487
β	Walter-Le vy, 1968[4]	$P2_12_12_1$	8.672	16.89	5.549	--	--	4	1.489
β	Kaminsky, 1997[14]	$P2_12_12_1$	8.694(7)	16.902(8)	5.549 (6)	--	815	4	1.490(2)
β	R. Fronczek, 2003[5]	$P2_12_12_1$	5.5381 (10)	8.580(2)	16.795 (5)	--	789.0 (3)	4	1.516
α	Walter-Le vy, 1968	$P2_12_12_1$	8.839	18.778	4.896	--	--	4	1.470
K	Kim, 1968[15]	$P2_12_12_1$	8.942	18.798	4.893	--	--	4	1.471
α	R. Fronczek, 2003	$P2_12_12_1$	4.8653 (10)	8.873(2)	18.739 (3)	--	809.0 (3)	4	1.496
δ	Walter-Le vy, 1968	$P2_1$ or $P2_1/m$	5.095	18.254	4.919	118.3 6	--	2	1.501
δ	R. Fronczek, 2003	$P2_1$	4.899(2)	18.268(6)	5.043 (2)	118.3 9(2)	397.0 (3)	2	1.524
δ	Botez, 2003[16]	$P2_1$	5.0894 1(5)	18.2504 (2)	4.9170 2(5)	118.3 03	402.1 03	2	1.377

* volume per molecule, and others are volume per unit cell.

¡è the author also mentioned the ¦$\hat{A}$ form they obtained was in agreement with Marwick¡¯s result.

2. Selective Nucleation of D-mannitol Polymorphs

The influence of the supersaturation on mannitol polymorphic nucleation is investigated in cooling crystallization, and the results are displayed in Figure 1 with a comparison to the solubility curves. As illustrated in Figure 1, it is quite interesting to find that the nucleation sequence is ¦Á, ¦Ä, ¦Â as the decrease of concentration while the solubility order is always ¦Ä > ¦Á > ¦Â along with temperature.

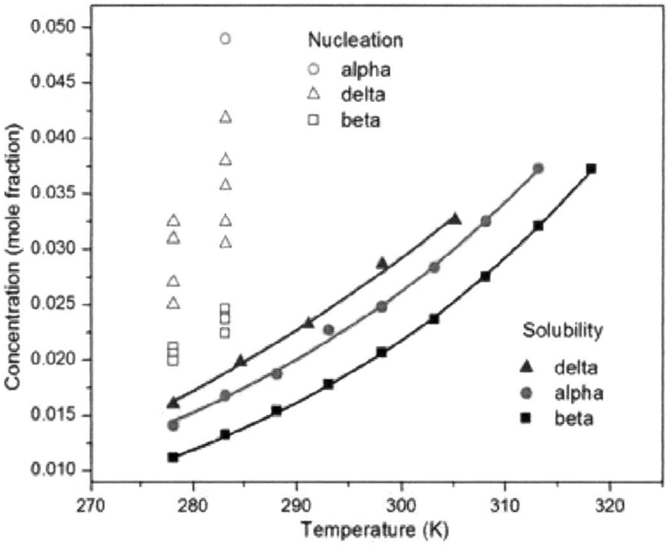

Figure 1. Polymorphic nucleation results of mannitol (open symbols, labeled as ¡°Nucleation¡±) at different initial concentrations and temperatures shown in aqueous solution with the solubility curves of the three polymorphs (solid symbols, labeled as ¡°Solubility¡±).

To clarify the nucleation zone for different polymorphs, the nucleation parameters (DG^*, r^*, and n^*) at various initial concentrations at 10 ¡ãC are obtained as shown in Table 2 based on the measured induction time and supersaturation data. It is apparent that when the initial concentration is higher than 0.0305, the critical excess free energy forming the ¦Ä form nuclei is lower than that of the ¦Â form. It indicates that the kinetic properties play a more important role than the thermodynamic properties. While the concentration is lower than 0.0246, the supersaturation is the dominant factor, that is why ¦Â mannitol is easier to nucleate in this region.

Table 2. Nucleation properties of two mannitol polymorphs at 10 ¡ãC, the subscripts ¦Ä and ¦Â refer to the polymorph ¦Ä and ¦Â forms respectively while *s* stands for the supersaturation

initial mole concentration	s_δ	s_β	ΔG^*_δ (10^3 J/mol)	ΔG^*_β (10^3 J/mol)	r^*_δ (10^{-10} m)	r^*_β (10^{-10} m)	n^*_δ	n^*_β
0.0418	2.22	3.16	*1.37*	3.12	*1.60*	2.05	*2*	3
0.0380	2.02	2.87	*1.76*	3.82	*1.81*	2.23	*3*	4
0.0357	1.90	2.70	*2.12*	4.32	*1.99*	2.37	*3*	4
0.0325	1.73	2.46	*2.92*	5.27	*2.33*	2.62	*5*	5
0.0305	1.62	2.31	*3.73*	6.09	*2.67*	2.82	*7*	7
0.0246	1.31	1.86	12.07	*11.02*	4.74	*3.79*	39	*16*
0.0239	1.27	1.81	15.31	*12.17*	5.34	*3.98*	55	*18*
0.0237	1.26	1.79	16.44	*12.52*	5.54	*4.04*	61	*19*
0.0234	1.24	1.77	18.43	*13.09*	5.86	*4.13*	73	*20*

The values highlighted in italics in Table 2 show lower critical excess free energy values, smaller radius and molecular number of the critical nucleus for either the ¦*Ä* or ¦*Â* form at certain concentrations, leading to the selective nucleation of this polymorph at those conditions. In this way, controlling to get the desired polymorphs can be fulfilled.

A reverse anti-solvent crystallization of adding saturated mannitol solution into cold ethanol is also performed. The nucleation results at -10¡æ are shown in Figure 2. It is similar to Figure 1 that high concentration favors the metastable ¦Â form and the least stable ¦Ä mannitol can only be obtained at relatively low concentration.

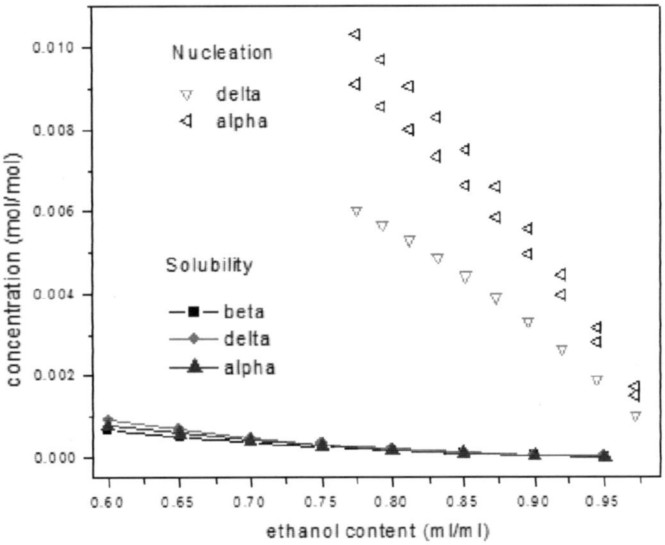

Figure 2. Polymorphic nucleation results of mannitol (open symbols, labeled as ¡°Nucleation¡±) at different concentrations in ethanol-water mixture shown with the solubility curves of the three polymorphs (solid symbols, labeled as ¡°Solubility¡±).

3. Crystal growth

After nucleation, the nuclei should grow to be crystals by consuming supersaturation. But in bulk aqueous solution, the solvent-mediated polymorphic transformation is quite obvious for mannitol metastable forms, which makes it even more difficult to clarify the growth mechanism. In order to enlarge the acquaintance to the growth of ¦Ä mannitol, the induction time are used to provide the growth mechanism following a reported method by Kuldiplumar et al[10]. In this method, a function F(s) is defined for normal, spiral, and volume diffusion-controlled growth mechanisms:

$$F(s) \equiv \ln\left\{s^{\frac{1}{n}}[f(s)]^{\frac{n-1}{n}} t_{ind}\right\} = \ln A_p + \frac{B}{n\ln^2 s} \quad (1)$$

Similarly for 2D nucleation-mediated growth, the function is defined as:

$$F(s) \equiv \ln\left[(s-1)^{\frac{2(n-1)}{3n}} s^{\frac{n+2}{3n}} t_{ind}\right] = \ln A_u + (n-1)\frac{B_{2D}}{3n\ln s} + \frac{B}{n\ln^2 s} \quad (2)$$

According to equation 1 and 2, the calculated F(s) against $1/(\ln^2 s)$ for the normal, spiral, and volume diffusion-controlled growth mechanism, as well as F(s) against 1/(lns) following the 2D nucleation-mediated growth mechanism are shown in Figure 3.

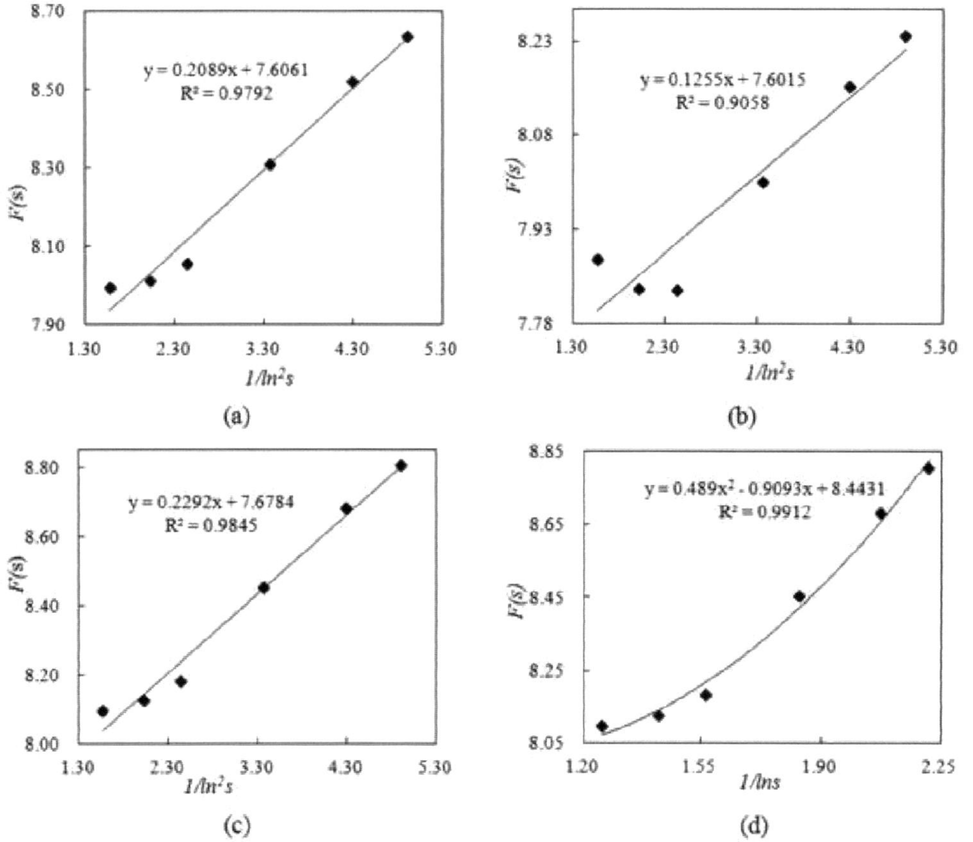

Figure 3. Plots of *F(s)* versus $1/ln^2s$ for ¦Ä form of mannitol under the normal growth mechanism (a), the spiral growth mechanism (b), the volume diffusion controlled growth mechanism(c), and *F(s)* versus *1/lns* under the 2D nucleation-mediated growth mechanism (d).

It is clear in Figure 3 that the largest R^2 of 0.9912 appears when the data are fitted to the 2D nucleation-mediated mechanism. Therefore it is reasonable to believe that the growth mechanism of the metastable ¦Ä form of mannitol in aqueous solution should be 2D nucleation-mediated.

Reference

[1] J. Bernstein, Polymorphism in molecular crystals, Oxford University Press, USA, 2002.

[2] S. Datta, D.J.W. Grant, Nat. Rev. Drug Discov. 3 (2004) 42-57.

[3] B. O'Sullivan, The application of in situ analysis to crystallization process development, University College Dublin, Dublin, 2005, p. 177.

[4] L. Walter-Levy, C. R. Acad. Sc. Paris, Ser. C 267 (1968) 1779.

[5] F.R. Fronczek, et al., Sect. C: Cryst. Struct. Commun. 59 (2003) o567-o570.

[6] A. Burger, etal., J. Pharm. Sci. 89 (2000) 457-468.

[7] C. Nunes, et al., J. Pharm. Sci. 93 (2004) 2800-2809.

[8] J. Cornel, et al., Industrial and Engineering Chemistry Research 49 (2010) 5854-5862.

[9] W Su, et al., J. Raman Spectrosc. 46 (2015) 1150-1156.

[10] A. Kuldipkumar, et al., Cryst. Growth Des. 7 (2007) 234-242.

[11] T.C. Marwick, Proc. R. Soc. London, Ser. A 131 (1931) 621-633.

[12] A. Rye, H. Sorum, Acta Chem. Scand 6 (1952) 1128-1129.

[13] H.M. Berman, et al., Acta Crystallogr., Sect. B: Struct. Sci B24 (1968) 442-449.

[14] W. Kaminsky, Z. Kristallogr. 212 (1997) 283-296.

[15] H.S. Kim, et al., Acta Crystallogr., Sect. B: Struct. Sci B24 (1968) 1449-1455.

[16] C.E. Botez, et al., Powder Diffr. 18 (2003) 214-218.

525689 Can the Solvation Manner Leads to a Nucleation Diversity for Polymorphic System? the Case of D-Mannitol

Abstract:

Can the solvation manner leads to a nucleation diversity for polymorphic system? The case of D-mannitol

There has been much interest in the role of solution chemistry plays in determining the polymorphic nucleation process, especially for the field requires a precise control and even design of the final crystalline product, such as pharmaceutical and fine chemical industries. In this study we focused on a quite challenging subject: D-mannitol (D-man). An interesting solvent-dependent nucleation behavior was found by us: in pure methanol the alpha form can be easily harvested regardless of the supersaturation and temperature, in water the alpha can also be sporadically observed but the solution mainly crystallized beta form, whereas for mixtures of water and methanol, interestingly, the pure delta form was exclusively yielded instead of either alpha or beta. We wondered the reason for this solvent dependent crystallization outcomes, especially the unexpected appearance of delta form, however, the nearly identical molecular conformations and extensive hydrogen bonds throughout the crystal structures making D-man a system difficult to resolve [1], as shown in Fig. 1.

Two points are mainly concerned in this study : 1. exploring is there any structural feature could differentiate three anhydrous polymorphs of D-man properly; 2. based on such difference, whether a ¡°link¡± could be detected between the nature of intermolecular interactions present in solution and the crystallization outcomes, especially the unexpected appearance of delta form.

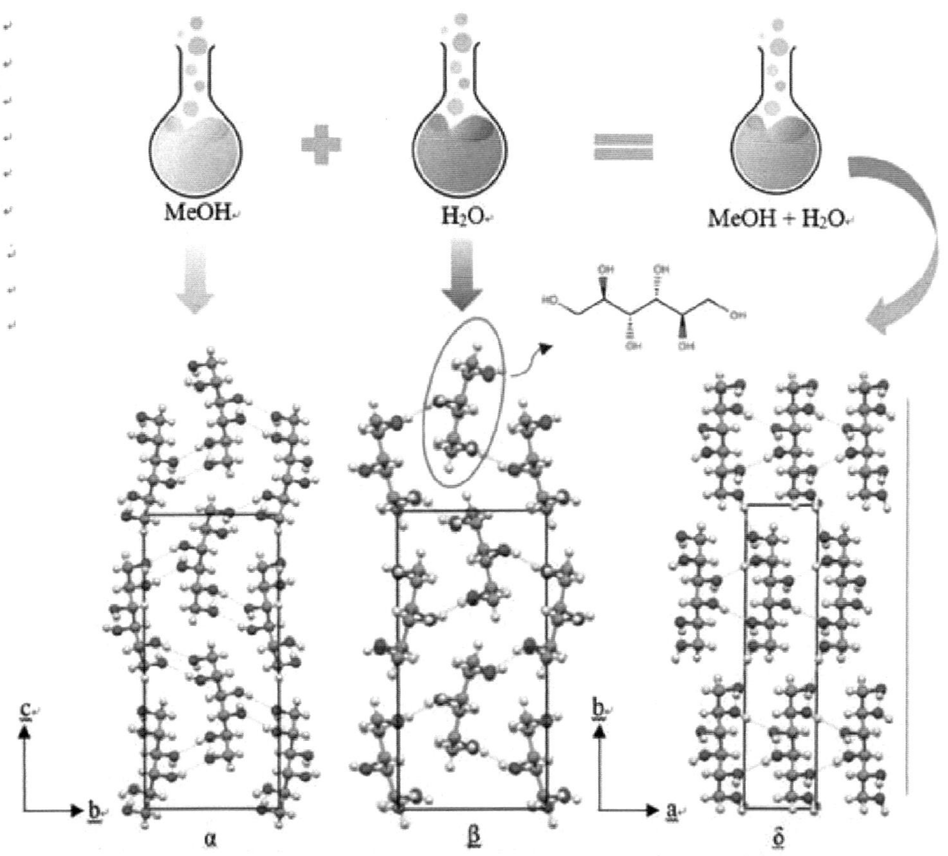

Fig. 1 Molecular structure and unit cell packings of three anhydrous D-man polymorphs (the dimers are overlapped in this view)

Though a Hirshfield Surface analysis [2], we found two important characteristics can be used to differentiate three polymorphs, namely the dimerization effect and terminal packing intensity, as illustrated in Fig.2 and TABLE 1.

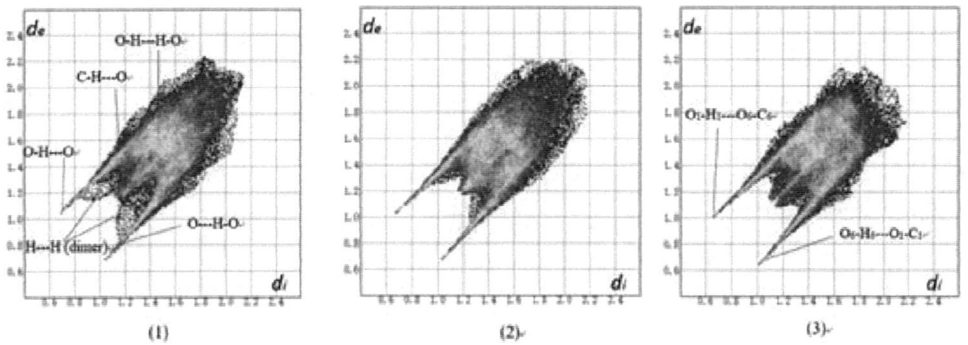

Fig. 2 2-D Finger plots for three D-man anhydrous polymorphs: α (1), β (2) and δ (3).

TABLE 1. Packing Characteristics of Three Anhydrous Polymorphs

	α	β	δ
Dimerization effect	high	medium	low
Terminal packing	medium	compact	loose

In light of above comparison, the nature of interactions in different D-man solutions was investigated in terms of the radial distribution function (RDF) calculation. The solvation effect on D-man¡¯s body parts (-O3H3) and terminal parts (-O1H1) were both investigated. The RDF results together with snapshots for the molecular distribution within different solutions are shown below:

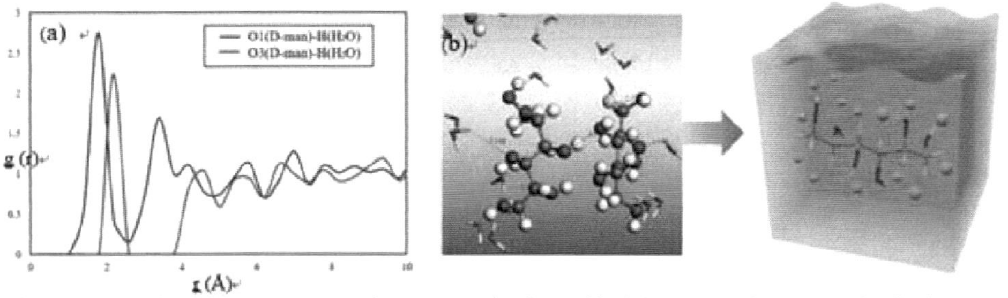

Fig. 4 (a) RDFs between oxygen atoms in D-man molecules and hydrogen atoms in water molecules in aqueous solution; (b) Snapshot of D-man molecules solvated in water.

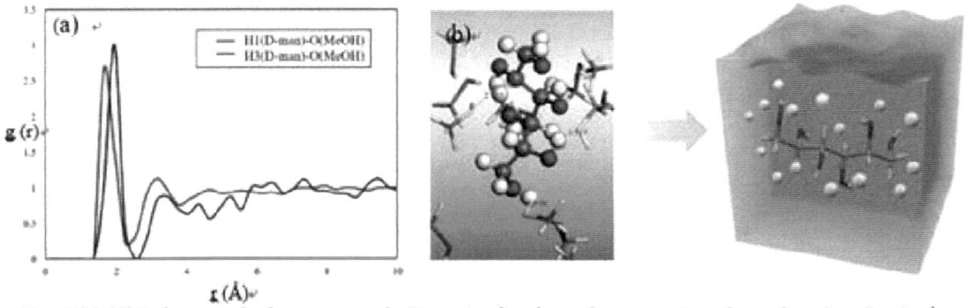

Fig. 3 (a) RDFs between hydrogen atoms in D-man molecules and oxygen atoms in methanol molecules in methanol solution; (b) Snapshot of D-man molecules solvated in methanol.

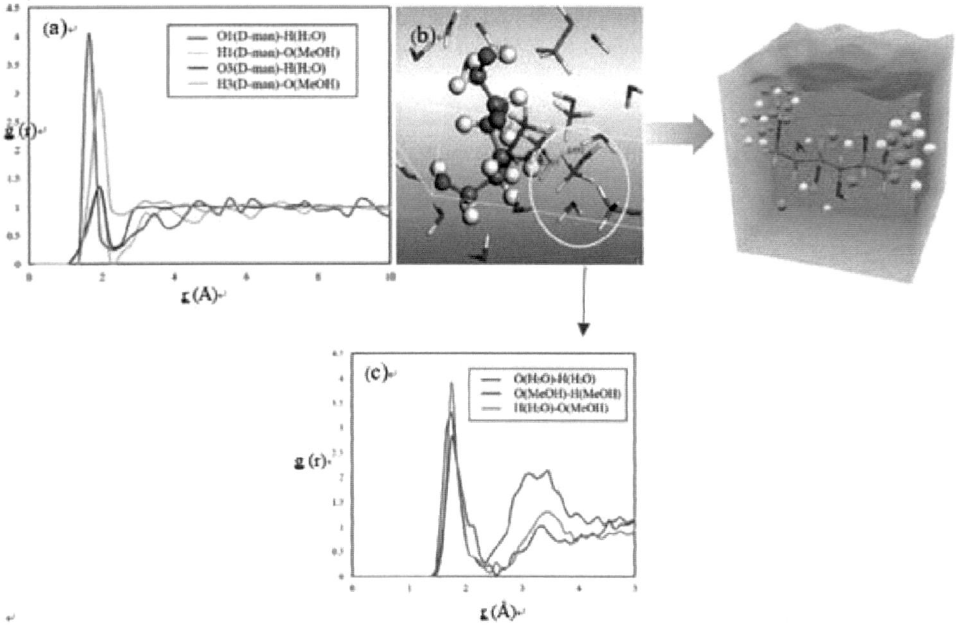

Fig. 5 (a) RDFs between atoms of D-man molecules and atoms in solvent molecules in binary solvent solution; (b) Snapshot of D-man molecules solvated in binary solvent; (c) RDFs between solvent molecules in binary solvent solution

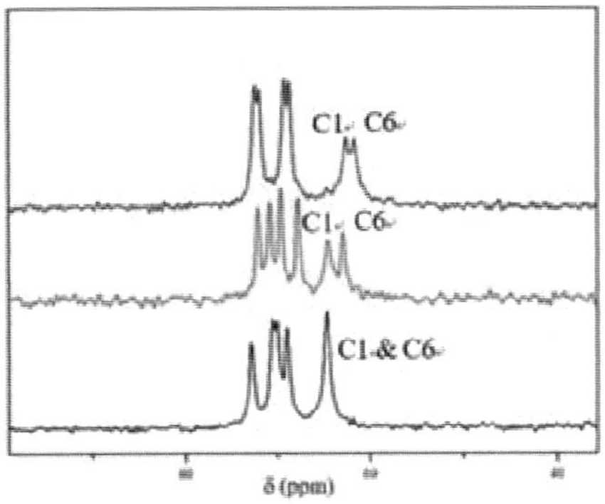

Fig. 6 ^{13}C solid-state NMR spectrum of three D-man anhydrous polymorphs.

From above results, no link was found between the solvation effect on D-man¡¯s body part and the dimerization effect in three forms, indicating that the dimerization strength cannot be a probe for the nucleation behavior explanation. While a good correlation was observed for terminal packing characteristics: in pure methanol and water, the solvation intensities at the end part are 3 and 2.7, respectively, while in the mix solvent, much higher solvating power on terminals was observed, with the terminal-water in 4 for g(r) and terminal-methanol in 3. A double-layer solvent shell with water molecules in the inside track while methanol in outsides was formed, considering the higher probability of methanol-water assemblies than the homologous solvent aggregates (as illustrated in Fig. 5(c)), we have reason to believe that those outside methanol may be bonded to the inside water shell to some degree. In the following process of solute molecule assembly, this distribution will serve as a shield for solute aggregation, create limited space for solute-solute interactions, thus facilitate a head-to-tail connection due to the relatively small volume of both ends, and meanwhile bring about a lower local packing compactness in final structure, consequently leads to the appearance of delta form. On the other hand, the light solvation degree in other two solutions can provide favorable condition for more solute molecules to occupy the area adjacent to terminals. As lots of solutes accumulate around the ends, one exposed hydroxyl group at the molecular terminal may be likely to bond with others in more orientations, contribute a lower selectivity for terminal groups, which could be supported by the separate peaks of C1 and C6 (two terminal carbon atoms) obtained from solid state NMR characterization, as shown in Fig. 6. The broader space in solution together with the optional bonding orientation finally stabilized an approximate herringbone linkage network and a dense packing motif exist in alpha and beta.

Thus we have found that the packing characteristics in the lattice of D-man should be a direct reflection of the solvation state of D-man in solution, the solvating power of different solvents on the both ends of D-man can provide distinct selectivity for terminal ¨COH groups to interact with each other. To verify the assumption, same experiments were carried out in the acetone-water system. The results showed that in water-acetone solvent, the delta form was not crystallized exclusively, instead a mixture of alpha and delta was usually yield. This phenomenon to a certain extent supports our conclusion, as acetone possesses lower HB donor and acceptor abilities than methanol, cannot construct a firm double-solvent-shell as observed in methanol-water, as a result a local environment being suitable for both alpha and delta was created.

This study, for the first time, investigated the solvent-dependent nucleation behavior of D-man which is one of the most important excipients in pharmaceutical and food industry [3, 4], and revealed the significant roles played by solvation manners in determining polymorphic outcomes of D-man. In this regard, the use of local packing compactness as a probe might constitute a potential way could be extended to other systems for polymorphic nucleation research.

Reference:

[1] Carespacheco, M. G., Vacamedina, G., Calvet, R., Espitalier, F., Letourneau, J. J., & Rouilly, International Journal of Pharmaceutics, 475 (2014) 69.

[2] M.A. Spackman, J.J. McKinnon, CrystEngComm. 4 (2002) 378.

[3] Wagner, C.M., Pein, M., Breitkreutz, Pow. Tech. 270 (2015) 470.

[4] Rowe, R.C., Sheskey, P.J.S., Owen, S.C., 2006. Handbook of pharmaceutical excipients, 449.

539029 Enzyme-Mimetic Luminescent Luminescent Nanoparticles As Hydrogen Peroxide Biosensors

Abstract:

Hydrogen peroxide (H_2O_2) is an abundant molecule associated with biological processes and reacts with natural enzymes, such as catalase. Thus, H_2O_2 quantification constitutes a powerful tool for detection of disease biomarkers linked to enzyme-based assays such as the plasmonic ELISA. However, the optical H_2O_2 biosensing without organic-dyes in biological media and at low, submicromolar, concentrations has yet to be achieved. The target of this work is to design biomimetic artificial enzymes based on antioxidant CeO_2 nanoparticles that become luminescent upon their Eu^{3+} doping [1,2]. CeO_2 nanoparticles have received a lot of attention recently due to their antioxidant enzyme-like (nanozyme) properties.

Here, europium-doped cerium oxide (CeO_2:Eu^{3+}) nanoparticles with well-controlled size (d = 4 – 16 nm) are prepared by flame aerosol technology and characterized in regards to H_2O_2 sensor response in physiologically-relevant solutions. Furthermore, we combine luminescent enzyme-mimetic CeO_2:Eu^{3+} nanoparticles with optically stable Y_2O_3:Tb^{3+} nanophosphors [1,2] that serve as concentration-independent reference value in the same nanoaggregates. Finally, the luminescence excitation is enhanced by the addition of Bi^{3+} within the same crystal matrix that improves its excitation in the near-UV wavelength region.

The developed biosensors are coupled to enzyme-based assays that consume or generate H_2O_2 aiming the detection of other important bioanalytes, such as alcohol and glucose, using alcohol and glucose oxidase enzymes, respectively [3]. Upon the addition of a luminescent material with no H_2O_2 sensitivity, ratiometric sensors can be made for the facile and fast H_2O_2 detection ***in vitro***. The sensing performance of the nanoparticles was further evaluated with hydrogen peroxide producing ***S. pneumoniae*** bacteria in biological media. Furthermore, the excitation spectrum engineering by the addition of Bi^{3+} ions renders these materials as suitable H_2O_2 biosensors with conventional fluorescent microscope filter sets utilizing the near-UV excitation. The biomimetic artificial enzyme developed here could serve as a starting point of sophisticated ***in vitro*** assays towards highly sensitive detection of disease biomarkers.

REFERENCES

[1] G.A. Sotiriou, M. Schneider & S.E. Pratsinis, "Color-Tunable Nanophosphors by Codoping Flame-Made Y2O3 with Tb and Eu", ***J. Phys. Chem. C*** **115**, 1084–1089 (2011).

[2] G.A. Sotiriou, D. Franco, D. Poulikakos, A. Ferrari, "Optically Stable Biocompatible Flame-Made SiO2-Coated Y2O3:Tb3+ Nanophosphors for Cell Imaging", ***ACS Nano*** **6**, 3888–3897 (2012).

[3] A. Pratsinis, G. A. Kelesidis, S. Zuercher, F. Krumeich, S. Bolisetty, R. Mezzenga, J-C. Leroux & G. A. Sotiriou. "Enzyme-Mimetic Antioxidant Luminescent Nanoparticles for Highly Sensitive Hydrogen Peroxide Biosensing" ***ACS Nano*** **11**, 12210-12218 (2017).

537134 Targeted Single-Walled Carbon Nanotubes for Photothermal Ablation of Breast Cancer Combined with Immunostimulation

Abstract:

The anionic aminophospholipid phosphatidylserine (Ptd-L-Ser) is actively sequestered to the inner leaflet of the plasma membrane of healthy cells. In a tumor, this tightly regulated Ptd-L-Ser membrane asymmetry breaks down, and Ptd-L-Ser is expressed on the outer leaflet of cancerous cells and the vascular endothelium. The presence of Ptd-L-Ser solely within the tumor serves as a tumor specific antigen which can be used in targeted therapeutics. We employ the natural ligand of Ptd-L-Ser, annexin V (ANXA5), in a protein-nanoparticle conjugate to sensitize tumors to photothermal ablation. The targeted ablation of the tumor in conjunction with anti-CTLA-4 monoclonal antibody immunotherapy generates a powerful in situ cancer vaccine creating a robust systemic immune response rejecting tumor metastasis.

Mice with well-developed orthotopic syngeneic EMT6 tumors (d â‰¥ 5 mm) were administered an intravenous systemic dose of the conjugate of single-walled carbon nanotubes (SWNTs) and ANXA5, which localized to tumor vasculature. Tumors were then irradiated with mild laser light using a Diodevet-50 NIR system for 175 s, at a power density of 1 W cm^2. Select mice additionally received three doses of 200 Î¼g of the immunostimulant anti-CTLA-4 in order to prime the immune system prior to photothermal therapy. The low energy NIR laser does not harm healthy tissue but rapidly generates temperatures in excess of 60Â° C within the tumor. This temperature is sufficient to instantly destroy cancerous cells. Healthy tissue fails to accumulate the SWNT-ANXA5 conjugate and remains unaffected by the NIR laser. The addition of anti-CTLA-4 immunotherapy in conjunction with this targeted photothermal ablation modality resulted in a significant synergistic increase in helper T cells (CD4+) and cytotoxic/suppressor (CD8+) in the spleen when compared to photothermal or immunostimulatory monotherapy. This increase in tumor effector cells led to tumor rejection. The combination of immunostimulation and targeted photothermal therapy led to significant increases in animal survival with more than half of all animals found to be tumor free at the conclusion of the study (120 days), at which time all of the animals in the control groups had died.

521835 Effect of Ethanol Solvent on Antimicrobial Efficiency of Magnesium Oxide Nanoparticles

Abstract:

Effect of Ethanol Solvent on Antimicrobial Activity of Magnesium Oxide Nanoparticles

Proma Bhattacharya, and Sudarsan Neogi

Indian Institute of Technology Kharagpur

Kharagpur 721302

Abstract: The emergence of different kinds of antibiotic resistant harmful microbes have led to extensive research for development of novel antimicrobial agents. It has been seen that different materials exhibit enhanced properties while in nano-scale and for exploiting these properties for combating pathogens, nanoparticles are very much appreciated. Different metal oxide nanoparticles such as zinc oxide, copper oxide, calcium oxide, titanium dioxide etc. have been investigated as potential antibacterial agents[1]. MgO is an inorganic compound with a significant band-gap, and it has already found wide scope of applications like catalysis, harmful waste treatment, development of refractories and adsorbents, different type of coatings with varied optical properties, superconducting thin films, and lithium ion batteries, etc. Recently, magnesium oxide nanoparticles have shown significant potential for application in biomedical fields, especially tumour treatment and wound healing due to its inherent antibacterial nature and ability to help in fibroblast adhesion after a skin injury [2]. The reaction parameters for synthesis of MgO nanoparticles play an important role in determining the morphology and properties [3]. In this work, the reaction solvent used is ethyl alcohol and its morphology and surface and antibacterial properties of the nanoparticles are compared to those synthesized in water. The particles were characterised by FESEM, EDX, XRD and FTIR for studying their morphology and chemical composition. It has been found that the size of the nanoparticles was greatly tuned by the solvent. The average diameter of the water based nanoparticles was found to be 25 nm while it changed to 10nm when the solvent was changed to ethyl alcohol. The surface properties were analysed by its zeta potential measurement where it showed marked change in the isoelectric point (point of zero charge) from 6.5 to 9 when the solvent is changed from water to ethanol. The evaluation of the antibacterial efficacy of the nanoparticles prove an unexpected increase in cell annihilation ability against both gram-positive (*Staphylococcus aureus*) and gram-negative (*Pseudomonus aeruginosa*) bacteria. Colony forming unit reduction tests reveal that a concentration of 5mg/ml of MgO prepared in ethanol was able to reduce bacterial count by almost 99%. This concentration was found to be haemocompatible (4% haemolysis), as indicated by the haemolysis tests. To understand the mechanism of the cell-nanoparticles interaction, FESEM imaging was done. The results infer that the ability of these nanoparticles to penetrate and rupture the cell membrane, leading to gradual leakage of cell components, to be the main reason behind their antibacterial nature. Amino acid leakage studies are in coherence with this proposed mechanism. Thus, magnesium oxide nanoparticles synthesized in ethyl alcohol are potential candidates for wound healing as it is biocompatible as well as highly antibacterial.

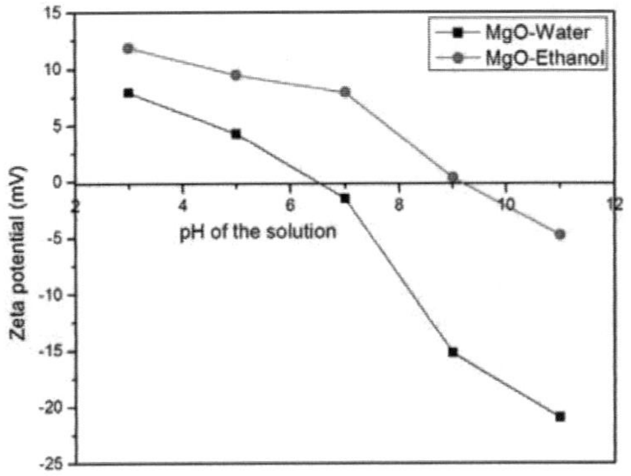

Figure 1 Variation of zeta potential of the nanoparticles synthesized in water and ethyl alcohol with pH of the solution

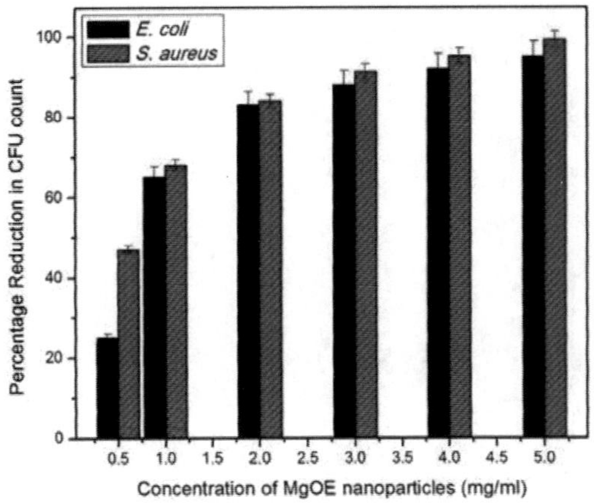

Figure 2 The percentage reduction of colony forming units of S. aureus and E.coli with increasing concentration of MgO nanoparticle synthesized in ethyl alcohol

(a)

(b)

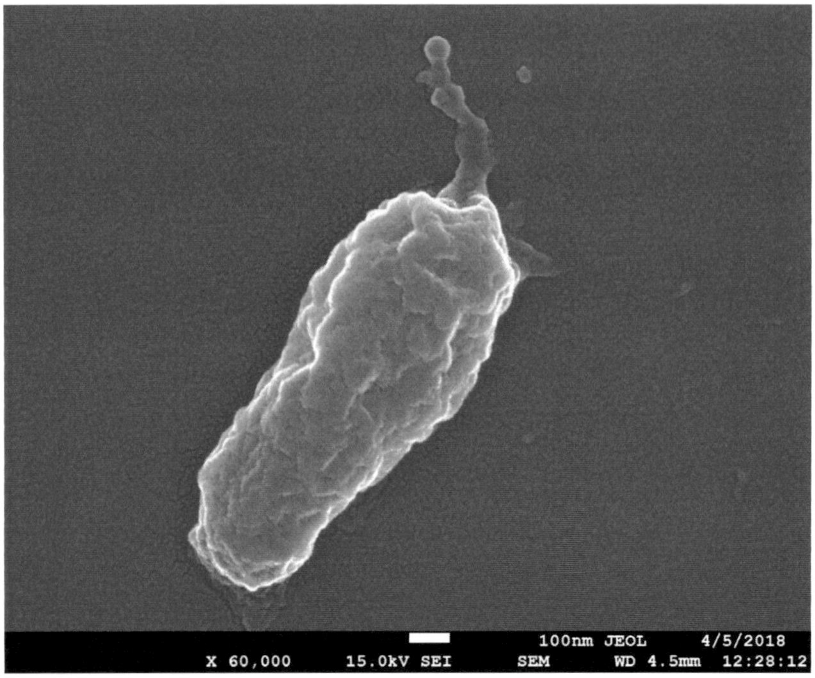

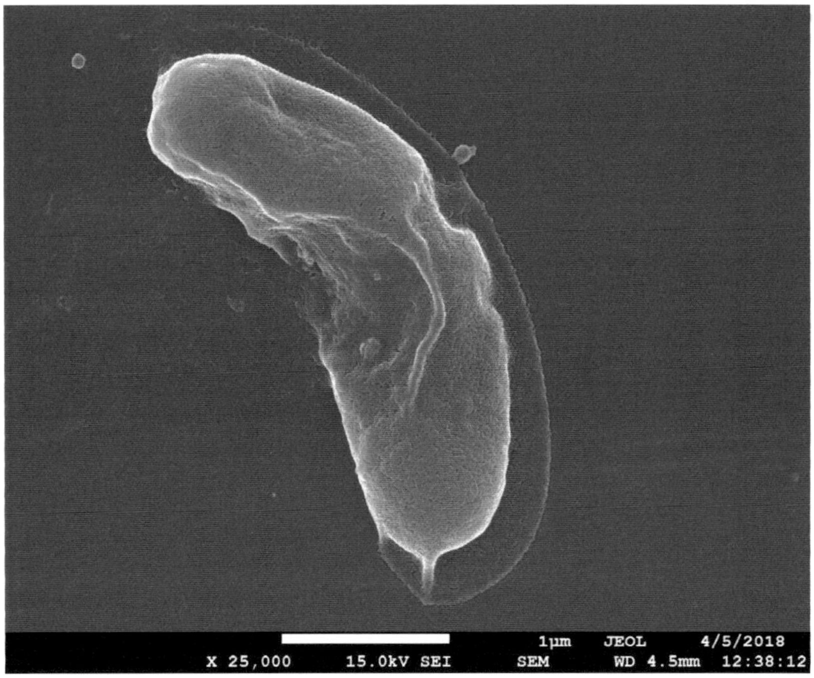

Figure 3 FESEM images showing (a) untreated E. coli and (b) E.coli treated with 5mg/ml of MgO in ethyl alcohol showing the rupture of the cell after nanoparticle treatment leading to the leakage of inter-cellular material

References:

[1] Manke A, Wang L and Rojanasakul Y 2013 Mechanisms of nanoparticle-induced oxidative stress and toxicity *Biomed Res. Int.* **2013**

[2] Jin T and He Y 2011 Antibacterial activities of magnesium oxide (MgO) nanoparticles against foodborne pathogens *J. Nanoparticle Res.* **13** 6877–85

[3] Samodi A, Rashidi A, Marjani K and Ketabi S 2013 Effects of surfactants, solvents and time on the morphology of MgO nanoparticles prepared by the wet chemical method *Mater. Lett.* **109** 269–74

525937 Synthesis and Optical Characterization of Gadolinium-Containing Scintillating Nanoparticles to Enable Neural Stimulation

Abstract:

Optogenetics is a method that combines optics and genetics, the two fields from which it is named, in order to monitor and control living tissue. This technique currently uses an invasive and potentially harmful process to control neuronal activity by implanting LEDs into brain tissue. These implantations are labor intensive, can lead to brain tissue damage, and some areas of the brain just are not accessible. It would be safer and potentially more versatile to improve this technique by systematically synthesizing and characterizing nano-scintillators that could be inserted into brain tissue via injection. These particles would serve as the visible light source to stimulate neuronal activity. These particles need to meet a number of strict requirements that would allow them to serve as local light sources to stimulate neuronal activity.

In order to achieve an inoffensive means to manipulating specific neurons, our particles will contain a scintillating inorganic crystalline material. The scintillating core must be able to effectively absorb X-ray radiation and emit photons with the desired energy required to activate the conformational switch in the chosen opsins. The particles have a stringent size restriction to be smaller than 100 nm to allow them access to the brain tissue.

We have been able to produce, through a core-shell synthetic approach, a number of species that are alluring candidates for such an application. $Ce:Gd_{4.67}(SiO_4)_3O$, $Ce:Gd_2SiO_5$, $Ce:Gd_2Si_2O_7$, and mixed Ce:GLSO core-shell nanophosphors were synthesized by the successful deposition of rare earth shells onto silica cores via sol-gel processing followed by the induction of a solid state reaction between the core and shell domains. SiO_2 nanoparticles were used as the templating core component to produce dispersible, spherical particles as well as providing the silicon source for the gadolinium silicate species. Careful monitoring of the resulting particles characteristicsâ€™ were accomplished by using powder X-ray diffraction (PXRD) to identify the reaction pathway, crystallinity, phase purity, and crystal domain size, scanning electron microscopy (SEM) to monitor particle size and morphology, and photoluminescence (PL) and radioluminescence (RL) measurements to assess optical behavior. PXRD results suggested that clean conversion of crystalline silicate phases occurred at 1000Â°C and increasing temperatures produced larger crystallite sizes. The final silicate phase of the rare earth shell was manipulated by varying the silicon to rare earth ratio and dopant concentration. The synthesized silicate particles were uniform spheres with an average diameter of 70nm.

521850 Synthesis, Characterization and Antibacterial Study of Copper-Nickel Bimetallic and Mixed Metal Oxide Nanocomposite

Abstract:

Synthesis, Characterization and Antibacterial Study of Copper-Nickel Bimetallic and Mixed Metal Oxide Nanocomposite

Debashri Paul and Sudarsan Neogi

Department of Chemical Engineering, Indian Institute of Technology,

Kharagpur - 721302, India.

E-mail: debashripaul@gmail.com, sneogiiit@gmail.com

Abstract:

Bimetallic nanoparticles are composed of two different metal elements. They gain more attention than the monometallic nanoparticles from both scientific and technological point of view due to their unique optical, catalytic, electronic properties that are not present in the corresponding monometallic nanoparticles. Generally, both Gram-negative and Gram-positive bacterial strains are thought to be harmful for the public health. For years, antibiotics and organic compounds have been used to control infectious diseases which result from both hospital environments and communities. Organic compounds which are being used as disinfectants are toxic to the human body; hence, the interest in inorganic disinfectants such as metal nanoparticles and metal oxide nanoparticles (NPs) are increasing.

Several studies have been made of anti-bacterial, anti-fungal and anti-microbial activity of copper and nickel nanoparticles individually along with their metal oxides. To improve the properties of pure copper and nickel nanoparticles along with their metal oxides, bimetallic nanoparticles have been synthesized by co-precipitation method. The nanoparticles were characterized by Transmission electron microscopy (TEM), Field Emission Scanning Electron Microscope (FESEM), X-ray powder diffraction (*XRD*) and Fourier Transform-Infrared Spectroscopy (FTIR). This work aims at studying the antibacterial activity of bimetallic nanoparticles and their mixed metal oxide along with mechanism and interaction of the nanoparticles with bacterial cell wall. The nanoparticle toxicity was assessed using two gram positive bacteria (S. *aureus* and B. *subtilis*) and two gram negative bacteria (E. *coli* and P. *aeruginosa)*. Disc diffusion assay, minimum inhibitory concentration (MIC) and time kill assay was carried out as the preliminary assessment of antibacterial activity of the bimetallic nanoparticles. A comparative study of these preliminary assessments was carried out between the individual metallic nanoparticles, metal oxides and bimetallic nanoparticles in order to understand the synergetic interaction of the bimetallic nanoparticles and how effectively they can kill the bacterial infection with a lower concentration of the nanocomposite in comparison to their individual metallic nanoparticle

Copper nanoparticles and nickel nanoparticles showed a zone of inhibition of 24.3 mm and 18.3 mm respectively in case of *B. subtilis* for a specific concentration of 1mg/ml whereas bimetallic copper/nickel nanoparticle showed a zone 30.3 mm thus showing the synergistic effect of bimetallic nanoparticles.

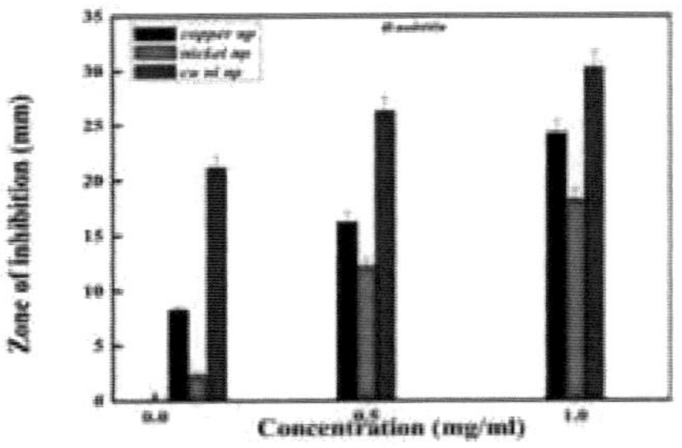

Fig.1 Zone of inhibition of different nanoparticles for B. *subtilis*

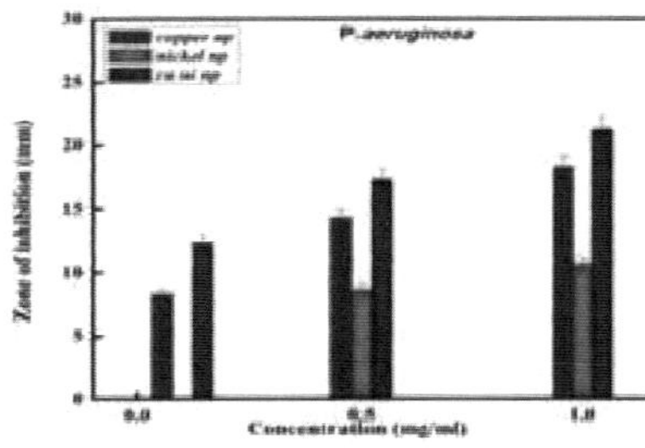

Fig.2 Zone of inhibition of different nanoparticles for *P. aeruginosa*

Copper nanoparticles and nickel nanoparticles showed a zone of inhibition of 18.3 mm showed a zone of inhibition of 10.6 mm in case of *P. aeruginosa* for a specific concentration of 1mg/ml whereas bimetallic copper/nickel nanoparticle showed a zone of of 21.3 mm thus showing the synergistic effect of bimetallic nanoparticles.

The bacterial growth study of E.*coli* and E.*coli* with bimetallic mixed metal oxide nanoparticles helped us to understand the effect of the particles on the growth rate kinetics. The lag, log, stationery and death phases were seen for both the control and treated bacterial strains. The exponential phase for the treated bacteria reduced to a great extent and the annihilation phase was achieved soon. The percentage decrease in absorbance with E.*coli* as control is 56.5% for CuO-NiO mixed metal oxide. It can be seen in the figure, that with an increase in the concentration of nanoparticle, there is a proportional effect on the cell toxicity leading to decrease in optical density at 600 nm.

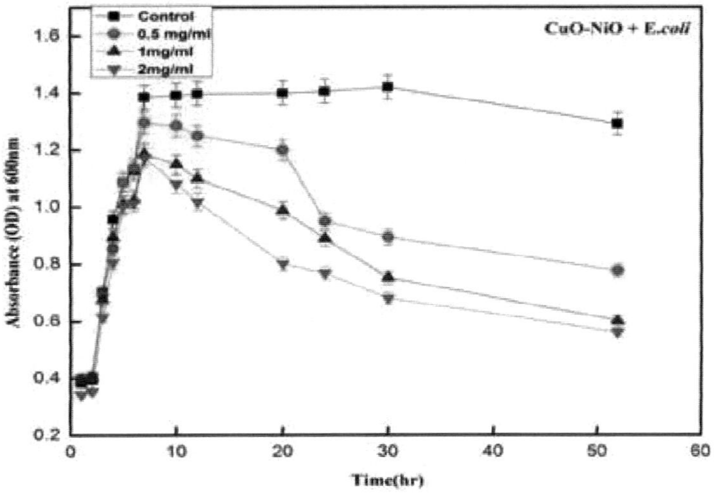

Fig 3: Growth curves for nanoparticle treated E.coli

References

[1] Zaleska-Medynska, Adriana, Martyna Marchelek, Magdalena Diak, and Ewelina Grabowska. "Noble metal-based bimetallic nanoparticles: the effect of the structure on the optical, catalytic and photocatalytic properties." *Advances in colloid and interface science* 229 (2016): 80-107.

[2] Azam, Ameer, Arham S. Ahmed, Mohammad Oves, Mohammad S. Khan, Sami S. Habib, and Adnan Memic. "Antimicrobial activity of metal oxide nanoparticles against Gram-positive and Gram-negative bacteria: a comparative study." *International journal of nanomedicine* 7 (2012): 6003.

[3] Hajipour, Mohammad J., Katharina M. Fromm, Ali Akbar Ashkarran, Dorleta Jimenez de Aberasturi, Idoia Ruiz de Larramendi, Teofilo Rojo, Vahid Serpooshan, Wolfgang J. Parak, and Morteza Mahmoudi. "Antibacterial properties of nanoparticles." *Trends in biotechnology* 30, no.

525048 Spatial Confined Capillary Flow and Precisely Controlled Crystallization Via 3D Printed Platform: A Comprehensive View

Abstract:

Spatial Confined Capillary Flow and Precisely Controlled Crystallization via 3D Printed Platform: A Comprehensive View

Mingguang Han, Gaohong He, Xiaobin Jiang*

State Key Laboratory of Fine Chemicals, Research and Development Center of Membrane Science and Technology, School of Chemical Engineering, Dalian University of Technology, Dalian, P. R. China

*Corresponding author: Email: xbjiang@dlut.edu.cn

Abstract

Precision crystallization control is crucial problem for crystallization progress design and industrial crystallization optimization in crystallizer designing, which was widely applied in pharmacy. Pacing the development of high precision manufacture method, micro-level or even nano level microfluidic chip and stable microwell array chip was developed and applied in recent state of the art crystallization processing research[1-4]. This chips-based crystallizer provides manageable environment control and repeatable crystal formation or nucleation results rapidly and accurately.

Here, we developed material-free micro pillar array chip as crystallizer utilizing high precise 3D printer (Digital Light Process (DLP)) with systematic dripping module (shown in Fig. 1) and dripped crystal solution drops onto the chip with different drop volume and different position in mild evaporation surrounding condition. We also found diversified initial contact angles when dripping different volume of drops, which possess different liquid-air interface initially and influence capillary flow inside system impactful for crystal formation (shown in Fig. 2).

For sessile drop, to understand evaporation condition and crystal formed position, concentration gradient, evaporation time and distance between formation and pillar center was encountered as variables to measure concentration change, vapor diffusion flux and nucleation energy barrier, and indicated the relationship of crystal form and capillary micro flow (shown in Fig. 3); For cover drops, we raised inside capillary micro flow by coffee ring inspiration to explain coffee-ring disappearance and guide the stable diffusion crystal formation in concave structure with observation of crystal form (shown in Fig. 3); For sitting drops and missing drip, some irregular and unexpected crystal form appeared without explanation, which need to be avoided. This method was well popularized in commercial crystal compounds with facile management and modification for high throughout crystal production. Meanwhile, lower concentration or even, polymorph system was probably suitable in crystallization process.

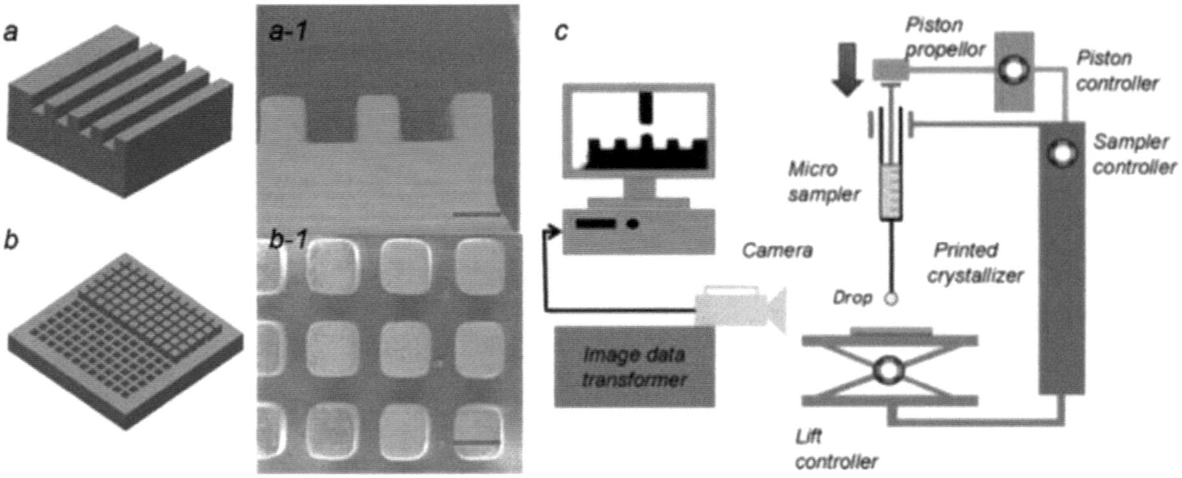

Fig 1: (a) The closed-up review of tunnel concave structure and (b) lattice concave structure of designed crystallizer. (a-1) SEM image of the cross section of (a). (b-1) SEM image of lattice concave from (b) (scale of the bar, 500 ¦Ìm). (c) The process of dripping drop system.

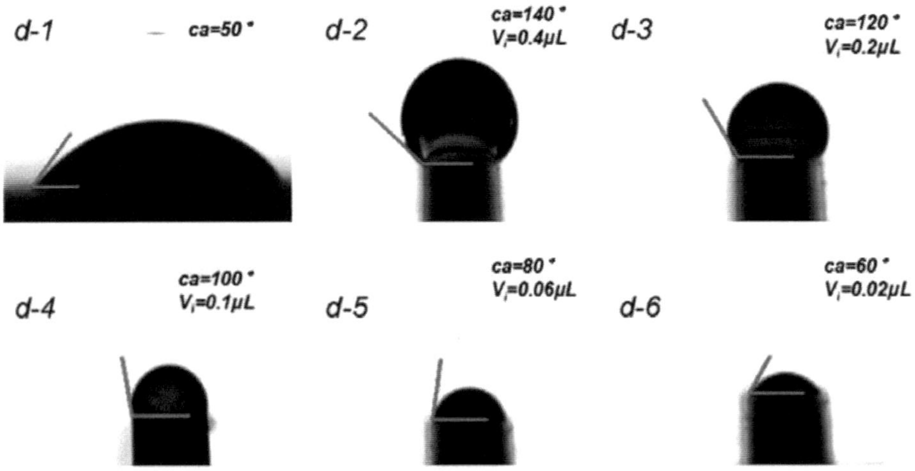

Fig 2: (d-1): The hydrophilic property of commercial 3D material shown in 50¡ãintrinsic angle. (d-2)~(d-6): gradient contact angle results when dripping numerous volume liquid drops on concave structure

Fig 3: The relationship of crystal form and capillary micro flow in sessile drop condition and cover drop condition

Acknowledgment

We acknowledge financial contribution from National Natural Science Foundation of China (Grant No. 21527812, 21676043, U1663223, 21606035), Changjiang Scholars Program (T2012049), the Fundamental Research Funds for the Central Universities (DUT16TD19, DUT17ZD203) and Education Department of the Liaoning Province of China (No. LT2015007).

References

1. Duffy DC, C MJ, A SOJ. Rapid Prototyping of Microfluidic Systems in Poly(dimethylsiloxane). *Anal. Chem.* . 1998;70:4974-4984.

2. Leng J, Salmon JB. Microfluidic crystallization. *Lab Chip.* Jan 7 2009;9(1):24-34.

3. Sultana M, Jensen KF. Microfluidic Continuous Seeded Crystallization: Extraction of Growth Kinetics and Impact of Impurity on Morphology. *Crystal Growth & Design.* 2012;12(12):6260-6266.

4. B Z, S RL, F IR. Screening of protein crystallization conditions on a microfluidic chip using nanoliter-size droplets. *Journal of the American chemical society.* 2003;125(37):11170-11171.

5. Bourgks-Monnier C, M. E. R. S. Influence of Evaporation on Contact Angle. *Langmuir.* 1995;11:2820-2829.

6. Shanahan MER, Bourg C. Effects of evaporation on contact angles on polymer surfaces. *INT. J. ADHESION AND ADHESIVES.* 1994;14(3):201-205.

7. Deegan RD, Bakajin O, Dupont TF, Huber G, Nagel SR, Witten TA. Capillary flow as the cause of ring stains from dried liquid drops. *Nature.* 1997;389(6653): 827-829

8. Cira NJ, Benusiglio A, Prakash M. Vapour-mediated sensing and motility in two-component droplets. *Nature.* Mar 26 2015;519(7544):446-450.

9. Mersmann A, ed *Crystallization technology handbook*. New York and Basel: Marcel Dekker, Inc.; 2001.

527642 Polymorph Screening of L-Glutamic Acid By Anti-Solvent Crystallization in Easy-to-Use Microfluidic Device

Abstract:

Most pharmaceuticals have polymorphs. Since different polymorphs exhibit distinct physical properties, such as solubility, bioavailability, density, mechanical properties, color and morphology, which result in difficulties in the downstream processing, including separation, handing and so on. Hence, it is crucial to control the crystallization of targeted polymorph. As an amino acid commonly used in pharmaceuticals and food, L-glutamic acid has two polymorphs: prismatic or granular shaped metastable Î± form and four different shaped stable Î² form, including need-, rod-, lozenge- and plate-shaped. In practical applications, Î± form has higher utilization efficiency. In previous work, to obtain a specific polymorph form of L-glutamic acid, the commonly adopted methods includes control of larger crystallization temperature difference, in quiescent and flow state through membrane crystallizers, adding amino acid additives, sonocrystallization and so on.

In this work, based on anti-solvent crystallization, solutions and antisolvent were loaded into separate syringes which were connected to channels. By varying the initial concentrations of solutions, temperature, and velocities of solution and antisolvent, the resulted crystals were flitered directly, and dried in an air oven. Then crystals were characterized by SEM and PXRD for morphology evaluation and form confirmation. The polymorph selectivity of L-glutamic acid was investigated. This method was simple in operation and low consumption was required. And the crystallization temperature was kept constant throughout the process. The results showed that this method was more flexible in the selectivity of L-glutamic acid polymorphs without other assistive technologies or additives which might introduce impurities.

533156 Thermodynamics of Co-Crystal Systems

Abstract:

Thermodynamics of Co-crystal Systems

Dipali. Ahuja,[1] and Åke. C. Rasmuson[1]

[1]*Synthesis and Solid State Pharmaceutical Centre, Bernal Institute, Department of Chemical Sciences, University of Limerick, Limerick, Ireland*

Introduction.

The pharmaceutical industry has a significant interest in co-crystals for their ability to alter the physical properties of the active pharmaceutical ingredient (API) without affecting its chemical structure *[1]*. Co-crystallization may enhance a wide range of pharmaceutical properties like hygroscopicity, thermal stability, solubility and bioavailability. Most research in this field has been focused towards discovering new co-crystals. Much less work has been done on studying the co-crystal thermodynamic properties and manufacturing techniques. Ternary phase diagrams are crucial for the design of the manufacturing process *[2]* as they highlight the stability regions for various solid crystalline phases. Factors like solvent and temperature significantly affect the shape of the phase diagram and the solubility differences between the co-crystal components can cause a system to be congruent or incongruent *[3]*. It is most likely that a large solubility difference causes a system to show incongruent dissolution. The main interest in the ternary phase diagram is the size of the region where co-crystal is the most stable solid phase.

Methods and Experiments.

For this work, thermodynamic studies for 1:1 sulfamethazine (SMT)-salicylic acid (SA) co-crystal system were carried out. SMT is a sulfonamide drug and is used an antibacterial and anti-infective agent; SA is used to relieve pain and fever and is used for anti-acne treatments. Ternary phase diagrams were constructed in three solvent systems (methanol, acetonitrile and DMSO-methanol) at three temperatures. The methodology behind the phase diagram construction involved determination of the pure component solubilities by gravimetric technique. The invariant points were determined by mixing the solid and solvent and allowing sufficient time for equilibration. Solids were then analysed by PXRD (Powder X-ray Diffraction) or DSC (Differential Scanning Calorimetry) and the solution composition was determined by HPLC (High Performance Liquid Chromatography).

Results and Discussion.

Solvent dependent congruent and incongruent dissolution was observed, highlighting how solvent choice can affect the phase diagram. The co-crystal system showed congruent dissolution in acetonitrile but showed an incongruent behavior in methanol and DMSO-methanol. The effect of temperature over the range studied was found to be weak. Based on the solubility data for the pure components and the co-crystal, the Gibb's free energy of co-crystal formation was determined, and ranged from -5.5 to -7.7 kJ/mol, changing with temperature. The negative values reveal the thermodynamic stability of the co-crystal and spontaneity of the co-crystal formation from SMT and SA. The enthalpy and entropy associated with co-crystallization were determined to be +23 kJ/mol and +0.1 kJ/mol, , respectively. From this data, co-crystal formation seems to be an entropy driven process. A relationship between the solubility ratio between API and coformer and the size of the region where the co-crystal is the most stable phase has been observed. In addition, it was found that the solubility ratio between the pure components does not allow for confident prediction of whether a system will show congruent or incongruent behaviour. The co-crystal yields and volumetric productivities in the three solvent systems were determined using the slurry crystallization technique.

References:

[1] J. Urbanus, C. P. M. Roelands, D. Verdoes, P.J. Jansens, J.H. ter Horst, Co-Crystallization as a Separation Technology: Controlling Product Concentrations by Co-crystals, Cryst. Growth Des., 10, 1171 (2010).

[2] D.M. Croker, M. E. Foreman, B. N. Hogan, N. M. Maguire, C.J. Elcoate, B. K. Hodnett, A.R. Maguire, Å. Rasmuson, Understanding the p-Toluenesulfonamide/Triphenylphosphine Oxide Crystal Chemistry: A New 1:1 Cocrystal and Ternary Phase Diagram, Cryst. Growth Des., 12, 869 (2012).

[3] S. Zhang, Å. Rasmuson, Cryst. Thermodynamics and Crystallization of the Theophylline–Glutaric Acid Cocrystal, Cryst. Growth Des., 13, 1153 (2013).

525830 Thermodynamic Properties of Paracetamol Impurities and Their Impact on the Crystallisation of Paracetamol from Solution

Abstract:

The impact of additives and impurities on active pharmaceutical ingredients is an essential yet poorly understood area in crystallization research. Paracetamol, which is one of the most widely used model systems in crystallization, is typically produced with trace amounts of unwanted impurities. However, it remains unclear how the main impurities 4-nitrophenol (NP) and 4â€™chloroacetanilide (CA) affect the solubility and polymorphism of paracetamol.

This work describes the solid- and solution characterisation of the main impurities of paracetamol. Temperature-dependent solid-liquid properties of 4-nitrophenol (NP) and 4â€™chloroacetanilide (CA) were obtained in four different alcohols and their effect as impurities on the crystallisation and solubility of paracetamol (PA) will be presented. HPLC measurements were used to determine the solution stability of these compounds. The solubility of NP appeared to be significantly higher than PA whereas the solubility of CA was lower than PA. The solubility difference between the impurities could be rationalised based on their molecular structure and hydrogen bonding interactions. Despite these significant differences between the impurities, their effect on the solubility of PA is the same, as each impurity slightly increases the solubility of PA. The solubility data was modelled using the empirical modified apelblat equation and thermodynamic models, which include the Margules, van-Laar, Wilson and NRTL models. The quality of fit of the tested thermodynamic models to the experimental data increases in the order Margules < van-Laar < Wilson < NRTL < Apelblat.

XRPD measurements were conducted to establish the polymorhpism of the compounds used in this study. Recrystallisation of PA from solution containing the highly soluble NP impurity resulted in small uniformly sized high purity PA form I crystals whereas the presence of the poorly soluble CA impurity induced the formation of large needle shaped crystals of PA form I. These differences in crystallisation are a consequence of the solubility difference and the different functional groups of PA and its impurities. The effect of the impurities on the nucleation kinetics of paracetamol was investigated using induction time measurements and showed that the nucleation process is inhibited for both impurities.

The presentation will cover experimental and modelling data and describes the effect of impurities on the solubility, crystallization and nucleation kinetics of paracetamol. Overall this work presents fundamental information for the development of crystallisation approaches for the purification of paracetamol from its main impurities.

525983 A Novel Approach into Secondary Nucleation and Crystal Growth, Controlled Particle Size Distribution, Using a Large Single Seed Crystal in Solution Crystallization

Abstract:

Secondary nucleation by means of seed crystals is extensively employed in crystallization process and remained a noteworthy consideration of research in recent years, but its in-depth understanding particularly at molecular level remained elusive until recently, molecular dynamic simulations provided an insight to it, i.e. crystal nuclei breeding. However, seed crystals are typically dispersed in solution, it becomes difficult to decouple nuclei breeding from attrition-enhanced secondary nucleation. In this work, a novel approach was used to experimentally investigate the mechanism of crystal nuclei breeding as an origin of secondary nucleation and its link with fluid shear, using a large single crystal of paracetamol (size = 15mm x 12mm) in propan-2-ol. The crystal was held stationary in the stirring vessel using a specially design crystal holder, to eliminate the possibility of enhanced secondary nucleation due to crystal attrition or breakage during agitation. The experiments were done in a low supersaturation ($S_o = 1.0$) at stirring speed of 300 and 370 rpm. Particle imaging velocimetry (PIV) was used to quantify the fluid shear rate at each stirring speed using a high-speed camera and laser illuminated sheet. Laser diffraction was used to perceive the onset of secondary nucleation using a focused beam of red laser passing through the vessel which was calibrated to detect the particle size as small as 20 ÆŽm. Focused beam reflectance measurement (FBRM) was used for real time tracking of particles which displayed a significant increase in particle counts with an increase in shear rate at higher rpm. At a high fluid shear, the MSZW exhibited a decrease in secondary nucleation threshold with a relative supersaturation ratio of 1.17 at 300 rpm and 1.08 at 370 rpm. Below this threshold, a growth only zone was apparent which was validated using the scanning electron microscopy (SEM). PSD data was measured on a volume basis with increasing number of fines at a higher share rate as expected. Since, no crystal attrition and breakage were involved in the onset of secondary nucleation, these results provided experimental evidence that support the hypothesized nuclei breeding mechanism. Further evidence from the SEM images of the seed crystal, showed the formation of alternate layered pattern of the surface growth and the existence of very fine crystallite layers attached to it. It is proposed that the surface of seed crystal served as a catalyst for the pre-existing clusters in the solution to attach and nucleate to form crystallites, which then sheared off by the fluid flow. The experiments presented an excellent repeatability over MSZW and growth only zone at specific shear rate, particularly the revelation of controlled particle size distribution elucidates the significance of fluid shear on the secondary nucleation using a single seed crystal. Controlled particle size distribution is an imperative quality factor in crystallization and the quality of crystalline product. This approach can offer a more precise modelling in process scale-up and manufacturing in industry.

533602 Crystal Growth Kinetics of Salicylamide Investigated Under Different Crystallisation Processes and Also Environmental Conditions

Abstract:

Crystal Growth Kinetics of Salicylamide Investigated under Different Crystallisation Processes and also Environmental Conditions

Crystallisation is a very important technique, especially in the Pharmaceutical Industry, as it is used in purification and recovery of the crystalline active pharmaceutical ingredients (API) [1]. Despite this, crystallisation fundamentals are still insufficiently understood. Efforts have been made by researchers to understand the kinetics' of salicylamide nucleation [2]. This research focuses on improving knowledge of crystal growth process and kinetics of salicylamide by investigating under rotating disk and also seeded desupersaturation agitated tank experiments. Observed the growth of individual crystals as well as populations of crystals. Crystal growth is also studied under different environmental conditions to detect any influence of the solvent, temperature and supersaturation.

The rotating disc experiments uses inverted optical microscope integrated with a camera, combined with a PC with image/ video capture and analysis software to capture micrographs of each crystal before and after growth. Multiple individual seed salicylamide crystals are fixed onto a rotating glass disc, and placed into a supersaturated solution for a growth period of 1 hour and rotated at 200rpm. Crystals are observed in-situ, so allows for the growth of specific faces to be measured and also view overall habit. The agitated tank experiments use a reactor vessel to perform seeded isothermal desupersaturation experiments. The concentration decay is recorded with a FTIR probe and first principles data treatment is applied to extract the concentration information from IR spectra; besides, FBRM is used for detection of nucleation during crystal growth. In the agitated tank experiments the average growth rate of a certain mass of seeds (better statistics) can be determined at hydrodynamic conditions closer to an industrial process.

The rotating disk crystal growth of salicylamide crystals at 15°C was measured as a function of relative supersaturation (S-1) in the range from 0.01 – 0.04, and in four different solvents (acetone AC, acetonitrile MeCN, ethyl acetate EA and methanol MeOH), rotating the disc at 200rpm. Different rotation speeds where tested in the range of 50 - 350rpm, and 200 rpm was found to be sufficient to essentially eliminate the film mass transfer resistance. The growth of a minimum of 25 crystals was determined for each system. The growth rates of the length and width of each individual crystal was measured using the maximum and minimum feret diameter respectively, also growth of the (200) face was measured. The studies revealed that salicylamide crystals grown from seed in different solvents, at different supersaturations and also within each solvent receive different shapes. There is a significant spread in the crystal growth rates obtained within each system, e.g. 0.05 µm/s up to 0.55 µm/s in one system; this is due to the crystal quality, where lower quality results in increased growth. The average growth rates increase with increasing supersaturation in each solvent. The average individual crystal growth rate behaviour, as a function of solvent and supersaturation, is similar for length and width. The average crystal growth rate of the (200) face of individual seed crystals decreases in the order MeOH>MeCN>AC>EA at supersaturation of 0.03. Salicylamide crystal surface was examined before and after the growth process using scanning electron microscopy; noticed growth steps and layers clearly present after growth indicating rough interfacial growth mechanism. The agitated tank experiments were also performed in all four solvents (AC, MeCN, EA and MeOH) and at 4 temperatures (10, 15, 20, 25°C). The empirical power law equation, spiral growth mechanism (BCF) equation and the birth & spread mechanism (B+S) equation were fitted to the data. In all four solvents, the growth rate increases with temperature as expected. The average crystal growth rate decreases in the order AC>MeOH>MeCN>EA at supersaturation of 0.03. The interfacial energy values for crystal growth depend on solvent, and where obtained by fitting the B+S model; the order is EA>AC>MeCN>MeOH. Due to the fact that low interfacial energy tends to lead to high crystal growth rate, which is in agreement with the result from the rotating disc experiments. The growth rate is faster in the agitated tank experiments as compared to rotating disc which possibly reflects an influence of mass transfer resistance in the latter.

In all growth processes, salicylamide was found to have the slowest growth kinetics in ethyl acetate. Molecular modelling was performed which high lightened the strong hydrogen bond propensity of ethyl acetate and due to its high molecular weight results in slow growth. These crystallisation studies with salicylamide in different solvents, supersaturations and temperatures have shown that crystal growth rates are affected significantly by the use of different environmental conditions.

Reference:

1. Lacmann, R., *Crystallization, Third Edition. J. W. MULLIN, Butterworth-Heinemann, Oxford 1997, 527 Seiten, zahlr. Abb. und ISBN 0-7506-3759-5.* Chemie Ingenieur Technik, 1998. **70**(11): p. 1468-1468.
2. Nordstrom, F.L., M. Svard, and A.C. Rasmuson, *Primary nucleation of salicylamide: the influence of process conditions and solvent on the metastable zone width.* CrystEngComm, 2013. **15**(36): p. 7285-7297.

539930 Fast Temperature Cycling via Microwave Heating Enables Enhanced Deracemization

Abstract:

Fast Temperature Cycling via Microwave Heating Enables Enhanced Deracemization

Christos Xiouras[1], Fabio Cameli[1], and Georgios D. Stefanidis[1]

[1] Process Engineering for Sustainable Systems, Department of Chemical Engineering, KU Leuven, Belgium;

Keywords: Deracemization, Crystallization, Temperature Cycling, Viedma ripening; Microwave Heating;

In 2005, Viedma [1] demonstrated a remarkable non-classical crystallization phenomenon according to which an initially racemic crystal population converts to a single chirality upon grinding the crystals in their saturated solution. The process was initially demonstrated for the achiral $NaClO_3$ which crystallizes as chiral crystals, but was later extended to intrinsically chiral compounds that crystallize as conglomerates (*i.e.* each individual crystal contains a single enantiomer) [2]. The underlying mechanisms of the phenomenon are still under discussion, but the process is considered to have high potential for application in the pharmaceutical industry, where separation of enantiomers is often required. However, the application of grinding at a larger industrial scale is rather complicated and could lead to problems in downstream processing. Recently, it was discovered that grinding could be replaced by periodic temperature fluctuations (temperature cycling), which is more convenient at a larger scale [3].

In contrast to conventional heating, which is slow and requires a heat transfer surface, microwave heating is rapid and volumetric in nature. Therefore, if combined with fast cooling, microwave activation can enable novel process operating windows for temperature cycling as compared to conventional heating. Figure 1 (left) shows that the application of such operation for the model system glutamic acid leads to an order of magnitude faster heating and cooling rates compared to conventional operation, which in turn leads to significant reduction in the total deracemization time required (Figure 2, right). Besides intensification of the deracemization process, the extreme operating windows applied here also lead to new insights into the obscure driving mechanisms of solid-state deracemization phenomena.

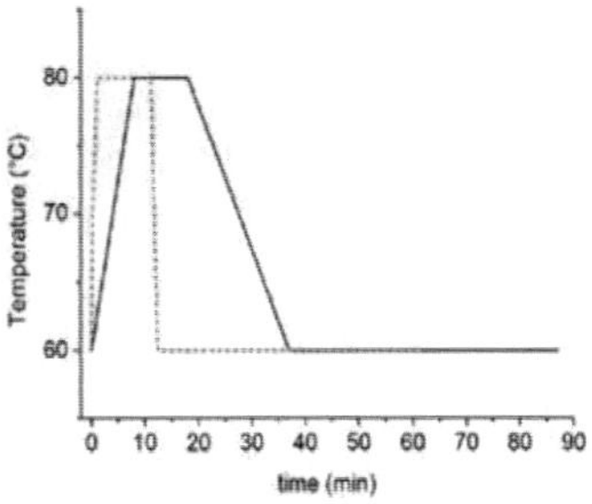

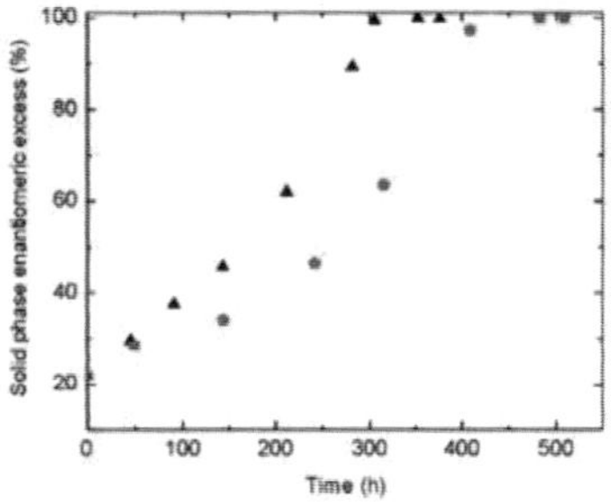

Figure 1. Left: Temperature profile using microwave heating (red dashed line) and conventional heating (black line). Right: Evolution of solid phase enantiomeric excess for in time in: microwave reactor (black points) and conventional heating configuration (red points). Model system: glutamic acid conglomerate crystals in acetic acid and catalytic salicylaldehyde.

References

[1] C. Viedma; Chiral Symmetry Breaking During Crystallization: Complete Chiral Purity Induced by Nonlinear Autocatalysis and Recycling *Phys. Rev. Lett.* 94 (2005), 3−6

[2] W.L. Noorduin, T. Izumi, A. Millemaggi, M. Leeman, H. Meekes, W.J.P. Van Enckevort, R.M. Kellogg, B. Kaptein, E. Vlieg, D.G. Blackmond, Emergence of a single solid chiral state from a nearly racemic amino acid derivative, *J. Am. Chem. Soc.*, 130 (2008), 1158-1159.

[3] K. Suwannasang, A. E. Flood., C. Rougeot, & G. Coquerel. Using programmed heating–cooling cycles with racemization in solution for complete symmetry breaking of a conglomerate forming system. *Cryst. Growth Des.*, 13 (2013), 3498-3504.

530887 Resolution of Conglomerates Using Preferential Crystallization and Enzymatic Racemization

Abstract:

Producing pure enantiomers is of great interest to pharmaceutical and fine chemicals industries, since these molecules interact differently with living organisms. Preferential crystallization (PC) is an important resolution technique to purify enantiomers. Nonetheless the separation can only reach maximum 50% yield. The addition of racemization of the non-desired enantiomer increases the theoretical process yield to 100%. Free and immobilized enzymes are attractive catalysts for this recycling step coupled with chiral separation processes. In this contribution, the separation of a model conglomerate forming system was investigated experimentally and the results were evaluated using a shortcut mathematical model for the coupling of enzymatic racemization with PC.

The amino acid racemase used was overproduced in bacteria and purified using affinity chromatography. The kinetic properties of the enzyme were investigated in preliminary studies. Low enantiomeric excesses produced by PC generate low driving forces for the racemization. This causes severe requirements related to the enzyme dosage. The optimal immobilization conditions were established to enable reuse of the catalyst and its easy separation from the products. Besides performing systematic experiments, a simple consistent mathematical model was developed assuming a lumped crystallization mechanism. The effective growth rate and the racemase kinetics were used to simulate the process combination. The performance of the biocatalyst was evaluated for the proposed application based on the experimental and simulated results.

Paper Number: 528336

ALKALINE CRYSTALLIZATION OF $CaCO_3$ IN A DIRECT AIR CAPTURE PROCESS

Caroline E. Giacomin[1], Thomas Holm[2], Luisa Burhenne[2], and Walter Mérida[2]

1) Chemical and Biological Engineering, University of British Columbia, Vancouver, BC, Canada
2) Clean Energy Research Centre, University of British Columbia, Vancouver, BC, Canada

Abstract

Crystallization of $CaCO_3(s)$ from solution in an alkaline environment was studied as a function of temperature, $CaCO_3$ pellet loading, and $CaCO_3$ pellet size. Occurrence of crystallization as spontaneous nucleation or growth on pellets are both possible where growth is the desired outcome. Growth is quantified as retention. The crystallization on pellets was found to be mass transport controlled, while spontaneous nucleation was found to be kinetically controlled. The highest reported retention, 73%, was found at the conditions where mass transport was favored by high pellet loading and small pellet diameter, and kinetics were hindered by low temperature.

Introduction

Crystallization of $CaCO_3$ from alkaline solution is relevant for a process capturing CO_2 directly from air. Recently, the economic feasibility of this process was assessed, and it was found to be possible at a cost of less than \$100/tonCO_2 [1]. A sketch of the overall industrial process is shown in Figure 1, and the pelletizer (central in Figure 1) is a fluidized bed reactor that precipitates dissolved carbonate as $CaCO_3(s)$. Precipitation can occur either on the solid $CaCO_3$ pellets or spontaneously in solution, where spontaneously nucleated particles are significantly smaller than the pellets and are referred to as fines. Precipitation on pellets is the desired outcome. Both forms of precipitation have the same driving force, the supersaturation, F_d, and this is defined in Equation (1) [2]:

$$F_d = [\mathrm{Ca}^{2+}][\mathrm{CO}_3^{2-}] - \frac{K_{sp}}{f_D^2} \qquad (1)$$

where, $[\mathrm{Ca}^{2+}]$ and $[\mathrm{CO}_3^{2-}]$ are the concentrations of calcium and carbonate in solution, K_{sp} is the thermodynamic solubility product, and f_D is the activity factor for divalent ions [3].

Crystallization of $CaCO_3$ has been studied in the water softening industry, where addition of dissolved carbonate is used to reduce the calcium content of water by precipitation on $CaCO_3$ pellets [3][4][5][6][7]. Tai et al. found that increased temperature, supersaturation, and pH increase the spontaneous nucleation rate and reduce pellet growth [4]. In works where the effect of surface area was studied, seed pellets with diameters between 0.6 and 4 μm were used, and it was found that the total available surface area significantly impacts calcium removal performance whereas pellet size showed no impact [6]. Also, Ševčík et al. used beaker experiments to show that increasing the probability of contacting a solid particle, via higher stirring velocities, increases pellet growth [5]. Therefore, conditions expected to be favorable for pellet growth are large total surface area, low temperature, low supersaturation, high stirring rate, and low pH.

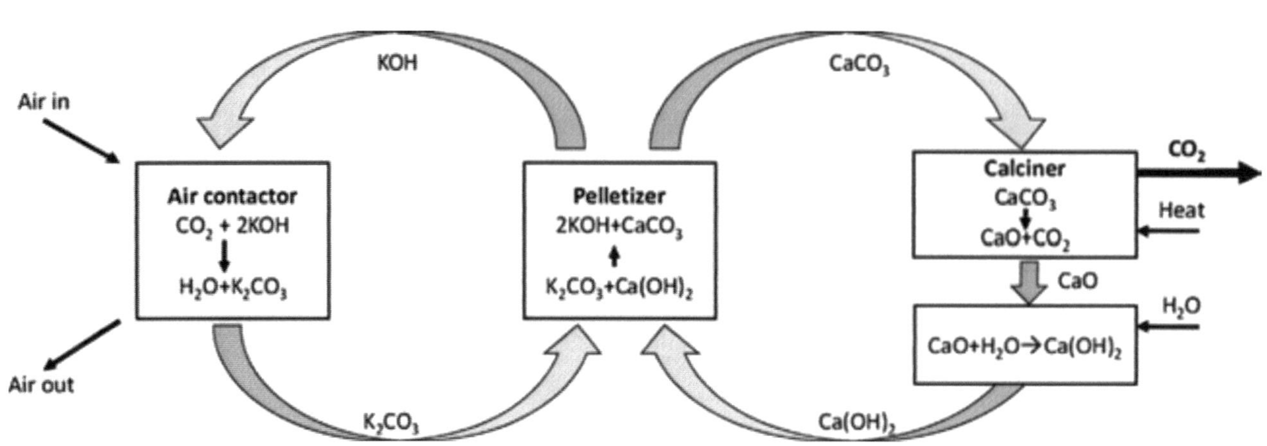

Figure 1: The direct air capture process and reactions of Carbon Engineering Ltd. [8].

The crystal growth on solid pellets was successfully modelled by van Schagen et al. [9] and the equation for the crystallization rate, C, is given in Equation (2):

$$C = \kappa A_s \left([\mathrm{Ca}^{2+}][\mathrm{CO}_3^{2-}] - \frac{K_{sp}}{f_D^2} \right) \tag{2}$$

where, κ is the rate constant, and A_s is the surface area. The reaction was described as a two-stage process where the first stage is the transportation of ions to the solid surface and the second stage is the crystallization on the pellets. Depending on the local conditions in the reaction vessel, the pellet growth reaction is limited by either mass transport, kinetics, or a combination of the two.

In contrast to the water softening process, the reaction studied in this work is operating at high pH (between 14-14.6), high carbonate concentration, and high supersaturation due to calcium being fed as solid $Ca(OH)_2$ particles. In addition, the pellets are large, more than 150 μm in diameter [8]. All these factors are expected to increase production of fines. However, high retention, of approximately 85%, has been demonstrated with high loading of $CaCO_3$ pellets even at high supersaturations [1][8]. Therefore, it is interesting to study pellet growth in a controlled environment to understand how the experimental parameters influence the retention.

In this work, $CaCO_3$ pellet growth was performed in beaker experiments controlled for pellet size, mass loading, total pellet surface area, and temperature in order to understand their effect on the overall retention. The most favorable conditions for high retention were for experiments with the smallest pellet size, highest mass loading and lowest temperature. Total pellet surface area had a limited effect on the retention. The experimental results suggest that crystallization on pellets is mass transport controlled, while spontaneous nucleation is kinetically controlled, and guides directions for further optimization of the direct air capture process.

Experimental Procedure

The reaction consists of three components: process solution, seed pellets, and lime slurry. The process solution used in these beaker experiments is a 1.0 M KOH (Sigma Aldrich 99% 45wt% solution) and 0.5 M K_2CO_3 (Sigma Aldrich 99% reagent grade) aqueous solution. The seed pellets consist of $CaCO_3$ and are sieved to yield a uniform size sample (see Table 1) and washed with deionized water to ensure all fine powder is removed (see Figure 2). Sieving is done using a Fritsch Analysette 3 PRO to shake approximately 150-200 g of pellets at an amplitude of 1.0 mm for 30 minutes. The lime slurry consists of a 2.0 M KOH solution that has $Ca(OH)_2$ powder (Sigma Aldrich, 95+%, A.C.S.) added to comprise 20wt% of the final slurry.

Table 1: These size groups represent the most common pellet sizes found within a bench scale reactor post-operation [8]. Seed is the average size fed; mid-size is about halfway grown and makes up the predominate size range in the reactor; and mature pellets are fully grown and ready for removal.

	Mesh	Diameter range [μm]	BET surface area [m^2/g]
Seed	60 - 50	250 – 300	39.7
Mid-size	45 - 40	355 - 425	27.6
Mature	25 - 14	700 - 1410	24.5

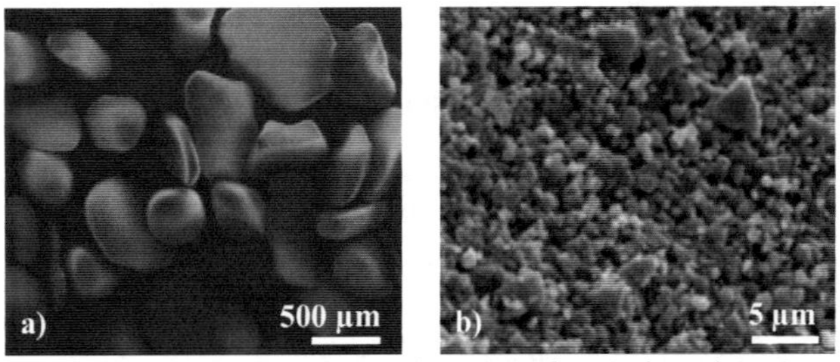

Figure 2: SEM images of mid-size pellets (a) and their surface (b) after being sieved to uniform diameters and washed of residual powder.

First, 250 mL of process solution is added to a 500 mL Pyrex beaker. The beaker was stirred using an overhead mixer (JJ-1 Precise Strength Power Mixer) with a two-bladed stainless steel impeller. A measured mass of calcium carbonate pellets is added from one size group listed in Table 1. Stirring is constant and sufficiently fluidizes the sample for all pellet sizes and loadings.

An aliquot of approximately 6.0 g of lime slurry was added. The precise mass and volume of lime slurry is recorded. After 30 minutes the beaker contents were poured through a wet sieving device consisting of a size 70 mesh (210μm) encased in PVC pipe. Effluent from the mesh was vacuum filtrated in a 9.0 cm Buchner funnel with a type 1 filter paper from Whatman. Solids found in the filter paper are the fines. By looking at the mid-sized pellets (see Figure 2) using a Nova NanoSEM their irregular shape and porous surface was observed. Surface area measurements and data for each size group were reported in Table 1.

Pellets and fines recovered are placed in separate glass petri dishes. These samples are dried overnight (at least 12 hours) at 105-110°C. Samples are allowed to cool to room temperature before mass gain is recorded. To account for relative humidity fluctuation, a control 50 g mid-size trial is run on all experiment days.

For temperature dependent trials, a type-K thermocouple attached to National Instruments LabView software was added to the reaction beaker and a Corning PC 420D heat plate was used as a heat source and temperature controller. The process solution was brought to the reaction temperature prior to pellet and slurry addition. Three trials were run for each set of conditions.

Results are reported as retention, R, calculated as in Equation (3):

$$R = \frac{m_{\text{pellets,after}} - m_{\text{pellets,before}}}{m_{\text{pellets,after}} + m_{\text{fines}}} \tag{3}$$

where $m_{\text{pellets,after}}$ is mass of pellets after reaction, $m_{\text{pellets,before}}$ is the original mass of $CaCO_3$ pellets added, and m_{fines} is mass of fines collected after reaction.

After reactions process solution concentrations are analyzed for their ion ratio, $[OH^-]:[CO_3^{2-}]$. SEM images were taken of both the pellets and fines produced. Powder X-ray diffraction (XRD) was also performed on both using a Bruker D8 Advance with a 2θ range of 10 to 90, step size 0.1, and scan time of 0.3s per step.

Results and Discussion

Ambient temperature

For all ambient temperature conditions reported, a positive retention rate was observed and $CaCO_3$ was formed both on pellets and as fines. In these experiments process solution concentration, amount of fed $Ca(OH)_2$ particles, and stirring rate were constant meaning that the supersaturation, F_d, was constant. Therefore, this set of experiments evaluates the influence of mass transport.

For varied mass loading with a constant pellet size (mid-size), Figure 3a, the retention varies from 52% at 25 g loading to 64% at 100 g loading. Higher loadings are expected to increase the surface area of $CaCO_3$ in the beaker and thus lead to higher retention. This trend was confirmed in our experimental data, but the effect is limited as lower loading did not seem to affect the retention. This suggests that other factors, such as pellet size, are more important.

When the pellet size was varied at constant mass loading, Figure 3b, the retention changed from 67% at seed size to 22% at mature size. The change of pellet from seed to mature increases the particle size and decreases the total surface area of $CaCO_3$ resulting in fewer particles present in the system. Both of these effects are expected to reduce the retention rate and therefore explains the trend observed in Figure 3b. In an attempt to deconvolute the two effects, a third set of experiments was done where the total surface area confirmed through BET analysis was kept constant (at 1,215 m^2). The results of these experiments are shown in Figure 3c. Here, it was evident that the pellet size has a large effect, and the retention changes from 58% at seed size to 22% at mature size. Notably, the seed size pellets and mid-size pellets give retentions within one standard deviation of each other.

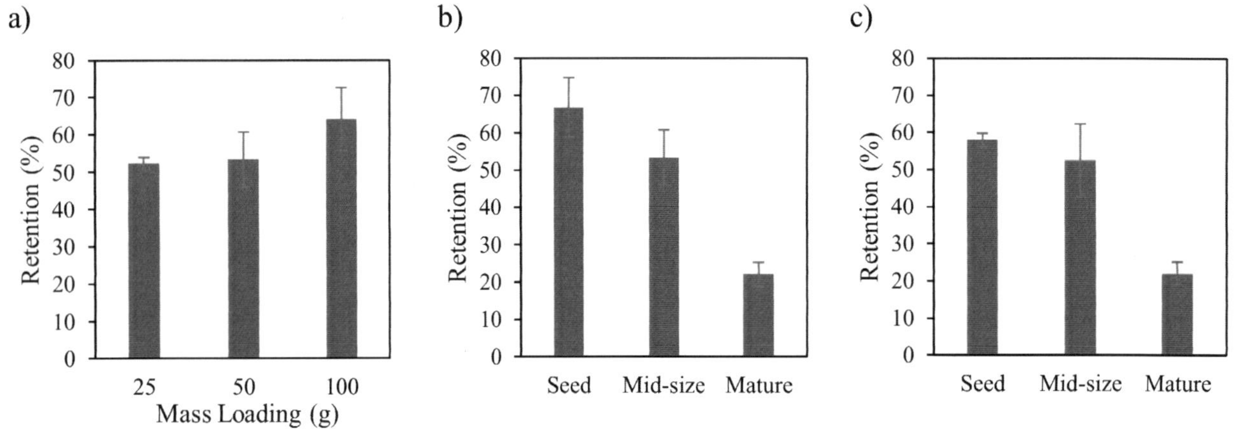

Figure 3: Experiments run at room temperature 20°C with the following variables being varied: a) mid-size pellets at different mass loadings of 25, 50, and 100 grams, b) constant mass loading with varied pellet size, and c) constant surface area with varied pellet size.

In order to understand this result and to separate the effect of pellet size and surface area, the retention rate was plotted as a function of surface area for all results reported in Figure 3. This is shown in Figure 4. Here, the trend in Figure 3c is confirmed where mature pellets give a low retention rate, while the seed and mid-size pellets get comparable retention rates. This can be expected if the actual pellet size is considered (Table 1), where the seed and mid-size pellets are close in diameter while the mature pellets have a diameter approximately 2.5-6 times the seed size and 1.75-4 times the mid-size pellets. From these results, it is evident that while Equation (2) can explain the overall reaction in some cases, care should be taken when interpreting the results solely with this equation. In particular, when $Ca(OH)_2$ is fed as solid particles, as is the case here, the actual supersaturation, F_d, is largely confined to the volume around the solid $Ca(OH)_2$ particles. Thus, to achieve high retention, one must ensure that solid $CaCO_3$ is present near the $Ca(OH)_2$ particles. From the results reported in Figure 3 and Figure 4, high mass loading and small pellet sizes is the best approach. To confirm this, an additional experiment was performed using 100 g of seed pellets resulting in 73% retention. This data is shown in Figure 4 as the rightmost triangle. Overall, this suggests that collisions, or near collisions, between the $Ca(OH)_2$ particles and $CaCO_3$ pellets are key to high retention and this gives a guideline for how to achieve an even higher retention rate at the pH and concentration used in this work.

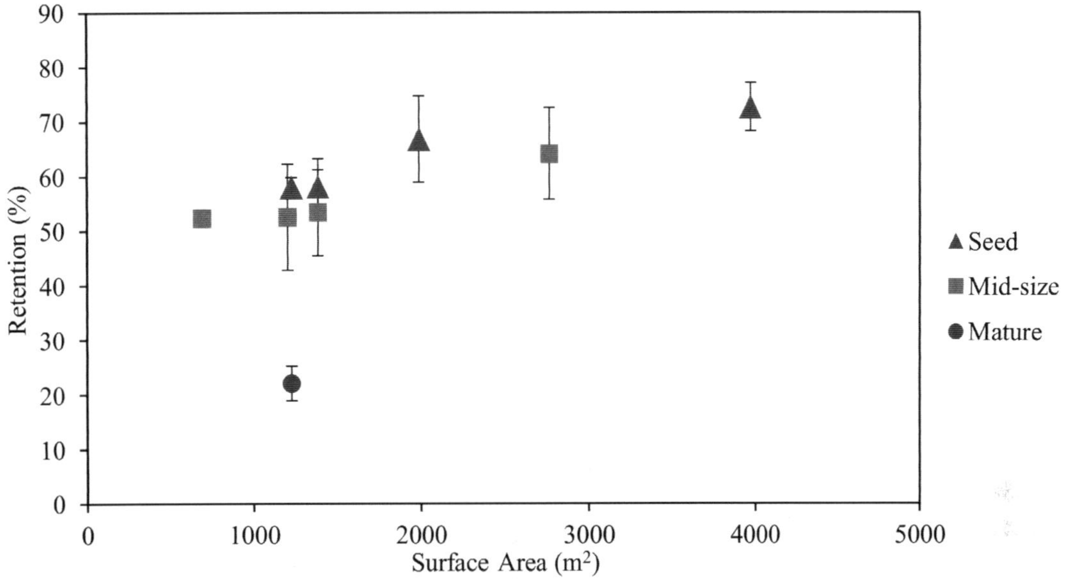

Figure 4: Retention for all trials shown in Figure 3 as well as best conditions experimental set. All shown with respect to surface area as calculated by BET isotherm measurements from the pellet size groups.

Elevated temperatures

Elevated temperature is expected to modify the kinetics of crystallization [3]. With mass transport variables held constant, increased crystallization rate is expected to increase spontaneous nucleation. In earlier work [5][10], at high temperatures, different $CaCO_3$ morphologies were reported and thus temperature can change the precipitation mechanism both for spontaneous nucleation and pellet growth. For example, the aragonite morphology has a needle-like structure that can help reduce the effect of increased temperature on the retention by assuring that a large surface area of $CaCO_3$ is present in the beaker. This morphology was reported at temperatures higher than 50°C [10].

In Figure 5, the retention rate at four temperatures is reported for mid-size pellets. Here, it is evident that the higher temperature gives significantly lower retention going from 53% at 20°C to -1.8% at 80°C. The negative retention at 80°C was within one standard deviation of no retention, and was assumed to be due to experimental uncertainty. If the limiting mechanism for pellet growth is physical proximity of $CaCO_3$ pellets to $Ca(OH)_2$ particles, an increase in temperature is expected to mainly influence the spontaneous nucleation rate. An increased spontaneous nucleation rate is expected to produce fines that are both more numerous and smaller in diameter than a slower spontaneous nucleation rate. In order to investigate this, SEM imaging was done on the fines at the various temperatures and the results are shown in Figure 6. Here, it is evident that the size of the fines decreases as the temperature is increased.

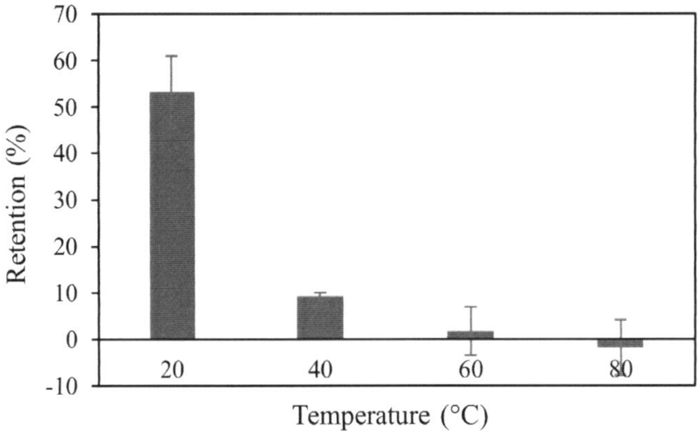

Figure 5: Percent of mass found on pellets over total mass collected after reaction for trials at different temperatures for mid-sized pellets.

Figure 6: $CaCO_3$ fines grown at a) 20°C, b) 40°C, c) 60°C, and d) 80°C.

SEM imaging was also done on the pellets after growth and is shown in Figure 7. No clear trend is visible in these images, and this can be expected as the amount of $CaCO_3$ deposited on a single pellet is relatively small compared to the overall pellet size (mass gain at 50% retention is 0.80 g or only 1.6% of total mass for a loading of 50 g). A long-term experiment might give different results and possibly different $CaCO_3$ morphologies, however that is outside the scope of this work but will be investigated in the future.

Figure 7: Pellet surfaces after reaction at a) 20°C, b) 40°C, c) 60°C, and d) 80°C.

The different $CaCO_3$ morphologies result in different shape of the particles so it may be visible in SEM images. However, if only small amounts of a morphology are present, it is hard to distinguish. Therefore, XRD was run to confirm the results of the SEM images. The XRD results are shown in Figure 8 for the a) pellets and b) fines and compared with reference spectra for three different $CaCO_3$ morphologies (calcite, vaterite, and aragonite) and $Ca(OH)_2$. For the pellets, this could be compared to the unreacted pellets. Here, it is evident that at all temperatures, the pellets are present mainly as calcite. Only a small peak appears in the 80°C results at a 2θ of 61. The peak position is not one of the main peaks for the vaterite or aragonite phases. Therefore, it is assumed that this is an impurity peak and that $CaCO_3$ is only present as calcite on the pellets.

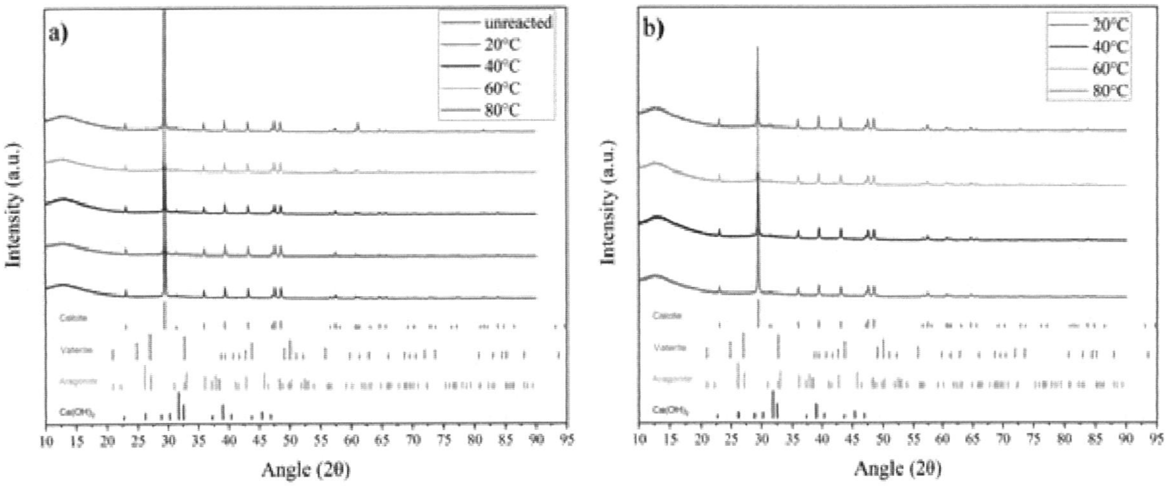

Figure 8: XRD results for (a) pellets, and (b) fines at different temperatures. Reference spectra for calcite (pink), vaterite (green), aragonite (orange), and $Ca(OH)_2$ (blue) are shown at the bottom.

For the fines, in Figure 8b, the XRD results were consistent with calcite at all temperatures. The fines are not seeded and consequently will form the morphology that is most kinetically and thermodynamically favorable. In our results, only calcite is observed, and this suggests that at the high pH used in this work, only calcite will be formed both through spontaneous nucleation and as precipitation on $CaCO_3$ pellets. Therefore, in contrast to work at lower pH values the temperature does not change the morphology of $CaCO_3$ produce, and only modifies the nucleation rate resulting in lower retention.

When results at ambient and elevated temperatures are compared, a framework for achieving high retention can be found. This builds on the understanding that high retention is achieved by improving the conditions for precipitation on $CaCO_3$ particles while hindering the spontaneous nucleation process. These two processes have equal driving force, the supersaturation, F_d. In the conditions used in this work, the supersaturation is high but locally confined around the $Ca(OH)_2$ particles. Thus, increased precipitation on the $CaCO_3$ pellets and resulting higher retention is improved by increasing the chance of a $CaCO_3$ pellet being present close to the $Ca(OH)_2$ particle. This is a pure mass transport constraint. On the other hand, the rate of spontaneous nucleation is increased by increasing the kinetic driving force, demonstrated here by the temperature. High retention is achieved when the mass transport conditions are favorable and the kinetic driving force is low. The highest retention reported in this work is 73% and is achieved for seed size pellets at 100 g loading and at 20°C. A further improvement of the retention can be expected for smaller particle sizes, higher mass loadings of $CaCO_3$ pellets, and lower temperatures.

Conclusion

Crystallization of $CaCO_3(s)$ from solution in an alkaline environment relevant for a direct CO_2 capture from air reactor was studied as a function of temperature, $CaCO_3$ pellet loading, and $CaCO_3$ pellet size. The results were evaluated in terms of retention: the fraction of $CaCO_3$ formed that grows on pellets. Higher mass loading and smaller pellet size resulted in higher retention. Higher temperatures resulted in lower retention. Higher temperatures also were shown to reduce the diameter of fines produced. The crystallization on pellets was found to be mass transport controlled, while spontaneous nucleation was found to be kinetically controlled. The highest retention, 73%, was found at high pellet loading, small pellet diameter, and low temperature.

The fluidized crystallization will next be looked at in a fluidized bed constructed at the lab scale to model the version used by Carbon Engineering Ltd. The work within fluidized bed reactor will provide longer reaction times, allowing larger net mass growth and conditions closer to the industrial process.

Acknowledgement

This work was supported by Carbon Engineering Ltd., the National Sciences and Engineering Research Council of Canada (NSERC), and Mitacs through the Mitacs Accelerate Program. The support of all colleagues and students at the University of British Columbia and Carbon Engineering Ltd. is greatly acknowledged, in particular we thank Jenny McCahill for many discussions pertaining to the direct air capture process of Carbon Engineering Ltd.

References

[1] D. W. Keith, G. Holmes, D. St. Angelo, and K. Heidel, "A Process for Capturing CO2 from the Atmosphere," *Joule*, 2018.

[2] N. Spanos and P. G. Koutsoukos, "Kinetics of Precipitation of Calcium Carbonate in Alkaline pH at Constant Supersaturation. Spontaneous and Seeded Growth," *J. Phys. Chem. B*, vol. 102, no. 34, pp. 6679–6684, 1998.

[3] H. N. S. Wiechers, P. Sturrock, and G. v R. Marais, "Calcium carbonate crystallization kinetics," *Water Res.*, vol. 9, no. 9, pp. 835–845, 1975.

[4] C. Y. Tai and H. P. Hsu, "Crystal growth kinetics of calcite and its comparison with readily soluble salts," *Powder Technol.*, vol. 121, no. 1, pp. 60–67, 2001.

[5] R. Ševčík, M. Pérez-Estébanez, A. Viani, P. Šašek, and P. Mácová, "Characterization of vaterite synthesized at various temperatures and stirring velocities without use of additives," *Powder Technol.*, vol. 284, pp. 265–271, 2015.

[6] K. L. Mercer, Y. P. Lin, and P. C. Singer, "Enhancing calcium carbonate precipitation by heterogeneous nucleation during chemical softening," *Journal / American Water Works Association*, vol. 97, no. 12. 2005.

[7] G. H. Nancollas and M. M. Reddy, "The crystallization of calcium carbonate. II. Calcite growth mechanism," *J. Colloid Interface Sci.*, vol. 37, no. 4, pp. 824–830, 1971.

[8] L. Burhenne, C. Giacomin, T. Follett, J. Ritchie, J. S. J. McCahill, and W. Mérida, "Characterization of reactive CaCO3 crystallization in a fluidized bed reactor as a central process of direct air capture," *J. Environ. Chem. Eng.*, vol. 5, no. 6, pp. 5968–5977, 2017.

[9] K. M. Van Schagen, L. C. Rietveld, and R. Babuška, "Dynamic modelling for optimisation of pellet softening," *J. Water Supply Res. Technol. - AQUA*, vol. 57, no. 1, pp. 45–56, 2008.

[10] Y. F. Ma, Y. H. Gao, and Q. L. Feng, "Effects of pH and temperature on CaCO3 crystallization in aqueous solution with water soluble matrix of pearls," *J. Cryst. Growth*, vol. 312, no. 21, pp. 3165–3170, 2010.

AUTHOR INDEX

AUTHOR INDEX

AUTHOR INDEX

AUTHOR INDEX

AUTHOR INDEX

AUTHOR INDEX

AUTHOR INDEX

AUTHOR INDEX